SOURCES AND EFFECTS OF IONIZING RADIATION

United Nations Scientific Committee on the Effects of Atomic Radiation

UNSCEAR 1994 Report to the General Assembly, with Scientific Annexes

UNITED NATIONS
New York, 1994

NOTE

The report of the Committee without its annexes appears as Official Records of the General Assembly, Forty-ninth Session, Supplement No. 46 (A/49/46).

The designation employed and the presentation of material in this publication do not imply the expression of any opinion whatsoever on the part of the Secretariat of the United Nations concerning the legal status of any country, territory, city or area, or of its authorities, or concerning the delimitation of its frontiers or boundaries.

The country names used in this document are, in most cases, those that were in use at the time the data were collected or the text prepared. In other cases, however, the names have been updated, where this was possible and appropriate, to reflect political changes.

UNITED NATIONS PUBLICATION
Sales No. E.94.IX.11
ISBN 92-1-142211-6

CONTENTS

UNITED NATIONS SCIENTIFIC COMMITTEE ON THE EFFECTS OF ATOMIC RADIATION

1994 REPORT

Report of the United Nations Scientific Committee on the Effects of Atomic Radiation to the General Assembly

---

CONTENTS

INTRODUCTION

1. During the last few years the United Nations Scientific Committee on the Effects of Atomic Radiation (UNSCEAR)[a] has undertaken a broad review of the sources and effects of ionizing radiation. Nine scientific annexes on particular subjects were issued in the UNSCEAR 1993 Report. Two further annexes have been completed, and these comprise the UNSCEAR 1994 Report. This is the twelfth substantive Report of the Committee, informing the General Assembly and the scientific and world community of its latest assessments[b]. The two reports, 1993 and 1994, are complementary and provide a coherent summary of the Committee's findings and programme of work.

2. The present report and its scientific annexes were prepared between the thirty-eighth and forty-third sessions of the Committee. Serving as Chairman, Vice-chairman and Rapporteur, respectively, at these sessions were: thirty-eighth and thirty-ninth sessions: K. Lokan (Australia), J. Maisin (Belgium) and E. Létourneau (Canada); fortieth and forty-first sessions: J. Maisin (Belgium), E. Létourneau (Canada) and L. Pinillos Ashton (Peru); forty-second and forty-third sessions: E. Létourneau (Canada), L. Pinillos Ashton (Peru) and G. Bengtsson (Sweden). The names of members of national delegations who attended the thirty-eighth to the forty-third sessions of the Committee are listed in Appendix I.

3. The scientific annexes of this report were developed at annual sessions of the Committee, based on working papers prepared by the secretariat. The Committee wishes to acknowledge the help and advice of a small group of consultants, appointed by the Secretary-General, who helped in the preparation of the material for this report. Their names are given in Appendix II. They were responsible for the preliminary reviews and evaluations of the technical information received by the Committee or available in the open scientific literature, on which rest the final deliberations of the Committee.

4. The sessions of the Committee held during the period under review were attended by representatives of the United Nations Environment Programme (UNEP), the World Health Organization (WHO), the International Atomic Energy Agency (IAEA), the International Commission on Radiological Protection (ICRP), the International Agency for Research on Cancer (IARC) and the International Commission on Radiation Units and Measurements (ICRU). The Committee wishes to acknowledge their contributions to the discussions.

5. In the present report, the Committee summarizes the main conclusions of the two scientific annexes, "Epidemiological studies of radiation carcinogenesis" and "Adaptive responses to radiation in cells and organisms". In addition, the Committee is reviewing the effects of radiation on the natural environment, and although the scientific annex has not yet been completed, a summary of this work in progress is given.

6. Following established practice, only the introductory part of the report is submitted to the General Assembly. The full UNSCEAR 1994 Report, including the scientific annexes, will be issued as a United Nations sales publication. This practice is intended to achieve a wider distribution of the findings for the benefit of the international scientific community. The Committee wishes to draw the attention of the General Assembly to the fact that the main text of the UNSCEAR 1994 Report is presented separately from its scientific annexes simply for the sake of convenience. It should be understood that the scientific data contained in the annexes are important because they form the basis for the conclusions of the report.

---

[a] The United Nations Scientific Committee on the Effects of Atomic Radiation was established by the General Assembly at its tenth session, in 1955. Its terms of reference are set out in resolution 913 (X) of 3 December 1955. The Committee was originally composed of the following Member States: Argentina, Australia, Belgium, Brazil, Canada, Czechoslovakia, Egypt, France, India, Japan, Mexico, Sweden, Union of Soviet Socialist Republics, United Kingdom of Great Britain and Northern Ireland and United States of America. The membership was subsequently enlarged by the General Assembly in its resolution 3154 C (XXVIII) of 14 December 1973 to include the Federal Republic of Germany, Indonesia, Peru, Poland and the Sudan. By resolution 41/62 B of 3 December 1986, the General Assembly increased the membership of the Committee to a maximum of 21 members and invited China to become a member.

[b] For the previous substantive reports of UNSCEAR to the General Assembly, see *Official Records of the General Assembly, Thirteenth Session, Supplement No. 17* (A/3838); *ibid., Seventeenth Session, Supplement No. 16* (A/5216); *ibid., Nineteenth Session, Supplement No. 14* (A/5814); *ibid., Twenty-first Session, Supplement No. 14* (A/6314 and Corr.1); *ibid., Twenty-fourth Session, Supplement No. 13* (A/7613 and Corr.1); *ibid., Twenty-seventh Session, Supplement No. 25* (A/8725 and Corr.1); *ibid., Thirty-second Session, Supplement No. 40* (A/32/40); *ibid., Thirty-seventh Session, Supplement No. 45* (A/37/45); *ibid., Forty-first Session, Supplement No. 16* (A/41/16); *ibid., Forty-third Session, Supplement No. 45* (A/43/45) and *ibid., Forty-eighth Session, Supplement No. 46* (A/48/46). These documents are referred to as the 1958, 1962, 1964, 1966, 1969, 1972, 1977, 1982, 1986, 1988 and 1993 Reports, respectively. The 1972 Report with scientific annexes was published as *Ionizing Radiation: Levels and Effects, Volume I: Levels* and *Volume II: Effects* (United Nations publication, Sales No. E.72.IX.17 and 18). The 1977 Report with scientific annexes was published as *Sources and Effects of Ionizing Radiation* (United Nations publication, Sales No. E.77.IX.1). The 1982 Report with scientific annexes was published as *Ionizing Radiation: Sources and Biological Effects* (United Nations publication, Sales No. E.82.IX.8). The 1986 Report with scientific annexes was published as *Genetic and Somatic Effects of Ionizing Radiation* (United Nations publication, Sales No. E.86.IX.9). The 1988 Report with annexes was published as *Sources, Effects and Risks of Ionizing Radiation* (United Nations publication, Sales No. E.88.IX.7). The 1993 Report with scientific annexes was published as *Sources and Effects of Ionizing Radiation* (United Nations publication, Sales No. E.94.IX.2).

I. EPIDEMIOLOGICAL STUDIES OF RADIATION CARCINOGENESIS

7. The Committee has paid particular attention to the review of results of epidemiological studies of human populations exposed to ionizing radiation, since these form the main basis for quantifying the risks of radiation-induced cancer in man. Several study populations are available, including the survivors of the atomic bombings of Hiroshima and Nagasaki, patients exposed in medical procedures, those exposed occupationally and inhabitants of high natural background or contaminated areas, and these groups are the subject of continuing investigations.

8. Estimates of the risks of cancer caused by radiation exposure were derived in the UNSCEAR 1972, 1977 and 1988 Reports and discussed in the UNSCEAR 1993 Report. Although all information was considered, the primary estimates of risk were derived from results of the main study population, the survivors of the atomic bombings. An objective of the Committee's present review of this subject is to consider the large number of additional epidemiological studies now contributing quantitative information on the effects in humans of ionizing radiation and to evaluate comparative risk estimates.

9. Studies of disease in human populations must adhere strictly to epidemiological principles in order to achieve valid quantitative results. These include sound case ascertainment, an appropriate comparison group, sufficient follow-up, an accounting for confounding factors and well-characterized dosimetry. Such epidemiological studies are able to provide clear-cut evidence of risks for various sites of cancer, and also to evaluate the factors that modify risks, following high radiation doses. However, at low doses epidemiological studies are not able to detect and quantify statistically significant radiation effects.

A. EFFECTS OF EXTERNAL EXPOSURES

10. The Committee has examined the epidemiological studies that could be used to derive risk estimates from external, sparsely ionizing (low-LET) radiation exposures at high and low dose rates. The Committee has summarized the main features of these studies, including their strengths and limitations.

11. The primary study for the estimation of risk of cancer induction is the Life Span Study of survivors of the atomic bombings of Hiroshima and Nagasaki. The study, which began in 1950, comprises a large population of all ages and both sexes exposed to a range of doses at high dose rate. Data on cancer mortality and new data on cancer incidence are now available up to 1987. Since most of the original survivors are still living, many more years of follow-up will be necessary to determine the complete lifetime cancer occurrence in this population. Consequently, lifetime risk estimation requires projection beyond the period of observation.

12. Cancers for which statistically significant excess risks have been determined from the Life Span Study mortality data are leukaemia, breast, bladder, colon, liver, lung, oesophagus, ovary, multiple myeloma and stomach. The incidence data are broadly similar, but two of the sites, oesophagus and multiple myeloma, do not show significant risks. The incidence data are probably more definitive than the mortality data. Two additional sites, namely thyroid and skin, have significant excess incident cancers.

13. Studies of other radiation-exposed populations such as cervical cancer patients, ankylosing spondylitics and children treated for tinea capitis serve to clarify and generally support findings from the Life Span Study. Some also provide information on issues that cannot be addressed by the atomic bomb survivor data, such as the effects of low chronic doses, highly fractionated exposures and variability among populations. For some sites of cancer, including breast, leukaemia and thyroid, there are a number of very useful results from studies other than the Life Span Study. In general, there are no great disparities in risk estimates between the Life Span Study and the other studies.

14. Although the Committee has presented risk estimates for specific sites from results of many studies, general estimates of lifetime mortality risks for all cancers must still be derived from the Life Span Study. For this report the Committee has analysed the data from 1950 to 1987 and made projections to the full life-span of the population in several ways. Using the constant relative risk model allowing for sex and age at exposure (a more refined analysis than in the UNSCEAR 1988 Report), the estimates of lifetime risk of mortality following an exposure to 1 Sv (weighted dose) is 11% for solid tumours and 1% for leukaemia. Using alternative projection methods allowing for some decline in relative risk with time (as suggested by some epidemiological studies), lifetime risk estimates for solid tumours are 20%-40% lower. The constant relative risk estimates in the UNSCEAR 1988 Report were 10% for solid tumours and 1% for leukaemia at 1 Sv.

15. The Committee indicated in the UNSCEAR 1993 Report that risk estimates derived at high doses and high dose rates should be divided by a small factor to obtain the risk at low doses (<0.2 Sv). If a factor of 2 is used, the risk derived from the UNSCEAR 1988 Report would be 5% per Sv and from this report 6% per Sv for a constant relative risk projection. If alternative projection methods are used, however, the risk would be 4%-6% in the Japanese population (the applicability to other populations involves some additional uncertainty). Consequently, the use of a nominal value of 5% per Sv for mortality due to leukaemia and solid cancers from irradiation at low

doses for a population of all ages (4% per Sv for an adult working population) still seems valid to the Committee.

16. The effects of low-LET radiation delivered at low doses or low dose rates have been examined in studies of occupational, natural background and environmental exposures. Occupational studies offer the most promise of providing results that are statistically significant because they are based on large populations with a range of individual dose estimates and long periods of observation.

17. The most comprehensive occupational study to date involves nuclear workers in the United Kingdom. This study reports a significant excess risk for leukaemia and a positive, but non-significant excess for all cancers as a group. A smaller study carried out in the United States found non-significant deficits of cases among exposed workers. In a combined analysis of these two studies, the results of which were statistically non-significant, there was excess incidence of leukaemia and all cancers, which were about half the estimates for the atomic bomb survivors. Initial findings in studies of workers in the atomic energy programme of the former Soviet Union with exposures of the order of several sievert accumulated over several years show clear excesses of cancer in the highest dose groups broadly consistent with the levels of risk seen in the survivors of the atomic bombings.

18. Comparisons of cancer incidence in areas of high and low natural radiation background have been undertaken in China, France, Japan, Sweden, United Kingdom and United States. None, including the largest, that in China, has produced statistically significant associations.

19. Populations exposed to environmental releases of radionuclides have provided little information on risk. However, one circumstance of special interest concerns releases of fission products into the Techa River in the former Soviet Union during 1948-1951. In the 28,000 people studied there was some evidence of an excess of leukaemia not inconsistent with results derivable from the study of the survivors of the atomic bombings.

B. EFFECTS OF INTERNAL EXPOSURES

20. Of the radionuclides emitting low-LET radiation that may enter the body, iodine-131 is the most important, since it is used to diagnose thyroid conditions and to treat hyperthyroidism and thyroid carcinoma. Environmental exposures to iodine-131 from fallout and from accidents at nuclear installations have also occurred. Iodine-131 appears to be less effective than external radiation in causing thyroid cancer, perhaps by a factor of 3-5. More studies are needed to clarify the possibly greater risks in children than in adults, as indicated by external radiation exposure. The Committee is aware of reports of thyroid cancer incidence in locally exposed individuals following the Chernobyl accident and intends to examine this issue in a future report.

21. More densely ionizing (high-LET) radiation exposures result from alpha-particle-emitting radionuclides, such as radon and its decay products and radium and thorium used in medical and industrial applications. High-LET radiation is more effective in causing damage in tissue than low-LET radiation. Alpha-radiation is not very penetrating, however, so exposures occur only when the radionuclides in air, food or water are taken into the body. The Committee has examined the few epidemiological studies that can provide risk estimates.

22. Radon is an important source of exposure of the public in houses and other buildings. The risk of lung cancer caused from exposure to radon is derived from studies of miners of uranium and other minerals. There is no consistent evidence that radon causes cancer in tissues other than the lung. The excess incidence of lung cancer from radon is concentrated in the period 5-14 years after exposure and decreases with time. The risks for low and protracted exposures are likely to be more appropriate for applying to exposure levels experienced by the public. The numerous studies of residential radon exposure have so far contributed little to radon risk estimation, mainly because of their low statistical power. Important issues that must be addressed include the impact of confounding factors such as smoking and arsenic-containing dusts in mines.

23. Estimates of carcinogenic risk in bone and liver have been derived from exposures to alpha-emitting radionuclides: radium-224 in the case of bone and Thorotrast, a thorium-based x-ray contrast agent, in the case of liver.

24. Long-lived radium-226 and radium-228 at high levels have caused bone sarcomas and carcinomas of the paranasal sinuses in radium dial painters, and the risk extends over the long periods in which these radionuclides are lodged in bone. Precise risk estimates have not been derived. No excess cancers were identified in workers exposed to small amounts of plutonium or to uranium dusts. Workers exposed in Russia to a combination of external radiation and plutonium did have excess lung cancers at the higher exposure levels.

C. OTHER RELEVANT STUDIES

25. In the last decade there were many studies of the incidence of leukaemia near nuclear installations in the United Kingdom following the identification of several leukaemia clusters. One report suggested paternal exposure as a cause. However, in the light of more recent reports it is unlikely that any of these clusters or excesses are due either to environmental radiation or to paternal exposure. A possible explanation is that the excesses are due to the spread of infection that occurs when populations from urban and rural areas mix. No such pattern of clusters was found in subsequent studies around nuclear installations in Canada, France, Germany and the United States.

26. Initial excesses in leukaemia were observed following a single nuclear test explosion in the United States and, following that, explosions carried out by the United Kingdom, but the observation seems to be due in the first case to chance and in the second case to an unusually low incidence in controls for the British participants in the tests and to unusual latencies in the cohort of New Zealand participants. No clear effect is evident.

27. People with certain recessive hereditary diseases, such as ataxia-telangiectasia and retinoblastoma, are known to be sensitive to radiation exposure and are more likely to develop second cancers if treated with radiation. There are indications that those who do not have the disease but are genetic carriers may also be more sensitive than normal individuals to cancer induction, possibly by radiation exposure, but studies so far are not definitive.

II. ADAPTIVE RESPONSES TO RADIATION IN CELLS AND ORGANISMS

28. The scientific community has been aware for many years of the possibility that low doses of radiation may result in changes in cells and organisms, which reflects an ability to adapt to the effects of radiation.

29. It has been suggested in recent years that conventional estimates of the risks of stochastic effects of low doses of ionizing radiation may have been overstated because no allowance was made for the process referred to as adaptation. This is the name given to the possibility that a small prior dose of radiation may condition cells in such a way as to stimulate cellular repair processes and thus reduce either the natural incidence of malignant conditions or the likelihood of excess malignancy being produced by radiation.

30. There is substantial evidence that the number of radiation-induced chromosomal aberrations and mutations can be reduced by a small prior conditioning dose in proliferating mammalian cells *in vitro* and *in vivo*. It seems likely that this effect is linked to an increased capacity for DNA repair. While it has been observed under specified and clearly defined conditions, it has not been seen with all cell systems.

31. There is increasing evidence that cellular repair mechanisms are stimulated after radiation-induced damage. It has to be resolved whether these are related to increased DNA repair. Whatever the mechanisms, they seem able to act not only on the lesions induced by ionizing radiation but also on at least a portion of the lesions induced by some other toxic agents. There appears to be similar overlap in regard to the type of DNA damage that induces adaptive response.

32. It remains doubtful whether the immune system plays any role in these processes. In the UNSCEAR 1993 Report, Annex E, "Mechanisms of radiation oncogenesis", the Committee concluded that the immune system may not have a significant influence on radiation carcinogenesis after low doses. In this Report, Annex B, "Adaptive responses to radiation in cells and organisms", that conclusion is not altered, although some transient effects on the immune system have been identified.

33. Extensive data from animal experiments and limited human data provide no evidence to support the view that the adaptive response in cells decreases the incidence of late effects such as cancer induction in humans after low doses. However, further experimental studies should be conducted.

34. As to the biological plausibility of a radiation-induced adaptive response, it is recognized that the effectiveness of DNA repair in mammalian cells is not absolute. The mechanisms of adaptation are likely to coexist with the mechanisms induced by low doses that may result in malignant transformations. An important question, therefore, is to judge the balance between stimulated cellular repair and residual damage. The Committee hopes that more data will become available and stresses that at this stage it would be premature to draw conclusions for radiological protection purposes.

III. EFFECTS OF RADIATION ON THE NATURAL ENVIRONMENT

35. All living organisms are exposed to radiation from natural sources (cosmic rays and the natural radionuclides present in all components of the terrestrial and aquatic environments) and from local, regional and global contamination arising from human activities.

36. The Committee has not previously attempted to review the effects of radiation on plants and animals in the environment. There is, however, a substantial body of information that can form the basis for such a review. The accumulation of radionuclides in plants and animals in the environment has been considered, particularly from the viewpoint of their transfer through food chains leading to man but also in terms of basic physiology. These data may be developed to provide estimates of the possible concomitant radiation exposure.

37. Previous reports of the Committee have presented summaries of the extensive laboratory studies of the effects of radiation on a variety of animals. In addition, data on radiation effects have been obtained from use of large, sealed sources of gamma rays in the environment and from investigations of the effects, actual or potential, in contaminated areas. Together, these data may be used to assess the relative radiosensitivities of a wide range of organisms and the effects of radiation exposure on those individual attributes (mortality, fertility, fecundity etc.) that are essential for the maintenance of healthy natural populations.

38. The Committee is in the process of reviewing these data and intends to provide a scientific assessment of the impact of increased radiation exposure on the natural environment in a future report.

Appendix I

MEMBERS OF NATIONAL DELEGATIONS ATTENDING THE THIRTY-EIGHTH TO FORTY-THIRD SESSIONS

ARGENTINA	D. Beninson (Representative), E. d'Amato, C. Arias, D. Cancio, A. Curti, E. Palacios
AUSTRALIA	K.H. Lokan (Representative)
BELGIUM	J. Maisin (Representative), P. Govaerts, R. Kirchmann, H.P. Leenhouts, P.H.M. Lohman, K. Sankaranarayanan, D. Smeesters, A. Wambersie
BRAZIL	J. Landmann-Lipsztein (Representative), E. Penna Franca (Representative)
CANADA	E.G. Létourneau (Representative), A. Arsenault, D.R. Champ, R.M. Chatterjee, P.J. Duport, V. Elaguppilai, N.E. Gentner, B.C. Lentle, D.K. Myers, R.V. Osborne
CHINA	Wei Kedao (Representative), Li Deping (Representative), Liu Hongxiang (Representative), Wei Lüxin (Representative), Leng Ruiping, Pan Ziqiang, Tao Zufan, Wu Dechang
EGYPT	F.H. Hammad (Representative), M.F. Ahmed (Representative), F. Mohamed (Representative), H.M. Roushdy (Representative), S.E. Hashish
FRANCE	P. Pellerin (Representative), E. Cardis, R. Coulon, H. Dutrillaux, A. Flury-Hérard, H. Jammet, J. Lafuma, G. Lemaire, R. Masse
GERMANY [a]	A. Kaul (Representative), W. Burkart, U.H. Ehling, W. Jacobi, A.M. Kellerer, F.E. Stieve, C. Streffer
INDIA	D.V. Gopinath (Representative), U. Madhvanath (Representative), N.K. Notani (Representative)
INDONESIA	S. Soekarno (Representative), S. Wiryosimin (Representative), S. Zahir (Representative), K. Wiharto
JAPAN	Y. Hirao (Representative), H. Matsudaira (Representative), T. Asano, Y. Hosoda, T. Iwasaki, A. Kasai, S. Kumazawa, K. Mabuchi, T. Matsuzaki, K. Nishizawa, H. Noguchi, K. Sato, K. Shinohara, N. Shiomitsu, S. Yano
MEXICO	E. Araico Salazar (Representative)
PERU	L.V. Pinillos Ashton (Representative)
POLAND	Z. Jaworowski (Representative), J. Jankowski, J. Liniecki, M. Waligórski, O. Rosiek, S. Sterlinski, I. Szumiel
RUSSIAN FEDERATION [b]	L.A. Ilyin (Representative), R. Alexakhin, A. Bakuniajev, R.M. Barhoudarov, Y. Buldakov, V. Bebeshko, N.A. Dolgova, A. Guskowa, D.F. Khokhlova, Y. Kholina, E. Komarov, I.S. Koshkin, O. Pavlovski, V.V. Redkin, G.N. Romanov
SLOVAKIA [c]	M. Klímek (Representative), D. Viktory (Representative),
SUDAN	O.I. Elamin (Representative), A. Hidayatalla (Representative)
SWEDEN	G. Bengtsson (Representative), L.E. Holm, J.O. Snihs, L. Sjöberg, J. Valentin
UNITED KINGDOM OF GREAT BRITAIN AND NORTHERN IRELAND	R.H. Clarke (Representative), J. Dunster (Representative), J. Denekamp, Sir Richard Doll, J.W. Stather
UNITED STATES OF AMERICA	F.A. Mettler (Representative), L.R. Anspaugh, J.D. Boice, C.W. Edington, J.H. Harley†, N.H. Harley, C. Meinhold, P.B. Selby, W.K. Sinclair, E.W. Webster, H.O. Wyckoff

[a] At the thirty-eighth and thirty-ninth sessions: Federal Republic of Germany.

[b] At the thirty-eighth, thirty-ninth and fortieth sessions: Union of Soviet Socialist Republics.

[c] At the thirty-eighth, thirty-ninth, fortieth and forty-first sessions: Czechoslovakia.

Appendix II

SCIENTIFIC STAFF AND CONSULTANTS COOPERATING WITH THE COMMITTEE IN THE PREPARATION OF THIS REPORT

B.G. Bennett
J.D. Boice
S.C. Darby
D.L. Preston
W.K. Sinclair
H. Smith

Scientific Annexes

ANNEX A

Epidemiological studies of radiation carcinogenesis

CONTENTS

INTRODUCTION

1. The results of epidemiological investigations form the main basis for quantifying the risks of cancer induction in man following exposure to ionizing radiation. Many such studies have been completed and published and many others are in progress to evaluate the effects of irradiation in groups of individuals exposed inadvertently as a result of conditions in the workplace, at home or in the environment or because of atomic detonations or intentionally for purposes of diagnosing or treating diseases.

2. In the UNSCEAR 1988 Report [U2] the results of epidemiological studies were examined to derive estimates of the risk of cancer induction from exposures to ionizing radiation. The estimates were based primarily on the extended follow-up with revised dosimetry of the life span study of survivors of the atomic bombings of Hiroshima and Nagasaki. The risk estimates for cancer referred specifically to the Japanese population exposed to absorbed doses of the order of 1 Gy of uniform whole-body low-LET radiation at high dose rate. The results of other studies were used primarily to support the principal results derived from the life span study.

3. In this Annex further review of epidemiological studies of radiation carcinogenesis is provided. One objective is to review the risk estimates in the UNSCEAR 1988 Report and assess their validity in the light of new information on all epidemiological sources. Further studies of the survivors of the atomic bombings in Japan provide new information on cancer incidence, mortality, dose-response relationships, the effects of exposure in childhood and the appropriateness of various models of radiation risk for use in risk projection. A large number of additional epidemiological studies are contributing quantitative or otherwise useful information on the effects of both high- and low-LET ionizing radiation in human populations. They include studies of patients exposed to radiation for diagnostic or therapeutic purposes, studies of workers

occupationally exposed in the nuclear industry, studies of populations exposed to terrestrial gamma rays at varying levels and to other environmental exposures, studies of individuals exposed to radon in homes and other buildings, and studies of the effects of local and global fallout from atmospheric tests of nuclear weapons.

4. Although it has been well established that radiation exposure after high doses causes an increased risk of cancer in many organs, there remain many important questions about the effects of high-dose and high-dose-rate exposure compared to the effects of low-dose chronic exposures; variation in risk with age at exposure, individual sensitivity and ethnicity, including transferral of risk from one population to another; changes in dose response with time since exposure; and the joint effects of radiation and other agents. As epidemiological data accumulate, it is possible to carry out more detailed investigations of these issues. Recent findings are reviewed in this Annex.

5. The data available from epidemiological studies allow estimates of radiation risk to be calculated. While it would be desirable to make joint analyses of all these data, differences in quality and lack of sufficient information in many of the studies present great difficulties. The Committee has, however, attempted to make a comparative study of risk estimates for specific types of cancer that can be separately derived from the various studies. The types or sites of cancer considered here include those for which data are available from one or more studies involving reasonably large populations with quantitative individual or population average dose estimates. These risk estimates should be interpreted cautiously because of differences between the various study populations, but they nevertheless provide a rough guide to the degree of consistency among the various studies. As an aid in the interpretation of the risk estimates derived here, descriptions are provided of the nature of the various populations considered and of the strengths and limitations of these studies.

I. METHODOLOGICAL CONSIDERATIONS

A. THE ROLE AND LIMITATIONS OF EPIDEMIOLOGY

6. Among the biological effects of concern when human beings are exposed to low doses of ionizing radiation, the most important long-term effect is the possible induction of cancer. Much can be learned about the nature of the cancer-induction process from laboratory experiments in cellular systems and animals. Even some insights into the mechanism of induction may be obtained in this way. Nevertheless, until now the most potent method of studying radiation-induced cancer is the epidemiological study of exposed human populations. Only epidemiology is able to describe the types of cancer induced in humans, their frequency as a function of dose and time after exposure and the many factors, such as age and sex, that modify their expression. Furthermore, only epidemiology is able to quantify these responses and, as a result, derive estimates of the risks of cancer induction as a function of dose. As successful as epidemiology has been in contributing to a broad understanding of radiation-induced tumours in humans and the quantitative risks of the process, it has some inherent limitations. When the dose is high, epidemiological studies have identified clear-cut responses, and risk estimates for many types of cancer have been derived and their dependencies on other factors explored. However, when the dose is low and the effect to be detected is very small compared with the natural occurrence of cancer in the irradiated population, quantification of the risks is very difficult and perhaps impossible. Additional strengths and weaknesses of epidemiology will be discussed further below.

7. Epidemiological studies can provide valuable information about the applicability to humans of dose response and other information suggested by data obtained in experimental systems. Although epidemiology is not intrinsically expected to contribute much to the overall knowledge of radiobiological mechanisms, the many parameters that epidemiology has established as modifying and influencing cancer induction must nevertheless be taken into account in any theory of the mechanisms of action of radiation-induced cancer. Radiation epidemiology has already contributed to general theories of carcinogenesis, as, for example, in breast cancer, where epidemiological studies have shown that at young ages the female breast is especially vulnerable to environmental insult.

8. Some information on the carcinogenic effects of radiation exposure is provided by clinical trials, but most epidemiological studies of radiation effects are inherently observational in nature. That is, they are arranged by circumstance rather than as a result of experimental design. This means that the conditions of exposure, the study population and the allocation of individuals to the various exposure levels are outside the control of the research worker. In fact, the exposures are often determined for reasons quite unconnected with the objectives of a particular epidemiological study. In the case of exposure to ionizing radiation, exposures occur, for example, as a result of the geographic location of an individual at a particular time (in studies of the survivors of the atomic bombings in Japan); the place of residence and type of dwelling (in studies of the effects of domestic exposure to radon and of exposure to terrestrial gamma-radiation);

occupation (in studies of radium dial painters, uranium miners, tin miners and workers in the nuclear industry); medical treatment of disease (in studies of patients with ankylosing spondylitis or cervical cancer); or simply prevalent medical fashion (in studies of children who received radiotherapy for a supposedly enlarged thymus and children whose mothers were given x-ray examinations during pregnancy).

9. The characteristics of epidemiological studies have the following consequences:

(a) the effects of interest cannot always be studied directly or with the desired precision;
(b) randomization procedures can rarely be used to ensure the absence of undesirable systematic differences between those exposed at different levels;
(c) studies cannot usually be repeated at the command of the investigator.

The observations that can be made are limited to the effects of exposures that have actually occurred. For example, while the effect of radon exposure on lung cancer incidence in the general population is of great interest, current risk estimates are determined only in studies of miners with occupational exposures to radon and its decay products. These populations consist of males who were between the ages of 18 and 65 at the time of exposure, and the conditions under which the exposures were received differ markedly from those in which general population exposures are received. In addition, there are often systematic differences between those exposed at different levels and the general population; appropriate comparison populations are therefore important. For example, in the thyrotoxicosis study carried out in the United States [S2], an excess of leukaemia was observed when rates for patients treated with ^{131}I were compared to rates for the general population. However, a non-exposed group of thyrotoxicosis patients was found to have leukaemia risks comparable to the patients treated with ^{131}I.

10. Because investigators lack control over many important aspects of epidemiological studies, observed associations between exposure and subsequent disease can be distorted by the presence of factors associated with the exposure and with the disease in the individuals studied. Thus, an important aspect of the design and analysis of any epidemiological study is to identify and eliminate such distortions. Factors such as age, sex and ethnicity are important, easily measured determinants of disease that should be taken into account in virtually every study worthy of serious consideration. Other factors, such as cigarette smoking, socio-economic status and employment history, are quantifiable in principle if accurate records are available or if interviews can be carried out with each study participant. When such information is obtained after the exposure or disease diagnosis, special care must be taken to ensure that the ascertainment is of equal quality for exposed and unexposed individuals or for cases and controls.

11. A common study design in radiation epidemiology is the cohort study. In a cohort study, a fixed, well-defined group of individuals not known to be suffering from the disease, or diseases, of interest is enrolled and followed forward in time. Information is collected on deaths from, and if possible, incidences of the diseases of interest in the enrolled population. Because the diseases of interest are often rare and the effects of exposure relatively small and because there is a need to determine how risks vary with sex, age at exposure, time or other factors, epidemiological cohort studies of radiation effects need to include thousands or even tens of thousands of individuals in the study group, and follow-up must continue for years or decades. Although prospective follow-up data are available in many cases, most cohort studies were begun with a retrospective review of historical data such as employment or medical treatment records. The size of a cohort study and, for a retrospective cohort study, its timing usually preclude personal contact between the investigators and individual cohort members. Thus, individual data on other than a few basic items are rarely available. An important element in the design of cohort studies is to ensure that the relationship between disease rates and the exposure is not distorted by unmeasured factors.

12. It is not possible at present to determine if a specific cancer was caused by radiation, although there are encouraging indications that advances in molecular biology will allow such a determination [T16, U1]. Therefore, the extent of radiation-induced cancer in a cohort must be inferred by comparing similar groups with different levels of exposure. In cohort studies this may involve the comparison of exposed and unexposed groups or, ideally, groups receiving a range of exposures. In many studies of radiation effects it is not possible to include an unexposed control group; therefore, inferences are based on comparison of the observed number of cases to an expected number derived from national or regional rates. While these rates are usually readily available for mortality, they are much less often available for cancer incidence. Thus, an internal comparison group is especially important for studies of radiation-induced cancer incidence, as illustrated by the thyrotoxicosis study mentioned above. When national or regional rates must be used, it is desirable to investigate, if possible, the validity of the comparison. This was done in the British ankylosing spondylitis study by identifying and following up roughly 1,000 individuals with the disease who were not treated with radiation [S21]. When internal controls are available, national rates may still be useful as a check on the validity of the control group and on the stability of inferences about dose effects. This has been done in some analyses of data on survivors of the atomic bombings in Japan [P13, P23].

13. When it is not possible or practicable to conduct a cohort study, case-control studies can provide valuable information on radiation effects. The strength of a case-control study lies in its ability to make effective use of

data on relatively small numbers of cases and controls. In this type of study, information is collected both on individuals who have already developed the disease and on disease-free controls. Controls are usually individually matched to the cases with regard to age, sex and other relevant factors. Case-control studies are often constructed from within previously defined cohorts in order to allow the use of more detailed data than could be obtained for the full cohort. In a case-control study, inference is based on a comparison of the level of various factors, e.g. radiation dose, in the cases and controls. The validity of a case-control study depends on the extent to which the controls are truly representative of the population from which the cases were drawn. This is not always easy to achieve. Case-control studies are appropriate for gaining insights into several areas in radiation epidemiology. Examples include the assessment of the effect of radon exposure in the home on lung cancer risk among people with different smoking habits [B37, P30, S5] and of the interactions of radiation and other risk factors for breast cancer in the survivors of the atomic bombings in Japan [L16] and cervical cancer patients [B20, B21].

14. An important consequence of the observational nature of epidemiology is that considerable caution must be exercised before assuming that the relationship between a disease and some environmental or occupational exposure is a causal one. Once the requirements for high quality data and appropriate analysis have been satisfied, criteria for evaluating causality include the following: the strength of the association between the disease end-point and the presumed causative factor; the consistency of data from a variety of studies; the existence of a dose-response gradient; and experimental evidence that similar effects can be produced in the laboratory. In the context of radiation, a good example is given by lung cancer and occupational exposure to radon [C6]. In the late nineteenth and early twentieth centuries, radon concentrations in the mines of central Europe were high, and it was eventually established that approximately 50% of the miners died from lung cancer [M14]. Elevated risks of lung cancer were later seen in a variety of other mining populations exposed to high concentrations of radon, including miners of iron ore in Sweden and Britain, fluorspar miners in Canada and uranium miners in Canada and the United States. A very recent analysis of 11 underground miner studies indicates that of almost 2,700 lung cancers found among 68,000 miners, fully 40% are attributable to radon exposure during the miners' working lives [L21]. Also, some of the more detailed studies have demonstrated a strong dose-response relationship, and similar results in laboratory studies on rats and dogs have been available since the 1970s. Doubts about the causality of the relation are reduced by the observation that miners in other types of mines, such as coal or nickel miners, which would have been equally dusty but where levels of radon are not increased, have been shown to have little or no increase in the risk of lung cancer or possibly even a decrease [A4, G12, M14, M16]. Despite these findings, the role of dusty conditions in the mines and especially the presence of arsenic are not yet fully understood [C21, L17].

15. Radiation exposures of the general population usually involve low doses delivered gradually over time. For example, in the UNSCEAR 1993 Report [U1], it was estimated that the average annual effective dose to adults from all natural sources of radiation is about 2 mSv and that the proportion of people receiving an annual effective dose in excess of 10 mSv is very low. Interest in assessing the effects of radiation for the purposes of radiological protection therefore centres on the effects of exposures to a few millisievert per year. The vast sample sizes that would be required for direct study of the effects of such exposures were discussed by Land [L7] and by Goss [G24]. That it is difficult for epidemiological studies to provide definitive results in such circumstances becomes evident.

16. The ability of radiation to act as a carcinogen in most organs and tissues of the body has been well established by studies at much higher doses (for example, of about 1 Sv or more, often delivered at a high dose rate) and in which the exposure has increased cancer morbidity rates by 40%-50% or more, i.e. studies in which the relative risk is about 1.5. As the results of numerous studies discussed in this Annex illustrate, epidemiological studies can detect and quantify effects of this magnitude and provide useful information on the factors that affect them, including sex and age at exposure. However, if a study is to address lower dose exposures (0.1 Sv or less), and if it is assumed that the response is linear in dose, the increase in cancer risks would be only 5% or less, i.e. the relative risk would be 1.05 or less. To have a reasonable probability of detecting a risk (i.e. statistical power) of this magnitude, an epidemiological study would require thousands or even hundreds of thousands of cases of the disease of interest among exposed subjects. Even if it were feasible to assemble cohorts (or case series) large enough to detect such low-dose effects, it would be difficult to interpret any findings because other undetected small differences between the control and exposed populations would easily mask existing effects or induce spurious effects of this size. In view of the lack of power of most studies to detect low-dose effects, it should be recognized that the failure of a particular study to demonstrate an effect (positive or negative) at low doses is not necessarily an indication that such effects do not exist. Furthermore, the fact that epidemiological studies lack the power to detect the small effects that might be seen at low doses does not mean that they provide no information on low-dose effects. Such studies can often provide upper limits for the magnitude of the risk. Also, it is important to note that epidemiological studies have been quite successful in detecting the effects of radiation exposure in sensitive subgroups of the population, e.g. prenatal exposure to x rays with exposures of about 0.01 Gy and childhood thyroid cancer at doses of roughly 0.1 Gy.

17. The term "statistical power" is relevant to the issue of significant results from radiation studies, especially low-dose studies. Statistical power depends on the sample size, the length of follow-up, the method of analysis and the average dose to which the group is exposed. The follow-up must extend for a substantial period beyond the latency period, which is the time between the exposure and the occurrence of disease. When doses are low (<0.2 Sv), the sample size must be very large in order to detect a small effect, as noted above [G24, L7].

18. It would seem that, for the foreseeable future, the important questions in epidemiology — the effect of fractionated low-dose exposures, the temporal pattern of the effect in a population, the relative magnitude of the effect among those exposed at different ages and the joint effect of radiation and other agents — must continue to be evaluated from studies on effects at high doses, with supporting laboratory experiments to guide the extrapolation from high to low doses. Nevertheless, it is essential that human populations exposed at lower doses be monitored to ensure that extrapolation from the results of studies on effects at high doses has not given risk estimates that are inappropriate. The most suitable groups for study in this way are those in which some individuals have exposures higher than those received by most members of the public from natural radiation or common diagnostic medical procedures. Examples of such groups include radiation workers and people living in areas with high indoor radon concentrations. Studies of the geographical variation in mortality with exposure to sources such as terrestrial gamma-radiation and radon are in principle also important. However, in practice it is difficult to interpret such studies owing to problems in obtaining adequate exposure data and in allowing for socio-economic and other factors that may influence disease rates to a greater extent than the variation in radiation exposures.

19. The interpretation of results from studies of populations exposed to low levels of radiation presents many problems. Inevitably, the studies will tend to be negative in the sense that major potential effects, such as the association between exposure and leukaemia rates may not reach statistical significance. However, they may still be of great value in so far as they provide upper bounds to the risk in situations of practical interest. Also, if several similar studies can be carried out, as is at present happening with studies of radiation workers and residential radon exposures, it may be possible to combine results, making it easier to quantify an effect.

20. There is still another consequence of studying populations exposed to low levels of radiation: if a large number of significance tests are carried out, some findings will reach nominal statistical significance purely by chance. To minimize the difficulties of interpretation in such situations, it is important to clearly specify which of the many hypotheses tested were of special interest a priori and which were singled out after the event. For example, in 1982 Smith and Doll [S13] reported a large, statistically significant standardized mortality ratio (SMR) for tumours of the central nervous system other than those of the brain among British patients who received radiotherapy for ankylosing spondylitis. With additional follow-up, however, the magnitude of this risk estimate decreased markedly, and the statistical significance of the effect disappeared [D5]. Similarly, earlier analyses reported statistically significant increased risks of multiple myeloma in survivors of the atomic bombings [S7] and in workers at the Hanford plant [G20]. However, in the life span study incidence data [P33] in which roughly 50% more cases are reviewed, evidence for increased myeloma risks in the cohort is weak and, in any case, the risk is likely to be much smaller than suggested in the earlier reports. A recent report on the Hanford workers [G16] reports lower risks that are no longer statistically significant.

21. The above discussion is formulated largely in terms of mortality and thus ignores the fact that some radiation-induced cancers are not fatal. The difference between incidence and mortality is sometimes small, as, for example, in cancers at sites such as the stomach, liver or pancreas or in adult leukaemia, which have low (<20%) relative survival rates. (The relative survival rate is the ratio of the probability of surviving for a fixed period, usually five years, after diagnosis to the survival probability in the general population [J6].) However, mortality data cannot be expected to provide complete or reliable risk estimates for sites with high relative survival rates, since such cancers are unlikely to be listed as the cause of death. Examples of cancers that have relative survival rates greater than 50% include thyroid, melanoma and non-melanoma skin cancers, female breast, uterine corpus, prostate and urinary bladder [M29]. The issue of incidence versus mortality is further complicated by the fact that survival rates vary widely from country to country (see Table 1) and by the fact that they vary with time within single countries for many reasons, including improvements in medical therapy (see, for example, [H30]).

22. Since many countries do not have general population cancer registries and radiation-exposed populations are not usually included in existing cancer registries, it is difficult to carry out incidence studies of radiation effects. However, despite the difficulties a number of important incidence studies are now available. These include the registry-based cervical cancer cohort and case-control studies [B12, B21], survey-based studies of breast cancer among women exposed to radiation for the treatment or diagnosis of various medical conditions [B1, B18, B31, M19, S9], analyses of breast cancer and, most recently, a series of general analyses of cancer-registry-based incidence data on the survivors of the atomic bombings in Japan [P33, T15]. As has been noted by Hoel and Dinse [H29], it is possible to construct models for cancer incidence at some sites by combining data on mortality

following radiation exposure with case-fatality data. However, these methods appear to be of limited utility as they require data on the time from diagnosis to death.

23. Doll and Peto [D19] discussed other issues related to the strengths and weaknesses of epidemiological studies. By the nature of epidemiological studies, the contribution of any single study, however excellent, is much more limited than that of an experimental study. Firm conclusions can be drawn only when a number of studies carried out in different circumstances provide similar results. Accordingly, it is especially important in this field to carry out critical reviews of all the relevant studies from time to time, and to attempt to integrate the available information into a meaningful composite.

B. QUALITY OF DATA

24. Careful attention to data quality is important in the conduct, interpretation and comparison of epidemiological studies. There are many factors that can affect the quality of the data in studies of populations exposed to radiation. These include incomplete follow-up or exposure-related differences in the completeness of follow-up; recall bias in the ascertainment of exposure information; and comparability of the exposed and control groups. In addition, when moving beyond simple ascertainment of the existence of an effect to more precise quantitative descriptions of the effect, there must be a good understanding of the nature and quality of the dosimetric data. In comparing the results of different studies or in applying risks from one population to another, in addition to all these factors that might affect the data quality, it is necessary to consider carefully other factors, such as the comparability of the populations under consideration with respect to age, sex, ethnicity, smoking, diet and possible specific demographic factors.

25. In cohort studies it is essential either to ensure that the follow-up is virtually complete and that all relevant occurrences of disease in the enrolled population have been identified or to understand the nature of incomplete follow-up and adjust for it. It is particularly important to obtain and make use of information on migration or loss to follow-up. Obtaining complete follow-up is generally much easier for mortality data than for incidence data. For example, in Japan the compulsory system of family registration *(koseki)* makes it possible to obtain death certificates for virtually all deceased members of the atomic bomb survivor cohort, while cancer incidence data are generally available only for survivors who reside in the areas covered by the Hiroshima and Nagasaki tumour registries at the time of diagnosis. In mortality analyses, *koseki* information on emigration from Japan is available and is used to identify people who are lost to follow-up. As will be described later, special measures were taken to allow for the effects of migration in analyses of the data on cancer incidence in the survivors of the atomic bombings. In some places (e.g. the Nordic countries), long-standing national cancer incidence registries exist, and cancer ascertainment is facilitated by record linkage procedures based on personal identification numbers.

26. Even when good sources of follow-up data are available, one must be concerned about the quality of the data and their interpretation. For example, the general underreporting of cancer on death certificates, which may be age-dependent, means that although reasonable estimates can be made of relative risks, absolute risks are underestimated from death certificate diagnoses. In site-specific analyses, especially those based on death certificates, the misclassification of cancer types on the death certificate (mainly underreporting of cancer) can also affect risk estimates. Sposto et al. [S66] investigated the impact of cancer/non-cancer misclassification and found that after allowing for cancer recorded as non-cancer and *vice versa*, cancer risk estimates from the life span study should be increased by more than 10% relative to estimates that ignore the effect of these recording errors.

27. Another concern in the conduct of epidemiological studies involves the comparability of the exposed and control groups. This is particularly important when an exposed population rate is compared to general population rates, in which case the availability of valid appropriate population rates is critical. The so-called "healthy worker" effect [H44], in which workers tend to have lower mortality rates for cancer and for other diseases than the general population especially in the first few years after selection for employment, is one illustration of this problem. The healthy worker effect is only one aspect of the general problem of comparability of rates in any selected cohort to rates for a general population. Cause-specific death or incidence rates can also exhibit great variability within the general population for reasons other than employment status. Thus, when making comparisons to general population rates it is important, whenever possible, to emphasize relative risks and trend tests rather than to focus simply on standardized mortality ratios. Even if an unexposed internal comparison group is available, there must be concern for the appropriateness of the comparisons being made and any unusual features of the control population.

28. It is also essential to avoid bias due to differing probabilities of disease or to differing exposure detection between exposed and unexposed individuals. Detection bias can be a serious problem for some outcomes (if, for example, surveillance is increased as a result of concern about radiation exposures). As a result, many occult cases that would otherwise go undetected may be diagnosed, as happens, for example, in the detection of thyroid cancer with enhanced screening procedures.

29. Another concern involves biases resulting from exposures that may have taken place because of symptoms

of a subsequent disease or from differential recall of exposure for cases and controls. As an example of the former, the elevated risk of neoplasms other than leukaemia or colon cancer seen in the first five years after treatment in the British ankylosing spondylitis study is discounted by the authors because "some of the tumours presenting soon after treatment may have caused the symptoms that were incorrectly ascribed to spondylitis" [D6]. The possibility of recall bias was raised in criticisms of the initial studies of prenatal radiation: it was felt that parents of cases might have been more likely to report prenatal x-ray examinations than parents of controls.

30. The important issue of adequate statistical power was discussed in the preceding Section. Indeed, it is well known that epidemiological studies are problematic when the effects being studied are much less than half the normal baseline risk. Thus, increasing power by increasing the sample size is unlikely to provide reliable estimates of risk at low doses and low effect levels, because of the possible presence of competing or confounding risk factors that might safely be ignored at high doses and effect levels. There may well be unknown factors that increase or decrease risk by a few percentage points, i.e. near the level of effect in a low-dose study. Such factors will not have come to attention earlier because they seldom if ever are present at levels high enough to be apparent in an exploratory study of modest size, and they might be confounded with exposure to low doses of radiation. The problem can be addressed by increasing the amount of information on each study subject, but that solution can be very expensive for a large study, and it would be guided by very little prior information.

31. A related question concerns the influence of chance in statistical observations in studies in which multiple comparisons are made. If, for example, tests of significance are conducted at the 5% level for a series of comparisons between control and exposed groups, 5% (1 in 20) of the tests will result in an apparent difference due to chance alone. Thus, in a study involving many such comparisons (e.g. in a low-dose environmental study with many different cancer end-points evaluated) erroneous positive or negative results are likely to occur with a frequency that depends on the level of significance in the test.

32. The Committee notes that reliance on studies that have demonstrated statistically significant effects can be misleading. First, journals are more likely to publish (and authors more likely to report) studies that have some statistically significant result. Secondly, point estimates of risks for statistically significant effects are, especially for small studies, biased upward and can be misleading if not put in the context of other results. In addition, since epidemiological studies of radiation effects cannot usually be designed to ensure adequate power to detect effects, the ranges of risk estimates from positive and negative studies can provide useful information on risk.

33. In epidemiological studies of radiation effects, suitable measurements of individual doses made at the time of exposure are usually unavailable. In such cases estimates are developed by reconstructing the radiation fields based on theoretical constructs or computational models of the original exposures (e.g. the life span study) or by carrying out post hoc measurements for representative exposures believed to be comparable to those that occurred originally (e.g. the tinea capitis study in Israel). Dose estimates may be assigned to groups of subjects believed to have been exposed under comparable conditions or, as was done for the ankylosing spondylitis study in the United Kingdom, individual dose estimates computed for a sample of subjects may have been used to estimate population-average doses to selected organs. These estimates were more accurate for some organs than for others. More precise estimates of organ dose were possible in the cervical cancer study [B21] and in the study of benign gynaecological disease [I8]. Average organ doses must be used with caution when the nature of the exposure varies widely between subjects. Even when individual monitoring measurements are available, as for example in many studies of nuclear workers, there are often questions about adequacy and completeness, since the data were collected for purposes other than the study of radiation effects and may not exactly suit the needs of an epidemiological study. Furthermore, the derivation of organ doses from personal monitoring measurements is subject to some uncertainties owing to the nature of the exposure and the orientation of the individual in the radiation field.

34. In epidemiological studies whose sole aim is the simple identification of an effect, such as cancer, with a causative agent, such as radiation, it can be sufficient to distinguish between exposed and unexposed individuals, i.e. a precise knowledge of individual exposure levels is not critical. However, in more quantitative studies, in which the goal is to provide an estimate of the dose-response function, the quality of the dosimetric data is as important as the quantity and quality of the data on the outcome of interest.

35. The impact of errors in dosimetry has been examined most extensively for the data of the life span study of survivors of the atomic bombings. Jablon [J1], who first considered this problem, concluded that uncertainties in the data on location and shielding for individual survivors would lead to errors of 35%-40% in individual dose estimates (T65D dosimetry system). Using data on discrete symptoms such as epilation and bleeding, Gilbert and Ohara [G10] found evidence of either systematic differences or differing amounts of random error in the methods used to construct the T65D dose estimates. Gilbert's results supported average errors in individual dose estimates comparable to those noted by Jablon. Following revision of the dosimetry (to DS86) for the survivors of the atomic bombings in Japan [R22], several

studies have examined the nature of random errors in individual dose estimates. Pierce et al. [P9, P31] developed methods to adjust risk estimates for the biases resulting from random errors in individual dose estimates. They concluded that random errors of 30%-40% can lead to risk estimates for solid cancers that are 7%-11% lower and risk estimates for leukaemia that are 4%-7% lower than they would be if "true" doses could be used in risk estimation. Thus, to correct for this effect, risk estimates would need to be increased by those amounts. Stram et al. [S22] found that for the same estimated doses epilation rates differed very significantly by shielding category. Sposto et al. [S55] compared chromosome aberration dose-response curves for epilators and non-epilators and concluded that 40%-50% random errors in individual DS86 dose estimates would be needed to explain the observed steeper slope seen for survivors who reported severe epilation.

36. The studies just noted have focused on random errors in individual dose estimates and the impact of such errors on risk estimates. They do not deal with possible systematic errors in the DS86 dosimetry, such as those described by Straume et al. [S56], which are discussed later (paragraphs 156-158). The attention is given to the impact of dose errors in the data on survivors of the atomic bombings does not imply that the dose estimates are worse than those in other studies of radiation effects; rather, it reflects the availability of a large amount of data on individual survivors and the concern of those involved in the Japanese studies to provide a thorough summary of the strengths and limitations of the data.

37. The quality of the dosimetry has also been examined for some other studies of radiation-exposed populations. For example, in a study of the estimated radiation doses to 12 bone marrow sites and various other organs of ankylosing spondylitis patients treated in the United Kingdom with x rays, Lewis et al. [L6] compared the mean marrow doses obtained from the Monte Carlo calculations with those obtained from experiments on a physical phantom. Although the results obtained using the two methods were very highly correlated, the Monte Carlo estimates were, on average, 19% higher. The most likely explanations for the difference are the different distributions of energies in the primary photon beam and the different compositions of bone marrow used in the two studies. The results of Lewis et al. suggest that the coefficient of variation, i.e. the standard deviation divided by the mean, for individual dose estimates ranges from 10% for organs close to the beam to about 50% for less directly exposed organs. More recently, Stovall et al. [S18] described efforts to reconstruct doses for medical exposures. Dose estimates developed by Stovall et al. are used in a number of epidemiological studies.

38. Better descriptions of the uncertainties in individual dose estimates are needed in most studies of radiation effects. The fact that such issues are not addressed in the presentation of results of a specific study should not be taken as an indication that the dose estimates are particularly accurate or that uncertainties in individual dose estimates would have a negligible impact on the reported risk estimates.

C. METHODS OF DATA ANALYSIS

39. The analysis of epidemiological studies should involve life-table methods to fit regression models that take into account sex- and age-specific death or incidence rates and any other factors influencing disease incidence that may be correlated with exposure and for which data are available. Statistical methods were summarized by Breslow and Day [B28, B29], and modelling issues in the context of epidemiological studies of radiation effects were discussed by Preston [P32] and Vaeth [V12].

40. For the most part, the questions of interest in the analysis of studies of the effects of radiation in persons known to have been exposed to appreciable doses have gone beyond answering the basic question of whether there is an association between exposure and disease. Interest now centres on issues of a more complex nature, such as the magnitude and duration of any effects, and the variation in risk among those exposed at different ages. For many studies data are now available to address these questions, at least in part. However, correct inferences can be made only if they are based on models that accurately reflect the variation in disease rates indicated by the data. If inappropriate or oversimplified models are used, highly misleading inferences can be made. For example, when the annual change in the excess relative risk of non-leukaemia cancer mortality among survivors of the atomic bombings in Japan after 1960 was calculated ignoring the effect of age at exposure, the excess relative risk was found to be increasing at 3.5% per year. However, when it was calculated adjusting for the effect of age at exposure, the excess relative risk was found to be decreasing slightly [S7].

41. In presenting the results of specific studies, investigators usually focus on relatively simple descriptive statistics or models of the experience of the study population. The results may average over or adjust for differences in the study population, such as age and sex, and may include extrapolation and projection to extend risk estimates to lifetime experience. To make results understandable and to allow comparisons to be made with other studies, it is necessary to give some attention to the definition and use of the risk estimates.

42. Radiation effects are usually described in terms of various simple measures of the excess risk. The most commonly used summary statistics are relative and absolute risk estimates. In simple terms, if O is the number of events observed in a population and E is the

number of events expected in the population in the absence of exposure, the relative risk (RR) is defined as

$$RR = \frac{O}{E} \qquad (1)$$

and the excess risk (ER) is defined as

$$ER = O - E \qquad (2)$$

When comparing relative risks or presenting risks per unit dose, as described below, it is preferable to summarize results in terms of the excess relative risk (ERR), defined as

$$ERR = \frac{O}{E} - 1 = \frac{O - E}{E} \qquad (3)$$

instead of the relative risk. In order to compare estimates derived from populations with different levels of exposure, it is useful to work with risks per unit dose, called risk coefficients. Thus, if D is the average dose received by an exposed population, the linear excess relative risk coefficient is defined as

$$ERR = \frac{O - E}{ED} \qquad (4)$$

The term dose and the symbol D are used in a general sense to represent absorbed dose (in gray) or weighted dose (in sievert) with neutron RBE = 10, for the atomic bomb survivors (see paragraph 64). Both quantities determined in equations (3) and (4) are commonly referred to as excess relative risk, the context making clear the distinction. To compute roughly comparable values for the excess risk in different studies it is necessary to consider both the dose and the amount of follow-up time. This is often done through the use of the excess absolute risk (EAR) per unit dose and per unit time at risk:

$$EAR = \frac{O - E}{PY\ D} \qquad (5)$$

where PY is the number of person-years of follow-up.

43. A variety of methods are used for computing the expected numbers of cases used in the above definitions. It is common practice in epidemiology to allow for the effects of age, sex and other characteristics of the study population when computing the expected numbers of cases. Estimates that are computed allowing for such effects are called adjusted risks. However, when (as current data on radiation effects suggest) excess risks vary with age at exposure, sex or follow-up time, the summary statistics defined above are weighted averages involving an implicit weighting that depends on the composition of the population, with the largest weight being given to groups that have the largest number of cases. This means, for example, that the risk estimates generally cited for the life span study, which were used as the basis for many calculations in the UNSCEAR 1988 Report [U2], are heavily influenced by the experience of the older survivors and that, owing to the nature of the study, risks for persons exposed as young adults contribute relatively little. When excess relative risks of cancers other than leukaemia from the life span study are computed by averaging risk estimates for a small number of groupings of age at exposure and sex using equal weights, the resulting excess relative risks are about 50% greater than the usual age-averaged estimate that gives the greatest weight to the older members of the cohort [P3]. Since the age-averaged summary risk estimates depend heavily on the demographics of the population on which they were based, similarities or differences in such simple average risk estimates should be interpreted with caution.

44. The most common approach to exploring variation in the risks of radiation-induced disease is to carry out separate analyses of subsets of the full data set. Examples include analyses of subsets defined by sex, age at exposure or time since exposure. In a subset analysis, a simple summary measure is computed, and separate tests of the null hypothesis of no excess risk are carried out for each subset. Such analyses tend to promote overreliance on significance tests and overinterpretation of sampling errors. Modern analytical methods and software make it easy to carry out generalized regression analyses that emphasize tests for heterogeneity in response functions, e.g. that determine whether or not risks differ significantly by age at exposure or sex and that estimate the size of the differences [P35]. These methods can be usefully employed even in studies without individual dose estimates.

45. In many studies, excess cancer believed to be associated with radiation is reported without regard to dose, usually because of the poor quality of individual dose estimates. Summarizing risks in this way complicates the comparison of results from different studies, since such comparisons are of little use in understanding the magnitude and nature of radiation risks. When dosimetric data are available and linear risk estimates are deemed to be appropriate, the risk per unit dose should be presented. When non-linear dose functions are used, they should be described clearly and accompanied by evidence for a lack of fit of the linear model. Any presentation of risks should be accompanied by a clear indication of the nature and quality of the dose estimates used.

46. In considering the risk of radiation-induced cancer projected in time, it is common to make use of what are often referred to as lifetime risk estimates. Several studies [P11, T14] have examined the definition, interpretation and presentation of lifetime risk estimates. Three principles can be identified that underlie the computation of lifetime risks:

(a) independence of the follow-up period in the studies contributing information;
(b) recognition of the impact of cultural and environmental differences such as smoking on normally occurring cancer rates;

(c) allowance, as necessary, for the modifying effects on radiation risks of sex and age at exposure.

47. The first principle, stated in (a), concerns what is called risk projection. Few epidemiological studies of radiation carcinogenesis have more than 40-50 years of follow-up information, so it is inevitable that the calculation of lifetime risks involves extrapolation beyond the time period covered by most studies, especially for persons exposed early in life. When presenting lifetime risks it is important to indicate explicitly the approach taken and the effects of projection beyond the period covered by the data on which the underlying risk models are based. In several reports, including the UNSCEAR 1988 Report [U2], the projection effects were assessed by comparing lifetime risks calculated for constant absolute risk models with those calculated for constant relative risk models. Since constant absolute risk models no longer appear to adequately describe the variation in risk of radiation-induced cancer with time, this comparison does not provide useful information on the effects of risk projection.

48. The principle stated in (b) involves using descriptions of risk estimated for one population to predict what might be seen in a population with different site-specific cancer rates. Extensive data on the variation in cancer rates around the world have been presented by the International Agency for Research on Cancer [M17, P10]. Some examples of high and low cancer rates in various populations are given in Table 1, from which it can be seen that site-specific cancer rates can differ by as much as two orders of magnitude, as do, for example, rates for nasopharyngeal cancer, and that such great differences are not uncommon. While some of the differences are due in part to reporting methods, they also undoubtedly reflect real differences in incidence rates. Of particular interest is the contrast between rates for cancers of the stomach, lung and breast in the population of the whole of Japan and in the populations of some countries, particularly western ones. Although Japanese stomach cancer annual rates have been falling in recent years, they are still much greater than those in western Europe or the United States (50-60 per 100,000 per year compared to 6-10 per 100,000 per year). Stomach cancer rates in the Russian Federation and other parts of the former Soviet Union average 40 per 100,000 per year but are quite variable across the country. On the other hand, despite the fact that breast cancer rates in Japanese women (about 25 per 100,000 per year) are increasing, they are still less than half those in Europe and the United States (55-90 per 100,000 per year). Lung cancer rates also tend to be lower in Japan (25 per 100,000 per year) than in many other countries (40-60 per 100,000 per year). The differences reflect variation in the levels of exposures to carcinogens other than radiation and, in some cases, variation in the susceptibility of different ethnic groups.

49. Such differences in background rates of cancer raise questions about the applicability of risk estimates to a population other than the one for which they were derived. Modern analytical methods allow the development of time-dependent models for excess absolute or relative risks that provide comparable descriptions of the current data for a specific population, but they can lead to different lifetime risk estimates when applied to populations with different cancer incidence. Despite the large number of studies of radiation and cancer in humans, it is still unclear how best to transfer risk estimates to different populations. This issue is discussed further in Chapter II.

50. The calculation of lifetime risks requires a set of normally occurring death rates from all causes and from the specific causes of interest for models of radiation effects. These are usually cross-sectional rates, i.e. rates for a specific year or averages over a limited number of consecutive years, for a specific national or international population. Since radiation-induced cancer risks vary with age at exposure and sex, the basic calculations are carried out for each age at exposure and sex. The resulting estimates are then averaged with respect to the population of interest, e.g. the current population of Japan. While lifetime risk estimates provide a convenient summary of radiation effects, it should be noted that cancer rates change with time. The lifetime risks are not therefore estimates of what would be expected in either the population(s) from which the risk models were developed or in the populations used to obtain the background rates for the lifetime risk computations.

51. In the UNSCEAR 1988 Report [U2] the primary measure of lifetime detriment was the risk that an individual would die from a cancer that arose owing to the exposure in question. In this Annex the term "risk of exposure-induced death" (REID) is used for this quantity. For an instantaneous dose D to the whole body or to an organ at age e, the lifetime risk of exposure-induced death from cause c (all cancers or a single cancer) is given by

$$\mathrm{REID}_c(D,e) = \int_e^{\infty} [m_c(a|D,e) - m_c(a)]\, S(a|D,e)\, da \qquad (6)$$

where $m_c(a|D,e)$ and $m_c(a)$ are the death rates from cause c at attained age a, with and without, respectively, instantaneous exposure to total dose D at age e, and where $S(a|D,e)$ is the probability that the individual survives to age a given that he or she was alive and received total dose D at age e.

52. The risk of exposure-induced death is not the only measure of lifetime detriment that can be used. Indeed, a different measure, the excess lifetime risk (ELR), has been used by the BEIR IV and BEIR V Committees [C6, C12]. The excess lifetime risk is the increase in the lifetime risk of the cancer in question experienced by an individual as a result of the specific exposure. For instantaneous

exposure to a dose D at age e, the excess lifetime risk for cause c is given by

$$\mathrm{ELR}_c(D,e) = \int_e^\infty m_c(a|D,e)\, S(a|D,e)\, da - \int_e^\infty m_c(a)\, S(a|e)\, da \qquad (7)$$

where S(a|e) is the probability that an individual survives to age a given that he or she was alive at age e and the other quantities are the same as in the definition of REID.

53. Although both the REID and the ELR are useful summaries of lifetime detriment, they measure slightly different quantities. The risk of exposure-induced death can be interpreted as the risk that an individual will die from a cancer that has been caused by the exposure in question, while the excess lifetime risk is the difference between the proportion of people dying of cause c in an exposed population and the proportion dying of this cause in an otherwise identical, but unexposed, population. The difference between the REID and the ELR concerns the counting of cases who would have died of cause c in the absence of exposure but who die of this cause at an earlier date following exposure. Such cases contribute to the risk of exposure-induced death but are ignored in the excess lifetime risk [T14]. The ratio of the ELR to the REID is approximately equal to (1 - B_r), where B_r is the lifetime risk for all exposure-induced causes among unexposed persons [V4]. This implies that if one considers lifetime risk for all cancers combined, the excess lifetime risk will be 15%-20% less than the risk of exposure-induced death, since the lifetime risk of dying of cancer is about 15%-20% in most populations.

54. The excess lifetime risk has some possibly undesirable properties, all of which are related to the fact that an exposure that increases an individual's chance of dying from one cause necessarily decreases his or her chance of dying from other causes. Thus the excess lifetime risk can be negative, even if the exposure increases the risk of cause c, if it increases the risk of other causes even more. If an exposure increases the risk of all causes of death to the same extent, then the excess lifetime risk for any cause would be zero. Individuals would then die sooner, but the distribution of causes of death would be unchanged. On the other hand, an advantage of the excess lifetime risk over the risk of exposure-induced death is that it has a more direct interpretation; it is, strictly speaking, the increase, attributable to the exposure, in the probability that a person's eventual death is due to the cause c. The undesirable properties of both the excess lifetime risk and the risk of exposure-induced death are largely a consequence of efforts to reduce a time-dependent quantity to a single value.

55. Since neither the risk of exposure-induced death nor the excess lifetime risk provide direct information on the times at which the exposure-induced events occur, it is useful to supplement estimates of these quantities with a statistic that describes these times. In the UNSCEAR 1988 Report [U2] this was done using a measure called "loss of life expectancy" (LLE). The loss of life expectancy is the difference between the expectation of life for an individual exposed at age e and the expectation for an unexposed individual. For an instantaneous dose D given at age e, it is given as

$$\mathrm{LLE}(D,e) = \int_e^\infty S(a|e)\, da - \int_e^\infty S(a|D,e)\, da \qquad (8)$$

The loss of life expectancy depends on the times at which exposure-induced events occur in the affected persons (the risk of exposure-induced death). However, if the loss of life expectancy is divided by the risk of exposure-induced death, the result is the years of life lost per radiation-induced case (YLC). This quantity is more independent of the risk of exposure-induced death and can thus be thought of as giving additional information about the risks. Since both the risk of exposure-induced death and the years lost per radiation-induced case combine information over all ages at exposure and also all ages at expression, they do not fully describe the impact of exposure on a population. Measures that display the impact separately for the different age groups may also be useful. One approach to this would be to present the excess lifetime risk as a function of attained age (A) and age at exposure (e) by, for example, plotting ELR(A|D,e) against A, where ELR(A|D,e) is given by equation (7) with the integrals evaluated between e and A.

56. Finally, it should be noted that even when the rate $m_c(a|D,e)$ is linear in D, as for solid tumours in the life span study, neither the REID nor the ELR is linear, since S(a|D,e) is not linear in dose, especially at high dose. As will be illustrated in the presentation of lifetime risk estimates later in this Annex, the non-linearity is such that linear extrapolation from 1 Sv to 0.2 Sv, for example, understates the low-dose estimates of risk of exposure-induced death and excess lifetime risk for solid tumours by 10%-20%.

57. Another issue in risk estimation is how to determine expected mortality rates following exposures to low-LET radiation at dose rates substantially lower than those that prevail in the majority of epidemiological studies on which estimates of risk following exposure can be based. Given the paucity of epidemiological data, the best approach at present is probably to carry out the risk projection for an exposure of interest, ignoring the possible need for a dose-rate effectiveness factor, and then to multiply the resulting estimate of lifetime risk by a factor based on radiobiological as well as epidemiological considerations (see UNSCEAR 1993 Report [U1], Annex F, "Influence of dose and dose rate on stochastic effects of radiation").

D. SUMMARY

58. Epidemiological studies provide the primary data on the association between late effects (induced cancer) in man and exposure to ionizing radiation. In interpreting these results it is important to remember the limitations imposed by the observational nature of such studies. Careful consideration should always be given to the quality of the data collected, to confounding factors and to other possible sources of bias (including random error in the estimates of exposure) that may be present. It is important also to appreciate the special problems that arise in studies involving low doses (0.2 Gy or less), in which the relative risks may not be much above unity and thus may be difficult to quantify. Even when a study has been carefully performed and appears satisfactory, it is important to obtain confirmatory evidence from other studies carried out in different circumstances before drawing firm conclusions.

59. As the questions of interest in the study of radiation effects become more refined, it has become necessary to make detailed and at times complex analyses to ensure that the full information content of the data is revealed and to ensure that misleading inferences are avoided. There is a need for a better understanding and increased use of modern analytical methods in radiation studies. Such methods allow researchers to obtain a clearer picture of the information provided by their data and to reduce the overreliance on subset analyses and inappropriate significance tests. More effort is needed to develop as-simple-as-possible methods of defining and presenting useful summaries of radiation effects for human cancer risk estimation.

60. In extrapolating from populations that have been the subject of a specific study to different populations exposed in other, and perhaps hypothetical, ways, it is essential that the assumptions made about variations in the risks of cancer with level of exposure, age at exposure, time since exposure and background incidence of the disease are in agreement with the variations seen in practice in human populations. Furthermore, any assumptions about projected effects at long times after exposure, where few data are as yet available, should be made explicit. Despite the large number of available studies, there is currently little useful information on the question of how to transfer risks estimated for one population for use with another population. Careful parallel analyses of existing data should be undertaken.

61. The measures of lifetime detriment from radiation exposure that were used in the UNSCEAR 1988 Report [U2], the risk of exposure-induced death (REID) and the loss of life expectancy (LLE) provide useful summaries of radiation effects. The risk of exposure-induced death and the years of life lost per induced case (YLC) will be used as the primary general descriptors in this Annex. It would be desirable to develop other measures that provide a more complete picture of the temporal nature of radiation-induced cancer. In addition to estimates of mortality-based risk, estimates of incidence-based lifetime risk are also needed, and these can be evaluated with data now becoming available.

II. CARCINOGENIC EFFECTS OF EXTERNAL LOW-LET RADIATION EXPOSURES

62. Epidemiological studies of populations exposed wholly or predominantly to low-LET radiation are of prime importance in evaluating risk in environmental and occupational settings. Effects have been demonstrated mainly in studies that involve exposures of relatively high dose (about 1 Gy). Increasingly, however, information is accumulating on effects from lower doses (0.2-1 Gy) and, in a few instances, from doses less than 0.2 Gy.

63. Many of the studies have been characterized not only by high doses but also by high dose rates (>0.1 mGy min^{-1}). These include the studies of survivors of the atomic bombings in Japan (0.01 Gy s^{-1} to several gray per second), patients treated with radiation and individuals who have had diagnostic or prenatal x-ray examinations. In recent years studies have also been undertaken involving exposures at low dose rate or exposures received as many fractions, each of low dose. These include studies of workers in the nuclear industry and individuals exposed to gamma rays from terrestrial radiation or fallout from nuclear explosions. However, it is noted in the UNSCEAR 1993 Report [U1], Annex F, "Influence of dose and dose rate on stochastic effects of radiation", that at low doses (0.2 Gy or less) the form of the dose response is expected to be the same (i.e. linear) whatever the dose rate. Thus, in some studies nominally at high dose rate the total dose received is so low that the studies should probably be classified along with low-dose, low-dose-rate studies.

64. When the dose is an absorbed dose of low-LET radiation (e.g. from x rays) it is specified in gray (Gy). When the dose refers to the life span study of survivors of the atomic bombings, the dose will usually be a weighted dose and will be specified in sievert (Sv), in which the absorbed dose from neutrons is multiplied by an RBE of 10 and added to the absorbed dose from gamma rays.

Furthermore, all low-LET radiations such as x rays, gamma rays and beta particles will be treated alike, i.e. they will be assumed throughout this Annex to have the same biological effectiveness, even though at very low doses well-known differences in effect occur [S67].

65. The epidemiological studies used to derive estimates of risk of cancer following low-LET radiation exposures are reviewed in this Chapter. The main features of these studies are given in Table 2. The nature of the exposures and the dosimetry are described, as are the age, sex distribution and follow-up of the population. Also indicated in the Table are the cancers that have been studied and those for which statistically significant excess risks have been reported. Although the studies have a wide variety of characteristics, an attempt is made to calculate summary risk estimates for selected sites. However, in view of the uncertainties in specific studies and the different natures of the populations considered, these estimates should be compared and generalized with caution. There is also some discussion of the specific effects, temporal trends and applicability of the results.

66. The diverse studies listed in Table 2 have different strengths and weaknesses, and some are more important than others in contributing to quantitative estimates of risk. To provide a perspective on these relative merits, some of the strengths and weaknesses are outlined in Table 3. The attributes considered include the composition of the population studied, the nature of the control group, the length of follow-up, the ascertainment of incidence or mortality, the range of doses to those exposed, the nature and quality of the dosimetry and any unusual or specific uncertainties. It is not possible to be completely uniform or comparable in making these assessments. The comments in Table 3 are somewhat subjective and are provided principally as general guidance.

67. All the studies to be considered by the Committee that are relevant to risk estimation for low-LET radiation, including external and internal radiation at high and low dose rates and dose, are included in Table 2. The numerous other studies in the literature, some of which have raised interesting points from time to time, but others of which have failed to make useful contributions to the estimates of risk, are not considered further here.

A. HIGH-DOSE-RATE EXPOSURES

68. Risk estimates based on the continuing follow-up of the life span study of survivors of the atomic bombings in Japan have been the primary source of information on the risk of high-dose-rate exposures to mainly low-LET radiation. However, there are now a large number of other epidemiological studies that can make important contributions to the understanding of cancer risks following high-dose-rate exposures to low-LET radiation.

69. The life span study has a number of features that have allowed it to play a central role in the estimation of radiation risks, notably the large number of exposed persons of both sexes and a broad range of ages, the whole-body exposures of individuals in the cohort, the inclusion of a large group of people who received very low doses, which can be used as an internal control group; the completeness of the mortality follow-up; and the availability of data on cancer incidence.

70. Despite the strengths of the life span study, a number of factors make it essential to consider other exposed populations. For example, the exposure was highly acute (a few seconds at most), and since the life span study is composed exclusively of Japanese, risk estimates derived from this cohort cannot directly provide information on the application of the risk estimates to other populations. This latter point is especially important because of differences between site-specific cancer incidence and death rates in Japan and in other countries. General cancer patterns in Japan have some notable features relative to patterns in European or American populations, including the high incidence of stomach cancer, the low rates of female breast cancer and the virtual absence of chronic lymphocytic leukaemia. Risk estimates from other studies can provide important additional information about risks for specific cancers or, in some cases, more general aspects of the risk of radiation-induced cancer, including factors such as age at exposure, sex, temporal patterns of the excess risk and the effects of fractionation.

71. The studies most useful for comparison with and as a supplement to the life span study involve populations exposed to x rays for diagnostic or therapeutic purposes. The features of the exposures and populations of the main studies are summarized in Table 2, as already noted. As a group, studies of medically exposed persons have a number of specific limitations not noted under strengths and weaknesses given in Table 3, including the following:

(a) the disease under treatment or observation may affect subsequent cancer risks;
(b) patients may have been exposed to other carcinogens as part of their medical treatment, e.g. alkylating agents in cancer therapy;
(c) the range of ages at exposure is often narrow and the data are often limited to, or include a very high proportion of, a single sex;
(d) the doses are in a limited range, the exposures are usually targeted to a specific tissue and the dose estimates to specific organs are sometimes not of good quality.

72. Another important group of studies relevant to the assessment of cancer risks following high-dose-rate, low-LET radiation exposures involves the highly fractionated exposures received by radiologists and other medical x-ray workers [A9, E7, M24, S38, W10, W14].

Dose estimates in earlier studies were often lacking, while in some more recent studies doses are assigned on the basis of year of exposure and knowledge of radiological protection practices in the relevant country at that time. Such studies cannot therefore be expected to provide precise information about risk estimates.

73. Despite the limitations of any single epidemiological study of radiation risk, studies of populations with appreciable exposures to low-LET radiation are important to an understanding of radiation risks because they offer information on risk in a number of ethnic groups and nationalities and because the conditions under which the exposures were received vary so greatly [D27].

1. Mortality and incidence in the life span study

74. The cohort of the life span study includes 93,000 survivors of the atomic bombings and 27,000 persons who lived at Hiroshima and Nagasaki in 1950 but who were not in the cities at the time of the bombings. The primary findings of the life span study have been used to clarify the association between radiation exposure and cancer mortality. The life span study cancer mortality data for the period 1950-1985 [S7, S69] were used as the basis of risk estimates for the UNSCEAR 1988 Report [U2]. In the following text, mortality data extended to 1987 or specified for the period 1958-1987 are described in order to compare with incidence data for 1958-1987 [R23, T15].

75. A complicating factor for estimates of low-LET cancer risks based on the life span study arises from the presence of a small, but not negligible, neutron component, especially at Hiroshima, where neutrons accounted for 1%-2% of the organ doses, with individual neutron dose estimates ranging from 0 to 100 mGy. The high correlation between a survivor's estimated gamma and neutron doses makes it impossible to use the life span study data to provide reliable estimates of the neutron relative biological effectiveness (RBE). Therefore, it has been common practice to assume a specific value or values for the RBE of neutrons in the computation of life span study risk estimates. Published analyses of the life span study data have usually made use of total absorbed dose, i.e. they have implicitly assumed an RBE of 1, or they have explicitly assumed a constant RBE of 10 or 20. While it has been demonstrated that risk estimates based on the DS86 dosimetry do not vary greatly with the assumed value of the RBE in this range [P3, S6], preference should be given to estimates based on values of 10-20. Questions raised recently about the magnitude of the neutron component [S56] are discussed in paragraphs 156-158.

76. Although important questions about the dosimetry of the survivors of the atomic bombings remain [S56], the availability of detailed data on location and shielding at the time of the bombings makes it possible to compute individual dose estimates. Uncertainties in the location and shielding data are a major component of the random uncertainty in the individual dose estimates. In recent years much attention has been paid to characterizing the nature of random errors in life span study dose estimates and to assessing their impact on risk estimates [P9, P31, R22, S55].

77. Recent reports of the life span study provide data on cancer incidence in the period 1958-1987 for solid tumours [T15] and in the period 1950-1987 for leukaemia, lymphoma and multiple myeloma [P33]. The life span study incidence data allow a more precise description of radiation-induced cancer risks both because of the larger number of cases in the incidence data than in the mortality data and because the registry-based diagnoses are of better quality than cause-of-death data from death certificates. The primary limitations of the incidence data are the lack of data on solid tumour incidence for the first seven years of follow-up (data on the first five years after exposure are unavailable for study for either mortality or incidence in the life span study) and the problems that arise owing to migration from the tumour registry catchment areas. Life span study migration probabilities depend on birth cohort, city and age. Among life span study members under the age of 16 years in 1945, roughly 25% of those from Nagasaki and 15% from Hiroshima were not resident in those cities in recent years. Older survivors were much less likely to have moved, and it appears that those who have left the area tend to return when they reach retirement age. In both cities, men were more likely to migrate than women. Migration probabilities were taken into account in the recent analyses of the life span study cancer incidence data [S23, T15]. The incidence and mortality data for solid tumours in the life span study are given in Table 4. While the life span study incidence data are important, they should be viewed as a supplement to, rather than as a replacement for, mortality-based risk estimates in the life span study. In particular, it is likely that both mortality- and incidence-based risk estimates will be important in the quantitative evaluation of radiation effects in human beings.

78. Estimates of the average excess relative and absolute risks for solid tumour incidence and mortality in the life span study are compared in Table 5 and illustrated in Figure I. The organ-dose-based cancer mortality risks derived by Shimizu et al. [S7, S69] for 1950-1985 were calculated assuming an RBE of 1. When organ-dose-based risk estimates were not available, approximate estimates were made by dividing the shielded-kerma-based values by estimates of the transmission factor for the target organ. The values for cancer mortality [R23] and cancer incidence [T15] for 1958-1987 are based on weighted organ doses (with the absorbed dose from neutrons given a weight of 10, i.e. a constant neutron RBE of 10), which are similar to equivalent doses and will be given in sievert. The 95% confidence bounds in the original papers were

transformed to approximate 90% intervals by reducing the original intervals to 84% of their original length in each direction.

79. Figure I shows that the results of the life span study incidence and mortality data are broadly similar, with both sets of data demonstrating statistically significant effects for all solid tumours as a group, as well as for cancers of the stomach, colon, liver, lung, breast, ovary and bladder. The incidence data also provide evidence of excess radiation risks for thyroid cancer and non-melanoma skin cancers. The incidence study also strengthens earlier findings of an increased risk of salivary gland tumours among exposed members of the life span study cohort. Oesophageal cancer, for which a statistically significant excess is seen in the mortality data [S7, S69], does not appear to have an elevated risk in the incidence data [T15]. The difference is due to the number of cases during 1950-1957 in the high-dose survivors. In view of the failure to observe an excess in subsequent years, the mortality results may be a statistical anomaly. Solid tumour sites for which statistically significant excess risks were not seen in either the incidence or the mortality series include cancers of the rectum, gall-bladder, pancreas, larynx, uterine cervix, uterine corpus, prostate, and kidney or renal pelvis.

80. The life span study incidence data [P33, T15] provide clear evidence of an excess risk of solid tumours in persons exposed to doses in the 0.2-0.5 Sv range. In addition, the solid tumour dose-response function appears to be remarkably linear (see paragraph 168). This result is noteworthy because the response is a composite for all tumours, which may include a variety of different responses for individual tumours. The incidence data include sites for which a radiation effect has not been convincingly demonstrated in other populations.

81. Risk estimates from the life span study mortality data for lympho-haematopoietic cancers [S7, S69] are compared in Table 6 with estimates for the same cancers from the life span study incidence data [P33]. The mortality data [S7] are based on absorbed dose in the bone marrow (neutron RBE = 1), while the incidence data [P33] are based on weighted organ doses as described above (neutron RBE = 10). Several leukaemia subtypes in the life span study incidence data [P33] were analysed, and strong evidence for an association with radiation exposure was found for acute lymphocytic leukaemia, acute myelogenous leukaemia and chronic myelocytic leukaemia but not for adult T-cell leukaemia. The risk estimates for the incidence of lymphoma parallel those for mortality, but both have confidence limits below zero. However, the myeloma risk estimates differ from earlier life span study reports on mortality [S7] or incidence [I5, I13, I14]. The life span study incidence data [P33] include a large number of newly diagnosed cases, and several of the cases used in the early analyses were excluded from the new analysis as a result of dose range restrictions or other considerations. In the new analyses there was no longer a significant association between radiation exposure and myeloma incidence.

82. Table 7 presents observed and expected numbers of cases by dose category in the life span study of solid tumour incidence and mortality and leukaemia incidence. The numbers in this table clearly indicate the effects of radiation exposure in survivors of the atomic bombings and provide some information about the shape of the dose response. The expected numbers were computed using information from the most recent reports on cancer incidence and mortality in the life span study [P33, R23, T15]. The numbers in the table suggest that between 1950 and 1987 roughly 300 of the 6,900 cancer deaths in the life span study are associated with radiation exposure above 0.2 Sv (weighted dose). Of the 231 first primary leukaemia cases diagnosed during this time period, about 75 are in excess of expectation. The incidence data indicate that about 500 of the 8,600 first primary solid tumours diagnosed among life span study cohort members between 1958 and 1987 are associated with radiation exposure.

2. Site-specific cancer risks

83. A large number of studies of cancer morbidity in populations exposed to low-LET radiation were listed in Table 2. The data from those studies have been used to calculate site-specific risk estimates that are presented and compared below. Studies were chosen for inclusion in these comparisons based on a number of criteria, including population size, data quality and the availability of some quantitative measure of either individual or population-average doses that allow the computation of estimated risks per unit dose. Many additional studies indicate excess risks in specific organs, but the dose information, particularly in earlier studies, is not adequate to derive risk estimates. An example is the study of Dickson [D21]. Comparative risk estimates for several cancer sites in the body have been calculated, and the results are summarized in Table 8. These estimates must be interpreted and compared with caution because of differences from study to study with respect to the nature and quality of the dosimetric information as well as differences in the sex, age at exposure and follow-up distributions. Supplementary data for sex- and age-specific risk estimates for incidence and mortality are available for the life span study. These more detailed life span study data are presented to facilitate comparison with studies in which the exposed group is restricted to a single sex or a limited range of ages at exposure.

84. For cohort studies, the data on the exposed portion of the population are summarized in terms of observed and expected numbers of cases. For the life span study,

expected numbers of solid tumours were derived from models similar to those described in [T15] and [R23]. Life span study mortality risk estimates are based on the 1950-1987 data presented in [R23]. Life span study solid tumour incidence risk estimates are derived from the 1958-1987 data used in [T15]. For the life span study lympho-haematopoietic cancer data, the time-dependent excess absolute risk (EAR) models for 1950-1987 presented in [P33] were used. For other cohort studies, the expected numbers were determined from published reports. If the study included a control group and if information about observed and expected numbers of cases in this group was available, a correction was made to the expected number:

$$E_e^a = E_e^u \times \frac{O_0}{E_0} \qquad (9)$$

where E_e^a and E_e^u are, respectively, the adjusted and unadjusted estimates of the expected number of cases in the exposed portion of the cohort and O_0 and E_0 are the observed and expected number of cases in the unexposed portion of the cohort. If the cohort included no internal controls, no adjustment was made. The estimates of the excess relative risk and the average excess absolute risk were calculated directly from these values. The linear excess relative risk coefficient is

$$ERR = \frac{O_e - E_e^a}{E_e^a \bar{D}} \qquad (10)$$

where O_e is the observed number of cases and $\bar{D}$ is the average dose in the exposed portion of the cohort. The linear excess absolute risk coefficient per person-year of observation is

$$EAR = \frac{O_e - E_e^a}{PY \bar{D}} \qquad (11)$$

Differences in time-averaged absolute risk coefficients for different age-at-exposure groups must be interpreted with caution, since current data suggest that excess absolute risk coefficients change with increasing time after exposure and follow-up is incomplete for the younger members of most cohorts used in the studies of radiation effects (see Section II.A.4).

85. Confidence intervals for studies other than the life span study were obtained using exact Poisson methods and assuming a linear dose response. These methods tend to understate the uncertainty in the estimates, especially when the average dose is much greater than 1 Sv or when the internal control group is small or non-existent. For the life span study, confidence intervals were taken from the relevant publications. This was done because confidence intervals for risks calculated from the data of the life span study given in Table 8 are too wide, since information on the low-dose group used in the computation of the expected number of cases is not included in this Table.

86. For case-control studies, risk estimates and confidence intervals were taken from the original papers whenever possible. In some cases, odds ratios were calculated based on unmatched analyses of published data. Excess absolute risk estimates are given for case-control studies in which the cases were obtained by complete ascertainment in a well-defined cohort and sufficient information was given to allow calculating the number of cases expected in the full cohort under the null hypothesis of no radiation effect.

87. The risk estimates for selected cancers are given in Parts I-XIV of Table 8. Listed in order of the international statistical classification of diseases, injuries and causes of death (ICD) code, each part is dedicated to one type or site of cancer. Figure II, which, in analogy with Table 8, also contains Parts I-XIV, illustrates the excess relative risks per sievert. Although both the Table and the Figure facilitate the comparisons of risk estimates, it must be repeated that the validity of the comparisons is impeded by the various exposure conditions, dosimetric evaluations and population-related factors. It is also important not to overemphasize the significance of differences in the risk estimates for specific cancers. While there are probably real differences in the magnitude of radiation effects for specific cancers, there is also a tendency to overinterpret the observed differences. Recent joint analyses of cancer-specific excess risks derived from the life span study mortality data [P20] suggest that differences in the excess relative risks for specific cancers are generally about what can be expected on the basis of random variation about a common level of risk for these specific cancers. The expected numbers of cases for the life span study were calculated using slightly different methods than were used in other estimations of the typical average relative risks (as given, for example, in Table 5). In some cases, this leads to slight differences between the estimates in Tables 5 and 8. These differences largely reflect the variability inherent in these data.

(a) Cancer of the oesophagus

88. Risk estimates for oesophageal cancer are presented in Part I of Table 8, and the excess relative risks are illustrated in Part I of Figure II. Data on the incidence of oesophageal cancer are available from the life span study [T15] and the cervical cancer cohort study [B11]. The risk estimates are similar in the two studies, but in neither case are they statistically significant. Data on mortality from oesophageal cancer are available from three studies. A statistically significant association between radiation exposure and cancer of the oesophagus is reported only in the ankylosing spondylitis study [D6, L6]. In an earlier

analysis of the life span study mortality data [S7], a significant excess risk of oesophageal cancer had been noted. Davis et al. [D14] noted a significantly elevated standardized mortality ratio for oesophageal cancer in the cohort of the Massachusetts tuberculosis fluoroscopy study [D14], but large oesophageal cancer standardized mortality ratios were also seen among unexposed members of this cohort. Once this is allowed for, there is no evidence of a radiation effect or a dose response. The authors speculated that the large standardized mortality ratios for both exposed and unexposed members of this cohort were associated with alcohol and tobacco consumption.

89. Variations in the risk of oesophageal cancer with age and sex, as seen in the life span study, are generally consistent with the pattern seen for many other cancers, i.e. a larger excess relative risk for females than males and higher average relative risks in those exposed before the age of 20 years. None of these differences, however, is statistically significant. In spite of some conflicting evidence, the mortality data of the life span study for 1950-1985 showed a significant risk for oesophageal cancer mainly because of its occurrence in the period 5-12 years after exposure. Thus the oesophagus may be recognized as having a positive risk for relatively early occurring radiation-induced cancer, but when the dose is fractionated (as in the Massachusetts tuberculosis fluoroscopy study) the risk is very small.

(b) Cancer of the stomach

90. Part II of Table 8 and of Figure II compares risk estimates for stomach cancer. A statistically significant association between radiation exposure and stomach cancer is shown in the incidence data of the life span study [T15] and cervical cancer case-control study [B21] and in the mortality data of the life span study [R23] and the peptic ulcer study [G11]. Despite the very large mean dose in the peptic ulcer study, the point estimates of the excess relative risk based on these data are only slightly less than those in the other populations. A non-significant estimate of risk of stomach cancer was observed 5-24 years after exposure in the mortality data of the ankylosing spondylitis study [D6, L6]. Although non-significant, the largest point estimates of risk were obtained from the mortality data of the metropathia haemorrhagica study [D23]. The relative risk of mortality from stomach cancer based on data of the benign gynaecological disease study [I8] is similar to that of the life span study but not significant.

91. Differences in the excess relative risk of stomach cancer with sex and age at exposure are statistically significant in the life span study data. Higher values are shown for females than for males (incidence and mortality) and for those less than 20 years old at exposure (incidence only). Differences in the excess absolute risks for males and females are more pronounced in the incidence data than in the mortality data of the life span study. Current results suggest that there is an association between low-LET radiation exposure and stomach cancer risk, with females having greater relative risks and perhaps also greater absolute risks than males.

(c) Cancer of the colon

92. Risk estimates for cancer of the colon are given in Part III of Table 8 and of Figure II. There is a clear association with radiation exposure in both the incidence and mortality data of the life span study [R23, T15] and in the mortality data of the benign gynaecological disease study [I8] and the metropathia haemorrhagica study [D23]. The excess absolute risk estimate derived from data of the benign gynaecological disease study [I8] is much higher than that derived from the life span study, reflecting the difference in normally occurring rates of colon cancer in the United States and in Japan. Two other studies provide data on cancer of the colon. These are the cervical cancer case-control study [B21] and the peptic ulcer study [G11]. Both involve either very high doses or a broad range of doses, and in both cases the estimated risks are low.

93. The sex-specific risk estimates for cancer of the colon from data of the life span study differ from those for cancer of the stomach, in that both the excess relative risk and the excess absolute risk are higher for males than for females. However, the difference in the excess relative risks is not statistically significant. (No formal test of sex-related differences in the excess absolute risk has been reported.) Although the difference in the excess relative risks for colon cancer with age at exposure is not statistically significant in the incidence data of the life span study [T15], statistically significant decreases in the excess relative risks with time have been reported for both the incidence and mortality data [R23, S7]. Taken together, the results of the studies considered here indicate an increased risk of colon cancer following radiation exposure.

(d) Cancer of the liver

94. Although an association between thorotrast exposure and liver cancer has been demonstrated in several studies (see Section III.B), the evidence for excess liver cancer following exposure to low-LET radiation is weaker. The risk estimates derived from several studies are given in Part IV of Table 8 and of Figure II. With the exception of the life span study, the numbers of observed cases are small in these studies. None of the medically exposed populations considered here suggests an association between radiation exposure and liver cancer. Evidence for an excess risk of liver cancer in the life span study has increased with additional follow-up time and the availability of cancer incidence data [T15].

95. Interpretation of the data on liver cancer in the life span study has been hampered by questions about the adequacy of the diagnoses. Because death certificate

diagnoses of liver cancer in the life span study do not usually allow for a distinction between primary and secondary tumours, analyses of the life span study mortality data [R23, S7] were limited to those cases in which the death certificate explicitly identified the cancer as primary. As indicated in Table 8, this greatly reduces the number of cases and, in contrast to an analysis that includes liver cancer deaths specified as primary or not specified as secondary or primary, no statistically significant radiation effect is found. A supplementary analysis of the incidence data of the life span study [T15] restricted to the 40% of tumour registry cases with histologically confirmed diagnoses resulted in an excess relative risk that was statistically significant, with the point estimate almost identical to that seen in the analysis of all liver cancer cases.

96. In contrast to most other cancers, the excess relative risk of liver cancer for females was less than that for males. Neither the sex nor age-at-exposure differences seen in the life span study were statistically significant. However, if females actually have lower excess relative risks than males, then the failure to see any evidence of excess risks in cohorts of the benign gynaecological disease study [I8] and in the cervical cancer cohort study [B11] is perhaps not surprising.

97. The life span study data, particularly those on cancer incidence, support the conclusion that liver cancers are induced by radiation exposure. Interpretation of mortality data of the life span study is complicated by the use of different criteria for case definitions; nonetheless the mortality results appear to be consistent with the incidence results. Studies other than the life span study provide little evidence for or against cancer risks owing to the small number of cases.

(e) Cancer of the lung

98. The life span study [R23, T15], the ankylosing spondylitis study [D6, L6] and the peptic ulcer study [G11] all indicate an association between radiation exposure and lung cancer (Part V of Table 8 and of Figure II). The lower excess relative risks in the last two studies do not seem particularly surprising because both of these cohorts have a relatively high proportion of males, and the life span study results indicate that excess relative risks are much higher for females than for males. The difference in sex-specific excess relative risk for lung cancer in the life span study almost certainly reflects differences in rates of lung cancer occurring normally; however, the differences in the sex-specific average excess absolute risks also appear to be large. Since the number of cases is relatively large, the failure to see a significant excess risk of lung cancer in the Massachusetts tuberculosis fluoroscopy study [D14] is surprising in view of the high excess relative risks for females in the life span study. In the Massachusetts tuberculosis fluoroscopy study, fractionation may have had a large effect. It is noteworthy that standardized mortality ratios for both the exposed and unexposed members of the tuberculosis cohort are very large. Recently published analyses of lung cancer data for the Canadian fluoroscopy cohort [H43] support the Massachusetts findings of no excess lung cancer risk for highly fractionated x-ray exposures. However, it should also be noted that an excess relative risk of 0.08 Gy^{-1} was found following radiotherapy to the breast, which involved average exposures to the lungs of about 9.8 Gy [I20, N8]. This positive but lower than expected risk may have been reduced because of cell-killing effects after high doses.

99. It has been suggested that the relative risk of lung cancer decreases with time since exposure, as indicated in the risk model used by the BEIR V Committee [C12]. The life span study incidence data [T15] indicate that the excess relative risk has decreased only slightly with the passage of time and that the excess absolute risk is continuing to increase. This contrasts with the finding of no apparent excess during the period beyond 25 years after exposure in the ankylosing spondylitis study. It is important in any study of respiratory cancer to indicate the smoking status of the population, as the joint effects of smoking and radiation alter the dose response (see paragraph 182).

100. In summary, an elevated risk of lung cancer is seen following acute exposure to low-LET radiation. The fluoroscopy data suggest that highly fractionated exposures may have little or no effect. Although descriptive models that allow for decreases in lung cancer risk with time since exposure have been proposed, current low-LET data do not appear to provide strong support for this notion. Therefore, lung cancer risk estimates presented later in this Annex are based on time-constant relative risk models with sex and age-at-exposure effects.

(f) Cancer of the bone and connective tissue

101. Although cancers of the bone and connective tissue are relatively rare, a number of studies have noted excess risks for these cancers following radiation exposure (Part VI of Table 8 and of Figure II). However, only for the skin haemangioma in childhood study [F12] and the childhood radiotherapy study [T6] are the results statistically significant. In the latter study, a dose response was evident over intervals of therapeutic doses. A study of children irradiated for retinoblastoma also found a very high risk of osteosarcoma and soft tissue sarcoma associated with radiotherapy exposures [E4]. A large, statistically significant excess risk for cancer of the bone and connective tissue was also reported for the children in the tinea capitis study in Israel [R10]. However, since the skeletal dose was not reported, it is not possible to present estimates of the risk per unit dose. Although the numbers

are small, the age- and sex-specific estimates based on data of the life span study and other populations suggest that survivors exposed before the age of 20 years may have high relative risks.

(g) Non-melanoma skin cancer

102. As can be seen in Part VII of Table 8 and of Figure II, elevated risks of skin cancer have been reported in a number of studies. The most detailed analyses of data on low-LET radiation and skin cancer come from the tinea capitis study in New York [A1, S8], from the tinea capitis study in Israel [R10, R19] and from the life span study on cancer incidence [T15]. Significant associations between radiation exposure and non-melanoma skin cancer incidence have been shown in these studies and in the Rochester thymic enlargement study, the lymphoid irradiation study and the thyroid irradiation study, all evaluated in the review by Shore [S62]. Non-significant risk estimates are derived from data of three other studies included in Part VII of Table 8. Those exposed at younger ages have been found to have significantly higher relative risks in both the life span study and the Israeli tinea capitis study. The issue of age was not investigated in the New York tinea capitis study. The New York studies suggest that persons with lighter skin are at higher risk of radiation-induced skin cancer, especially in areas of skin exposed to sunlight, e.g. the face. None of these studies indicates that the excess relative risk depends on sex. The life span study results suggest that the dose response is non-linear with little or no effect at doses below 1 Sv.

103. The similarity of excess relative risks for skin cancer evaluated from the main studies in Part VII of Table 8 should not be overinterpreted, since the life span study includes older survivors who have lower risks while the tinea capitis studies include different proportions of light- and dark-skinned patients. For the younger persons in the life span study, the excess relative risk appeared to be greater than that in the Israeli tinea capitis study. In Japan, the incidence of skin cancer in normal circumstances is quite low. In the cervical cancer cohort study, only 206 non-melanoma skin cancers were observed, with 220 expected. The absence of a risk following high-dose exposures to the skin of these patients is noteworthy. Comparison of absolute risk estimates for skin cancer is problematical because of differences in the exposed area of skin in the various studies. Excess risk to medically irradiated patients occurred only to areas of skin also exposed to ultraviolet radiation [S62].

104. Present results suggest that elevated skin cancer risks are associated with low-LET radiation exposures. Relative risks for children are greater than those for adults. Questions about the shape of the dose-response function and of a possible interaction between high- and low-LET radiation exposure warrant further examination.

(h) Cancer of the female breast

105. Elevated risk of breast cancer following exposure has been demonstrated in several studies, including the life span study [T2, T15], the Massachusetts tuberculosis fluoroscopy study [B31], tuberculosis fluoroscopy studies in Nova Scotia and other Canadian provinces [M19] and the acute post-partum mastitis study in New York [S9], as can be seen in Part VIII of Table 8 and of Figure II. Wide confidence intervals are indicated for the scoliosis patients study [H24], the Rochester thymic irradiation study [H22] and for women in the ankylosing spondylitis study [D6, L6]. A number of studies have considered radiation exposure and breast cancer but provide little or no convincing evidence of an association: the contralateral breast studies in Denmark [S60] and in the United States [B18]; the cervical cancer case-control study [B11, B20, B21]; the Israeli tinea capitis study [M20]; and for women in the Swedish skin haemangioma study [F12].

106. The failure to detect an increased breast cancer risk in the cervical cancer case-control study is somewhat surprising in view of the large number of cases and moderate doses (average 0.3 Gy). However, as suggested by the very different estimates of excess relative risk for the younger and older populations (see Part VIII of Table 8), breast cancer risks following radiation exposure exhibit a reasonably consistent decrease with increasing age at exposure (see paragraph 137). Thus, the failure to detect radiation effects in the cervical cancer case-control study and in the contralateral breast studies [B18] is possibly due to the fact that these cohorts include a large number of women who were over 35 years of age at exposure. Indeed, a significant excess risk of breast cancer was found in a subset analysis of younger women in the contralateral breast study in the United States [B18]. It has also been suggested that the lack of an increased risk of breast cancer in the cervical cancer case-control study is related to the cessation of ovarian function caused by irradiation [B20]. However, when the data on cervical cancer were restricted to cases among women whose ovaries had been removed before or shortly after the diagnosis of cervical cancer, a positive, though not statistically significant, effect was seen. The Hodgkin's disease patients study clearly indicates a radiation effect with the highest risk at young ages [H2, T8]. Also, the risk per unit dose appears to be lower after high doses than is suggested in studies at low doses, which could be due to cell-killing, limiting the number of cells at risk (see paragraph 171).

107. Relative risks appear to be especially high among women who were exposed to radiation as children, as indicated by the results of the life span study, the Rochester thymic irradiation study and the scoliosis study. The risk estimates for the cohort of scoliosis patients are based on only 11 cases, with 6 expected, and are probably biased upward as a consequence of factors unrelated to

exposure, such as a greater rate of nulliparity among women with severe spinal curvature. A relatively large effect (an excess relative risk of 1.11) at very low doses (16 mGy) has been reported in the Israeli tinea capitis study [M20]. This finding should be viewed with some scepticism, since it is based upon results for a specific subgroup, which had an unusually low number of cases in the control group. Compared with general population rates there was no excess.

108. In general, constant relative risk models that have a linear dose response and that depend on age at exposure appear to describe the breast cancer data quite well within cohorts. Based on a parallel analysis of the incidence data from the life span study and two cohorts from the United States (the Massachusetts tuberculosis fluoroscopy study [B31] and the New York acute post-partum mastitis study [S9]), it was concluded that although relative risk models describe the data well within a single population, average excess absolute risks appear to be similar across populations [L2]. In the analysis carried out by the BEIR V Committee [C12], it was suggested that a common relative risk model for the three cohorts fit the data better than did a common, time-dependent absolute risk model. In the same report, a similar conclusion was reached based on a comparison of breast cancer mortality in the life span study and the Canadian tuberculosis fluoroscopy study [M19]. These results suggest that fractionation has little effect on breast cancer risk. This is in marked contrast to the results for lung cancer in the tuberculosis cohorts [D14, H43], in which no increased risk was seen. Although the simple summary statistics shown in Table 8 cannot be used for detailed formal comparisons, because of differences between cohorts in age at exposure and other factors that might affect the risk, the results of the studies suggest there is a need to carefully consider the most appropriate risk model for transfer between populations. Breast cancer risks are clearly elevated following low-LET radiation exposures. Relative risks are higher for women early in life, including women exposed prior to menarche.

(i) Cancer of the urinary bladder

109. Estimates of risk for bladder cancer from several studies are given in Part IX of Table 8 and of Figure II. Statistically significant excess risks have been derived for incidence data [T15] and for mortality data [R23] of the life span study, the metropathia haemorrhagica study [D23] and the benign gynaecological disease study [I8]. Although the doses are considerably higher in the last two studies and in the cervical cancer case-control study [B21], the risk estimates are about the same as the non-significant risk estimate in the ankylosing spondylitis study [D6]. Within the life span study the effects of age and sex on the risks are unclear. In particular, the incidence data exhibit a statistically significant sex difference, with the excess relative risk for females exceeding that for males by a factor of about five but the average excess absolute risk showing no significant difference; in the mortality data, the point estimate of the excess relative risk for males is higher than that for females, although the difference is not statistically significant. Neither the mortality data [S7] nor the incidence data [T15] in the life span study exhibit statistically significant variation with age at exposure for either the excess relative or absolute risks; however, there is a suggestion of some variation with age in the cervical cancer case-control study [B21].

110. In summary, statistically significant excess risks of cancer of the urinary bladder are seen in several populations exposed to low-LET radiation. The life span study risk estimates are somewhat greater than those in other studies, however, since those other studies involve much higher doses, the differences may reflect cell-killing.

(j) Tumours of the brain and central nervous system

111. Risk estimates for tumours of the brain and central nervous system are presented in Part X of Table 8 and of Figure II. Fairly large and statistically significant excess relative risks have been demonstrated in both tinea capitis studies [A1, R10] and in the ankylosing spondylitis study [D6, L6]. A lower, but still significant, risk is estimated from the data of the pituitary adenoma study in the United Kingdom [B9]. Significant risks were not seen in either the life span study or in the skin haemangioma study [F12]. Except for the ankylosing spondylitis study and the pituitary adenoma study, the studies showing significant risks involved children. It has been noted that some of the cases in which brain tumours were recorded on death certificates in the ankylosing spondylitis study may have involved secondary tumours from lung cancer [D6]. The tinea capitis study in Israel included both benign and malignant tumours of the brain, with 42 of 60 tumours being benign (mostly nerve sheath tumours). The life span study also included both benign and malignant brain tumours.

112. The mortality data [R23] and incidence data [T15] of the life span study provide weak evidence of an association between radiation exposure and cancers of the central nervous system excluding the brain. The estimates from the incidence data [T15] suggest the possibility of differences related to sex and age at exposure, but these differences are not statistically significant. A number of other studies have reported significant increases in brain tumours and tumours of the central nervous system associated with various types of radiation exposure, including prenatally exposed individuals [B23, M9], persons receiving radiation therapy for childhood cancers [A2, R14], persons treated for enlarged tonsils [S32] and individuals who received radium implants in the nasopharynx [S3]. Preston-Martin et al. have reported an association between dental x rays and brain tumours [P5, P6, P14] and between malignant brain tumours and other factors in a study in the Los Angeles area [P21, P22].

113. The results of these studies suggest that low-LET radiation may be associated with increased risks of tumours of the brain and the central nervous system and that risks for those exposed as children may be greater than for those exposed later in life.

(k) Cancer of the thyroid

114. Risk estimates for thyroid cancer are presented in Part XI of Table 8 and of Figure II. The compilation is based on a review by Shore [S61]. Excess risks have been found in a large number of populations. Because of the strong association between age at exposure and the excess relative risk for thyroid cancer seen in most studies, results for childhood and adult exposures are separated in both the Table and the Figure. Since the detection of thyroid cancer is heavily dependent on the degree of screening, average excess absolute risk estimates are not comparable even for studies that have equal follow-up intervals and similar age and sex distributions. This is illustrated by the results for the Michael Reese tonsil study [S63]. An extensive thyroid cancer screening programme was started in this cohort in 1974, after which thyroid cancer rates in the cohort increased almost 10-fold. The impact of this screening programme is apparent in Table 8. While the excess relative risk estimate for this study is roughly comparable to that seen in other populations, the absolute risk estimate is much larger than that in any other population considered. Schneider et al. [S63] estimated the screening-adjusted excess absolute risk to be $1.7\ 10^{-4}$ $(PYSv)^{-1}$ (weighted dose, RBE = 1 for x rays).

115. Excess relative risks for incidence of thyroid cancer in children based on data of the life span study [T15] were 10.3 for those of age 0-9 years and 4.5 for age 10-19 years. Values from other studies are comparable, with the exception of the tuberculosis adenitis screening study [H1] and the Israeli tinea capitis study [R15]. The very large risk estimate and narrow confidence interval for children in the tuberculosis adenitis screening study [H1] is a consequence of the small expected number of cases. Case ascertainment was very good in the Israeli tinea capitis study [R15], and unlike many of the other thyroid cancer studies, there is an internal control group. Risk estimates in this study appear to be quite large and, as noted in paragraph 169, there are some uncertainties regarding thyroid doses in this cohort. Shore et al. [S54] found significantly greater excess relative risks for Jewish than for non-Jewish children in their study of thymic irradiation in infancy.

116. Age-at-exposure effects are apparent in all studies in which this factor can be evaluated, such as the tinea capitis studies, the life span study and even some studies that involve only a limited age range. Sex-specific excess relative risk estimates for thyroid cancer are similar, despite the fact that the rate of naturally occurring thyroid cancer in females is typically three times the rate in males. This can be seen from age- and sex-specific estimates in the life span study, but is equally apparent from the Israeli tinea capitis study [R15] and other studies. While neither the life span study nor the Israeli tinea capitis study shows any statistically significant variation in age-at-exposure specific relative risks over time, data from the somewhat smaller Rochester thymic irradiation study suggest that the relative risk decreases with time.

117. In summary, low-LET radiation exposures, possibly at doses as low as 0.1 Gy, are associated with significant excess risks of thyroid cancer in children. The excess risk for adults appears to be lower. In contrast to the pattern seen for many other solid tumours, excess relative risks are similar for males and females despite large differences in the sex-specific background rates in most populations.

(l) Leukaemia

118. Risk estimates for leukaemia are presented in Part XII of Table 8 and of Figure II. Estimates can be made from a large number of studies. The life span study incidence data [P33] suggest that risks are highest in younger survivors, but there is little difference in excess relative risk or excess absolute risk estimates by sex. Most studies in which the data allow examination of temporal patterns suggest that the excess relative risks and the excess absolute risks of leukaemia peak within 3-10 years of exposure and then decline with time. For those exposed as children, the risk after 25 years is very low. In the life span study incidence data [P33] it was found that the temporal pattern of the leukaemia excess absolute risk depended on age at exposure, with the initial peak being higher and the decrease more rapid among those exposed as children. Figure III illustrates this pattern. There has been much speculation about whether or not excess leukaemia risks disappear 20 or more years after exposure. The data from both the ankylosing spondylitis study [D6, L6] and the life span study suggest that elevated risks are seen 25 or more years after exposure. The most recent analysis of the data of the benign gynaecological disease study [I12] suggests that risks for acute leukaemias vary little over time among people exposed as adults. This finding seems to be consistent with the recent results of the life span study [P33].

119. Although temporal variation in leukaemia risks following radiation exposure complicates the direct comparison of risk estimates, these estimates can provide some useful information about radiation-induced leukaemia risks. In particular, the estimates from the life span study incidence data [P33] suggest that risks per unit dose from high doses (in excess of 3 or 4 Gy) are lower than those seen in studies of populations exposed to lower doses. The results of detailed analyses of leukaemia risks in the cervical cancer cohort [B12, B17] as well as a recent

study of radiotherapy of the uterine corpus [C36] and, to a lesser extent, the life span study high-dose data suggest that cell-killing tends to reduce excess leukaemia risks per unit dose following high-dose exposures. It is likely that cell-killing can explain the relatively low risk estimates in the ankylosing spondylitis study [D6, L6] and in the childhood cancer treatment study [T7].

120. Many analyses of leukaemia risks have investigated the risks for leukaemia subtypes. In general, radiation exposure leads to increased risks for chronic myelocytic leukaemia and for acute myelogenous or lymphocytic leukaemias. There is no convincing evidence of excess risk for chronic lymphocytic leukaemia. As adult T-cell leukaemia is endemic at Nagasaki because of the unusually high prevalence of the HTLV-1 virus, recent life span study analyses considered the risk of T-cell leukaemia in adults but found no evidence of an elevated risk associated with radiation exposure. Ichimaru and Ishimaru [I5] also investigated this issue and found no evidence of excess adult T-cell leukaemia risks among survivors of the atomic bombing of Nagasaki. Subtype-specific dose-response functions considered in the life span study incidence data [P33] show that the leukaemia dose-response function is concave upward and that there is evidence of significant leukaemia risk at doses below 0.5 Gy. A joint analysis of the three leukaemia subtypes (chronic myeloid, acute myelogenous and acute lymphocytic) found that while there were statistically significant differences between subtypes with regard to the temporal patterns and the effects of age at exposure and sex on the excess absolute risk, there was no evidence of significant differences in the shape of the dose-response function.

121. Numerous studies not included in Part XII of Table 8 and Figure II have investigated leukaemia risks in other groups of people exposed to radiation. Groups in which an association between radiation exposure and leukaemia risks has been reported include workers at the Mayak facility in the Russian Federation [K23], adults who received radiation treatment for non-Hodgkin's lymphoma [G8, T4] and patients with cancer of the uterine corpus [B16, C8, C10, C14]. Radiotherapy for Hodgkin's disease was associated with subsequent leukaemia in some studies [K16, T8] but not in others. No excess leukaemia risks were found in a study of women treated for ovarian cancer [K15]. Leukaemia was found to be associated with postnatal exposure to diagnostic radiation in several studies [G15, G18, P17, S44] but not in others [B38, L4]. A 30-year pilot study of almost 5,000 children estimated to have received 0.2-0.3 Gy during cardiac catheterization failed to show a significant excess risk of leukaemia, based on three observed cases compared with 1.9 expected [M47, S45]. However, the excess relative risk estimate from this small number of cases was 2.4, which is similar to that based on data from other studies included in the Table.

122. In summary, the incidence of acute leukaemias or of chronic myelogenous leukaemia exhibits strong associations with low-LET radiation exposure. Risks, especially for those exposed as children, appear to peak 3-10 years after exposure and then decline with time. This pattern is less pronounced for those exposed as adults, in which case there is little evidence for an early peak in the excess relative or absolute risk and the decline in risks, if it occurs, is less pronounced. Recent work also suggests that the temporal pattern of risk of leukaemia following radiation exposure depends on sex. While leukaemia risks have generally decreased with time since exposure, small excess risks persist 25 or more years after exposure (Figure III). Despite the strong evidence for radiation effects on several subtypes of leukaemia, no study has found evidence of an association between radiation exposure and chronic lymphocytic leukaemia.

(m) Malignant lymphoma (non-Hodgkin's)

123. As summarized by Boice [B19], the evidence of an association between radiation exposure and morbidity from non-Hodgkin's lymphoma is weak. Risk estimates from a number of studies are presented in Part XIII of Table 8 and of Figure II. While none of these studies offer convincing support of a radiation effect, the excess relative risk estimates for several of the larger studies are similar to those seen for other cancer types. The life span study incidence data [P33] do not provide strong support for a radiation effect.

124. Non-Hodgkin's lymphoma has been considered in a number of other studies for which it is difficult to estimate the risk per unit dose. No evidence of an excess of non-Hodgkin's lymphoma was found among persons who received diagnostic x ray examinations for various reasons [B38]. Data from the Oxford survey of childhood cancers in children exposed prenatally to radiation suggested an elevated risk of lymphoma [B23], while a study of twins in the United States provided no evidence of an excess [I15]. Occupational studies of radiologists and x-ray technicians [M24, W10] and of nuclear industry workers in the United Kingdom [K21] and in the United States [G17] do not find significant excess risk of non-Hodgkin's lymphoma. In general, there is no convincing evidence that non-Hodgkin's lymphoma is associated with radiation exposure.

(n) Multiple myeloma

125. Studies of the survivors of the atomic bombings in Japan and of people exposed to radiation for medical purposes provide little evidence of an elevated risk of multiple myeloma following radiation exposure. Of the studies included in Part XIV of Table 8 and of Figure II, only the mortality data of the metropathia haemorrhagica study [D23] show a statistically significant excess risk of myeloma. Significant excess myeloma risks were found in

earlier analyses of the mortality data [S7] and incidence data [I13] of the life span study. In the most recent life span study mortality report [R23], the point estimate of the excess relative risk for multiple myeloma [2.3 (1.7-6.3)] was higher than that for any other non-leukaemia cancer type. The life span study incidence data [P33] involved not only an extension of the follow-up period but also a review of the data for each myeloma case. Although several cases included in earlier reports were excluded on the basis of the reviews, the new life span study data set includes about 20 more cases than either of the earlier reports. Persons with dose estimates in excess of 4 Gy and cases diagnosed as second primaries were excluded from the routine analyses. In this analysis the estimated excess relative risk is much lower [0.4 (<0-1.7)] and is not statistically significant. As noted in the analysis of the incidence data [P33] of the life span study, when both the survivors with high doses and the second primary cases were included in the analysis, the point estimate of the risk increased and approached significance.

126. In an earlier review, Cuzick [C20] examined a large number of cohorts and concluded that radiation exposure was associated with an increase in myeloma risk. About 40% of the cases in this review were obtained from the life span study. There have been several recent studies of the association between multiple myeloma and exposure to diagnostic x rays. In a case-control study of 208 cases from a United States health maintenance organization, Boice et al. [B38] failed to find a statistically significant association, but they noted that the risk appeared to increase with the number of x-ray exposures. In a study of 399 myeloma cases and matched controls, Cuzick and de Stavola [C9] found no evidence of an association with diagnostic x-ray exposures.

127. An association between low-dose radiation exposures (average cumulative dose 0.03 Sv) was initially reported for workers at the Hanford site [G20], but later studies cast doubt on this finding. Weak evidence of an increase in risk of multiple myeloma with external dose was seen in the study of workers in the United Kingdom [K20, K21]. In view of more recent findings for the Japanese cohort [P33] and in the Hanford study [G16], an association between radiation exposure and multiple myeloma is not clear at this time.

3. Differences by sex and by ethnic origin

128. The life span study data provide clear evidence of sex differences in the excess relative risks. For solid tumours, the excess relative risk for females is generally about twice that for males. However, the ratios of naturally occurring age-specific cancer rates in females in Japan are about half those for males, which implies that absolute excess risks are roughly equal for males and females. The joint analysis of the life span study mortality data for three broad categories of cancer (digestive system, respiratory system and other non-sex-specific solid tumours) by Pierce and Preston [P20] adds support to the hypothesis that sex effects in the excess relative risks largely reflect differences in sex-specific rates of normally occurring cancer.

129. Sex-specific estimates of the excess relative risk and excess absolute risk for the life span study mortality data and for the life span study incidence data [T15], as well as sex ratios for each of these summary measures of the excess risk and for the background rates are given in Table 9. For solid tumours as a group, the excess relative risks for females are larger than those for males. However, in accordance with the hypothesis described above, the excess absolute risk sex ratios are generally closer to one than are the excess relative risk ratios. Results for cancers of the liver and thyroid and for leukaemia appear to deviate from this pattern. For liver cancer the excess risks, both relative and absolute, appear to be substantially lower among females than among males despite the fact that in the life span study, as in other populations, naturally occurring liver cancer rates for males are higher than those for females.

130. In the life span study and other populations, thyroid cancer occurs only about one third as often in males as in females. Despite this difference in background rates, sex-specific excess relative risk estimates are not significantly different. The sex ratio seen for leukaemia background rates is about the same as that seen for other cancer types, but the sex-specific excess relative risks are similar.

131. Because many of the medically exposed populations used in studies of radiation effects are composed wholly or primarily of a single sex, the information on sex effects on radiation risk from populations other than the Japanese is limited. In the New York tinea capitis study [S4, S8], thyroid cancer excess absolute risk estimates were found to be two to three times greater in females than in males, and it was noted that this ratio is similar to the female-to-male ratio in background rates. The findings in the Israeli tinea capitis study [R15] were similar: a significant difference in the sex-specific excess absolute risk estimate, but no significant difference in the excess relative risks. Sex differences were also considered in the analysis of skin cancer in the Israeli tinea capitis study [R19]. Despite differences in the estimated risks per unit dose, sex effects in the two tinea studies are quite similar to those seen in the life span study.

132. Since only 16% of the members of the ankylosing spondylitis study population are females, there is little power to detect sex differences in this population. However data presented in [D6] allow some comparisons of sex-specific risk estimates. For cancers other than leukaemia and colon cancer there is no significant difference in the estimated excess relative risk per sievert for

males (0.15) and females (0.13) or in the estimated sex-specific excess absolute risk (5.8 cases per 10,000 PYSv for males and 4.3 for females). Excess absolute risk and excess relative risk estimates for leukaemia among females in the ankylosing spondylitis study are about one third of those for males. However, since the number of leukaemia cases among females is small (3 observed versus 1.8 expected), the standard errors are large, so these differences are not statistically significant.

133. Understanding how radiation risks differ between ethnic groups is an important problem for radiation research and protection. Although formal comparison is difficult, the risk estimates presented in the last Section suggest that there is no simple characterization of interpopulation variation. Relative risk summaries are emphasized in many studies; however, one must be cautious in applying these risks to a population with very different background rates. That there is no consistent, simple characterization of sex differences in radiation risks suggests that ethnic or cultural differences in radiation risks are likely to be complex or difficult to distinguish.

134. Some differences in sensitivity to radiation by ethnic origin have been found in the few studies that have included substantial numbers of more than one ethnic group. For example, in the study of about 2,200 children who received x-ray treatment for tinea capitis in New York and a comparable group of 1,400 treated without x rays, 41 persons in the irradiated group had one or more basal cell carcinomas of the face or scalp compared with only 3 in the controls. However, despite the fact that 25% of the irradiated group was black, all 41 skin cancer cases were in whites [S8]. In the study of subjects irradiated in Israel for tinea capitis, the risk per gray of radiation-induced skin cancer was intermediate between the high risk found in the New York whites and the absence of risk found in New York blacks. In the Israeli tinea capitis study, a difference by country of birth was also noted, with subjects born in North Africa experiencing much greater relative risks of radiogenic leukaemia and thyroid cancer than subjects born in Asia or Israel [R4, R5]. (It should be noted that all the subjects born in Israel in this study had fathers born in Asia or North Africa.) However, country of birth did not appear to modify the risk of radiation-related tumours of the brain and central nervous system [R10]. The possible effect of the prevalence of other carcinogenic agents is considered in Section II.A.9. Risks of radiation-induced thyroid cancer in Jewish and non-Jewish subjects were compared in the study of the effects of thymic irradiation noted above [S35]. After adjustment for the effects of sex and time period, the risk of radiation-induced thyroid cancer was 3.5 times higher ($p = 0.01$) in Jewish than in non-Jewish subjects. The population living in the contaminated areas along the Techa River in the southern Urals, which includes both Russians and Tartars, who have different natural cancer rates [K18], might be able to provide useful data on ethnic differences.

4. Effect of age at exposure

135. Although the age range in populations used in studies of radiation effects is often restricted, data from various studies provide a fairly consistent picture of the effect of age at exposure on the risk of radiation-induced cancer. As with sex effects, the nature of variation in risk with age at exposure depends on whether one considers relative or absolute risks. In particular, it appears that for solid tumours at a given time after exposure, persons exposed early in life have higher average excess relative risks and lower excess absolute risks than those exposed later in life. This point is illustrated by comparing the mortality data for the period of observations from the life span study [R23] and the observed data from the ankylosing spondylitis study [D6] (Table 10). The data from the life span study are given for both sexes combined and for males only. Because of the high proportion of men in the ankylosing spondylitis study, the life span study male risks are likely to be more comparable with the estimates obtained in the ankylosing spondylitis study than are the pooled estimates of the life span study. Since these studies involve different follow-up times and since the exposure patterns both in the body and in time, as well as many other factors, differed [U12], the excess absolute risk estimates are not directly comparable. Nevertheless, for leukaemia and for all solid tumours these were compared in the UNSCEAR 1988 Report [U2] (Annex F, Table 56) and in [U12]. Considering the many differences between these studies, the agreement between risk estimates is not unreasonable. Differences in time-averaged excess absolute risks for different age-at-exposure groups should be interpreted with caution because current data suggest that excess absolute risks increase with increasing time since exposure and because follow-up is incomplete for the younger members of most cohorts used in the studies of radiation effects.

136. An increase in average excess absolute risks with increasing attained age would be a natural consequence of a model in which the risk of radiation-induced cancer is simply proportional to background cancer rates. However, findings in the Israeli and New York tinea capitis studies [R19, S8], the Massachusetts tuberculosis fluoroscopy study [B31] and a Canadian tuberculosis fluoroscopy study [M19] support the inference drawn from the life span study that solid tumour relative risks are larger for those exposed at younger ages than for those exposed at older ages. As discussed earlier, for most cancer types the excess relative risk decreases with age at exposure. The large excess relative risk for those exposed as children is not easily interpreted but probably reflects, at least in part, the fact that background rates for those exposed as children have been low for most of the follow-up to date. Kellerer and Barclay [K12] recently noted that a model in which the excess relative risk is allowed to decrease with increasing attained age without any additional dependence on age at exposure describes the life span study data about

as well as an age-at-exposure-dependent relative risk model. Under the fitted Kellerer-Barclay attained-age-dependent excess relative risk model, there is virtually no difference in attained-age-specific excess absolute risk estimates. At the present time, data are complete only for the older members of the population, so models are difficult to test. Shimizu et al. [S7] noted a tendency for excess risk estimates for solid tumours at a given attained age to be greater for those exposed at younger ages than the corresponding estimates for those exposed at older ages. This is an important question that needs to be considered further.

137. Studies of breast cancer incidence provide the clearest evidence of age-at-exposure effects on the risk of radiation-induced cancer. Table 11 contrasts data on age-at-exposure-specific excess relative risk estimates based on breast cancer incidence data from several studies. Although the numbers might suggest differences between cohorts in the level of risk, in virtually every case the time-constant excess relative risk estimates decrease with increasing age at exposure. This is shown more clearly when the data are plotted as in Figure IV. The studies used for the plot are the life span study incidence data [T15], the Massachusetts tuberculosis fluoroscopy study [B31], the New York acute post-partum mastitis study [S9], the Rochester thymic irradiation study [H22], the Swedish benign breast disease study [B1], the Stanford Hodgkin's disease study [H2], the United States study of the contralateral breast among women who received radiotherapy for breast cancer [B41] and the Canadian tuberculosis fluoroscopy study [M19]. All studies used data on incidence, except the Canadian study, which considered mortality data.

138. Despite the limited age range, data on thyroid cancer from the Israeli tinea capitis study [R15] and the Michael Reese tonsil irradiation study [S63] demonstrate highly decreasing relative risks with age at exposure. A similar pattern is seen among younger members of the life span study cohort [A13, T15]. The life span study, which is the largest source of data on adults, suggests that thyroid cancer risks are quite small among those irradiated as adults. Age-at-exposure-specific risk estimates for thyroid cancer from the life span study, the Israeli tinea capitis study and the Michael Reese study are compared in Table 12. The relative risks for some other cancer types estimated for various age-at-exposure groups in the life span study cohort are given in Table 13.

5. Prenatal exposure

139. Studies of medical exposure to diagnostic x rays during pregnancy are important because of the possibility that the developing fetus may be more susceptible to the carcinogenic effects of ionizing radiation than the adult or the young child. The Oxford survey of childhood cancers [B22, M22, S52] was the first (and is still the largest) study to find an association between obstetric x rays and childhood cancer. Since its publication in 1958, numerous other studies have reported similar relative risks of childhood cancer of about 1.4 without taking the fetal dose to the embryo and fetus into account. These results are listed in Table 14. Doubts about the causal nature of the association have been raised, however. They were summarized in the UNSCEAR 1986 Report [U3] and, more recently, by others [M25, M33] and will be discussed further below. It should be stressed that the poor quality of the information on the dose to the embryo or fetus and the possibility that some of the excess risk might be due to factors unrelated to the x-ray exposure preclude an accurate estimate of risk per unit dose. Any such estimates must be interpreted cautiously.

140. Evidence of a causal association includes the consistency of findings in many studies conducted over a period of 35 years [B22]. Case-control studies based on personal interviews were initially criticized because of the possibility of response bias, but subsequent large studies based only on reviews of obstetric records confirmed the association [M9]. The possibility of unidentified conditions that select mothers for x-ray examination whose children were destined to develop cancer was addressed in a reanalysis of twin data within the Oxford survey of childhood cancers [M23, M32] and in two case-control studies of twins in population-based tumour registry regions in Connecticut [H3] and Sweden [R16]. Twins were likely to have been x-rayed for reasons unrelated to the mother's health, such as to diagnose the twin pregnancy or to determine fetal position before delivery, and all three evaluations found an association between prenatal x-ray exposures and childhood cancer. Controlling for maternal illness and drugs taken during pregnancy in the Oxford survey of childhood cancers did not change appreciably the estimates of relative risk [M22].

141. Arguments against a causal association between low dose, obstetric x rays and childhood cancer have been raised. *In utero* exposure to the atomic bombings in Japan (mean uterine dose: 0.18 Gy [Y2]) were not linked to an increase in childhood cancer. During the first 14 years of life, no leukaemias occurred in the *in utero* group, and only two cancers were diagnosed: one case of liver cancer at age 6 years and one case of Wilms' tumour at age 14 years [J2]. The sample, however, is not large. Most studies of prenatal x-ray exposures found the relative risk of leukaemia to be 1.4 and the relative risk of all the major groups of solid tumours also to be 1.4. The similarity between these relative risks is strange, because these cancers are known to have dissimilar origins and very different excess relative risks following postnatal exposures. Such a correspondence is uncharacteristic of any other human or experimental animal exposure, raising questions about biological plausibility [M25, M33]. Furthermore, while Japanese children who were less than

10 years old when exposed to the atomic bombings were at a very high risk of developing childhood leukaemia (14 cases occurred), they did not develop any other childhood cancers in excess [Y1]. Finally, it is noteworthy that excess relative risks have been found only in case-control studies. No cohort study has detected a significant increase in childhood cancer following prenatal irradiation [C15, D16, I15, J2, O5, R13]; indeed, relative risks of about 1.0 or less have been found. The cohort studies can be criticized, however, because of the small numbers of childhood cases and associated low statistical power to detect an excess. Furthermore, one author later concluded that there was a strong possibility that ascertainment was incomplete in his investigation [D25]. Despite the presumed higher frequency of prenatal x-ray exposure among twins than among singletons, cohort studies of twins in Sweden, Norway, Finland, California and Connecticut have also failed to find an increase in childhood cancer, compared with the general population, and several studies report significant deficits [I15, R13]. The exact values of prenatal exposure in the twin studies, however, were not known, and the proportion irradiated may not have been as high as believed. Furthermore, it is not necessarily correct to expect twins to be at the same risk of childhood cancer as singletons. Nevertheless, on the basis of the above discussion the possibility of a causal association must be recognized, although quantification of the risk remains uncertain.

142. If it is assumed that fetal doses from medical prenatal exposures were about 10 mSv, then the estimated excess relative risk per sievert derived from the various case-control studies is roughly 40 for leukaemia and 40 for other cancers. This is in marked contrast to the prenatal or postnatal risk in children derived from the atomic bomb survivor studies. As noted above, there were no leukaemias in the prenatal group. Averaged over the first 35 years of follow-up (1950-1985), the life span study mortality data suggest that the excess relative risks per sievert in survivors of the atomic bombings who were less than 10 years old when exposed are approximately 17 for leukaemia and 2 for other cancers [S7]. Other studies of childhood irradiation find excess relative risks per sievert for leukaemia that are generally smaller than seen in survivors of the atomic bombings (Table 8, Part XII) [H21, R5].

143. Despite the considerable uncertainty about the mean fetal dose per obstetric examination, estimates of risk have been made by UNSCEAR [U2, U4, U6] and others [B6, K6, M22, M30, M32, S37]. These estimates vary widely. Some estimates [K6] failed to account for the decrease in fetal dose per obstetric examination that occurred over the years [M22]. Others relied upon the concept of dose per film [U2] rather than the more appropriate dose per examination, since more than one film may be taken per examination [M30, M32]. Some considered only fatal and not incident cancers; some were computed for the first 10 years of life and others for the first 15 years. One recent analysis of the Oxford survey of childhood cancers with adjustments for many of the previous inadequacies in former computation indicates a decrease with year of birth in the excess absolute risk of incident cancers over ages 0-14 years, perhaps corresponding, at least in part, to a decrease in fetal dose [M22]. The results are summarized in Table 15. Based on these more recent analyses of the Oxford survey of childhood cancers, the excess absolute risk of mortality of all cancers for the first 14 years is about 5 10^{-2} Sv^{-1} [M30].

144. In the only study that has evaluated adult onset cancers following prenatal exposure, Yoshimoto et al. [Y2, Y5] reported a statistically significant increased risk of adult cancer among 1,630 prenatally exposed survivors of the atomic bombings followed through 1984. However, no additional cancers occurred in the most recent follow-up interval (1985-1989), and although the point estimate of the excess relative risk per unit dose remains quite high (0.9), it is no longer statistically significant [Y1] (95% CI: −0.12-3.81). The small number of cancer deaths (24) and the limited follow-up during adult life contribute to the uncertainties in the result. Further follow-up will clearly be important.

145. Studies of *in utero* exposure have thus given a wide range of risk estimates, from relatively high risk to essentially small undetectable risk, including (possibly) none at all. Since there is no biological reason to assume that the embryo or fetus is resistant to the effects of ionizing radiation, and in particular to leukaemogenesis, the finding of a positive risk after appreciable doses is to be expected. However, on the basis of present data the exact quantification of effects is subject to much uncertainty.

6. Temporal pattern of radiation-induced risk

146. The temporal pattern of radiation-related excess cancer risks is a central issue in radiation effects studies. It is clear that temporal patterns for leukaemia and solid cancers differ greatly. Excess absolute risks for leukaemia generally peak 3-10 years following exposure and decline thereafter. Excess absolute risks for solid tumours increase more gradually, possibly increasing monotonically for the entire lifetime following exposure. In the life span study data, for a given age at exposure and sex, this increase with time in excess absolute risk for solid tumours is remarkably proportional to the increase with age in the background cancer rate for a given sex and age at exposure. The data are most usefully and simply described by time-constant excess relative risk models, at least over limited follow-up periods. This should not, however, be interpreted to mean that there is something fundamentally and biologically true about such a model. In particular, it

should not be uncritically assumed that the model can be used to project beyond current follow-up for the life span study or other studies. For the life span study this is mainly an issue for those exposed as children, who are only now attaining ages at which the background rates of cancer become large. Indeed, in the absence of complete lifetime follow-up, the strong dependence of excess relative risks on age at exposure may raise questions about the adequacy of time-constant excess relative risk models. It is important to realize that even if excess relative risk decreased substantially with time, the excess absolute risk would still increase, since background cancer risks increase very rapidly with age. As has been noted in recent papers [K12, P20], reasonably simple time-increasing excess absolute risk models can describe the life span study data as well as constant excess relative risk models. These absolute risk models often predict relative risks that decrease with increasing time since exposure, and they predict smaller differences in attained-age-specific relative risks than are predicted by time-constant relative risk models with age-at-exposure effects. However, time-constant absolute risk models, which were widely considered in the past, do not fit current data and are no longer useful for projection purposes.

147. There is convincing evidence of an excess risk of leukaemia among survivors of the atomic bombings in Japan as early as 1948 [F4] (all major subtypes except chronic lymphocytic leukaemia), but formal description of the risks is possible only from late 1950, when follow-up of the fixed life span study cohort began. When the life span study data for all age groups are considered together, the relative risk was greater during the first five years of follow-up than during later periods, and the subsequent decline in relative risk is statistically significant [P33, S7]. However, the temporal pattern of the excess leukaemia risk depends on both age at exposure and sex and on the type of leukaemia induced. In particular, as noted earlier and illustrated in Figure III, leukaemia excess absolute risks have a higher initial peak and decline more rapidly for those exposed as children than do the excess risks for those exposed as adults. For a given age at exposure, the life span study results also suggest that the initial peak value for excess absolute risk is lower and the decline less rapid for women than for men. Indeed, the pooled leukaemia model described in [P33] indicates a slight increase in excess absolute risk estimates with time since exposure for women over age 40 at exposure. It is also noted in [P33] that a statistically significant leukaemia risk was seen between 30 and 40 years after exposure.

148. Studies of the effects of medical irradiation on the induction of leukaemia generally support the picture emerging from the life span study data. The relatively short interval between irradiation and the maximum relative risk of radiation-induced leukaemia has been confirmed in several studies. For example, in the studies of women irradiated for cervical cancer [B12], patients irradiated for ankylosing spondylitis [D6] and children in Israel given radiotherapy for tinea capitis [R5], the relative risk of leukaemia was at its highest within the first five years after irradiation. In the study of women irradiated for metropathia haemorrhagica [D23, S12], the maximum relative risk occurred 6-10 years after irradiation. The study of ankylosing spondylitis patients is also consistent with the life span study cohort in finding a continued risk of radiogenic leukaemia more than 25 years after irradiation, although the relative risk was considerably lower than at earlier periods (an excess relative risk of 1 for the period from 1 to 14 years after irradiation and 0.2 for 15 or more years after irradiation [D6]).

149. Although it is often stated that radiation-induced solid cancers are not seen for 10 or more years after exposure [D23, S12], a radiation-related increase in solid tumour mortality is apparent in the life span study cohort during the first five years of follow-up, that is at 5-10 years after exposure (excess relative risk per gray shielded kerma: 0.24 (90% CI: 0.05-0.5) in 1950-1955 [S7]). This small relative risk for all cancers, involving a total number of excess cases of about 15, cannot be further subdivided into meaningful risks for individual cancers. A more detailed study of the types of cancer involved in these early deaths might be useful, however. These excess cancers were evident 5-10 years after the exposure in prior evaluations of the life span study, and early excess deaths are also evident in the study of the ankylosing spondylitis patients in the United Kingdom. Attention has not previously been drawn to this period because the number of cases and the excess risks are small. They are pointed out now merely to indicate that the 10-15 year latency period sometimes quoted should be interpreted with caution. At present, the Committee continues to use a 10-year latency period for solid tumours.

150. With increasing time since exposure, the excess absolute risk in the life span study cohort following a given dose and within a given age-at-exposure category has increased dramatically: by factors of 9, 3 and 10 for age-at-exposure groups 0-19, 20-34 and 35 years or more, respectively, between the periods 1956-1965 and 1976-1985 (see Table 16). In contrast, the relative risk of mortality from all cancers other than leukaemia has remained relatively stable, so that when follow-up to 1985 is considered, models that allow the excess relative risk to depend on age at exposure and sex but not on time since exposure are often used to describe the data [P3, P4, S7, T15]. However, as noted earlier, there are alternative models in which the excess relative risk is not constrained to be constant in time that describe the data at least as well as time-constant excess relative risk models with sex and age-at-exposure effects [K12].

151. Although models in which the relative risk is constant with increasing time since exposure are used to summarize the data on mortality from all cancers other

than leukaemia in the life span study cohort up to 1985, such models conceal some differences in the temporal patterns of the radiation-related increase among different age-at-exposure groups. The report on life span study cancer mortality through 1982 [P4] noted statistically significant differences in temporal patterns for survivors with different ages at exposure. In particular, it was found that the excess relative risk tended to decrease in time for those who were younger at exposure and to be constant or perhaps increase slightly for those who were older at exposure. Pierce et al. [P23] carried out an extensive examination of the more recent life span study mortality data and concluded that the evidence for a decrease with time in the non-leukaemia cancer mortality was based on the experience of the youngest survivors. A similar conclusion was reached in the life span study solid tumour incidence analysis [T15]. Figure V illustrates age-at-exposure-specific relative and absolute risks based on the life span study incidence data for 1950-1987 [T15]. The excess relative risk model is linear in weighted dose (RBE for neutrons = 10) and allows for sex-, age-at-exposure and time-since-exposure dependencies with different slopes for those over and under age 20 years at the time of exposure. While the relative risks differ significantly with age at exposure, the decrease in the excess relative risk with time is not statistically significant overall or for either age group. Excess absolute risks were computed as the product of the fitted excess relative risks and the fitted background mortality rates. While the excess relative risks for females are about 40% greater than those for males, the excess absolute risks for males and females are similar. From this figure it can be seen that despite a large estimated decrease in the excess relative risk for survivors who were less than 20 years old at the time of exposure, excess absolute risks have risen at about the same rate as those for other age groups. The plots suggest that as the youngest survivors reach the ages at which there are rapid increases in background cancer incidence, excess relative risks may decline to those seen for older life span study survivors.

152. A number of studies have paid attention to differences in temporal patterns of the excess relative risk for specific non-leukaemia cancers. The life span study mortality and incidence reports [R23, S7, T15] provide trend tests for most types considered. A statistically significant increase in the risk with time (without allowance for age-at-exposure effects) is reported for all cancers other than leukaemia. Statistically significant time trends were noted for one or two specific cancer types; however, in view of the large number of tests carried out and the relatively small number of cases at a particular site, these tests fail to provide convincing evidence of a significant variation between sites with respect to temporal patterns of the excess relative risk. The BEIR V Committee [C12] developed separate models for cancers of the digestive system, the respiratory system, the female breast and other non-leukaemia tumours and elected to base risk projections for cancers of the respiratory system and breast on models in which the excess relative risk decreases with increasing time since exposure. However, a recent joint analysis of the life span study data [P20] using the three non-sex-specific, non-leukaemia categories considered by the BEIR V Committee [C12] found that no evidence of significant differences between these categories with respect to age-at-exposure or temporal patterns. It also found that the category-specific, time-dependent excess relative risk models used by the BEIR V Committee [C12] did not provide a statistically significant improvement in fit relative to a simple, two-parameter, age-at-exposure-dependent, constant excess relative risk model. As described in the life span study cancer incidence report [T15], point estimates of a time trend for the breast cancer data were negative, indicating a decreasing excess relative risk with time, but the decreases were not statistically significant [P20].

153. In the study of ankylosing spondylitis patients the relative risk for all cancers other than leukaemia or colon cancer (40% of which are lung cancers) decreases significantly with increasing time since exposure, with the estimated excess relative risk (0.07) for the period 25 or more years after exposure being only about 20% of the average excess relative risk (0.38) during the preceding 20 years [D6]. Only for cancers of the oesophagus, liver, larynx, bladder and skin and for multiple myeloma were the relative risks 25 or more years after exposure greater than those seen during the 5-25 year period, and only for cancer of the oesophagus was the excess relative risk in the later period statistically significant. The overall decline in relative risk could not be attributed to the variation with time in the types of cancer observed or to effects of age at exposure or to inadequate ascertainment of death in the oldest age groups in the study. There was no evidence that patients who survived and remained in the study for 25 years after the initial treatment received doses that were lower than those who did not [D6]. A decrease in the relative risk for all cancers other than leukaemia with increasing time following exposure has also been seen in three groups of children given medical irradiation [L18]. In contrast, results from the cervical cancer case-control study suggest that relative risks for heavily irradiated sites, which do not include the lung or the breast, tended to be highest 20 or more years after exposure [B21], although there were few cases beyond 30 years after exposure.

154. Temporal variation in excess relative risk estimates has been considered in a number of other recent studies of medically irradiated populations. In the most recent analysis of the data on breast cancer incidence among women given frequent chest x-ray fluoroscopy in Massachusetts, it was found that the risk remains high for more than 50 years [B31]. The relative risk of thyroid tumours following thymus irradiation among infants in New York appears to decrease smoothly with increasing time since exposure 15 or more years after exposure, with

the relative risks for the periods 5-14, 15-24, 25-34 and >34 years since exposure estimated to be 11.0, 4.8, 2.0 and 1.8, respectively [S10]. There is also a reduction in relative risk with time in the tinea capitis study in New York [L18]. In contrast, the extended follow-up of subjects in the tinea capitis study in Israel has found that the relative risk remains approximately constant for 30 years after exposure [R15]. Apparent differences such as these require reconciliation. Additional statistical analyses of all the available data may be helpful.

155. In studies of the effects of exposure in childhood, no increase is seen in some types of cancer until the population has reached an age when such tumours begin to appear in the absence of unusual exposure to radiation. For example, in the study of infants irradiated for supposedly enlarged thymus, the first breast cancer was diagnosed 28 years after irradiation [H22]. Similarly, in the study of children in New York irradiated for tinea capitis the minimum latency period for radiation-induced skin cancer was about 20 years [S8]. However, the interval between irradiation and the subsequent excess is not always greater in children than in adults, especially for tumours that are common in childhood and early adult life. For example, in the study of children irradiated for tinea capitis in Israel, tumours of the brain and nervous system began to appear only 6 years after irradiation [R10].

7. Dose response

156. In any discussion of the dose response, the dose itself is of paramount importance, and because the life span study is so important, the dosimetry of the survivors of the atomic bombings of Hiroshima and Nagasaki is especially relevant. The formal dosimetry scheme DS86 has been used as the basis of risk estimation in the life span study since 1986. It has generally been regarded as satisfactory, but recently some important questions have been raised about a potentially larger contribution to the dose from fast neutrons than was specified in DS86. This may have come about either because the RBEs for neutrons are very high at low doses, as has been suggested [R26, S33], or because the presence of thermal neutrons [S56] is an indication of greater numbers of fast neutrons, especially at distances in Hiroshima, than were allowed for in DS86.

157. When the DS86 was adopted (and even before that) physicists were aware of some discrepancies for both Hiroshima and Nagasaki between thermal neutron fluxes based on activation measurements and values derived from the fast neutron transport equations. These discrepancies have become more evident for Hiroshima (but not, apparently, for Nagasaki [S72]) with the measurements of ^{152}Eu and ^{154}Eu in soil and measurements of ^{36}Cl in concrete, both activated by thermal neutrons and reported in [S56]. The ratio of measured thermal neutron activation to that derived by calculation increases from less than 1 at 800 m or less from ground zero to about 2 at 1,000 m and about 10 at 1,600 m. (The average DS86 fast neutron organ absorbed doses for survivors at those distances are 1%-2% of the gamma ray absorbed doses.) The origin of these extra thermal neutrons that seemed to have been present at Hiroshima is not known. If they come from fast neutrons not accounted for, presumably in the source term for Hiroshima, the total equivalent dose at greater distances would need to be increased. The matter is still under investigation. In the meantime some calculations have been made of the possible maximum impact of potentially increased numbers of fast neutrons on risk estimates. The increased equivalent dose would probably reduce the risk estimates by perhaps 10%-20%, reflecting the fact that the change could not be large at dose categories that contribute most to risk [P15]. The fact that risk estimates based on the incidence for all cancers at Nagasaki alone differ very little from estimates based on that at Hiroshima and Nagasaki together would appear to confirm that the impact cannot be large. It should also be noted that adding more neutrons at greater distances in Hiroshima will tend to increase the disagreement between the two cities, which DS86 improved over TD65. On the other hand, measurements of chromosome aberration frequencies (which do not necessarily relate to cancer induction) induced by gamma rays and a range of neutron energies were recently used to reanalyse aberration data for the survivors of the atomic bombings. A neutron component of about 5% of the absorbed dose was derived for Hiroshima, compared with about 2% in DS86 [S53]. Yet another recent reanalysis of stable chromosome aberrations among the atomic bomb survivors at Hiroshima and Nagasaki examined the differences between the two cities. It concluded that, in DS86, the RBE for neutrons is very high or the neutrons at Hiroshima have been underestimated or the gamma rays at Nagasaki have been overestimated [S33]. The last-mentioned reason seems unlikely, since TLD (thermoluminescent dosimetry) measurements of the gamma rays at Nagasaki confirm calculations.

158. Thus, evidence appears to be mounting that DS86 does not fully account for the neutrons that were probably present at Hiroshima, especially at the greater distances. Present indications are that in analyses based on linear or linear-quadratic models, the overall reduction in risk estimates resulting from increases in neutron doses would not be large [P15]. There are also other features of DS86 that may, in time, require amendment. These include correction for biases in risk estimates resulting from random errors in individual dose estimates (paragraph 35) and for the effect of misclassification of cancer on death certificates (paragraph 26), both of which would tend to increase risk estimates [S1].

159. There are a number of other important issues related to the shape of the dose-response function for radiation-induced cancer. These issues revolve around the degree of curvature in the cancer dose-response function

at low doses and the impact of cell-killing on the response, especially at high doses. In a recent monograph, Kondo [K25] summarizes much of the experimental and epidemiological data on low-dose effects and offers a criticism of the linear no-threshold model of dose response. The question of the shape of the dose-response function at low doses is important since the doses of interest in radiation protection are often much lower than the doses in the epidemiological studies that contribute most of the information to risk estimates. Another issue, which arises in part because of the neutron component of the life span study exposures at Hiroshima, involves description of the RBE of neutrons for cancer induction. Experiments and epidemiological studies involving the direct effects of low-LET radiation, e.g. the induction of chromosome aberrations in lymphocytes, have generally found that a linear-quadratic model describes the low-LET radiation dose response better than a simple linear model. These experiments also indicate that cell-killing leads to a flattening of the dose-response function following high-dose-rate exposures to high doses. While these experiments are crucial to an understanding of radiation effects, it is important to remember that cancer induction is a lengthy, multi-step process, and it is not clear that the results for specific, observable cellular events should apply directly to radiation-induced cancer risks (see UNSCEAR 1993 Report [U1], Annex E).

160. Without a better understanding of the issues involved in transferring risk estimates from one population to another, comparison of risks for populations exposed to different doses is of little use in clarifying the shape of the dose-response function. Thus, the most useful information on dose response, especially at low doses, comes from studies of a given population with a range of exposures and individual dose estimates. The lack of individual dose estimates limits the usefulness of several large studies, including the ankylosing spondylitis study in the United Kingdom and the cervical cancer cohort study. There are, however, a number of important studies that do have individual estimates. These include the life span study, the cervical cancer case-control study, the various tuberculosis fluoroscopy cohorts, the benign gynaecological study and the Israeli tinea capitis study.

161. As was discussed in Chapter I, errors in individual dose estimates affect the nature of observed dose-response relationships. As has been well documented [G4, J1, P3, P9], random errors in individual dose estimates lead to a downward bias in the estimated risk per unit dose. These errors also distort the shape of the observed dose-response curve. If, as is often the case in radiation effects studies, errors in individual doses are roughly proportional to dose, the apparent dose response will tend to have less upward curvature than the true response.

162. Because of the interest in effects at low dose, the radiation effects literature is replete with two-group comparisons involving low-dose groups and unexposed groups; these comparisons are used to address the question of the lowest dose for which effects can be detected. Such tests should be interpreted with caution, since in most cases the failure to detect effects is more likely to be a consequence of the lack of power of the test than an indication of the absence of an effect. Furthermore, positive results can occur by chance only. While such comparisons should not be ignored completely, once a dose effect has been found, emphasis should be placed on tests for departures from linearity (based on linear-quadratic or spline models) rather than on specific dose-group comparisons.

163. Figure VI illustrates the relationship between excess relative risk and weighted dose to the bone marrow or large intestine for mortality up to 1987 from leukaemia and cancers other than leukaemia, respectively, in the life span study cohort, using the DS86 dosimetry [P11, P13, R23]. For leukaemia, the estimates of excess relative risk were derived from data for both sexes and all ages at exposure. For cancers other than leukaemia, the estimates of risk illustrated are those for a male aged 30 years at the time of exposure; the risks for females in the life span study cohort were roughly twice those for males. There is an apparent non-linearity in the dose responses for leukaemia over the range of doses, while the dose response for solid tumours exhibits less curvature. The issue of the non-linearity of dose-response curves from the life span study for cancer mortality (and for incidence) will be discussed further below.

164. Because of concerns about the role of cell-killing and the impact of errors in individual dose estimates on the shape of the dose-response curves at high doses, and because the life span study risk estimates are primarily used for effects at low doses, recent analyses of the life span study data have often focused on individuals with shielded kerma of less than 4 Gy. Even when the analyses are restricted to survivors with less than 4 Gy shielded kerma, the risk estimates from the life span study are biased by dosimetry errors. Calculations presented in [P9] suggest that bias corrections based on an assumption of 35% errors in individual dose estimates increase linear risk estimates by about 6% for leukaemia and 9% for cancers other than leukaemia [P9].

165. Statistically significant risks for solid tumours in the life span study (see Table 7) are presently seen only above 0.2 Sv, i.e. the relative risks for solid tumours in the lower dose categories, namely 0.01-0.05, 0.06-0.09 and 0.10-0.19 Sv, are not significantly different from unity. However, these dose categories all have positive nominal risk estimates, and the slope of the dose response for doses lower than 0.5 Sv, while lower than the slope for all doses up to 4.0 Sv, does not differ significantly from it [S65] (see also Table 17). For some specific cancer sites, point estimates of the excess risk are negative in some

low-dose categories, although not significantly so. However, while a linear relationship with dose for solid tumours is consistent with the life span study data, other dose-response models that exhibit curvature in the low-dose region cannot be ruled out.

166. The shape of the life span study cancer mortality dose-response curve has been studied in detail by Pierce and Vaeth [P11, P13] for individuals whose shielded kerma was 0-4 Gy. Separate calculations were made for leukaemia and for all other cancers, with and without an adjustment for the effect of errors on the estimated risks. The fitted models are of the form

$$\text{ERR} = (\beta \alpha_{sex} \delta_{exp}) \times (D + \theta D^2) \qquad (12)$$

where D is the weighted dose assuming a neutron RBE of 10, α_{sex} and δ_{exp} represent possible sex- and age-at-exposure-dependent modifications in the linear risk per unit dose β, and θ represents the degree of curvature in the dose-response relationship. The reciprocal of θ is the weighted dose at which the linear and quadratic parts of the dose-response relationship intersect and is often referred to as the crossover dose. For cancers other than leukaemia, θ is estimated to be 0.1 Sv^{-1} or 0.2 Sv^{-1}, depending on whether or not the doses are adjusted for the effect of errors. These values are not significantly different from zero. (A value of zero corresponds to a linear dose response.) For leukaemia, θ is significantly greater than zero, with estimated values of 0.4 Sv^{-1} for unadjusted doses and 0.8 Sv^{-1} for adjusted doses.

167. The quantity θ can be related to the dose-reduction factor, which is the factor by which the slope of a linear model should be divided to give the slope at low doses, i.e. the linear term in a linear-quadratic dose-response model. The value of the dose-reduction factor is equal to $1 + \theta D_0$, where D_0, which depends on the distribution of doses in the data under analysis, takes the value 1.5 for the life span study data [P11, P13]. The dose-reduction factor for mortality from cancers other than leukaemia is estimated to be 1.2 or 1.3, depending on whether dosimetry errors are taken into account. These values are not significantly different from unity. The best estimate of the dose-reduction factor for mortality from leukaemia is 1.6 if errors in the doses are ignored. This estimate increases to 2.2 after allowance for the effect of dosimetry errors. In addition to carrying out significance tests, Pierce and Vaeth [P11, P13] calculated confidence intervals for both θ and the dose-reduction factor. The 95% upper confidence limits (one-sided) for the dose-reduction factor for cancers other than leukaemia are 2.4 and 3.6 before and after adjustment for dosimetry errors, respectively. The lower confidence limits were not given explicitly but were less than 1.0 in both cases. The corresponding upper 95% limits for leukaemia are 3.6 and 6.0. Without adjustment for dosimetry errors the lower bound is less than 1.0, but after adjustment for dosimetry errors the lower bound is 1.1.

168. In the analysis of the life span study solid tumour incidence data [T15], the dose response was found to be approximately linear for survivors with DS86 weighted doses between 0.2 Sv and 4 Sv, as can be seen in Figures VI and VII. Below 0.2 Sv, limitations of sample size etc., discussed elsewhere, render quantification difficult. The pooled leukaemia data (Figure VI) [P33] were best described by a linear-quadratic model with upward curvature. Although subtype-specific leukaemia analyses failed to find evidence against linearity for acute lymphocytic or chronic myelogenous leukaemia, a joint analysis that included these two subtypes along with acute myelogenous leukaemia failed to find statistically significant differences in the curvature for the three subtypes. For all leukaemias considered as a single group and for acute myelogenous leukaemia and chronic myelogenous leukaemia considered separately, statistically significant excess risks were seen at doses below 0.5 Sv. For chronic myelogenous leukaemia and acute lymphocytic leukaemia, the slope of the dose-response function below 0.5 Sv did not differ significantly from that for survivors with larger doses. Vaeth et al. [V14] presented preliminary results of an analysis of cancer incidence data on the shape of the life span study dose response for cancer incidence similar to that carried out for the mortality data. The results of this analysis generally parallel those for mortality described above [P11, P13] and are shown graphically in Figure VIII. The difference in curvature between the solid tumour and the leukaemia dose-response functions is significant, and with a bias correction for dosimetry errors, the estimated value of the low-dose reduction factor for solid tumours is about 1.05 with a 90% confidence interval of 0.6-1.6. The corresponding leukaemia point estimate is 2.5 with a 90% confidence interval of 1.3-8.4. Figure VIII summarizes the results for low-dose extrapolation factors for leukaemia and for solid tumours for both adjusted and unadjusted doses.

169. Dose-response relationships have been considered in several studies of the effects of medical irradiation. Table 18 summarizes the results of an analysis of the dose-response relationship for cancers and benign tumours of the thyroid in the tinea capitis study in Israel [R15]. Relative risks increased appreciably even within the limited range of thyroid doses received by study subjects. After adjustment for sex, ethnic origin and attained age, the relative risk for cancers rose from 3.3 at the lowest dose category (mean dose: 0.062 Gy) to 6.1 at the highest dose category (mean dose: 0.214 Gy), while in a linear fit the relative risks rose from 3.0 to 8.0 for the same dose categories. For benign tumours, the dose-response curve was not as steep. However, the adjusted relative risks still increased from 1.1 to 4.5, while the fitted risks increased from 1.9 to 4.0. It should be noted that while some aspects of the dosimetry in the tinea capitis study in Israel seem to be sound, the values of the doses to the thyroid gland and to the breast may have greater uncertainties than those of the doses to organs directly in the x-ray field. In addi-

tion, it is possible that the irradiation of the pituitary gland may influence the response of the thyroid, and in the tinea capitis study in Israel the pituitary doses were high.

170. Dose-response relationships for breast cancer incidence have also been examined in several tuberculosis fluoroscopy studies. In neither the Massachusetts incidence study [B31] nor the Canadian mortality study [M19] was there any evidence of significant non-linearity in the dose response. The data in Tables 19 and 20 describe the dose response seen in the Massachusetts data. The original Canadian tuberculosis fluoroscopy data set included only women treated in Nova Scotia. The most recent report [M19] included data on women treated in other provinces as well. Both the average doses and the estimated excess relative risk at 1 Sv were lower for women treated in other provinces than for those treated in Nova Scotia. This finding had been interpreted as suggesting a non-linear dose response. However, results described in [M19] indicate that a linear dose-response model that allows for a region-dependent slope fits significantly better than a linear-quadratic model that does not allow for regional differences. Adding a quadratic term to the linear model did not improve the fit, however, and a pure quadratic model was found to fit significantly worse than either the linear or linear-quadratic alternatives.

171. If one examines the studies at higher doses in the risk comparison tables presented earlier in this Chapter, the comparisons often suggest that estimates of the risk per unit dose at higher doses are low. These apparent lower risks are consistent with the idea that cell-killing at high doses reduces the pool of cells at risk of developing cancer. As noted earlier, it is difficult to separate cell-killing from the effects of errors in individual dose estimates in the life span study data when both probably contribute to a levelling off or downturn at higher doses (Figure VI). However, a number of studies of medically exposed populations can provide information on the effect of high doses. Extensive analysis of data on leukaemia following exposure to high doses has been carried out using the cervical cancer patient data [B8, B12]. For this population, in which the average total bone marrow dose was about 7.5 Gy, it was found that relative risks increased with dose for doses up to about 4 Gy but decreased at higher doses. The breast cancer data from the acute post-partum mastitis study in New York [S9] suggest that relative risks levelled off at doses in excess of 3 Gy and appeared to decrease after about 5 Gy (see Figure IX). In the study of patients treated with x rays for ankylosing spondylitis in the United Kingdom, the excess absolute risk of leukaemia was found to vary erratically with radiation dose over a range of mean marrow doses up to about 7 Gy [S13], but a model with linear and quadratic terms for the leukaemia induction rate and exponential cell-killing at higher doses fitted the data reasonably well. In a case-control study of cancers following radiotherapy for childhood cancer, it was found that exposures as high as 60 Gy were associated with a high risk of thyroid cancer [T5]. While there was no downturn in the risk at high doses, the results are not inconsistent with a hypothesis that the slope of the dose response is lower at high doses than at low doses (relative risks in the dose ranges <2, 2-10, 10-30, 30-60 and >60 Gy were 1.0, 7.5, 7.6, 9.2 and 12.0, respectively) [B8, T5]. The first cancers in this study were neuroblastoma (seven cases), Hodgkin's disease (five cases), Wilms' tumour (four cases), non-Hodgkin's lymphoma (four cases), brain cancer (two cases) and other cancers (three cases). Finally, the excess relative risks of leukaemia derived from high-dose studies (7 Gy or more) such as those involving breast radiotherapy [C24] tend to yield lower values of excess relative risk per unit dose than for the survivors of the atomic bombings, indicating a probable effect of cell-killing. Thus the evidence for cell-killing effects varies, but it is probably a factor in most higher dose responses and may even have an impact at lower doses [S1]. Furthermore, cell-killing effects may vary according to the site of cancer considered.

172. The life span study is the only important epidemiological study in which people were exposed to both gamma rays and neutrons. It is not possible to compute useful estimates of the neutron RBE from the life span study data because of the high correlation between gamma and neutron doses. However, it has been shown [P3] that risk estimates are not greatly affected by changes in the assumed (constant) value of the neutron RBE for RBEs in the range 1-20 because the neutron doses are small. Although the life span study data do not allow precise estimation of the RBE, it has been demonstrated [T15] that RBE values developed on the basis of experimental data on chromosome aberrations [D20, S53] do not fit with those derived from the data for solid tumours of the life span study. This lack of fit is not related to the value of the limiting RBE of neutrons; instead, it results from the fact that there is curvature in the gamma dose response in the experimentally derived chromosome aberration dose-response models but not in the solid tumour incidence or mortality data.

8. Fractionation

173. In many situations of interest in radiation protection and risk estimation, such as those involving occupational or diagnostic exposures, the risk associated with any given total dose to a particular organ may depend on the dose rate and on the number of fractions in which it is delivered. Exposures in almost all of the studies considered in Section II.A were received at high dose rates. However, the degree of fractionation varied from a single fraction for the survivors of the atomic bombings in Japan to a few fractions for patients given thymus irradiation to large numbers of fractions for the exposure of radiologists and patients with pulmonary tuberculosis receiving air lung collapse treatment.

174. Some tuberculosis patients were monitored with chest fluoroscopy every two weeks for five years or longer. Their exposures are thus much more comparable to current occupational or diagnostic medical exposures than the experience of persons exposed to high single doses. Two major studies of tuberculosis patients have been carried out. One included 4,940 women who were treated in Massachusetts between 1925 and 1954 and followed for an average of 30 years [B31]. Among 2,573 women who were examined an average of 88 times, the mean breast dose was 0.79 Gy; 142 breast cancers were observed compared to 107.6 expected based on incidence rates for the general population used as controls. In contrast, no excess was found for 2,367 women treated by other means and not regularly irradiated (87 observed versus 100.9 expected). The relation between dose and relative risk of breast cancer was consistent with linearity up to 4 Gy, and for those aged 20 years at exposure the excess relative risk at 1 Gy was 0.70. This is lower than the value seen for breast cancer incidence in the life span study, which is about 2 (Table 11). However, much of the difference in these excess relative risks may be due to the large differences between background breast cancer incidence rates in the United States and those in Japan. After allowing for these it is likely that the excess absolute rates in the Massachusetts tuberculosis fluoroscopy cohorts and the life span study are quite similar. The variation in excess relative risk with age at exposure and time since exposure was also consistent with that seen in the life span study (see Table 11).

175. The second study of tuberculosis patients given fluoroscopies included 31,710 women treated at sanatoria in Canada between 1930 and 1952 [M19]. About a quarter of these women received doses of 0.10 Gy or more, and their relative risk of death from breast cancer as compared with those exposed to less than 0.10 Gy was 1.36 (95% CI: 1.11-1.67). Women in Nova Scotia experienced excess relative risks per unit dose that were approximately three times higher than women in other provinces. The data are consistent with a linear dose-response model, and the relative risk appeared to decrease with increasing age at exposure. The BEIR V Committee [C12] reported that the temporal pattern of the relative risks was similar to that seen in the life span study. However, recent analyses of the life span study incidence data with five to seven additional years of follow-up [T15] suggest that relative risks have remained fairly constant in the survivor cohort. This seems to differ from the findings reported in [M19] and [C12]. Within the Canadian cohort, differences in the slopes of the dose-response curve for women treated in Nova Scotia and in the other provinces is puzzling. Although the mean numbers of fluoroscopic exposures were similar, the dose rate was more than an order of magnitude greater in Nova Scotia than elsewhere [M19], and a possible explanation suggested by the authors is that this might be a dose-rate effect. While this may be the case, the estimated relative and absolute risks are much higher than those in other populations of women with similar background rates, e.g. the acute post-partum mastitis series in New York, whose radiation exposures involved relatively high dose rates and little fractionation. Another possible factor is that of better ascertainment in the well-studied Nova Scotia population as well as more precise dosimetry because all the patients faced the x-ray tube.

176. The breast is not the only organ for which there are human data on the effect of dose fractionation. There has also been an opportunity to examine fractionation effects in patients irradiated in infancy for supposedly enlarged thymuses [S10]. Patients treated with x rays in Rochester, New York, from 1926 to 1957 and their unirradiated siblings were contacted by mail questionnaire and pertinent medical conditions were verified with the doctor or the hospital noted on the questionnaire. To eliminate the possibility that cell inactivation at high doses might mask any sparing effect of dose fractionation, the fractionation analyses were limited to the 2,358 subjects with total thyroid doses less than 6 Gy. Analyses were carried out of the dose per fraction, the number of fractions and the interval between fractions; the results are summarized in Table 21. The excess absolute risk of thyroid cancer per gray was numerically greater in the group with the lowest dose per fraction, but there was no significant trend in excess absolute risk with increasing number of fractions or with a greater interval between fractions. A Cox regression analysis considering relative risk also revealed no evidence of a sparing effect of any of the three fractionation variables.

177. The effect of fractionation on subsequent lung cancer risk has also been examined in patients with tuberculosis [D14, H43]. In an extension of the Massachusetts study, 6,285 patients examined by x-ray fluoroscopy an average of 77 times during lung collapse therapy were followed for an average of 25 years. The average dose to the lung was 0.84 Gy. Based on United States national mortality rates, there was no evidence of increased mortality from lung cancer (SMR: 0.8, based on 69 deaths). It is by no means clear that national rates in the United States for lung cancer form an appropriate basis for calculating expected numbers for patients who are undergoing lung collapse therapy. In addition there was no evidence of a dose-response relationship despite a wide range of doses to the lung. Lung cancer risk did not vary with time since exposure or age at exposure, and adjustment for smoking habits and the amount of lung tissue at risk did not appreciably modify these findings. The authors interpreted these data as suggesting that, in contrast to breast cancer, fractionated exposures as experienced by this cohort may be less effective in causing lung cancer than a single exposure of the same total dose. This interpretation is supported by a recent study on the Canadian fluoroscopy experiences, in which no excess of lung cancer was detected after fractionated exposures averaging about 1 Gy [H43]. However, a small risk of lung cancer was found after radiotherapy to the breast [I20, N8].

178. Another group who received highly fractionated high-dose-rate x-ray exposures is formed by radiologists and other medical x-ray workers. The results of studies of individuals in these groups were reviewed recently by Carpenter [C25] and are summarized in Table 22. For many of the studies, dosimetry information is very limited. However, although many of the early radiologists had occasional single doses that were high enough to cause overt radiation effects [M24], those joining the profession after the introduction of radiation protection guidelines would have been exposed repeatedly only to low-dose x or gamma rays. For most of the groups listed in Table 22, guidelines were introduced during the 1920s (for example in 1921 in the United Kingdom), although they were not issued until the late 1950s in China [W14]. Thus, the increase in mortality with increasing time since entry into the profession seen in the United Kingdom study for radiologists entering the profession after 1920; the excesses in the study in the United States for those entering in 1940-1969; and the results obtained for recent follow-up periods in the studies in Japan and Denmark provide strong evidence of the carcinogenicity of very highly fractionated exposures to low-dose low-LET radiation. However, there are reasons to interpret these studies with some caution. In particular, the dose estimates are crude, and it is possible that some workers, especially those employed before radiation protection guidelines had been introduced, might have received high doses. In addition, since there is no internal comparison group, results are based on comparison with national rates. Furthermore, some of the differences may reflect factors other than radiation, such as alcohol or tobacco.

179. A tendency for higher relative risks among those exposed at younger ages is also apparent among diagnostic x-ray workers in China [W14], who received highly fractionated exposures. In this group the relative risks for leukaemia and thyroid cancer decreased with increasing age at first employment (see Table 23). The relative risks for all cancers combined, exclusive of leukaemia, was highest among workers first employed before the age of 20 years, but otherwise the data suggest little variation in relative risks across age-at-employment categories. The highest relative risks for breast cancer were seen among those aged 25-29 years at first employment. Some of the variation in relative risks by age at first employment may be attributable to the varying length of follow-up or the different total doses accumulated to date for the different groups. (Average observation times ranged from 19.9 years among workers who were less than 20 years old when they started work to 11.7 years among those who were 40 years old or over.)

9. Joint effects of radiation and other agents

180. The effects of joint exposure to more than one carcinogen can be studied in depth only if detailed information can be obtained on the extent of exposure to each of them. There are relatively few examples where such information is available for human populations. However, detailed studies have been carried out of cigarette smoking and asbestos exposure in relation to lung cancer among insulation workers, and also of cigarette smoking and alcohol consumption in relation to cancers of the oral cavity, pharynx, oesophagus and larynx, and these have been summarized by the International Agency for Research on Cancer [I9]. In these examples, it was clear that the two carcinogens did not act independently and that the relative risks among those exposed to both were similar to the product of the relative risks observed among individuals exposed to only one or the other. Further evidence that different carcinogens do not act in isolation from one another is provided by the increase in excess absolute risks with increasing age seen in many populations exposed to radiation in adult life, presumably the result of cumulative causative factors (see Section II.A.4).

181. The topic of joint effects was last reviewed by UNSCEAR in 1982 [U4]. Since then, a number of additional studies have been published. Several epidemiological surveys, conducted between 1963 and 1970, ascertained cigarette smoking status for various subsets of the life span study cohort. In interpreting the results of analyses based on these studies it should be noted that current cigarette smoking status, rather than smoking status at the time of the bombings, was ascertained and that cigarettes were very scarce in Japan towards the end of the Second World War. Prentice et al. [P12] examined cancer mortality among 40,498 such individuals in relation to total T65DR radiation dose and cigarette smoking habits at the time of the survey. Thirty-three per cent of lightly exposed persons (<0.10 Gy) were cigarette smokers as compared with 41% of heavily exposed persons (>1.00 Gy). There thus appears to be some potential for cigarette smoking habits to confound and distort the relationship between radiation dose and cancer mortality in the life span study cohort. In the analysis of Prentice et al. [P12], the at-risk period for each individual extended from the month of the first survey in which the individual took part to the end of 1978 or to death, if this occurred earlier. Cancer mortality for sites reported at that time to be related to radiation exposure [K9] was found to be still significantly related even after taking into account cigarette smoking status. An exception was multiple myeloma mortality, where there were not enough cases to allow study. Further analyses using relative risk models that were either multiplicative or additive in the two exposures (smoking and radiation) or models that took more general forms were also carried out for some sites of cancer. Persons heavily exposed to both cigarette smoke and radiation were found to have significantly lower cancer mortality than multiplicative relative risk models would suggest for all non-haematologic cancer, stomach cancer and digestive cancer other than stomach cancer. Surprisingly, the relative risk function for those cancers appeared not only to be submultiplicative (i.e. the excess relative risks were lower

than those predicted by the multiplicative relative risk model) but also to be subadditive (i.e. the excess relative risks were less than those predicted by an additive model). The lung cancer relative risk function, on the other hand, could not be said to be either multiplicative or additive, but there was no suggestion that it was subadditive.

182. A further study of the joint effects of radiation exposure (T65DR) and cigarette smoking status during the 1960s in relation to lung cancer incidence up to the end of 1980 was carried out by Kopecky et al. [K8] in a subset of the life span study cohort. The data were found to be well fitted by a model in which the excess relative risk was the sum of two components:

(a) a radiation-related term that increased with decreasing age at the time of the bombing;
(b) a smoking-related term that increased with daily cigarette consumption, as measured at the beginning of follow-up, and with attained age.

In this model there was no significant synergism between the effects of radiation exposure and smoking, suggesting that the two factors might combine to increase the relative risk of lung cancer in an additive rather than a multiplicative fashion. The data are illustrated in Figure X. One consequence of this is that the radiation-related relative risk is greatest among non-smokers and decreases with increasing tobacco consumption. Adjustment for the effect of smoking by means of this additive relative risk model substantially reduced the apparent sex differences in the radiation-related relative risk of lung cancer, since few women and most men were smokers. Thus, differences in smoking habits between men and women may largely account for the generally higher relative risks seen for women than men in the life span study cohort for cancers other than leukaemia (see Section II.A.3 and Table 9).

183. The interplay between risk of radiation-induced skin cancer and degree of skin pigmentation was discussed in Section II.A.3. In the study of children in New York irradiated for tinea capitis [S8], the skin cancer risk was particularly pronounced on exposed areas such as the face. The excess skin cancers occurring on the fringe areas of the face were estimated to yield a risk about 4.6 times greater than that estimated for skin cancers on the hairy scalp [H46]. This strongly suggests that exposure to ultraviolet radiation has increased the carcinogenic effect of the x rays on the face, relative to their effect on the scalp, which is protected from ultraviolet rays.

184. Another factor that alters the risk from radiation was observed in women given radiotherapy for breast inflammations/infections associated with childbirth or lactation [S9, S46]. Even after controlling for age at treatment, women who were irradiated shortly after their first childbirth subsequently had a greater excess risk of breast cancer per gray than women who were irradiated after their second or later pregnancies. This finding complements those of other epidemiological studies and confirms that age at first childbirth seems to be an important risk-modifying factor for breast cancer [L11].

185. Patients with cancer treated with radiation and chemicals provide an opportunity to evaluate whether the two exposures interact in causing secondary leukaemias and, possibly, other cancers. In the largest series to date [C24], patients with breast cancer who were treated with radiotherapy and alkylating agents were found to have a significantly higher risk (RR = 17.4) of developing leukaemia than women receiving radiotherapy alone (RR = 2.4, mean dose to bone marrow: 7.5 Gy) or alkylating agents alone (RR = 10.0). Ninety secondary leukaemias were evaluated in a cohort of 82,700 women. High doses to bone marrow in the chest wall appeared to add appreciably to the risk of leukaemia after chemotherapy. Other smaller studies disagree regarding the risk of leukaemia after combined therapy. Two studies of Hodgkin's disease [K16, T8] concluded that the leukaemogenic effect of chemotherapy was not affected by concomitant radiotherapy; another reported the opposite [V8]. Breast cancer risk following Hodgkin's disease treatment with radiation may have been enhanced due to chemotherapy with alkylating agents [H2]. Evaluations of patients with ovarian cancer have consistently reported that combined chemotherapy and radiation is not associated with a greater risk of leukaemia than chemotherapy alone [G6, K15]. Previous studies, however, were somewhat limited by small numbers of leukaemia cases in each treatment category and by the absence of a large comparison group of patients who received neither radiation nor chemotherapy. In the peptic ulcer study [G11] it was noted that radiotherapy in combination with surgery greatly enhanced stomach cancer risks. The authors, while acknowledging that this may have been a chance finding, hypothesized that the joint effect might be due to radiation damage to the stomach mucosa.

186. The possible joint effects of oncogenic virus infection and radiation exposure are of increasing interest. As discussed in paragraph 120, recent incidence data on adult T-cell leukaemia (ATL) in the survivors of the atomic bombings show no association between ATL and radiation exposure [P33], providing evidence for a lack of joint effects between radiation exposure and HTLV-1 infection, which is endemic in Nagasaki. Those liver cancers found to be increased as a result of radiation exposure from the atomic bombings are largely hepato-cellular carcinomas, for which hepatitis-B, and also recently hepatitis-C, virus infections are major risk factors in certain areas, including Japan. The possible interaction between radiation and hepatitis-B virus infection is suggested by findings that hepatitis-B surface antigen positivity is increased in persons exposed to 1 Gy or more, although exposed and non-exposed individuals seem to be infected by the virus equally frequently [K24].

187. Additional interactions of interest involve the influence of reproductive history, oestrogen levels and other

known breast cancer risk factors on the risk of radiation-induced breast cancer. Recent studies [L19] found no evidence of such interactions. The hypothesis that there exists a small population of women with a genetic susceptibility to radiation-induced breast cancer (suggested to explain the large relative risks for early onset breast cancer among women in the life span study who were exposed as children [L20]) should be further investigated.

188. An excess of leukaemia has been reported among children who received therapy with pituitary growth hormone. Of the six cases of leukaemia, five had also received radiotherapy. Growth hormone is known to stimulate proliferations of both normal and leukaemic human lymphocytes. The authors speculate that growth hormone might be a promoting agent in the subgroup of children who had been exposed to therapeutic radiation [F19].

189. The joint effects of radiation and other factors on the induction of cancer is an important and interesting subject that can only be addressed through epidemiological studies. Unfortunately, epidemiological studies will rarely have enough power to provide definitive answers to these questions.

10. Lifetime risk estimates

190. Lifetime risk estimates are often used to summarize radiation risks. A number of issues must be considered in the definition and presentation of lifetime risk estimates. These issues, which were discussed in Chapter I of this Annex and which are reviewed at length in [T14] and [P11], include the following:

(a) the model(s) to be used to describe the excess incidence or mortality for the cancers of interest;
(b) the method of projecting risk beyond the period for which follow-up data are available;
(c) the background age- and sex-specific death rates for the reference population;
(d) the definition of lifetime risk to be used;
(e) the method of transferring risks estimated for one population to another population.

Lifetime risks have generally been summarized by a single number that represents the experience of a specified population following a specific type of exposure. While having such a number is convenient, simple summaries can conceal substantial variation with regard to sex and age at exposure. Estimates stratified by age at exposure and, when necessary, by sex are provided in the analysis in this Section.

(a) Models for lifetime risk estimation

191. Despite the large number of studies of radiation and cancer in humans, the life span study remains the main source for developing models for risk estimation. While it would be useful to be able to develop models from other studies for use in lifetime risk estimation, this is not usually possible owing to insufficient sample size, the highly non-uniform distribution of doses to different tissues within an individual, inadequate dose estimates in some cases, short follow-up periods, and, in many cases, the restrictions on the age and sex distribution. The risk estimates to be computed here are based on models that describe the life span study data.

192. Since the temporal pattern for radiation-induced leukaemia differs markedly from that for other cancers, it is appropriate to consider separate models. One could also attempt, as was done by the BEIR V Committee [C12], to develop separate models for specific non-leukaemia cancer types or categories. While differences in excess risks for specific cancer types undoubtedly exist, observed differences in these risk estimates largely reflect sampling errors because of the relatively small number of excess cancers of specific types. The results in [P20], discussed in Section II.A.7, show that even using the broad categories in the report of the BEIR V Committee [C12], it is difficult to detect statistically significant differences between the category-specific dose-response models. The non-leukaemia risk estimates presented below were computed using a model developed from the pooled life span study data on solid tumour cancer mortality for 1950-1987 [R23].

193. With the exception of the youngest survivors, excess relative risks for solid tumours in the life span study have so far been remarkably constant during the follow-up. Even for those exposed as children, a constant relative risk model provides a useful summary for the current follow-up, and the primary question for this group involves the projection of risks beyond the current follow-up period. The much larger excess relative risks to date for those exposed as children compared with adults may be due to the low background cancer rates for the younger group thus far, or it may reflect a general increase in the sensitivity of younger survivors. This uncertainty suggests that a reasonable approach to the estimation of age-at-exposure-specific lifetime risks would be to use a time-constant excess relative risk model for a period corresponding to the current follow-up period and to pay explicit attention to the effects of different methods for projecting the risks beyond the current follow-up period. This approach makes clear the extent to which extrapolation uncertainties vary with age at exposure. In particular, for those exposed after about 30 years of age little extrapolation is currently necessary, while for the youngest survivors the choice of projection method has an appreciable impact on lifetime risk estimates. Because it is based on more adequate summaries of the current data, this approach to the assessment of sensitivity to modelling is preferable to comparing lifetime risk estimates based on models of time-constant excess relative risk and time-constant excess absolute risk.

194. While it seems reasonable to use time-constant excess relative risk models to describe the current life span study solid tumour data, life span study excess relative risk estimates exhibit statistically significant variation with age at exposure and sex. Thus, the basic models used in the lifetime risk computations for solid tumours are excess relative risk models of the following form:

$$ERR(D,e) = \alpha_s D e^{\beta(e-25)} \qquad (13)$$

where α_s is the sex-specific linear excess relative risk per sievert, D is the weighted dose, e is the age at exposure in years and β is an age-at-exposure effect. Separate models were used for cancers of the oesophagus, stomach, colon, liver, lung, urinary bladder, breast, ovary and all other sites as a group. Parameter estimates are based on models fit to the life span study mortality data for the period 1950-1987 [R23]. These estimates, along with parameter estimates for a single model as in equation (13), fitted to the data for all solid cancers as a group are given in Table 24. Excess relative risks predicted by these models for selected ages at exposure are given in Table 25.

195. The solid tumour lifetime risk estimates presented below were computed using cancer risks obtained by summing the estimated instantaneous site-specific risks. A 10-year latency period was assumed in the computations for all solid tumours. This approach differs in two respects from that used for the primary risk estimates in the UNSCEAR 1988 Report [U2]. First, the excess risk models allow explicitly for the effects of sex and age at exposure. Secondly, site-specific risk coefficients were used for all estimates.

196. Leukaemia excess risks exhibit substantial sex- and age-at-exposure-dependent variation with time since exposure on either a relative or an absolute scale. In the computations summarized below, the pooled life span study excess absolute risk model for leukaemia from 1950 through 1987 [P33] was used. The excess absolute risk function for this model with units $(10^{-4}\ PYSv)^{-1}$ is given by the following equation:

$$EAR(D,t) = \alpha_{s,e}(D + \theta D^2) e^{\beta(e-25)} \qquad (14)$$

where D is the weighted dose in bone marrow (neutron RBE = 10), t is the time since exposure (years), θ is 0.79 Sv^{-1}, and α and β are sex- and age-dependent parameters (for age at exposure of 0-19 years: $\alpha = 0.33\ Sv^{-1}$, β = 0.17 per year for males and $\alpha = 0.66\ Sv^{-1}$, β = 0.07 per year for females; for age at exposure of 20-39 years: $\alpha = 0.48\ Sv^{-1}$, β = 0.13 per year for males and α = 0.66 Sv^{-1}, β = 0.03 per year for females; for age at exposure of ≥40 years: $\alpha = 1.31\ Sv^{-1}$, β = 0.07 per year for males and $\alpha = 2.64\ Sv^{-1}$, β = 0.03 per year for females). This model involves a non-linear time-dependent dose-response function in which the excess absolute risk depends on time, sex and age at exposure. Figure III illustrates the temporal pattern of the excess risk for men and women of different ages. Table 26 contains point estimates of the excess absolute risk for two doses at several different times after exposure. This model differs from the model used in the UNSCEAR 1988 Report [U2] in that it allows for a non-linear dose response and for temporal variation in the excess risk. Since chronic lymphocytic leukaemia is rare in Japan, this model should be taken as a model for leukaemia risks excluding chronic lymphocytic leukaemia.

197. Data on the survivors of the atomic bombings in Japan and from other exposed populations indicate that significant increases in deaths from leukaemia can be seen as early as two years after exposure. Since follow-up of the life span study did not begin until slightly more than five years after exposure and excess risks, especially for younger survivors, were very high at that time, it is important to allow for some excess leukaemia risk during the period from two to five years after exposure. For the main lifetime risk computations (deaths per unit population per sievert) it was assumed that the excess rate for the period before 5 years after exposure was equal to half the fitted excess rate at 7.5 years after exposure.

(b) Projection methods

198. Three projection methods were used to examine the impact of different assumptions about the future course of the solid tumour excess risk. In all three, the excess relative risk values predicted by the model given above were assumed to hold for the period from 10 to 45 years after exposure. For the first projection method, it was assumed that the excess relative risk would remain constant throughout the life of the survivor. The second and third methods both involved the assumption that for survivors who were less than 45 years old at exposure the excess relative risk would decrease linearly starting 45 years after exposure. Under the second method the rate of decrease was defined so that when the survivor reached age 90 years the excess relative risk would be equal to the average excess relative risk for a survivor who was 50 years old at exposure. For the third method, the rate of decrease in the excess relative risk was defined so that when the survivor reached age 90 years the excess relative risk would be zero. Figure XI illustrates the projection procedures for survivors aged 10, 30 and 50 years at the time of exposure. Life span study leukaemia risks are small and appear to be decreasing for the youngest life span study survivors. Thus, there appears to be no need to consider alternative projection methods for leukaemia risks.

(c) Transfer of risk estimates between populations

199. Despite the relatively large quantity of data on radiation risks, the question of how to apply risk estimates derived for one population to a different population remains unanswered. The data that are available suggest that there is no simple solution to this problem. For

example, the limited data (Table 8, Part II) on stomach cancer following radiation exposure in different populations suggest, albeit weakly, that relative risks may be more similar than excess absolute risks in populations with disparate background stomach cancer rates. The BEIR V Committee [C12] concluded that for breast cancer the excess relative risks are more similar than the excess absolute risks. However, although no new parallel analysis of the various breast cancer data sets has been reported, based on the data presented here (Table 8, Part VIII), the current evidence for such a conclusion seems less compelling than it appeared to be five years ago. In particular, the updated life span study breast cancer data do not suggest that age-specific relative risks are decreasing with time. Also, the time-averaged excess relative risk per sievert for women exposed at any age at exposure is larger than that in most of the other breast cancer cohorts. The observation that sex differences in solid tumour excess relative risks are generally offset by differences in sex-specific background cancer rates might suggest that absolute risks are more similar than excess relative risks.

200. Since current data provide no clear evidence of how risks estimated for the life span study should be applied to another population, lifetime risk estimates presented in this Annex are calculated only for a specific Japanese population, which in itself requires a transfer from the actual population exposed in 1945. Age- and sex-specific national death rates in Japan in 1985 [J5] from all causes, all cancers except leukaemia and leukaemia are given in Table 27. Although age-at-exposure-specific estimates are emphasized, population average lifetime risks were computed using age- and sex-weighting based on the 1985 Japanese population. It should be noted that calculations of risk in other populations have been made in the UNSCEAR 1988 Report [U2] for the populations of Puerto Rico and the United Kingdom using a multiplicative transfer, and the results for lifetime risk based on multiplicative projection were quite similar for all three countries, differing by only 20% overall. ICRP [I10], in a more extensive comparison of five populations (including the Chinese population) with rather different demographic characteristics, used both a multiplicative and an additive transfer [L12]. Lifetime risks varied at most by a factor of 2 for the multiplicative transfer (less for additive transfers) and by only 30% if the Chinese population was not included. Thus the effects of transfer between populations, while very important, are not so large for total cancer risk. For the transfer of risk in individual organs, as ICRP points out, some much larger differences can arise [I10]. Results presented in paragraph 212 indicate that sex-specific risks for individual sites can differ greatly within a single population, even when there is little difference in total excess cancer risk. These results underscore further the difficulties of transferring risks between populations.

(d) Methods and results

201. As noted in Chapter I, issues related to the definition of lifetime risk have received considerable attention in recent years [M42, P11, T14]. Thomas et al. [T14] identified three different measures of lifetime risk that have been used in recent analyses. The lifetime risks presented later in this Section are estimates of the risk of exposure-induced death (REID) given as the percentage of the exposed population (paragraph 51). Since the lifetime risks given in the UNSCEAR 1988 Report [U2] were expressed per 1,000 people, they should be divided by 10 to allow comparing them with those given here. When comparing UNSCEAR lifetime risks with those in the report of the BEIR V Committee [C12], it is important to note that the estimates given by the BEIR V Committee are based on a different measure, excess lifetime risk. While risk of exposure-induced death and excess lifetime risk estimates are similar for leukaemia, estimates of the risks of exposure-induced death for solid tumours are 10%-20% greater than excess lifetime risks, as discussed in Chapter I. The two risk measures are compared, using the same underlying population and cancer death rates, in Table 28.

202. Another important issue regarding lifetime risks concerns the lack of linearity in dose. Even if the underlying risk model is linear in dose, the risk of exposure-induced death is a non-linear function of dose. Because of the non-linearity, simple linear extrapolation from values at 1 Sv underestimates the value at 0.2 Sv by about 15% for the young but less at older ages. The results at both doses are given in Tables 29 and 30.

203. Loss of life expectancy is an important supplement to risk of exposure-induced death or excess lifetime risk, since neither of the last two measures adequately reflects the time at which the excess cancers occur. In other reports, the loss of life expectancy has been defined as the expected life-shortening per exposed individual. This is ordinarily a very small number, since it reflects both the life-shortening among those dying of radiation-induced disease and the small chance of this occurring. In this Annex loss of life expectancy is expressed as the average years lost per exposure-induced case (YLC), which is calculated as the average years lost per exposed individual divided by the risk of exposure-induced death. The UNSCEAR 1988 Report [U2] provided estimates of the years lost per 1,000 exposed individuals. The years lost per exposure-induced case corresponds to those estimates divided by 1,000 multiplied by the risk of exposure-induced death, which is the expected number of exposure-induced deaths per 1,000 persons.

204. Tables 29 and 30 present the age-at-exposure-specific risk of exposure-induced death and years lost per exposure-induced case for leukaemia and solid tumours, respectively, following acute exposures of 0.2 Sv and

1 Sv. These lifetime estimates were calculated using the sex- and age-at-exposure-specific model described earlier. Since sex differences between the estimates of risk of exposure-induced death and years lost per exposure-induced case for solid tumours were small, single averaged values are given for each age at exposure. For example, under the assumption of constant relative risk, the lifetime risk of exposure-induced death from solid tumours for a person exposed to 0.2 Sv at birth are 5.4% and 6.2% for males and females, respectively, for an average of 5.8% (Table 30). Corresponding estimates of sex-specific years lost per exposure-induced case are 12.5 and 16.3 years for an average of 14.3 years (Table 30). Sex differences are even smaller for those exposed later in life. Table 31 contains estimates for each of the three projection methods outlined above. As can be seen, risk estimates for the alternative projection methods are 20%-30% lower than those given for the constant relative risk projection. For the lifetime risk of leukaemia (Table 29), no projection is needed (see paragraph 198).

205. The constant relative risk projection predicts that 6% of those exposed at early ages to 0.2 Sv will die from radiation-induced solid tumours, with an average of about 14 years of life lost per case (Table 30). Estimates of risk of exposure-induced death and years lost per exposure-induced case decrease with increasing age at exposure, falling by age 60 years to about 13% and about 8 years, respectively. Comparison of the estimates for the three projection methods shows that the values of the risk of exposure-induced death based on declining relative risk projections for those exposed early in life are about half those predicted by a constant relative risk projection and that the effect of the projection method decreases for those who were older at exposure. In contrast to risk of exposure-induced death, the years lost per exposure-induced case increase when declining risk projections are used. This is to be expected, since the number of exposure-induced cases later in life is reduced by the declining risks.

206. The estimates of leukaemia risk are given separately in Table 29 for males and females. The estimates of risk of exposure-induced death for those exposed to 0.2 Sv early in life are about 0.2% for males and about half of that for females. The large values of years lost per exposure-induced case are a consequence of the short latency period and the generally decreasing risk with time since exposure. Because of the interaction of age at exposure with sex and with time in the leukaemia risk model, estimates of the risk of exposure-induced death for persons aged 40 years at exposure are higher than those for persons exposed at age 35 years. Risks of exposure-induced death at 1 Sv are greater than would be predicted by proportionality with 0.2 Sv risks because of the non-linearity in the leukaemia dose-response function used for these projections. As is the case for solid tumours, estimates of years lost per exposure-induced case are relatively independent of dose.

207. Population-weighted risks of exposure-induced death and years lost per exposure-induced case have been calculated for comparison with the estimates given in the UNSCEAR 1988 Report. The comparisons are given in Table 31. For a 0.2 Sv exposure, the risk estimate of exposure-induced death and the years lost per exposure-induced case for solid tumours are 2.4% and 11.2 years, respectively. The corresponding estimates for a 1 Sv exposure for solid tumours are 10.9% and 11.6 years. The age distribution of the 1985 Japanese population was used to provide weights for averaging the age-at-exposure-specific risks. The most relevant estimates from the UNSCEAR 1988 Report [U2] are those calculated using the multiplicative risk projection model with age-specific risk coefficients. The 1 Sv estimates (from Table 62 of Annex F of the UNSCEAR 1988 Report [U2]) are 9.7% and 11.4 years.

208. A non-linear dose-response model is used for the pooled leukaemia data. For a 0.2 Sv exposure, the estimate of risk of exposure-induced death from leukaemia and years lost per exposure-induced case are 0.16% and 34 years for men and 0.11% and 27 years for women. The corresponding estimates of risk of exposure-induced death for a 1 Sv exposure are 1.3% for men and 0.9% for women. The estimates of years lost per exposure-induced case are unchanged from the values at 0.2 Sv. The sex-averaged estimates in the UNSCEAR 1988 Report [U2] for an exposure of 1 Sv are 1% and 26 years. The estimated number of years lost per exposure-induced case in the UNSCEAR 1988 Report is less than the number given in this Annex, because the earlier estimates did not adequately allow for time-dependence in the leukaemia risks.

209. For all cancers, solid tumours plus leukaemia, the risk of exposure-induced death is 2.5% after exposure to 0.2 Sv and 12.0% after exposure to 1 Sv. The latter number compares with the value of 10.7% determined for the risk of exposure-induced death after exposure to 1 Gy (organ absorbed dose) in the UNSCEAR 1988 Report [U2] (see Table 31).

210. The solid tumour risk model used for the calculations presented above is for fatal cancers. An excess relative risk model for first primary tumours in the life span study is also available [R23]. With this model it would be possible, in principle, to calculate lifetime incidence risks based on methods similar to those used above. In particular, solid tumour incidence rates could be substituted for the mortality estimates and used together with death rates for all causes other than solid tumours in the basic calculations. The main practical problem in carrying out such computations is the need for appropriate age- and sex-specific cancer incidence rates. In many countries, such rates are not readily available.

211. As has been noted several times in the Annex, models for site-specific risks have large uncertainties, and

the differences in site-specific risks that are seen are greatly affected by sampling errors in the data as well as by details of the models chosen. With these caveats in mind, a number of site-specific risk estimates have been determined and are listed in Table 32. The non-leukaemia estimates were calculated using site-specific time-constant relative risk models for the life span study data in which risks were allowed to depend on age at exposure and sex. As with the earlier estimates in this Annex, these models were applied to 1985 Japanese site-specific cancer mortality rates. Age-at-exposure-specific REID estimates were averaged using weights determined from the Japanese population. The life-table computations for lifetime risk estimation involve decrements in the population at risk at any age that are due to the exposure. Thus, the lifetime risk for a given site depends on whether other sites have been exposed. For example, the lifetime risk for lung cancer will differ according to whether the exposure is to the lung only or to the whole body. The results presented here (Table 32) assume that only the target organ was exposed to 1 Sv weighted dose.

212. Table 32 provides sex-specific and averaged estimates of the REID for several sites and also values of the average years lost per exposure-induced case (YLC). Although specific values in this Table should be interpreted with caution, it is interesting that while the total risks for each sex are similar, there are striking differences in the distribution of the site-specific excess risks for men and women. The differences for liver and stomach seem particularly noteworthy. The large liver cancer risk for men reflects the unusual sex difference in the relative risk for this site. Overall, however, given the uncertainties in the data and some arbitrary aspects of choice of models, the site-specific risks of Table 32 should be regarded as representative of a given choice of models and data rather than as definitive of the relative values of risks for different sites.

11. Summary

213. Studies of the carcinogenic effects in humans of external exposures to low-LET radiation delivered at high dose rates, even though varying widely in scope and quality, provide convincing evidence that these exposures increase the risk of cancer in many organs and tissues of the body. In particular, for some cancers other than leukaemia, lymphomas and multiple myeloma, it seems reasonable to conclude that excess risks occur within about 5-10 years after exposure and that the elevated risks continue to increase, on an absolute scale, for most or all of the remaining lifetime. This must be qualified by the fact that virtually no lifetime follow-up data are yet available on populations of persons exposed before the age of 20 years. Analyses of data on radiation carcinogenesis in humans, including the analyses presented in this Chapter, often focus on results for specific solid tumour types. Indeed, site-specific analyses are all that can be carried out for many studies. There is strong evidence for an association between radiation exposures and many solid tumours, especially for the most common solid tumour types such as those of the lung, stomach and female breast. However, there is little or no evidence of radiation induction for such cancers as those of the cervix, uterus and gall-bladder or for chronic lymphocytic leukaemia. In between these two extremes is a middle group of tumour end-points for which indications of excess risk occur in some studies, possibly or even probably suggesting an association with radiation exposure, but the results fall short of statistical significance. In those epidemiological studies of solid tumours in which adequate dosimetric data are available, the dose-response function for all solid tumours taken together is consistent with linearity at weighted doses between 0.2 to 4 Sv.

214. The pattern of excess leukaemia risks differs from that seen for solid tumours in several respects. First, excess risks are seen within 2-3 years of exposure. Secondly, leukaemia excess risks (relative or absolute), especially in those exposed as children, appear to peak and then decline with the passage of time. Despite the decrease in risk with time, current data do not indicate that excess leukaemia risks ever disappear completely. Again, as with the solid tumours, it is important to note that there is virtually no lifetime follow-up data on persons exposed under the age of 20 years. Although investigated in numerous studies, there is no evidence that radiation exposure increases the risk of chronic lymphocytic leukaemia.

215. Results for malignant lymphomas and multiple myeloma are less clear than those for leukaemia or solid tumours. Studies of Hodgkin's lymphoma following radiation exposure have consistently failed to demonstrate an excess. The data on non-Hodgkin's lymphoma incidence in the life span study [P33] provide no clear evidence of an effect, although some evidence of an excess is seen in the mortality data of the ankylosing spondylitis study [D6] and the peptic ulcer study [G11] based on comparisons with population rates. Recent data from the life span study [P33] suggest that evidence of an excess of multiple myeloma is not as strong as had been suggested in earlier analyses of the mortality data. At this time there is no conclusive evidence of an increased incidence of lymphoma and myeloma following high-dose low-LET radiation exposures.

216. In this Annex site-specific risk estimates for a number of sites and for leukaemia from a number of studies have been presented in a common format. It has been emphasized that direct comparisons of the simple summary estimates must be interpreted with caution because of differences in the nature of the exposures, the quality of the dose estimates, the demographic characteristics and the length of follow-up time. However,

some overall impressions can be formed based on the risk estimates provided in Table 8. These estimates highlight the central role of survivors of the atomic bombings in risk estimation. The life span study is the most comprehensive single source of data on the effects of external exposure to high-dose-rate low-LET radiation and one of the few studies that has enough excess cases along with a broad range of doses and individual dosimetry and with variability in age, age at exposure and sex to support efforts to model dose response, effect modification and temporal patterns of risk. Despite the importance of the survivors of the atomic bombings in the characterization of the risk of radiation exposure, studies of other radiation-exposed populations help to clarify, confirm or qualify the findings of life span study. More importantly, they provide information on issues that cannot be addressed directly by the Japanese data, in particular the effect of chronic low-dose and highly fractionated exposures and the variability of risk in different populations. Also there are a few cancers, including leukaemia and breast and thyroid cancer, for which a number of useful results from studies other than the life span study are available. For most cancers, including such important ones as the stomach and the lung, the number of useful results is more limited. For some other cancers, such as cancers of the prostate, pancreas, cervix and testes, as well as Hodgkin's disease and chronic lymphocytic leukaemia, no estimates of radiation risk exist because no study has convincingly related these sites to radiation exposures. While the results that are available do not suggest great disparities between risks in the life span study and those in other studies, questions of the comparability of risk estimates have not and, unfortunately for most sites, cannot be addressed in detail using current epidemiological data.

217. The Committee has attempted to derive roughly comparable risk estimates of excess relative and absolute risks per unit dose for a broad range of studies. Although there are many problems with the calculation, interpretation and comparison of the simple summary measures used here, the Committee feels that presenting estimates in this way is useful because it moves away from an over-emphasis on the results of statistical tests of the hypotheses of no effect and examines instead the differences and similarities of risk estimates for different cancer types in various studies.

218. The Committee has derived lifetime estimates of the risk of mortality from all solid tumours and from leukaemia from the data of the life span study for the period 1950-1987 using three different projection procedures. Although the methodology differs somewhat from that used for the lifetime risk estimates of mortality presented in the UNSCEAR 1988 Report [U2], which were based on data for the period 1950-1985, comparison is possible for the constant relative risk (or multiplicative) model. The risk of mortality from all radiation-induced cancers is slightly higher in this Annex: 12.0% following exposure of 1 Sv, compared with 10.7% in the UNSCEAR 1988 Report [U2]. Using alternative projection methods that allow for uncertainties in the projection for those exposed at younger ages, the lifetime risk estimates are 20%-30% lower. Lifetime risks are also derived for induced cancer in individual organs and tissues, but since these depend on many assumptions, some of them arbitrary, they are much less precise than the total lifetime risk of fatal cancer.

B. LOW-DOSE-RATE EXPOSURES

219. For more than 30 years, estimates of the number of radiation-induced cancers that might be expected to occur among populations exposed to external low-LET radiation received at low dose rates have been based on risk coefficients derived from data on populations irradiated at high dose rates. Usually these studies have also been at high doses, although some, such as the survivors of the atomic bombings, cover a very broad range of doses with many persons exposed at low doses as well. In general, low-dose studies have been less useful for risk estimation because they are of low statistical power, and many of the published studies have severe limitations. Uncertainties in epidemiological low-dose studies can be due to several factors, including inadequate dosimetry, narrow dose ranges, inadequate or inappropriate controls and extraneous effects from confounders that are potentially more troublesome when the effect being detected is expected to be small. However, by far the most consistent problem in low-dose studies is their lack of power, i.e. the number of persons involved, the length of the follow-up and the doses to which they were exposed may collectively be too small to have any chance of detecting a radiation effect.

220. In recent years, several reports have appeared from which some estimates of risk may be extracted. These reports, which present data on persons occupationally exposed to radiation at low dose rates, on the effect of varying levels of terrestrial gamma-ray exposures and on the effect of exposure to environmental releases, are summarized below. In order to have reasonable power to detect any radiation-related effects, such studies must be very large. However, even if these low-dose studies are not sufficiently large, they may be indicative of an effect and perhaps even able to provide crude risk estimates, but with wide confidence limits. It seems quite unlikely, however, that studies of low-dose-rate exposures to low-LET radiation will be able to provide a means of examining detailed patterns of risk, such as variations with age at exposure, time since exposure and other relevant parameters.

1. Occupational studies

221. For the purpose of quantifying risk, occupational studies offer the most promise of providing informative results because of the large number of subjects who have

been exposed occupationally in many countries and whose exposures have been individually monitored and recorded. By far the most important occupational data come from detailed observations on radiation workers in the nuclear energy industry who have received cumulative doses varying from less than 1 mSv to over 1 Sv in the course of many years. Observations have thus far been published on about 20 groups of workers, totalling well over 100,000 subjects, mostly men, in the United Kingdom, the United States, Canada, India and the former USSR [A10, B4, B5, B27, B30, C3, C16, C22, D10, D22, D26, F11, F16, G3, G5, G9, G16, G17, G19, H17, H32, H45, K20, K21, K23, M15, M41, N1, N2, N3, N4, N6, P24, R1, S14, W3, W7, W20, W21, W22, W23, W24]. Total mortality in these groups was generally less than that in the general population for all men in the same ages over the same period in the country in which they worked (see Table 33). This establishes that they did not suffer a gross hazard, such as has been observed in coal miners and asbestos workers, but it does not exclude the possibility that a small or moderate number of radiation-induced deaths have occurred in these groups. As noted in paragraph 27, it is quite common for mortality rates in selected populations to differ from general population rates, and in such situations it is important to make use of internal comparisons to the extent possible and to try to understand and explain any differences between the exposed cohort of interest and the population from which the rates were drawn. In occupational studies, working populations are often observed to have lower age-specific death rates for all causes and for cancer than the general population. This is known colloquially as the healthy worker effect. An important source of uncertainty in some worker studies is possible confounding due to exposure to other workplace agents, such as chemicals, that may elevate the cancer risks.

222. To assess the effects of radiation exposures in these occupational groups, it is necessary to make comparisons within the industry, comparing the mortality of workers (mainly men) by level of occupational radiation exposure. This has been done in the United States for workers at the Hanford plant [G5, G16], the Rocky Flats nuclear weapons plant [W7] and the Oak Ridge National Laboratory [C3, W21], the Mound facility [W23, W24], the Oak Ridge Y-12 plant [C22] and the United Nuclear Corporation [H32]; in the United Kingdom for workers at the Sellafield plant [S14], the Atomic Energy Authority [B4, F11], the Atomic Weapons Establishment [B5] and the National Registry of Radiation Workers [K20]; and in Canada for workers at Atomic Energy of Canada Ltd. [G3, H45]. Results of studies that have provided risk estimates are summarized in Table 34. It should be noted that in most of these studies a large number of associations were examined. Some of the significant observations may therefore have been due to chance. It should also be noted that in these studies risks are presented with different lag intervals in the associations of doses and effects that may represent subjective decisions based on posterior views of the data. The results should thus be interpreted cautiously.

223. The study of workers at the Sellafield plant [S14] included 10,000 monitored workers with an average cumulative recorded dose from external exposure of 124 mSv. The average length of follow-up was 22 years. When exposures were lagged by 15 years (but not necessarily for other lag periods), there were significant associations between dose and mortality from leukaemia ($p = 0.04$), cancer of the bladder ($p = 0.02$), multiple myeloma ($p = 0.01$) and all lymphatic and haematopoietic cancers ($p = 0.03$).

224. The study of workers at United Kingdom Atomic Energy Authority plants included 22,000 workers with a radiation record, who were followed for an average of 23 years [F11]. The average cumulative recorded dose from external exposure was 40 mSv. When exposures were lagged by 10 years, there was a significant trend in mortality with radiation dose for cancer of the uterus ($p < 0.05$) and cancer of the prostate ($p = 0.01$). A case-control study of cancer of the prostate was carried out in the Atomic Energy Authority workforce [R25]. Although the risk of prostate cancer in that study increased significantly with increasing level of external exposure, this association was limited to subjects who were probably contaminated with radionuclides.

225. The study of United Kingdom Atomic Weapons Establishment workers included 9,000 employees with a radiation record, who were followed for an average of 18.3 years [B5]. The average recorded cumulative whole-body dose was 8 mSv. When exposures were lagged by 10 years, there was a significant association between mortality and dose for all malignant neoplasms ($p < 0.05$), as well as for cancer of the lung. A significant association between radiation dose and cancer of the prostate was also observed among workers monitored for exposure to radionuclides. The trend for all cancer mortality was particularly strong in workers monitored for radionuclide exposure ($p < 0.001$) and was mainly attributable to the dose-related increase in lung and prostate cancer mortality.

226. The study of workers at the Chalk River Laboratories of Atomic Energy of Canada Ltd. in Canada [G3, H45] covered 8,977 male workers who were monitored for external radiation and who had an average cumulative dose of 52 mSv. The most recent publication evaluates mortality in 1956-1985. When exposures were lagged by 10 years, there was a significant association ($p = 0.06$, based on four deaths) between cumulative radiation dose and mortality from leukaemia, excluding chronic lymphocytic leukaemia.

227. For the Hanford study, the internal analyses are based on data from 33,000 monitored workers employed

for six months or longer, with an average cumulative dose of 26 mSv above background. The most recent report on mortality was for the period 1945-1986 [G16]. Of the 24 specific cancer categories evaluated, only cancer of the pancreas and Hodgkin's disease, which do not seem to be associated with radiation in other studies, showed significant positive correlations with radiation dose, with one-tailed p values of 0.03 and 0.04, respectively. In previous analyses of the Hanford data, multiple myeloma was the cancer most strongly linked with radiation dose [G5]. However the findings have changed with additional follow-up, and the one-tailed p value obtained in the most recent analysis was 0.10. In addition to the analyses described above by Gilbert et al., the Hanford data have also been analysed by other authors (see, for example, [K5]). These authors have reached conclusions very different from those of Gilbert et al., primarily owing to their use of non-standard and inappropriate statistical methods, which failed, among other things, to allow properly for age at death, calendar year of death, occupation and length of employment. They will not, therefore, be discussed further here.

228. The study of workers at the Rocky Flats plant [W7] included 5,000 white males employed for at least two years and followed for an average of 14 years. The average recorded cumulative dose from external exposure was 40 mSv. No overall association with external radiation exposure was found.

229. The studies of workers at Oak Ridge National Laboratory included 8,000 white male workers hired between 1943 and 1972. Their mean cumulative dose from external exposure was 17 mSv. In an initial report, which included mortality to the end of 1977 [C3], no consistent gradients of cause-specific mortality were detected for radiation exposure. However, in a subsequent report, which included mortality to the end of 1984 [W21], a significant association was found between mortality from all cancers combined when exposures were lagged by 10 years ($p = 0.01$) and 20 years ($p = 0.001$). It is noted that these results are based on non-standard statistical methods in which, for example, a group of men with an unusually low mortality from cancer were included in the unexposed groups. A further analysis of these data [G9], using methods similar to those applied to the Hanford data, derived estimates of excess relative risk per sievert for all cancer mortality similar to those given in [W21] but with much wider confidence intervals. The excess relative risks were 2.9 (90% CI: 0.4-6.5) based on a 10-year lag and 6.1 (90% CI: 1.5-12.0) based on a 20-year lag [G9]. In a further analysis in the same report [G9] it was shown that the significant associations were largely due to smoking-related cancers. In addition it was shown that there were significant positive correlations of non-cancer mortality with radiation dose for diseases known to be related to smoking. It thus seems likely that confounding by smoking may have contributed to the significant positive correlations for cancer mortality seen in this study, although the original authors disputed this and pointed to other complexities [W20].

230. Although the significant trends noted above may represent specific hazards peculiar to those cohorts, there is of course the danger of chance results appearing when many types of cancer are studied separately in many cohorts, and overall conclusions may be strengthened if the available data can be grouped in one or more ways. Two partial groupings have so far been carried out. The first is a joint analysis of combined mortality data on workers at three installations in the United States (the Hanford site, Oak Ridge National Laboratory and the Rocky Flats nuclear weapons plant) [G17]. The second is based on the United Kingdom National Registry for Radiation Workers, which comprises workers at sites operated by British Nuclear Fuels Ltd., the Ministry of Defence, the United Kingdom Atomic Energy Authority, and Nuclear Electric [K20]. Table 35 shows estimates of excess relative risks with 90% confidence limits, as calculated in these two comprehensive analyses. In the combined analysis of data from the United States, no significant association was seen between radiation exposure and all cancer or leukaemia mortality, but it was seen for multiple myeloma ($p = 0.04$) [G19]. Since the maximum likelihood estimate for leukaemia occurred at a value that would have led to negative excess relative risks for workers with the largest doses, the estimate has been reported as negative without assigning a value. Estimates based on a further analysis of risk estimates from data of the combined studies in the United States and of the United Kingdom registry have also been calculated [K21] and are also given in Table 35. For comparison, risk estimates based on analyses of an appropriate subsection of the survivors of the atomic bombings in Japan are shown as well. The studies of nuclear workers covered only individuals exposed in adulthood, they included very few women and the follow-up period after exposure was shorter for nuclear workers than for survivors of the atomic bombings in Japan, so that risk estimates for the entire life span study are not directly comparable with those based on the nuclear workers. In addition, dose estimates for the workers tend to be estimates of whole-body dose and are not directly comparable with the organ dose estimates in the cohort of survivors of the atomic bombings. Another uncertainty arises from the possible confounding by carcinogenic chemical exposures in nuclear facilities, particularly in reprocessing facilities such as Sellafield, one of the main sources of leukaemia risk. A low-dose-rate reduction factor has not been included in the figures for the life span study. For all cancers, the point estimate from the combined workers studies is close to the estimate from the Japanese data, with the 90% confidence interval extending from less than zero to about twice the upper confidence limit for the Japanese data. For leukaemia, the point estimate from the combined workers studies is about the same as the estimate based on the

Japanese data, without a low-dose-rate reduction factor applied. The confidence intervals extend from less than zero to several times the upper confidence limit from the Japanese data. Results from the three studies are summarized in Table 36.

231. Most of the studies of workers in the nuclear industry have not been able to take account of data on smoking habits, a serious problem if tobacco use is somehow related to radiation dose. It was possible to carry out a case-cohort study of lung cancer among white male radiation workers at the Hanford plant in which data on tobacco use were obtained from occupational medical records [P25]. The analysis took account both of the case-cohort design and changes over time in the quality of the data on tobacco use. Tobacco use was not strongly related to the level of radiation exposure, and adjustment for tobacco use did not greatly modify the results. With or without adjustment for tobacco use, the estimated risks per unit of cumulative equivalent dose were negative, and the 95% confidence intervals were wide and included values several times those estimated from populations at higher doses and higher dose rates. A cross-sectional survey of tobacco habits has also been carried out within the workforce of the United Kingdom Atomic Energy Authority [C32]. There was little evidence for an association between level of tobacco smoking and cumulative radiation dose.

232. It is noted that formal combined analyses of updated mortality data from Sellafield, the Atomic Energy Authority and the Atomic Weapons Establishment in the United Kingdom as well as formal combined analyses of the data from Canada, the United Kingdom and the United States have recently been completed; the data were not, however, available for the Committee to review at this time. In addition, an international collaborative study of cancer risk among nuclear industry workers in 14 countries is under way, using a common core protocol, under the auspices of the International Agency for Research on Cancer [C33]. This study should result in more precise direct estimates of cancer risk associated with low-dose-rate external radiation exposures in the future.

233. Information has recently become available concerning the radiation doses and mortality of workers at the reactor and reprocessing plants at the Mayak nuclear weapons facility in the Russian Federation [B27, B30, N4]. In the initial period of intensive activity, during start-up and the early years of operation, average radiation doses to the staff were substantially higher than would now be permissible. The average annual doses received by workers in the nuclear reactor (plant A) for the years 1948-1956 are estimated to be have been 0.20, 0.40, 0.31, 0.18, 0.15, 0.20, 0.09, 0.10 and 0.05 Sv. From 1949-1962, the average annual doses received by workers in the radiochemical processing plant (plant B) were 0.48, 0.94, 1.13, 0.66, 0.31, 0.20, 0.21, 0.16, 0.18, 0.11, 0.15, 0.15, 0.11 and 0.08 Sv. The proportion of workers with an annual dose >0.25 Sv in 1949 and 1950 was 50%-80% in plant A and 35%-90% in plant B. From 1968 onwards in plant A and 1974 onwards in plant B, the annual radiation doses to staff did not exceed 0.05 Sv on average.

234. So far, the health of more than 80% of the workers who began working in these unfavourable conditions has been followed for over 40 years. During the early years, chronic radiation sickness (defined as chronic fatigue, depression and an altered blood picture [B30]) was common, and the mean total dose among affected cases was about 3 Sv. It was subsequently determined that such cases rarely occurred in persons whose annual dose remained below 0.25 Sv or whose total dose remained below 1 Sv. The cessation of exposure or the reduction of its intensity to levels close to those more recently regarded as permissible limits (<0.05 Sv a^{-1}) led to a recovery process that resulted in the normalization of the blood count for 80% of the individuals affected by chronic radiation sickness. Also, persons transferred, for prophylactic reasons, away from contact with radiation once their total doses of gamma-radiation had reached a certain level did not subsequently develop radiation sickness. At both plants, workers whose total external gamma dose exceeded 1 Sv and those who received over 0.25 Sv in any one year had substantially higher cancer mortality than other workers (Table 37). The cancer mortality in other workers with lower doses was shown to be similar to that of the population as a whole.

235. Individual film-badge dosimetry data are available for 3,098 men and 1,453 women who started work at the two installations between 1948 and 1953 and for an additional 1,987 men and 362 women who started work between 1954 and 1958 [K23]. The average cumulative doses for the men in these four groups vary from 0.49 to 2.45 Sv, and the doses for the women are also thought to be high. Data on total cancer and leukaemia mortality in men to the end of 1989 are shown in Table 38. When compared with national rates in the former Soviet Union for 1970-1986, an elevated risk is apparent for all cancer mortality in the subgroup with the highest average exposures. This effect appears to be primarily due to a large excess of leukaemias (see Table 38). The excess relative risk of leukaemia per gray has been calculated only for the subgroup in which there was a significantly increased risk; it is estimated to be 1.4 Gy^{-1}. The death rate in this subgroup was 25 per 100,000 person-years compared with 5.5 per 100,000 person-years expected from national rates. The paper states that these results agree well with the data on the survivors of the atomic bombings if a low-dose-rate reduction factor of 2.7 is applied to the survivor data. In interpreting these figures it should be noted that expected values are based on general population rates, which may not be appropriate in this selected population. In addition, the national rates for 1970-1986 may be inappropriate for years before 1970. The relevant data for earlier periods are

often not available. No data on women are available at present.

236. In a separate analysis [D26], mortality patterns during a 40-year period have been analysed for 931 persons who worked at the first Soviet nuclear facility during 1949-1953 and were exposed to radiation under the most unfavourable conditions. The average accumulated dose from gamma-radiation was 3.26 ± 0.13 Gy. All individuals were subject to individual dosimetric control and systematic medical monitoring. The total number of deaths during 1952-1989 was 177, of which 22.6% were due to malignant tumours and 6.2% to leukaemia. The proportion of deaths from different causes and the average age of deaths varied over the decades. During the first decade, seven persons died of leukaemia (26.9% of deaths), two persons of aplastic anaemia and nine of accidents (30.8% of deaths). During the second decade the proportion and frequency of solid tumours rose from 18% to 23%. As would be expected in an ageing population, the beginning of the final decade saw the relative and absolute number of deaths from cardiovascular causes rise (the average age of death was about 60 years). Therefore, the carcinogenic effect has been detected only for acute leukaemia, which developed in the first 5-10 years after the first exposure at high total doses (>3 Gy).

237. In another Russian occupational study, the frequency of lung cancer was investigated among 2,346 workers in the Mayak radiochemical plant who were exposed both externally and internally to plutonium [H17]. External exposures were determined by film-badge dosimetry, and lung doses from inhaled plutonium were determined for all individuals using measurements of the excretion of radionuclides in urine and a lung clearance model. The relative biological effectiveness of plutonium alpha particles was taken to be 16. Mortality was studied during 1970-1989, at the end of which time the total collective combined dose to the lung in the group of 2,346 workers was 4,812 Sv. Of that combined dose, 3,327 Sv was from external exposure and 1,485 Sv from inhaled plutonium (RBE for alpha particles = 16). The number of deaths observed and the number expected from national rates in the former Soviet Union are shown in Table 39. The risk of mortality increased with increasing equivalent dose and was significant among workers in the highest dose group. Overall, the relative risk was 1.8 Sv^{-1} compared with 1.63 Sv^{-1} for the survivors of the atomic bombings ([S7], Table 4: both sexes). The authors also compared their estimate of the lifetime risk for lung cancer in these workers, 0.033 Sv^{-1}, using an internal control and applying a low-dose-rate reduction factor of 2, with the ICRP-derived value from the survivors of the atomic bombings, 0.0068 Sv^{-1}. They pointed out that against national statistics the result would be 0.014 Sv^{-1}, closer to the ICRP-derived value. Even better agreement with the ICRP value would have resulted if the low-dose-rate reduction factor had not been applied (equivalent to a reduction factor of 1). It should be noted that a substantial part of the occupational exposure (about 25%) is from plutonium alpha particles, and for these a low-dose-rate reduction factor would not be expected to be applicable. Also, the contribution from the alpha particle portion depends on the value of RBE selected (16 was used). A more complete study of these workers would include information on the smoking habits of the workers, on the possible influence of the healthy worker effect and on the precision of the dosimetry.

238. Some information on the effects of low-dose-rate exposure to external gamma-radiation on subsequent breast cancer risk is available from a study of about 1,200 women who worked from 1939 to 1961 in the United Kingdom with paint containing radium [B26, B35]. The study is fully described in Section III.B.2. These women accumulated small body contents of ^{226}Ra. However, they also received external irradiation to the breast at a rate of 0.05-0.2 Gy a^{-1}, resulting in accumulated doses up to more than 1 Gy (average 0.4 Gy). The breast cancer mortality in these women to 1 January 1986 was greater than in the controls (28 deaths compared with 20.5 expected) but was not statistically significant (Table 40). The women were subdivided into two groups, aged <30 and ≥30 years at the start of their luminizing work, and in both groups the ratio of observed to expected deaths among those with gamma-ray absorbed doses <0.2 Gy was higher than that for women with gamma-ray absorbed doses of at least 0.2 Gy (Table 40). For women less than 30 years of age, the time distribution of these deaths as a function of years of employment shows a significant increase in excess deaths from breast cancer only after 30-40 years of exposure. While this timing is consistent with an effect of radiation, the numbers are too small to draw conclusions. No leukaemias were observed in this population. In analysis of dial painters from the United States, Stebbings et al. [S42] did not find excess breast cancer that could be attributed to the radiation exposure.

239. Additional studies of radiation workers were carried out in India, where there are two studies: one of workers in the atomic energy community and their families in Bombay [N3] and the other in Tarapur [N2]. In both cases the numbers involved are small, about 5,000 people, and the length of follow-up is short (1975-1988 in Bombay, 1971-1988 in Tarapur). In Bombay the standardized mortality ratios were 80 and 65 for male employees and female family members compared to national statistics, and there was no evidence of a radiation effect. In Tarapur the standardized mortality ratio for employees and their families was 160 for all cancer, which was not significantly different from 100. The standardized mortality ratios for leukaemia were 238 for employees and 222 for families, but again neither was significant. These Indian studies are not at this point able to make any contribution to risk estimation, but they may provide a baseline for more comprehensive studies in the future.

2. Exposure to natural background radiation

240. External exposure of the population to natural sources, such as cosmic rays and terrestrial gamma rays, varies with geographical location, and several authors have attempted to examine the risk following low-dose-rate exposures by comparing cancer mortality or incidence rates in different geographical areas. Despite the obvious attraction of having large populations available for study, such attempts are fraught with difficulties. These include uncertainties in the doses actually received by individuals, geographic variability in the accuracy of cancer diagnoses, and confounding with the numerous other environmental factors that may well have a much greater influence on cancer risk. Furthermore, when geographical areas are compared, exact matching of control groups is difficult. It is not surprising, therefore, that studies that have attempted to correlate exposure to background gamma-radiation with cancer mortality or incidence without carefully considering these other factors have at times produced erratic and implausible results.

241. Recent calculations have been made of the proportion of deaths from various cancers caused by natural background radiation [D7]. These calculations used the models of the BEIR V Committee [C12] for leukaemia and for solid tumours, which were derived from data on the survivors of the atomic bombings in Japan. They predict that about 11% of deaths from leukaemia might be caused by postnatal exposure to natural background sources other than radon, while for other fatal cancers the figure is about 4% or less [D7]. In addition to the higher relative risk from leukaemia, the influence of other environmental risk factors on subsequent leukaemia risk is less than for many other types of cancer risk, and the interval between exposure and any subsequent risk is generally much less for leukaemia than for other types of cancer. Thus the effect of exposure to background radiation is much more likely to be assessable for leukaemia than for most other fatal cancers. However, studies in Connecticut [W2], Japan [I2, I11, S47], France [T12], the United States [I134, J3], Sweden [E1] and China [H35, W8, W15, W16] found no significant association between leukaemia and background radiation. An early study in Scotland found a significant association between natural background radiation and leukaemia (excluding chronic lymphocytic leukaemia) but concluded that even for leukaemia a proper interpretation of geographical variations can be made only when the effect of social and economic factors can also be taken into account [C27]. A later study in the United Kingdom on childhood cancers also reported a statistically significant effect [K7], but the statistical methods used were obscure; the result is not only difficult to interpret, but it was not confirmed in a subsequent report [M13].

242. Few of the above studies have attempted to relate the observed leukaemia rates to realistic estimates of dose, taking into account migration and differences between indoor and outdoor exposure rates. One exception is a Chinese study, in which two neighbouring regions having different levels of exposure to background radiation (owing to the high thorium content of the monazite sands in one of them) were selected for study [H35, W8, W15, W16]. In both regions the overwhelming majority of the inhabitants had lived in the area for their entire lives. Exposures were determined by measurements not only at fixed locations indoors and outdoors but also in a sample of individuals in each area who wore personal dosimeters for two months. The average annual dose in the red bone marrow was estimated to be 1.96 mSv in the high background area and 0.72 mSv in the control area. During 1970-1985 the age-adjusted mortality rates for leukaemia in males were 3.32 and 3.82 10^{-5} PY^{-1} in the high background and control areas, respectively (16 and 17 deaths), while in females the rates were 2.21 and 3.56 10^{-5} PY^{-1} in the high background and control areas, respectively (10 and 16 deaths), and consequently not significant. This study thus provides no evidence for a radiation effect following low-dose protracted exposure throughout life, since leukaemia rates were lower in the population with higher exposure living on radioactive monazite sands. However, the low doses and small numbers of cases are such that the study had low statistical power to detect an excess risk, given that one existed. For example, a typical individual in the high background area would, by age 50 years, have received a red bone marrow dose about 60 mSv greater than one living in the low dose area. Based on estimates from the life span study cohort (Table 8, Part XII), doses of this magnitude would be expected to result in a relative risk of leukaemia of the order of 1.2. Relative risks of this magnitude are very difficult to detect epidemiologically, so even if observation is extended over a lifetime it will be difficult to obtain a definitive result. Similar considerations apply to many of the smaller occupational studies, although some of these low-dose studies can be expected to provide an upper limit to the risk and therefore make a useful contribution to radiation-induced cancer risk evaluation.

243. Thyroid disease was also examined in a Chinese study of high background radiation [W17]. Cumulative thyroid doses at age 50 years were estimated to be about 0.14 Gy among those living in the high background area and 0.05 Gy among those living in the control area, a difference of 0.09 Gy. Approximately 35% of the accumulated exposure was received during childhood, which appears to be a sensitive time for thyroid carcinogenesis (Table 8, Part XI). Physical examinations were carried out on approximately 1,000 women from each area in the age group 50-65 years [W17]. For all nodular thyroid disease, the relative risk for residents of the high background area compared with that of residents of the control area was 1.02 (95% CI: 0.76-1.35), while for single nodules it was 1.13 (95% CI: 0.82-1.55), and the difference, while positive, was not significant. The authors concluded that an excess thyroidal exposure

totalling about 0.1 Gy delivered over a lifetime, much of it during the adult years of life, did not result in a significantly increased risk of clinically detectable thyroid tumours. Chromosome aberration frequencies were also studied and found to be significantly higher in the high background region than in the low background region [W17].

244. Where the use of different building materials results in differences in the rates of exposure to background gamma-radiation within a single geographical area, the case-control study provides an alternative method of investigation. This approach has been adopted in two studies of acute myeloid leukaemia in Sweden [F7, F8, F18]. In both studies the exposure to background gamma-radiation was assessed using an index that took into account the time spent indoors and outdoors and the building material (stone, wood etc.) in the homes and workplaces of the subjects. In each study there was a marginally significant association between their risk of myeloid leukaemia and exposure to higher background radiation (0.04 Gy in 20 years compared to 0.02 Gy in 20 years) after adjusting for confounders, with relative risks of 2.2 (95% CI: 1.0-4.9) in one study and 2.0 (95% CI: 0.9-4.3) in the other. However, the index of exposure was crude, and in a case-control study of multiple myeloma by the same authors, using a similar design, significantly fewer cases were found in the high exposure category of background gamma-radiation than in controls [F17].

3. Exposures from environmental releases

245. Several geographical studies have reported an association between the amount of fallout in southwestern Utah from nuclear tests in 1952-1958 and subsequent leukaemia rates (see, for example, [M1]). Accordingly, a case-control study with 1,177 individuals who died of leukaemia and 5,330 who died of other causes (controls) was conducted using estimates of dose to bone marrow determined from fallout deposition rates and the location of the individual's dwelling [S48]. For all leukaemias other than chronic lymphocytic leukaemia, when all ages and all time periods after exposure were combined, there was a weak association with bone marrow dose that was not statistically significant. However, a significant association was seen for those individuals who died when they were less than 20 years old (p = 0.02), in the period of highest exposure, 1952-1957 (p = 0.04), and for those individuals who died of acute leukaemia, especially acute lymphocytic leukaemia (p = 0.009) (see Table 41). Subgroups combining the modifying factors shown in Table 41 were defined before the analysis. In addition to the results given in Table 41, a significant association with dose was found for acute leukaemia discovered from 1952 to 1963 among individuals who were younger than 20 years at exposure, with odds ratios of 1.33 (95% CI: 0.64-2.74) in the 3.0-5.9 mGy dose category and 7.82 (95% CI: 1.90-32.2) in the 6-30 mGy dose category. A non-significant association with dose was also found for chronic lymphocytic leukaemia. The dose was predominantly from external low-LET radiation. No significant associations were seen in other a priori defined subgroups, including those *in utero* during the fallout exposure period. The observed risks in this study were approximately double those predicted by the estimates based on the life span study cohort, but the difference was not statistically significant.

246. A series of releases and accidents occurred at the Mayak nuclear weapons facility at Chelyabinsk-65 in the southern Urals. In the early 1950s, liquid wastes with a total activity of 100 PBq (in beta emitters) were discharged into the Techa River. As a result, the 124,000 people living along the Techa and Iset Rivers received internal and external radiation exposures. The long-term consequences of this exposure are being investigated in the most heavily exposed portion of this population, consisting of 28,000 people who lived on the banks of the Techa, of whom about a quarter were Tartars or Bashkirs and the remainder were Russians. Efforts have been made to reconstruct internal and external doses for this cohort. Individual external dose estimates are not available for cohort members; however, average external doses have been estimated on a village by village basis, while internal contamination has been measured using a whole-body counter on over 12,000 individuals [D27, K26]. Another issue of concern is the possibility of confounding effects from other liquid wastes containing toxic chemicals. The cohort was defined as all people living in the defined area surrounding the Techa River on 1 January 1950. It was assembled in 1968 using taxation and medical records, supplemented by lists of persons compensated during evacuation. Data are available on the vital status of the cohort up to 1983, by which time about 8,000 people were known to have died and 6,000 to have moved away. Emigration is clearly a special problem as is the variable quality of the data. Although efforts are being made to determine vital status for all cohort members, death certificates are routinely available only for those deaths occurring in the local area, so information on the cause of death is available for only about two thirds of the known deaths. Of those for whom the cause of death is known, about 15% had cancer indicated as the underlying cause of death. Special efforts have been made to collect data on cases of leukaemia, and ascertainment is believed to be high. Data on morbidity from other types of cancer has been obtained from cancer centres in the Chelyabinsk region; these data include about 17,000 of the most heavily exposed members of the cohort. Results to date indicate an increasing risk of leukaemia, especially in high-dose groups, and of total cancer mortality with increasing dose (Figure XII). Analyses of data for several cancers other than leukaemia have shown excess risks that are not significant [K18]. Table 42 summarizes data for several site-specific analyses, while Table 43 provides more detailed data on the leukaemia dose response.

247. In 1957 a radioactive waste storage tank at the Mayak facility exploded. The explosion, called the Kyshtym accident, contaminated a large area of the southeastern Urals, known as the East Urals radioactive trace [K19]. The resuspension of activity from a storage pond at the Mayak facility in 1967 resulted in small exposures in an area largely included within the East Urals radioactive trace. More than 270,000 people were living in the area at the time of the 1957 accident. Marrow dose estimates for the 20,000 most heavily exposed people are estimated to have ranged from 0.03 to 0.25 Gy. Mortality and morbidity in this group are being monitored. Little information is currently available on the cohort.

248. Since the cohorts of environmentally exposed individuals in the southern Urals are a valuable source of information and potentially important with respect to low-dose radiation risks, efforts should be made to improve the quality of the individual exposure estimates. In addition, the lack of detailed information on data collection methods and the incompleteness of the mortality data continue to limit the interpretation of risk estimates derived from currently available data.

249. To investigate the effects of the accident at the Chernobyl nuclear power station in 1986, a detailed study was carried out of cancers diagnosed in the period 1981-1990 in residents of the three districts of Ukraine that were most heavily contaminated with radioactive caesium and that had not been evacuated [P34]. This involved a search of oncology, haematology and other records in the regional and district hospitals and in specialist institutes. Annual age-specific incidence rates for leukaemia and for all other cancers, excluding thyroid cancer, are shown in Table 44. The numbers are small, but at ages 0-14, 15-44 and 45-64 years, the age-specific incidence rates are broadly similar before and after 1986. At ages ≥65 years, the observed incidence of leukaemia increased abruptly in 1987 and in subsequent years remained two or three times higher than the pre-accident rates, while the incidence of all other cancers, excluding thyroid cancer, increased in the year after the accident by about one third and remained at about that higher level in subsequent years. These increases do not follow the usual pattern for radiation-induced cancers, in that they are confined to the oldest group and, for cancers other than leukaemia, appeared before the usual latency period. It seems likely that they are attributable to local concern about cancer after the accident, which led to a more thorough investigation of elderly patients for cancer. Before 1985, the incidence rates of leukaemia and of all cancers combined were lower at ages ≥65 years than in other eastern European countries. Since 1986, the rates have been more in line with those in other countries, suggesting under-diagnosis in the elderly before the accident. This survey also included data on thyroid cancer, and these are discussed in Section III.A.1. It should be noted here that the international Chernobyl project [I3] did not find any evidence that the incidence of leukaemia, thyroid cancer or other cancers had increased in the short period up to late 1990 in people living in contaminated zones compared to those living in uncontaminated zones in Ukraine, Belarus or the Russian Federation. Also, cancer registry studies to date have not linked increases in childhood leukaemia with the Chernobyl accident [P38] although the follow-up may not yet be sufficient for an effect to be detectable [L22].

4. Summary

250. The risks of low-dose-rate external exposure to low-LET radiation are currently obtained by extrapolation from high-dose-rate data. The quantitative observations already available for workers at nuclear plants in the United States and the United Kingdom now provide some direct evidence that these estimates of risk may be reasonable. The study of workers in the United Kingdom shows a trend with dose for leukaemia; however, taken together, these studies have such broad confidence intervals that they could also be consistent with no risk at all. Firmer conclusions should be possible once the international overviews of these occupational data that are under way have been completed. The implications of the available results for dose-response relationships and the effect of protraction of exposure were discussed in the UNSCEAR 1993 Report [U1], Annex F, "Influence of dose and dose rate on stochastic effects of radiation".

251. The current observations on Russian workers at the Mayak plant demonstrate a carcinogenic effect of exposures to high doses of external gamma-radiation delivered at low dose rates and are also broadly consistent with the results derived from the survivors of the atomic bombings. The data are indicative of a dose response for groups of workers with different exposure levels. It is hoped that further study of these workers and, possibly, of additional cohorts will improve the estimates of risk from these potentially important data.

252. The study of radioactive fallout from the Nevada test site provides some indication of the ability of external doses of low-LET radiation in the range 6-30 mGy delivered at low dose rate to cause leukaemia, especially in persons exposed in childhood. The risks derived from the association seen in this study are somewhat higher than, but consistent with, the risks derived from exposures delivered at high dose rate. The great uncertainty in estimating exposures 30 years later, however, renders any conclusions doubtful.

253. The data on the populations exposed to radiation as a consequence of releases into the Techa River and the Kyshtym accident are also potentially important sources of risk estimates, especially for leukaemia, and of information on the nature of risks in different ethnic groups. As with

the studies of the Mayak workers, the interpretation of risk estimates in these populations is complicated by the presence of both internal and external exposures as well as by possible exposure to other carcinogenic substances that may be highly correlated with radiation dose. The Committee recognizes the potential importance of data on the exposed populations in the southern Urals. It hopes that efforts will be made to make available more information on these populations and to support improvements in the quality of dose estimates and follow-up data, which might lead to more reliable risk estimates. In the meantime, it can be said that quoted risk estimates are in broad agreement with those derived from high-dose-rate studies.

254. At present there is no evidence that external low-LET radiation from the Chernobyl accident has increased cancer rates in the populations living in the three most heavily contaminated districts of the former Soviet Union that were not evacuated [I3, M21]. Studies of the workers ("liquidators") engaged in the clean-up at Chernobyl are potentially valuable for risk estimation because the numbers are large (several hundred thousand) and the doses appreciable (0.1-0.4 Gy or more).

255. The study of thyroid nodularity in women in areas of high background radiation in China has provided evidence that external exposure to low-LET background radiation of about 0.1 Gy in adult life does not cause an appreciable risk of nodular thyroid disease. Overall, studies of cancer risk in areas of high background radiation have yet to provide definitive information on radiation effects, in large part because of the very low exposures involved and the lack of statistical power to detect small effects. These same limitations apply also to many occupational and environmental exposures.

III. CARCINOGENIC EFFECTS OF INTERNAL RADIATION EXPOSURES

256. Internal radiation exposures are of primary importance in considering risks to the specific organs in which ingested or inhaled radionuclides predominantly deposit. Studies of such exposures include effects in the thyroid and other organs from ^{131}I, effects in the bone and liver from isotopes of radium and thorium and effects in the lungs from radon. Exposures from a few other radionuclides, such as polonium, uranium and plutonium, have been considered in more limited studies.

A. LOW-LET RADIATION EXPOSURES

1. Iodine-131

257. Iodine-131 as sodium iodide has been used medically in three ways. Very large amounts of ^{131}I, usually 2,000-5,000 MBq, have been given for the treatment of thyroid cancer, with the aim of destroying the thyroid gland and/or metastatic tissue. Somewhat smaller amounts, 200-400 MBq, have been given to treat hyperthyroidism; the resulting radiation doses are sufficient to impair thyroid function, and the majority of patients treated in this way eventually became hypothyroid. Much smaller amounts, 0.4-4 MBq, have been given to diagnose thyroid diseases. Exposure to ^{131}I has also occurred as a result of fallout following explosions of nuclear weapons, and nuclear installations have discharged ^{131}I into the atmosphere, especially during accidents.

258. Follow-up of patients treated for thyroid cancer has been reported in Sweden [H31], the United Kingdom [E6], Denmark [B24], Finland [T11] and Connecticut [T13]. The most detailed study is from Sweden, where second cancer risks have been reported in 834 thyroid cancer patients given ^{131}I. The mean total administered amount was 4,550 MBq, and the mean observation period was 14 years. A total of 99 new cancers developed in the patients treated with iodine therapy, and when compared with national data the standardized incidence ratio was 1.43 (95% CI: 1.17-1.75). There were significant increases for cancers of the salivary glands, female genital organs, kidney and endocrine glands other than the thyroid. When organs were grouped by estimated radiation dose, the standardized incidence ratio was raised for organs receiving <0.1 Gy (SIR = 1.57; 95% CI: 1.07-2.21), lower for organs receiving doses 0.1-0.6 Gy (SIR = 1.18; 95% CI: 0.87-1.56) and raised among organs that had received doses of at least 1 Gy (salivary glands, stomach, small intestine and bladder) (SIR = 2.59; 95% CI: 1.53-4.09). The bladder and stomach received on average 2.1 Gy and the salivary glands and small intestine 1.9 and 1.3 Gy, respectively. For the bone marrow and the breast, organ doses were estimated to be between 0.1 and 0.6 Gy, and the standardized incidence ratio for leukaemia was 2.44 (95% CI: 0.66-6.25) based on four cases, while the standardized incidence ratio for breast cancer was 0.74 (95% CI: 0.34-1.40). Although some of these SIRs appeared significant, they did not increase after 10 years of exposure. In addition, second cancers were increased in patients not treated with ^{131}I, suggesting that the general population may not be an appropriate comparison group.

259. In the United Kingdom, Edmonds and Smith [E6] followed 258 patients who were treated with an initial ablation amount of 2,900 MBq of ^{131}I, followed, when

necessary, by further treatment with one or more 5,500 MBq amounts at intervals of a few months. The mean follow-up time was just over 10 years. A total of 20 incident cancers was observed compared with 13.2 expected (p = 0.05), and there were significant excesses for leukaemia (3 observed, 0.25 expected; p = 0.002), breast cancer (6 observed, 2.53 expected; p = 0.04) and bladder cancer (3 observed, 0.46 expected; p = 0.01). There were no cases of cancer of the central nervous system. The mean dose to the bone marrow for those developing leukaemia was 3.5 Gy, while for patients with breast and bladder cancer the mean doses to the relevant organs were 2.1 Gy and 22.9 Gy, respectively.

260. Some evidence of an excess of leukaemia following ^{131}I treatment for thyroid cancer has also been seen in Denmark, where 2 cases of myeloid leukaemia were found compared with 0.097 expected (0.05 > p > 0.01) in a series of 194 patients, representing all patients with thyroid cancer treated with ^{131}I in Denmark between 1948 and 1972 [B24]. Record linkage studies of cancer registry data have shown an increased incidence of leukaemia following thyroid cancer in Finland, Connecticut and Denmark [O1, T11, T13], although treatment with ^{131}I exclusively was not evaluated.

261. Several studies have been carried out of patients treated with ^{131}I for hyperthyroidism. One of the most recent studies included about 10,500 patients treated between 1950 and 1975 at seven hospitals in Sweden [H11, H40]. The patients received a mean total amount of approximately 500 MBq of ^{131}I, and radiation doses to the thyroid were between 60 and 100 Gy. Doses to organs other than the thyroid were relatively low. An average dose of less than 0.1 Gy was estimated for the colon, liver, pancreas, lungs, breast, uterus, ovaries, testis, kidneys and bone marrow. The dose to the bladder was 0.14 Gy, to the salivary glands 0.20 Gy and to the stomach about 0.25 Gy. The follow-up was continued for a mean time of 15 years and was very complete. Cancer incidence data were obtained from the Swedish Cancer Registry. Overall, the number of tumours occurring one or more years after therapy (1,543) was almost exactly equal to the number expected (RR = 1.06). When the data were subdivided into 20 different sites of cancer, there were significant increases of tumours of the lung (RR = 1.32; number of cases = 105), kidney (RR = 1.39; n = 66) and parathyroid gland (RR = 1.78; n = 30). In contrast, there was no increase in leukaemia (RR = 0.94; 95% CI: 0.65-1.31; n = 34), and the increase in thyroid cancer was not significant (RR = 1.29; 95% CI: 0.76-2.03; n = 18). Among 10-year survivors, significantly elevated risks were seen for cancers of the stomach (RR = 1.33; n = 58), kidney (RR = 1.51; n = 37) and brain (RR = 1.63; n = 30). Only the risk for stomach cancer increased over time (p < 0.05), and it tended to increase with increasing amount administered, although the trend was not statistically significant.

262. The cooperative thyrotoxicosis therapy follow-up study in the United States involved evaluating patients with hyperthyroidism for subsequent cancers. The original study included 36,000 patients, about 19,200 of whom were exposed to ^{131}I, and the radiation dose to the thyroid was always more than 20 Gy (mean 90 Gy). They were followed for an average of 8 years, but no increase in leukaemia or thyroid cancer was seen [D9, S2]. The follow-up of about 1,000 of these patients (all female) who attended the Mayo Clinic was extended to an average of 15 years. When their mortality was compared with that in a similar sized control group of women treated surgically, no increase in cancer risk was observed [H8]. An initial analysis of cancer incidence in the same population reported an elevated risk of cancer (RR = 1.8; 95% CI: 1.1-3.2) of organs that concentrate ^{131}I (salivary glands, digestive tract, kidney and bladder) [H9], but a later report that ranked organs according to the estimated doses received concluded that no clear patterns had emerged [H7]. When cancer incidence in the ^{131}I treated patients was compared directly with that in surgically treated patients, adjusting for differences in age at treatment and duration of follow-up, relative risks were 1.0 for all malignant neoplasms, 0.6 for leukaemia, 9.1 for cancer of the thyroid, 0.8 for cancer of the breast, 3.4 for cancer of the kidney and bladder and 0.3 for cancers of the central nervous system. Only the relative risk for thyroid cancer was significantly increased. It is likely that this was because the surgery group was at exceptionally low risk (much of the thyroid had been removed) and not that the ^{131}I-treated patients were at an increased risk. However, when compared with thyroid cancer incidence rates in Connecticut, the standardized incidence ratio for thyroid cancer was 3.8 (3 cases observed compared with 0.8 expected).

263. Follow-up has also been carried out for 1,762 hyperthyroid women who were treated at the Massachusetts General Hospital thyroid unit, many of whom were in the original study. These women have now been followed for an average of 17.2 years [G7]. Comparison was based on age- and calendar-time-specific incidence rates for Connecticut white women; the standardized incidence ratios for all malignant neoplasms in women never treated with ^{131}I and in those receiving ^{131}I amounts of 4-240, 240-369 and >370 MBq were 1.0, 0.9, 0.9 and 1.0, respectively (see Table 45). There was an excess of brain cancer among those receiving at least 370 MBq of ^{131}I, but there was no consistent trend with increasing dose. The excess reported in the initial analysis of the Mayo clinic series [H9] for organs that concentrate ^{131}I (salivary glands, digestive tract, kidney and bladder) was not repeated in this group (SIR = 0.9; 95% CI: 0.7-1.2). After stratification for age and year of treatment for hyperthyroidism, the standardized incidence ratio for developing breast cancer for women treated with ^{131}I was 1.9 (95% CI: 0.9-4.1) compared with women never treated with ^{131}I.

264. The effects of the diagnostic use of ^{131}I have been evaluated by Holm et al. [H12, H27, H41]. A total of 35,074 patients examined for suspected thyroid disorders between 1951 and 1969 in Sweden were followed for an average of 20 years. The mean age at the time of examination was 44 years. The average amount of ^{131}I administered to the patients was 2 MBq, and the radiation dose to the thyroid gland was approximately 0.5 Gy, while for other organs the dose was <10 mGy. Record linkage with the Swedish Cancer Register identified 3,746 cancers occurring five or more years after the initial ^{131}I examination, and the standardized incidence ratio for all malignant neoplasms, based on general population rates, was 1.01 (95% CI: 0.98-1.04). There were 50 cases of thyroid cancer, and the standardized incidence ratio was initially estimated as 1.27 (95% CI: 0.94-1.67) [H12], but this was reduced to 1.18 (95% CI: 0.88-1.56) after adjustment for region of residence [H27]. There was a significant excess of thyroid cancer 5-9 years following irradiation (SIR: 1.95, based on 23 cases; 95% CI: 1.24-2.92). However, women and those observed for 10 years or more showed no evidence of an excess or of a dose response, so it was concluded that the increase during years 5-9 was probably due to a high level of medical surveillance, which led to an increased level of detection of indolent tumours. Furthermore, a portion of this early excess may have been related to the underlying conditions being evaluated. Also, among 2,000 patients less than 20 years old, no increased risk of thyroid cancer was observed. For the other cancers, there were significant excesses of leukaemia (SIR: 1.34, based on 119 cases; 95% CI: 1.11-1.60), but only after 15 years, and both chronic lymphocytic leukaemia and non-chronic lymphocytic leukaemia showed similar significant excesses. Endocrine tumours other than the thyroid (SIR: 1.93, based on 130 cases; 95% CI: 1.62-2.29) and cancers of the nervous system (SIR: 1.19, based on 135 cases; 95% CI: 1.00-1.41) also showed significant risks. For breast cancer and cancers of the kidney and bladder there was no evidence of any increase (breast cancer SIR: 0.98, based on 739 cases; 95% CI: 0.91-1.06; kidney and bladder cancers SIR: 1.08, based on 259 cases; 95% CI: 0.95-1.22).

265. A combined analysis of the nearly 47,000 Swedish patients given ^{131}I for thyroid cancer, for hyperthyroidism or for diagnostic purposes and included in the studies above has also been reported. Interpretation of this analysis is complicated by differences in the demographic characteristics of the populations studied and the non-overlapping dose ranges. Bone marrow doses were estimated for individual patients based on total administered ^{131}I activity, 24-hour thyroid uptake and ICRP tables [H26]. Standardized incidence ratios for all leukaemias, all leukaemias excluding chronic lymphocytic leukaemia, and chronic lymphocytic leukaemia are shown by sex, absorbed dose in bone marrow, age at exposure and years after exposure in Table 46. The administration of ^{131}I did not appear to influence the subsequent risk of leukaemia in either males or females. Nor was there evidence for a higher risk among those exposed at age <40 years or in years 2-9 after exposure. The risk of chronic lymphocytic leukaemia, which has not been associated with ionizing radiation, was similar to that of other types of leukaemia. For the whole population (mean absorbed bone marrow dose: 14 mGy), the standardized incidence ratio of all leukaemia, excluding chronic lymphocytic leukaemia, was 1.09 (95% CI: 0.91-1.29); for those in the <100 mGy dose group (mean marrow dose: 48 mGy) it was 1.09 (95% CI: 0.91-1.30) and for those in the >100 mGy dose group (mean absorbed bone marrow dose: 221 mGy) it was 1.04 (95% CI: 0.28-2.67). Thus, although the combination of these three cohorts is not without difficulties, no clear association of cancer induction by radiation was evident in this analysis.

266. Several studies have been carried out of the effects of radioiodine in fallout. Over 200 inhabitants of the Marshall Islands were exposed to a mixture of radioiodines (^{131}I, ^{132}I, ^{133}I, ^{134}I and ^{135}I), tellurium and gamma-radiation in fallout from the 1954 Bravo thermonuclear test [C28, R21]. The whole-body gamma-ray dose was estimated to be 0.11-1.9 Gy to the Marshallese on three atolls. The mean thyroid dose from gamma rays and radioiodines was estimated to be 3-52 Gy in children and 1.6-12 Gy in adults. Many of these individuals subsequently received anti-thyroid drugs or had thyroid surgery. Thyroid cancer appeared in 7 of 130 women and in 2 of 113 men over a follow-up period of 32 years. The earliest thyroid lesion appeared 9 years after exposure in a girl who was 3 years of age at the time of the exposure. The calculated risk coefficient for thyroid cancer was 1.4 10^{-4} Gy^{-1}, but this estimate is uncertain for several reasons. The high thyroid doses may have resulted in cell-killing and decreased the number of cells at risk for cancer development. Frequent thyroid surgery may have reduced the tissue at risk, and prophylactic thyroid hormone medication might have influenced the risk. Furthermore, the effects of gamma-ray exposures could not be distinguished from those of internal radioiodines. The dose to the thyroid from the short-lived and more energetic radioiodines was 2-3 times greater than that from ^{131}I, and thus the contribution of ^{131}I to the development of thyroid cancer could not be determined.

267. The possibility of thyroid effects in people exposed to radioiodine releases during and after the Chernobyl accident has received much attention. In the international Chernobyl project [I3], nearly 600 children aged 5 and 10 years from contaminated and control areas were examined in 1990. The children had been between 1 and 6 years old at the time of the accident. The proportion of children having thyroid nodules was the same in the exposed and unexposed children [M21]. More recently, however, an increase of thyroid cancer in children living in areas contaminated by radionuclide deposition from the accident, where doses from short-lived

radioiodines may have been substantial, has been reported from the Ukraine [P34] and Belarus [K22] but not from the Russian Federation. In Belarus an increase was observed in children only four years after the accident. In all other studies of thyroid cancer an increased risk has never been seen until more than five years after radiation exposure [S61]. No information on the individual thyroid doses received by these children is available. It is not yet possible to determine whether the reported increases are truly associated with radiation exposure or whether they are a result of increased surveillance for thyroid disease, which is known to result in apparent increases [B15, B42, C5, R11, S63, S70].

268. A survey of thyroid disease among children in Utah and Nevada exposed to fallout radiation from weapons testing in the 1950s and control groups living in Utah and Arizona has found no significant differences the incidence of any type of thyroid disease [R12]. No cases of thyroid carcinoma were found in 1,378 children thought to have been exposed, while two cases occurred among 3,443 children who were not living in the primary fallout area during infancy. In a recent paper, Kerber et al. [K1] report an association between ^{131}I exposure and thyroid neoplasms among a cohort of about 2,500 children exposed to fallout from nuclear weapons tests in Utah. A total of 19 cases were observed and about 13 expected during 20 years of follow-up (1965-1986). The average thyroid dose, based on individual dose estimates, was 0.17 Gy. The estimated excess relative risk per gray was about 7. When the analysis was limited to the eight thyroid cancers included in this series, the point estimate of the excess relative risk per gray was slightly greater than that for all neoplasms but was not statistically significant ($p = 0.10$). As indicated by the authors, factors related to the study design might have led to biases in this estimate. These included issues related to selection and dietary recall used in the dose reconstruction. The authors concluded that such effects were unlikely to have had a large impact on their finding. While this estimate is not out of line with risks seen in children exposed to external low-LET radiation, it is substantially higher than risks seen in other studies of populations exposed to ^{131}I. In particular, the risks derived in the Kerber study [K1] are much larger than those seen in the Swedish study [H12] of patients, primarily adults, who had received an average dose of 0.5 Gy for whom the (non-significant) estimated excess relative risk per gray was 0.4. Based on only two cases among children in the Swedish study, the non-significant estimated excess relative risk per gray was 1.4.

269. The incidence of thyroid cancer has also been studied in Cumbria, United Kingdom, near the Sellafield nuclear site, which has discharged both ^{129}I and ^{131}I into the sea and atmosphere since the early 1960s [B25]. Measurements of ^{129}I were made on 130 thyroid glands taken opportunistically at necropsy from adults who died of various causes in Cumbria between November 1984 and September 1987. Levels of activity decreased with the distance of the patients' homes from Sellafield ($p < 0.001$). In contrast, age-standardized thyroid cancer registration rates for the 288 parishes in Cumbria in 1969-1986 showed a positive correlation with distance from Sellafield ($p < 0.001$), with the higher incidence at greater distances. It was estimated that most of the sample population received less than 0.67 μSv a^{-1} from ^{129}I, and none were likely to have received more than an additional 1.5 μSv a^{-1}. Because of its short half-life, ^{131}I could not be measured in the sampled thyroid glands, but the authors estimated that total cumulative doses from the two radionuclides were probably about equal. Such small exposures would not be expected to result in detectable effects of any kind, and indeed none were found.

270. Comparisons between the carcinogenic potential in the thyroid of external radiation and radioiodines such as ^{131}I have frequently been attempted. In view of the scant and uncertain data on increases in cancer risk following exposure to ^{131}I outlined above, such comparisons are difficult [C12]. The National Council for Radiation Protection and Measurements [N5] found that, based on human experience, the relative effectiveness of ^{131}I compared with external radiation is between zero and one half; data from animals suggest a value between one tenth and one. A value of one third was recommended in the NCRP report. Shore [S61], in a more comprehensive and recent review of the human data, found only 8.3 observed excess cancer cases from all ^{131}I studies combined compared with 37.0 cases expected based on risk estimates for external exposure. The ratio, 0.224, is about one quarter. Evidently the protracted dose to the thyroid that results from ^{131}I exposure is one reason for this difference, but the distribution of ^{131}I in the thyroid gland and the resulting non-uniformity of dose may be another [S68].

2. Other radionuclides

271. Phosphorus-32 is a high-energy beta emitter that tends to concentrate in the skeleton. Repeated injections of ^{32}P have commonly been used to treat polycythemia vera, a disease characterized principally by the overproduction of red blood cells. In a study of patients with polycythemia vera diagnosed in 1937-1953 in the United States and followed to the end of 1961, the cumulative incidence of leukaemia by 15 years after diagnosis was 20% in the ^{32}P-treated group but only 2% in the non-irradiated polycythemia vera patients who had been treated with phlebotomy, sometimes in combination with chemotherapy [M34]. Detailed dosimetry calculations have not been performed, but crude dosimetry suggests that the average skeletal dose to these patients was about 3 Gy, which appears to have resulted in an excess cumulative incidence of leukaemia of 18%. This risk is high compared to that seen in patients exposed to external

low-LET radiation. In a randomized, controlled trial, patients treated with ^{32}P also appeared at high risk: 6% (9 out of 156) patients treated with ^{32}P developed leukaemia compared with only 1% (1 out of 134) treated by phlebotomy [B33]. Extrapolation from these data to other situations may not be justifiable, however, as patients with polycythemia vera may be unusually susceptible to radiation-induced leukaemia owing to the nature of their underlying disease.

272. Sulphur-35 is a low-energy beta emitter. Treatment of chondrosarcoma and chordoma with ^{35}S injected as sulphate caused severe marrow damage, as expected, but it may also have caused the three leukaemia cases that occurred in the nine subjects who survived at least two years [M31, M37]. It is possible that the marrow damage would have materially increased the risk of radiation-induced leukaemia.

273. As part of a study of iron metabolism in pregnancy, the effects of oral administration of ^{59}Fe (a beta-gamma emitter) were examined in 679 pregnant women; 705 control women chosen from the same clinical population did not receive ^{59}Fe [H25]. Frequency and type of malignancy in the next two decades did not differ for the two groups of women. Among the 634 exposed children of these women, one case of leukaemia and two cases of sarcoma were discovered, while no malignancies occurred in the 655 children in the comparison group, representing a small but statistically significant increase ($p = 0.03$). Among those exposed, the fetal doses were estimated to be 0.05-0.15 Gy.

274. River water heavily contaminated with fission products from fuel processing caused large body burdens of ^{90}Sr in people living along the upper reaches of the Techa River in the southern Urals, who were using the water for drinking and irrigation. Tooth and whole-body counting based on Bremsstrahlung yielded cumulative bone marrow doses of up to 3 Gy [D24]. Dose contributions from other radionuclides were small, but external dose rates near the river banks were considerable (up to 1 mGy h^{-1}). In a cohort of the 28,000 persons most heavily exposed, an excess risk for leukaemia was reported (see paragraph 246).

3. Global fallout

275. Estimates of the average annual absorbed doses in the bone marrow from low- and high-LET radiation received from various sources by children born in Thurso near the Dounreay nuclear installation in northern Scotland have been published by the National Radiological Protection Board [D8]. For both low- and high-LET radiation the annual dose from fallout from atmospheric nuclear weapons tests peaked sharply in about 1964. For low-LET radiation the maximum annual dose in the bone marrow was estimated to be 650 μGy, mostly from ^{137}Cs and ^{90}Sr, while for high-LET radiation the maximum annual dose in the northern hemisphere was estimated to be much lower, 2.6 μGy, mostly from ^{239}Pu but with some ^{238}Pu [F20]. The dose from strontium would have been entirely internal from beta radiation, while for caesium about two thirds of the dose would have been internal from a mixture of beta and gamma-radiation. The temporal pattern of fallout throughout northwest Europe was similar to that estimated for Dounreay: doses from high-LET radiation were similar but doses from low-LET radiation varied depending on rainfall. In the Nordic countries the low-LET radiation doses were likely to have been about twice those experienced in Britain. If doses of this magnitude of either high- or low-LET radiation cause an appreciable childhood leukaemia risk, the incidence of leukaemia in northwest Europe should have increased at about the same time in all affected countries. Data from the Nordic countries have been presented [D2]. Surprisingly, in view of the small doses involved, a small but statistically significant increase in the relative risk of leukaemia was noted in children in the period following the highest fallout as compared with the preceding and following periods of more moderate fallout.

4. Summary

276. The data summarized above provide a substantial amount of information about the effects of ^{131}I on the risk of thyroid cancer for a very wide range of doses, from the very high doses delivered in the treatment of hyperthyroidism to the low doses received by persons exposed to diagnostic procedures or to radiation from fallout. The information relates mainly to individuals exposed in adult life, and it provides little evidence that exposure to ^{131}I is associated with an increased risk of thyroid cancer, although in some cases the follow-up periods have been comparatively short. Significant excesses, all of which must be qualified, were reported in four studies, one of them in children. This lack of a clear effect (when such effects are well known to follow external radiation exposure of the thyroid) may be due to the lesser effectiveness of ^{131}I because of dose protraction and inhomogeneous distribution of the beta emitter in the thyroid gland. Thyroid cancer risks are dependent on age at exposure. The risks of childhood exposure to ^{131}I are, however, less well known than those of adult exposure, so further studies of children are needed.

277. Comparisons of the carcinogenic effectiveness in the thyroid of ^{131}I with that of external radiation are difficult. Gross estimates from human experience, supported in part by animal studies, suggest that ^{131}I is about one fifth to one third as effective as external radiation.

278. While there have been excesses of cancer of sites other than the thyroid in cohorts exposed to high levels of

^{131}I, the findings have not been consistent or reproducible and may well be due to chance. A large study in the United States of patients with hyperthyroidism did not find any difference in the risk of leukaemia in patients receiving ^{131}I therapy and in those treated surgically. An extensive study of leukaemia in Swedish patients given ^{131}I in adult life indicated that the administration of ^{131}I did not influence the subsequent risk of leukaemia. Doses were generally low, but even at bone marrow doses >100 mGy no increased risk of leukaemia was observed.

279. Few studies have as yet been carried out of the effects of internal exposures to low-LET radiation from radionuclides of iodine other than ^{131}I, especially short-lived radioiodines. Such studies could provide valuable information on the effects of mixtures of radioiodines released to the environment, and the Committee recommends that exposed populations should be identified and studied.

B. HIGH-LET RADIATION EXPOSURES OTHER THAN RADON

280. The risk of cancer resulting from the exposure of human beings to high-LET radiation must be established either by studying the direct effects of such radiation in humans or by knowing, as has been outlined earlier in this Annex, the risk of cancer for low-LET radiation and establishing the relative biological effectiveness (RBE) of high-LET compared with low-LET radiation in a variety of other biological systems. There are few direct measurements of the ability of external high-LET radiation to induce cancer in human beings. Neutron exposures at Hiroshima (and Nagasaki) are largely overwhelmed by the effects of gamma rays, as assessed in DS86 [R22], and their impact may not be large in terms of possible adjustments (paragraphs 156-158). This makes it impossible to derive either an RBE or separate risk estimates for neutron-induced cancers from these studies. The only distinct human data on the effects of neutrons come from two criticality accidents in which the neutron doses exceeded the gamma-ray doses substantially. Two of the seven long-term survivors died from acute myeloid leukaemia 19 and 33 years later, after neutron doses of 0.12 Gy and 0.08 Gy, respectively [H36]. Gamma doses were only 0.04 Gy and 0.001 Gy, respectively. It is difficult to construct a useful risk estimate from such a small sample.

281. In the absence of human information, many studies comparing high-LET radiation with low-LET radiation have been undertaken in a wide variety of biological systems [I4, N7, S67] to determine the relative biological effectiveness. These have allowed the calculation of values of quality factors or radiation weighting factors for high-LET radiation [I10] (see Table 47). The application of these factors to derive appropriate high-LET equivalent doses (in sievert) enables the risks for low-LET radiation to be applied to high-LET radiation. End-points involving tumour induction, life-shortening and cell transformation are of special relevance, particularly if the RBEs are determined at low doses of both the high-LET and the low-LET radiation, where responses are linear and a maximum value of the RBE, namely RBE_M, results [I4, N7, S67]. Throughout this Section, absorbed doses of high-LET radiation are quoted in gray wherever possible. Equivalent doses are obtained by applying the radiation weighting factors (w_R) of Table 47.

282. There is one type of high-LET radiation to which human beings have been widely exposed internally. This is alpha-particle radiation, mainly from naturally occurring radionuclides. The large body of data on the exposure of human beings to alpha particles from radon and its progeny are discussed in Section III.C. Additional information on the effects of alpha particles in human beings comes mainly from studies of patients either treated with ^{224}Ra or injected with the diagnostic contrast agent Thorotrast; from studies of occupational exposure to ^{226}Ra, ^{228}Ra, plutonium, or uranium dust; and, to a very small extent, from studies of radium in the natural environment. Reviews have been carried out by Mays [M4], the BEIR IV Committee [C6] and Taylor et al. [T1].

283. It is necessary to examine this material to see to what extent an excess of cancer induction has been observed and whether there are any unusual features of its induction compared with induction by low-LET radiation. More importantly, if both the number of excess cancers per unit population and the dose causing them is sufficiently well known, a direct measure of risk may be derived for the specific organ or organs in which the majority of the alpha-emitting material becomes lodged. As will be seen below, this is possible for both bone and liver cancer; indeed, present knowledge of risk in these two organs derives primarily or exclusively from exposures to high-LET radiation from which or with which low-LET risks are inferred or compared.

1. Radium-224

284. After the Second World War, about 2,000 children and adults in the Federal Republic of Germany were treated for tuberculosis, ankylosing spondylitis or other diseases with repeated intravenous injections of ^{224}Ra as radium chloride [M3, S39, S40]. Bone is the principal repository for radium that is incorporated into the body, but as ^{224}Ra has a half-life of 3.6 days, almost all the radiation dose is delivered to the bone surface, before the radium can be incorporated throughout the bone volume. In beagle dogs, injected ^{224}Ra has been found to deliver an appreciable alpha-particle dose to the kidneys, liver and eyes as well [L13].

285. It was not possible to follow the entire group of 2,000 patients, but a subgroup of about 900 patients was studied from 1952 onwards. Until 1988, by which time 58% of the patients were known to have died, the average length of follow-up was 24 years (range 0-43 years) [M4, M7, S41]. The average injected amount of ^{224}Ra was 0.67 MBq kg^{-1}, leading to an estimated average skeletal dose of just over 4 Gy [S40]. Those less than 20 years old at first injection received on average nearly twice the injected amount per kilogram as those who were at least 20 years old at first injection, and because of the enhanced uptake of radium by growing bone, their estimated average endosteal dose was nearly five times that of adults (10.6 Gy compared with 2.1 Gy). The injections were given over periods ranging from 1 to 45 months, with average values of 5.5 months for adults and 10.9 months for children and juveniles.

286. The most notable feature of the subsequent health of these patients is that 54 developed a total of 56 primary bone sarcomas, mostly osteosarcomas [M4, S41]. Cancer incidence rates give an expected number of only 0.2, and there is no evidence that this would have been increased by the disease for which patients were being treated, as neither ankylosing spondylitis nor tuberculosis is associated with an increased incidence of bone cancer in non-irradiated patients. Thus all, or nearly all, even allowing for some possible misdiagnoses, of these bone sarcomas are likely to have been radiation-induced.

287. The first bone sarcoma appeared only 3.5 years after injection, the peak frequency was after 6-8 years, and the most recent sarcoma appeared after 33 years, with the temporal pattern of expression being similar for those first irradiated in childhood and as adults [M5]. Only one new tumour appeared in the most recent 10 years of follow-up (i.e. 30-40 years after treatment), and it seems likely that most, if not all, of the radiation-induced bone sarcomas have already appeared, so that the temporal distribution is unlikely to be altered much in the future. The excess absolute risks per gray appear to be roughly equal among men and women, among those irradiated for tuberculosis and spondylitis and among those who were irradiated at less than and more than 20 years of age [M7]. However, the estimated risks are not conclusive, as they depend on the estimated individual skeletal doses, which are not very accurate.

288. Initial examination of the dose-response relationship indicated that a simple linear model did not fit the data very well and that it might be necessary to include a quadratic term [C4]. Some evidence was also noted of a tendency towards an increased risk per gray when the injection span was increased, causing the period of irradiation to be protracted [M5]. A recent reanalysis of the data has examined the dose-response relationship and also taken into account the varying exposure periods of the individual patients [C17]. This analysis confirmed the increased risk per gray when the injection span was increased ($p < 0.01$). This might be due to an inverse dose-rate effect (a dose-rate effect is called inverse if instead of producing a greater effect per unit dose at higher dose rate it produces a lesser effect), in which, for example, an increased injection span would allow more cell division between successive injections, thus maintaining the number of cells at risk in the endosteum at a higher level during the course of injections [M26]. It can also be explained by cell-killing, which reduces the slope at higher doses [M26]. The lowest average skeletal dose that was followed by the development of a bone sarcoma was 0.90 Gy, but the analysis suggested that for a constant exposure time the data were consistent with a linear no-threshold dose response. The apparent linear-quadratic dependence that had earlier been inferred was, in fact, the combined result of a linear relationship between bone sarcoma incidence and dose for constant exposure time and of an increase in the incidence with longer exposure times for constant dose. The model of choice was of the following form:

$$R(D,t) = 0.0055\,D(1+0.18t) \qquad (15)$$

where R(D,t) is the cumulative incidence of bone sarcoma following an average skeletal dose D, in gray, delivered over a period of t months. There was no indication that the dependence on delivery period applied only to the larger doses.

289. It is possible to derive a lifetime risk estimate for the induction of bone cancer by alpha particles from these studies with ^{224}Ra. In the report of the BEIR IV Committee ([C6], page 208) juveniles and adults were treated as slightly different in sensitivity and as risk estimates for bone sarcomas of 188 10^{-4} Gy^{-1} for juveniles and 133 10^{-4} Gy^{-1} for adults were derived (it is noted that if dose protraction were taken into account, the difference between juveniles and adults would disappear, in agreement with Mays et al. [M7]). The weighted average yields an absolute risk coefficient of about 150 10^{-4} Gy^{-1}, and for fatal cancer (lethality 0.70 [I10]) it is about 100 10^{-4} Gy^{-1}. Thus the low-LET risk derivable for a w_R of 20 is 5 10^{-4} Gy^{-1}, and it applies to low dose and low dose rate. It should be noted that these values are derived from high-dose information and require extrapolation from high dose to low dose and from high-LET to low-LET. It is for these circumstances that the w_R of 20 is presumed to apply. Stated more generally, for radiation of high-LET and low-LET, the lifetime risk of fatal bone cancer is 5 10^{-4} Sv^{-1}. This is in very good agreement with the low-LET value of 0.2 10^{-4} Sv^{-1} annual risk rate (Table 8, Part VI) for a 25-year period at risk.

290. In addition to an increase in bone sarcomas, there have also been statistically significant increases in breast cancer in these patients (14 cases observed, 4.1-6.1 expected; $p < 0.005$), liver cancer (6 observed, 1.1-1.2 expected; $p < 0.002$) and leukaemia (6 observed, 2 expected;

$p < 0.05$) and a non-significant increase in kidney cancer (5 observed, 2.4-2.6 expected; $p < 0.13$) [S41]. The expected values were based on data from cancer registries in Hamburg and the former German Democratic Republic [S41]. The calculated organ doses averaged 2.4 Gy for the kidney and 1.4 Gy for the liver, but doses for the breast or bone marrow have not been published. The increase of leukaemia has been attributed to the frequent use of phenylbutazone by the patients with ankylosing spondylitis, but the evidence that this drug is leukaemogenic is weak [F14, I6]. In contrast to findings from studies of the effects of ^{226}Ra, no carcinomas of the paranasal sinuses or mastoid air cells have been observed [M7].

291. It was shown by Spiess that ^{224}Ra was ineffective in the treatment of bone tuberculosis [M3]. For ankylosing spondylitis, however, there was acceptable evidence of its therapeutic value. The original high-dose treatment was discontinued after about 1950, but treatment involving much lower amounts of ^{224}Ra continued to be used. In order to monitor the recent ^{224}Ra therapy for ankylosing spondylitis, Wick et al. [W4, W5, W18] identified about 1,500 adult patients treated mainly between 1948 and 1975. The majority of the patients had received a single series of 10 weekly injections, each of 1 MBq of ^{224}Ra, although some patients received more than one series, and for all the patients the resulting mean endosteal dose was estimated to be 0.65 Gy. The injection span averaged about 12 weeks. By 1988, the date of the most recent published results, the average length of follow-up was 17 years. By that time only one patient had developed a definite bone tumour: a fibrosarcoma of the ileosacral joint appearing 11 years after an average skeletal dose of 0.67 Gy. Wick et al. [W5, W18] referred to three tumours, but two were not bone tumours in the conventional sense. One was a reticulosarcoma of the bone marrow and might be more properly described as a lymphoma; the other was a multiple myeloma. The number of spontaneous bone tumours expected from national rates is about 0.5, and the predicted number of radiation-induced bone sarcomas based on the Spiess series is 6. Thus, there is some evidence that for a given injection span the cumulative risk per gray may be lower for this group than for patients treated with the original high-dose therapy. As mentioned above, one patient in this group also developed multiple myeloma, and six patients developed leukaemia. However, expected numbers have not yet been published. No carcinomas of the paranasal sinuses or mastoid air cells are known to have occurred so far [W5].

2. Radium-226 and radium-228

292. Populations exposed to skeletal alpha radiation from internally deposited long-lived isotopes of radium have also been studied [B26, E8, R6]. The best known group comprises over 4,000 dial painters, radium chemists and patients given ^{226}Ra or ^{228}Ra therapeutically in the United States before 1950. Treatment with internally deposited radium isotopes has been discontinued. Some of these individuals were internally contaminated with pure ^{226}Ra and others with various mixtures of ^{226}Ra and ^{228}Ra. Both isotopes have long half-lives (^{226}Ra: 1,600 years; ^{228}Ra: 5.77 years). Some 87 bone sarcomas are known to have occurred in 85 persons among the 4,775 persons "for whom there has been at least one determination of vital status" [C6]. Among the 2,403 subjects for whom there is an estimate of skeletal dose, 66 sarcomas occurred in 64 subjects (fewer than 2 would be expected from national rates). In addition, some 35 carcinomas of the paranasal sinuses and mastoid air cells are known to have occurred in 31 people for whom estimates of skeletal dose were made [C6]. In unirradiated populations these tumours are exceptionally rare. The sinus and mastoid air cell carcinomas seem to have been caused mainly by ^{222}Rn gas and its progeny. Radon-222, produced by the decay of ^{226}Ra, escapes from bone and collects in the frontal sinuses and mastoid air cells.

293. Cancer mortality has also been studied in about 1,200 women who worked in the United Kingdom with paint containing radium from 1939 to 1961, whose average level of contamination was much less than that of the United States group and who received much lower doses [B2, B26, B35]. By the end of 1985, one fatal bone sarcoma had occurred, compared with an expected number of 0.17, and there had been no deaths from head sinus carcinoma. The ^{226}Ra content of 470 luminizers was measured about 20 years after they had entered the industry. Forty-five had more than 0.05 μg, 20 had more than 0.1 μg (the greatest being 0.6 μg), and 264 had less than 0.01 μg. No person who had luminized after 1941 had more than 0.05 μg. Like the United States dial painters, each woman was also irradiated externally by gamma rays emitted from the container of radioactive paint that stood beside her as she worked.

294. In the group in the United States, there is little evidence to suggest that age at exposure influenced osteosarcoma risk [P29]. The first bone sarcoma was diagnosed five years after exposure, and the first tumours started to appear at about the same time after exposure had begun, regardless of the level of exposure (see Figure XIII). Owing to the protracted exposure from ^{226}Ra (half-life: 1,600 years), tumours have continued to appear. Thus, in contrast to the patients treated with ^{224}Ra, where the risk appears to be minimal more than about 30 years after exposure, tumours have continued to appear more than 60 years after first exposure. Analyses have not yet been carried out to learn whether the rate of occurrence of tumours per person-year at risk is increasing with increasing time since first exposure, remaining constant or decreasing. For head sinus carcinomas, the first tumour appeared nearly 20 years after first exposure, and tumours have continued to appear until 60 years after first exposure. By 1967 no bone sarcomas or head sinus

carcinomas had occurred in the patients in the United States who had an estimated average skeletal dose less than 12 Gy, and it was suggested that a practical threshold might exist. However, this now seems unlikely, as since then three head sinus carcinomas have occurred in individuals with estimated doses of 11.8, 7.1 and 1.2 Gy, while bone sarcomas have occurred in patients with estimated doses of 8.9 and 4.6 Gy. In addition, a bone sarcoma occurred in a British dial painter with an estimated skeletal dose of 0.85 Gy [B2]. In all six cases the majority of the dose was from ^{226}Ra. There are about 300 dial painters in the United States and United Kingdom with measured average skeletal doses between 0.8 and 12 Gy, and by this time about 0.2 naturally occurring bone sarcomas or head sinus carcinomas would be expected to have occurred among them. Thus it is likely that most or all of these six cancers can be ascribed to radiation [M4], and any practical threshold, if it exists at all, cannot be greater than about 1 Gy. No other bone sarcomas occurred in the United Kingdom dial painters.

295. Rowland et al. [R6, R7] studied dose-response relationships in the United States series. An equation of the form $I = (\alpha + \beta A^2) \exp(-\gamma A)$ was found to provide an acceptable fit to the osteosarcoma data. I, the incidence rate, is the observed number of bone sarcomas per person-year at risk and A is the activity of radium entering the blood during the exposure period. In contrast, a linear form, $I = \alpha + \beta A$, fitted the head carcinoma data. Marshall et al. [M38, M39] developed an elaborate two-target model, proposing that two successive initiating events and, later, a promoting event are required for osteosarcoma induction. The initiation events remove the ability of a cell to stop dividing; the promotion event is a signal to divide associated with natural remodelling of bone. The model also allowed for the competitive effects of cell-killing. A third model was that developed by Raabe et al. [R20], which used time to death and average skeletal dose as the parameters and is relevant to the question of a practical threshold (see also [C6]). All these analyses should, however, be interpreted with caution: the intake of radium was estimated many years after the event and may be inaccurate; the distribution of radium in bone is probably non-uniform, and hot spots capable of extensive cell-killing may have occurred; the dose responsible for tumour induction cannot be distinguished from any dose received after initiation; the relative contribution of the alpha-particle emissions cannot be separated from the other radiations accompanying radium decay; and the fraction of the total dose received by the endosteal cells cannot be specified precisely [B34].

296. In the United Kingdom study of the dial painters, there was an elevated risk of mortality from breast cancer that was not statistically significant except at one time after exposure [B26]. However, this risk is more likely due to external irradiation by gamma rays than to internally deposited ^{226}Ra (Section II.B.1) [B35, B36]. In the dial painters in the United States, mortality from breast cancer was increased when compared with national rates, but the excess was not significant when compared with local rates [S42]. There was also an excess of multiple myeloma in the United States study (6 deaths observed, 2.15 expected; $p = 0.045$ [S42]), but it correlated with duration of employment (a surrogate for gamma-ray exposure) rather than body content of radium. Fatal brain tumours and other tumours of the central nervous system in the United States series, recently reviewed, were not found to be in excess [S50]. Leukaemia may have occurred in early cases, which were characterized by very high radium burdens [C6, M27, P18], but the inadequate diagnosis and classification of leukaemia in the 1920s and 1930s prevents any detailed conclusion. A review of leukaemia incidence in the dial painters in the United States found 10 cases of diagnosed leukaemia compared with 9.24 expected on the basis of rates in the general population [S15]. No leukaemia was reported in the British dial painters exposed to relatively low levels of ^{226}Ra, who averaged 37 years at risk, compared to their natural expectation of 1.93 leukaemia cases [B35]. Thus, except perhaps in individuals with very high body burdens, it is concluded that ^{226}Ra and ^{228}Ra deposition in the skeleton from dial painting does not result in a risk of leukaemia.

297. Three studies of the geographical association between the level of ^{226}Ra in groundwater or drinking water and cancer incidence or mortality have been reported. The first study examined mortality between 1950 and 1962 among almost 908,000 residents of 111 communities in Iowa and Illinois whose water supplies contained at least 0.1 Bq l^{-1} [P19]. Those exposed to more than 0.1 Bq l^{-1} had a death rate from bone cancer of 1.41 10^{-5} per year, not significantly higher than the rate in control cities (<0.04 Bq l^{-1}) of 1.14 10^{-5} per year ($p = 0.08$). The second study examined nearly 64,000 residents of 28 towns where water was obtained solely from deep wells [B3]. The water supplies were divided into three groups by radium level. Only for bladder cancer in males was the highest ^{226}Ra level associated with a cancer rate higher than that observed for the residents of towns in Iowa with surface water supplies [C6]. The third study examined the association between leukaemia and ^{226}Ra in groundwater in Florida [L14]. The ^{226}Ra and ^{228}Ra contents of 50 private wells in 27 selected counties were measured. In 17 counties classed as high-exposure counties, leukaemia rates were greater than in the 10 counties classed as low-exposure counties. An association with increasing levels of radium concentration was suggested. A later study carried out over a longer time also found an apparent excess of leukaemia in these counties [O2].

298. The findings of these three studies are not convincing and, in fact, are typical of geographical correlation studies. The exposure levels were not very different and were not based on individual exposure

measurements. In one study there is an indication of a significant association, but the effect noted is not in agreement with the results of the occupational studies of higher levels of ^{226}Ra. It is noteworthy that none of the geographic studies found excesses of bone sarcoma, suggesting that the positive findings for other cancers may only be a consequence of the large number of comparisons and of chance, as indicated in paragraph 31. Further studies of the effects of low-level exposure to radium based on individual measurements might be useful. Except for the bone sarcomas and head sinus carcinomas, no excess of leukaemia nor of any other type of cancer is presently ascribed to the internal deposition of long-lived radium isotopes.

3. Thorium

299. Thorotrast, a 25% colloidal solution of thorium dioxide, was a commercially prepared contrast agent used in medical radiography. It was used in several countries from about 1930 until after the Second World War. Usually given by intravascular injection it was used extensively for the radiological diagnosis of war injuries of servicemen in the Federal Republic of Germany and Japan and for cerebral angiography, although it was also used for other purposes and administered by other routes.

300. Thorotrast is retained in the body more or less indefinitely, since the excretion rate is very low [F21], and although some of its decay products (e.g. ^{224}Ra), are normally mobile within the body they have difficulty in diffusing out of the deposits of ThO_2 in which they are created. The colloidal material injected rapidly becomes granular, and Thorotrast granules are taken up by phagocytic cells anywhere in the body, allowing transport to other sites such as lymph nodes. Alpha irradiation around phagocytes containing ThO_2 granules, as in the liver, is localized to aggregates assumed, for dosimetric purposes, to be of spherical shape and to have a radius up to 100 μm. The average organ dose cannot be casily specified because of self-absorption of alpha particles in the aggregates. It depends, among other things, on the amount of Thorotrast injected but is not proportional to it. In addition, tissue dosimetry is complicated by the activity ratios between ^{232}Th and its decay products.

301. Thorotrast is deposited at various sites in the body in approximately the following proportions: liver 59%, spleen 29%, red bone marrow 9%, calcified bone 2%, lungs 0.7% and kidneys 0.1% [K3, M48]. Thorium-232 has a long half-life (more than 10^{10} years) and decays through a series of alpha-emitting decay products, including ^{228}Th and ^{224}Ra. Persons with burdens of Thorotrast administered for radiodiagnostic purposes are thus continuously irradiated by ^{232}Th and its alpha-emitting decay products. The estimated typical dose rate of alpha-radiation following an injection of 25 ml of Thorotrast in these patients is about 0.25 Gy a^{-1} in the liver, 0.09 Gy a^{-1} in the bone marrow, 0.70 Gy a^{-1} in the spleen, 0.16 Gy a^{-1} in the endosteal layer of bone, 0.13 Gy a^{-1} in the bronchi and 0.004 Gy a^{-1} in the kidneys. (For further details, see [C6, M48].)

302. A number of epidemiological studies of patients injected with Thorotrast have been carried out in the Federal Republic of Germany [V1, V2, V9], Japan [I1, M10, M11, M12, M36], Portugal [D15, H13], Denmark [A12, A15, F1, F2, O3, O4] and the United States [F3]. The largest was the German study, which was started in 1968. Slightly more than 5,000 patients known to have received Thorotrast were identified, as was a similar sized control group drawn from the same hospitals and with a similar age- and sex-distribution. To reduce the influence of any underlying diseases in the patients, Thorotrast patients and controls known to have died within three years of injection or hospitalization were excluded from further study. Follow-up of the remainder was poor, chiefly because of the disruption caused by the Second World War, with about 1,100 Thorotrast patients and 2,700 controls untraced. Nevertheless, the results that are based on the remaining 2,334 exposed and 1,912 control patients who could be followed until 1988 are striking. Among the Thorotrast patients there were 396 deaths from liver cancer, compared with only 2 in the controls (see Table 48). These figures correspond to a relative risk of liver cancer of about 200. The majority of the liver cancers were carcinomas, primarily of the cholangiocellular type, and haemangiosarcomas. The shortest interval between injection of Thorotrast and death from liver cancer was 16 years, and it seems likely that the full burden of disease in the group has not yet become apparent, as a substantial number of the liver cancer deaths occurred in the most recent period of follow-up, with intervals of more than 40 years between injection and subsequent death having been observed.

303. In the German study the cumulative rate of malignant liver tumours did not depend on age at injection. However, although there was no difference between male and female patients with regard to age at injection, mean volume of injected Thorotrast or exposure time, the estimated cumulative rate of liver cancers in males was about twice that in females, and the difference was statistically significant. Since the amounts of Thorotrast injected in the patients differed by a factor of no more than two, no internal dose-response relationship could be derived. In the Thorotrast patients with liver cancer, the three lowest doses at 10 years before diagnosis were 1.88, 1.98 and 2.06 Gy. Risk estimates for the German study have, however, been calculated by taking the cumulative number of liver tumours up to 40 years after injection as the numerator and the cumulative dose of all patients up to 30 years after injection as the denominator. Patients who died within the first 15 years of exposure were, however, excluded from the evaluation, as

they were not considered to be at risk for Thorotrast-induced malignant liver tumours. The cumulative risk of malignant liver tumours was estimated to be about 300 10^{-4} Gy^{-1}.

304. The other study of Thorotrast-exposed patients for which detailed results have been presented is from Denmark [A12, A15, F1, F2, O3, O4]. A group of 999 patients exposed for cerebral angiography, mostly between 1935 and 1947, was followed until 1 January 1989 and numbers of incident cancers determined through linkage of records with the national cancer registry in Denmark. Many of the patients had been given angiographies because they presented symptoms suggestive of intracranial tumours, but even when these tumours were excluded the cumulative incidence of cancer at all other sites was estimated to reach 86% 50 years after injection with Thorotrast. A total of 79 cases of primary liver cancer had occurred, giving a standardized incidence ratio, compared with national age-, sex- and calendar-year-specific rates for Denmark, of 126 (see Table 49). Other cancer sites for which there was a statistically significant increase were liver not specified as primary, including haemangiosarcoma (SIR: 30); gall-bladder and extrahepatic bile ducts (SIR: 14); leukaemia (SIR: 10); nasal cavity (SIR: 10); peritoneum, i.e. malignant mesothelioma, (SIR: 8.6); multiple myeloma (SIR: 4.6); ovary (SIR: 2.4); lung (SIR: 2.3); breast (SIR: 1.8); metastatic sites (SIR: 12); and cancer of unspecified site (SIR: 11). Cancer risks were also increased at most other sites, although the individual increases were not statistically significant.

305. In a recent report on the Danish studies [A15], 16 cases of acute myelogenous leukaemia (AML) and 7 cases of myelodysplastic syndrome (MDS) were diagnosed 8-40 years after injection of Thorotrast. A multivariate model was used to describe the response of these dyscrasias with dose (mean alpha-particle bone marrow dose, 1-34 Gy). A risk estimate of 173 10^{-4} Gy^{-1} (AML and MDS) or of 248 10^{-4} Gy^{-1} (all haematological types) was derived. Compared with low-LET estimates of leukaemia risk, this would be consistent with an RBE value closer to 1 than to 20. It should be noted, however, that in many experimental circumstances, summarized for example in [N7] and [M43], values of RBE for alpha radiation of about 20 are often found, but not for myeloid leukaemia in mice, for which low RBE values, around 3, are more often found [U14]. Although explicit risk estimates had been calculated only for leukaemia [A15], a more recent report [A16], citing 127 liver cancer cases among 1,003 patients, derives a risk estimate of about 710 10^{-4} Gy^{-1}, higher than for other studies.

306. The standardized incidence ratio for liver cancer was strongly related to injected volume, time since injection and cumulative radiation dose (Table 50). A positive trend in the standardized incidence ratio with young age at injection was also seen, although the cumulative frequency of primary liver tumours showed no difference between ages 0-25 years and 26-45 years (see Figure XIV). At older ages the cumulative incidence of primary liver cancer was lower, although the difference was not statistically significant. In contrast to the German study, the cumulative frequency of liver cancer relative to the estimated dose in the liver was not different for males and females (see Figure XIV), and the standardized incidence ratio for primary liver cancer was higher in females (SIR: 165; 95% CI: 117-227) than in males (SIR: 102; 95% CI: 73-139) [A15]. There was a clear trend for the standardized incidence ratio for leukaemia, with the estimated cumulative radiation dose to red bone marrow similar to that observed for liver cancer (Table 50), and there were suggestions of positive trends for the standardized incidence ratio for leukaemia with age at injection, injected volume and time since injection, although they were not statistically significant (Table 50). There was no obvious relation between the standardized incidence ratio and injected volume of Thorotrast or time since injection for lung cancer or breast cancer, and the authors concluded that confounding factors were probably responsible for the increased risk at those sites. No excess lung or breast cancer was found in the other studies.

307. The Portuguese Thorotrast study involved about 2,500 patients injected with an average of 26 ml of Thorotrast between 1929 and 1955 and 2,000 controls, all followed for 30 years [A14, D15, H13]. Of 1,244 traced Thorotrast-exposed patients, 955 had died, 137 from malignant tumours (of which 87 were primary liver tumours), 12 from leukaemia, 8 from stomach carcinoma, 5 from carcinomas of the lung, 2 from carcinoma of the larynx and 5 from primary bone tumours. For liver cancers the average latent period was 29-34 years (for leukaemia, about 20 years), and the risk estimated in the report of the BEIR IV Committee ([C6], page 269) was 275 10^{-4} Gy^{-1}, quite similar to the result from the German study.

308. The study of Thorotrast exposures in Japan [I16, I17, M10, M11, M12, M36] is smaller (262 cases injected intravascularly, with an average of 17 ml per injection). Follow-up has now been extended to 42-50 years post injection. There have been 56 cases of liver cancer, 6 cases of blood disease, 4 cases of lung cancer, 1 osteosarcoma, 26 other malignant tumours and 18 cases of liver cirrhosis. With a linear dose-response model, the risk of liver cancer was estimated to be 330 10^{-4} Gy^{-1}. The findings, time relationships, dosimetry etc. were similar to those of the German study. The relative risk for lung cancer is estimated to be 1.7 [I16].

309. A much smaller study was undertaken in the United States [F3] of 26 cases of hepatic angiosarcomas found between 1964 and 1974 in Thorotrast-exposed patients. Latent periods ranged from 20 to 40 years. The hepatic tumour incidence is still increasing. No risk estimate was derived for this study.

310. Exposure to thorium and its daughter products via inhalation has also occurred in the production of thorium from monazite sand and its processing for the manufacture of products such as incandescent gas mantles. The distribution of thorium is very different from that in Thorotrast exposure. Polednak et al. [P28] studied a cohort of men employed at a plant in Illinois between 1940 and 1973. Among 592 men who worked for at least 1 year in selected jobs indicative of high exposure to thorium and its decay product thoron, there was a significant increase in mortality from cancers of all types combined (SMR = 1.75; 95% CI: 1.26-2.39; based on 38 deaths) (see Table 51). There was a significant excess of pancreatic cancer (SMR = 4.13; 95% CI: 1.34-9.63; based on 5 deaths), and for several other sites the number of deaths observed exceeded that expected, but since the numbers were small, the individual increases were not statistically significant.

311. The cumulative risk of fatal malignant tumours of the liver induced by the alpha particles from Thorotrast was estimated in the report of the BEIR IV Committee ([C6], page 269), using the results of the German, Japanese and Portuguese studies, to be about 300 10^{-4} Gy^{-1}. If this is divided by a w_R of 20, the expected low-LET risk is then 15 10^{-4} Sv^{-1}. This figure compares with estimates in the report of the BEIR V Committee ([C12], page 304) of all liver cancers in the life span study of about 18 10^{-4} Gy^{-1} for 40 years at risk. However, possibly only a third of these are actually primary tumours. More recent incidence data and revised modelling were presented earlier in this Annex (Table 8, Part IV), and they provide improved low-LET risk estimates. The risk of fatal primary liver cancer (all cases) was estimated from the life span study mortality data to be about 1.2 10^{-4} $(PYSv)^{-1}$ or 48 10^{-4} Sv^{-1} for 40 years at risk. The liver dose-response curve for the survivors of the atomic bombings appears to be linear; however, non-linearities at doses below 0.2-0.5 Sv cannot be ruled out. If a low-dose-rate reduction factor of 2 is applied, the low-LET risk estimate would be 24 10^{-4} Sv^{-1} for 40 years at risk. The risk estimate derived from the alpha particles (using w_R = 20) is presumed to be for low-dose-rate low-LET exposure. Thus, this value 15 10^{-4} Sv^{-1} or some value intermediate between 15 10^{-4} and about 40 10^{-4} Sv^{-1} may still be the most reasonable estimate. An additional complication in the comparison that may be important concerns the age of the exposed: all ages are included in the life span study and only adults in the Thorotrast studies.

312. The above risk estimates must be regarded as highly tentative, for several reasons. First, the calculations assume linearity of risk with dose throughout except where a low-dose-rate reduction factor other than 1 is applied. Secondly, the dosimetry of thorium in the Thorotrast material is uncertain. Thirdly, the liver is structurally abnormal as a consequence of liver damage in virtually all patients, and it is possible that damage caused by the chemical or foreign body toxicity of thorium may have contributed to the risk of liver cancer. Recent animal experiments suggest, however, that any such toxic effect is likely to be small [D17, W11, W12], although promotional effects cannot be discounted.

313. In addition to the increase in liver cancer, there also appears to be a marked increase in deaths from myeloid leukaemia in the German study, as well as in the Danish, Portuguese and Japanese studies, as noted by Mole [M27, M28]. In the German study 35 deaths were observed in Thorotrast patients and only 3 among the controls. There were similar differences in the other studies. These excess cases have been estimated in the report of the BEIR IV Committee ([C6], page 272) to lead to a lifetime excess risk of leukaemia of about 50 10^{-4} Gy^{-1}, with a latency period of five years. This estimate, assuming an RBE of 10 or 20, is much less (2.5-5 10^{-4} Gy^{-1}) than the low-LET risk of leukaemia, which is also about 50 10^{-4} Gy^{-1} [I10], i.e. an RBE of about 1 would fit better for leukaemia induction from Thorotrast. However, doses are high in this study, and it is possible that cell-killing and other factors play a role. Three cases of erythroleukaemia have been reported among those exposed to Thorotrast, and it has been suggested that high-LET radiation might specifically cause this disease [M40].

314. For cancers other than liver and leukaemia, elevated risks at a number of sites (see Tables 48 and 49) have been cited in each of the four main studies. For example, although relative risks and significance levels have not been calculated for the German study, there are possible increases in deaths from cancers of the extrahepatic bile ducts, pancreas, oesophagus and larynx and for non-Hodgkin's lymphoma, bone sarcoma, plasmacytoma and mesothelioma of both pleura and peritoneum. For several other sites of cancer, including chronic lymphocytic leukaemia, acute lymphocytic leukaemia, Hodgkin's disease or lung cancer, no increase was seen (see Table 49). Risk estimates could not be derived for any of these sites except bone. For bone, in the report of the BEIR IV Committee ([C6], page 272) a lifetime risk of about 100 10^{-4} Gy^{-1} or, using a w_R of 20, 5 10^{-4} Gy^{-1} was estimated for low-LET radiation (5 10^{-4} Sv^{-1} for all radiation), in good agreement with earlier quoted values.

4. Plutonium

315. Animal experiments have shown that plutonium is poorly absorbed by the oral route but slowly dissolved in the alveolar macrophages after inhalation, which means that the highest doses are delivered to the lung. When plutonium isotopes enter the blood they are concentrated in the liver and by various mechanisms at the bone surface

[V3], but their longer half-lives cause many of the alpha emissions to be made after the radionuclides are more widely distributed within the relatively acellular calcified bulk of the bone.

316. In two of the studies of nuclear workers exposed to plutonium and mentioned in Section II.B.1, a substantial proportion of the radiation workers was monitored for plutonium: for workers at the Sellafield plant [S14], detailed dose estimates of plutonium are not yet available; for workers at the Rocky Flats plant, individual plutonium body contents were estimated from health physics records based on periodic urine bioassays [W7]. The Rocky Flats study included 5,413 white males employed for at least two years, and follow-up was to the end of 1979. The average external dose (low-LET radiation) for the entire cohort was 41 mGy, and the average plutonium content of the body was 65 Bq. Mortality was compared with United States national rates for the entire group, and no excess was found for all causes or for all cancers. There were no deaths from bone cancer. When mortality in employees with body burdens greater than 74 Bq was compared with that in employees with smaller body burdens, using lag times of 2, 5 and 10 years, elevated rate ratios were found for all causes of death and for all lymphopoietic tumours. For all causes of death these increases were not statistically significant for any of the lag times considered, and for lymphopoietic tumours (which included one myeloid leukaemia) the increase was significant only for a five-year lag time. Some other cancer sites also showed elevated rates, but none were significantly elevated. There were no elevated rate ratios for either bone or liver cancer. There is no clear evidence of an effect of plutonium in this study.

317. Twenty-six men who worked with plutonium during the Second World War under very crude conditions at the Los Alamos Laboratory have also been studied [V5, V7, V11]. Inhalation was the primary mode of plutonium exposure. Current estimates of the systemic plutonium depositions in these individuals range from 52 to 3,180 Bq, with a median value of 500 Bq. The last published follow-up included the period up to 1990. By that time, seven individuals had died. The causes of death were lung cancer (two cases), myocardial infarction, arteriosclerotic heart disease, accidental injury, respiratory failure due to pneumonia/congestive heart failure and osteosarcoma of the sacrum. Three men also reported a history of skin cancer. The man who died from osteosarcoma did not suffer from Paget's disease, which is usually associated with bone sarcoma at older ages. In a separate study, all workers at the plant who had estimated plutonium depositions of 370 Bq or more on 1 January 1974 were identified [V6]. Of the 224 white males in the cohort, 43 had died at the time of the last follow-up compared with 77 expected based on United States national rates; 8 men had died from cancer compared with 15 expected.

318. Workers in a radiochemical plant at Mayak [H17] received exposures from both external gamma-radiation and plutonium, about one quarter of the total exposure being from the latter. Those in the highest dose group had over 4 Sv total (Table 39), and a significant excess of lung cancer was noted (see paragraph 237).

5. Uranium

319. Miners of uranium and other ores are exposed to high concentrations of radon gas and its progeny, and they are considered separately in Section III.C. Among those working above ground in the processing of uranium, exposures to radon gas and its progeny are low, but there may be exposure to alpha, beta and gamma-radiation from uranium compounds and progeny. A number of cohort studies have been carried out of the health of men exposed to uranium mainly in the form of UCl_4 and UO_3 in the nuclear energy industry [C22, P24]. These studies were considered together with other studies of nuclear energy workers in Section II.B.1. No hazards specific to uranium were identified in the cohort studies, apart from a suggestion in the study of workers employed at a uranium processing plant in Tennessee during the Second World War that men first exposed to uranium dust at the age of 45 years or older might be at increased risk of developing lung cancer [P24]. A case-control study of men who had worked at the plant and subsequently died of lung cancer was performed, in which cumulative radiation lung doses were evaluated in detail and smoking status, x-ray examinations received during employment, history of lung diseases and potential for diagnostic radiation were determined from employee medical records [C29]. The study included 330 cases and 641 controls. When all the men in the study were included in the analysis, there was no suggestion of an increase in risk with increasing exposure. However, among those aged over 45 years when first exposed, the relative risk of lung cancer was found to increase with increasing level of exposure after controlling for age and smoking status. Among men with high exposures to the lung (0.20-0.75 Gy), for which a quality factor of 10 was suggested in the paper (mean equivalent dose: 2.75 Sv) [B32], there was a fourfold increase in lung cancer risk (based on 11 cases and 6 controls), and this increase was statistically significant. It is possible that these findings indicate an increased risk of lung cancer among men exposed to uranium dust. However, it is also possible that the increase is a chance finding based on subset analysis or inadequate control for smoking.

320. Uranium millers are also exposed to airborne dust containing ^{234}U, ^{238}U and ^{230}Th. Two studies of uranium millers have been carried out [A11, W19]. In both studies the number of deaths observed from lung cancer was less than expected, and neither study reported an excess of bone cancer. However, both studies reported an excess of

cancers of lymphocytic and haematopoietic tissue other than leukaemia (4 observed, 1 expected in the entire period [A11]; 6 observed, 2.6 expected in the period more than 20 years after commencement of employment [W19]). The observed deaths comprised giant follicular lymphoma (two), multiple myeloma (two), lymphosarcoma (three), and Hodgkin's disease (three).

6. Polonium

321. A cohort study of 4,402 white males employed by the Mound facility in the United States during the period when polonium operations were conducted (1944-1972) has been carried out [W13]. Overall, no excess mortality was observed, but among workers initially hired during the Second World War, mortality was elevated, especially for deaths from all cancers, cancers of the lung and cancers of the rectum. However, these increases did not appear to be related to exposure to ^{210}Po. Among workers monitored for ^{210}Po, mortality was significantly less than expected, although more lung cancers were observed than expected. In this subcohort, no significant dose-response trends were identified for all causes combined, all cancers combined or cause-specific cancers.

7. Summary

322. There is overwhelming evidence that bone sarcomas can be induced by both ^{224}Ra and the long-lived isotopes of radium incorporated into the body after injection or ingestion. Because of the short half-life of ^{224}Ra (3.6 days) and the relatively short duration of administration, the study of patients injected with it has been informative in revealing the time pattern in which the radiation-induced tumours occur. The excess risk appears to start about 3 years after exposure and to last for about 30 years, with the excess absolute risk reaching a maximum 6-8 years after exposure. Excess absolute risks per gray do not appear to depend strongly on sex or age at exposure. Further analyses of the data in terms of the risk of osteosarcoma per person-year at risk in conjunction with dose, treatment period, time since exposure and other factors would be helpful in enabling comparisons with the risks seen in other populations following other types of exposure. The excess absolute lifetime risk of fatal bone cancer derived from these high-LET exposures is estimated to be 5 10^{-4} Sv^{-1} for both high- and low-LET radiations.

323. Studies of dial painters and others who accumulated body burdens of ^{226}Ra and ^{228}Ra confirm the findings of the ^{224}Ra series of a relatively short minimum induction period after first injection or ingestion. Dose-response relationships have been more difficult to evaluate than among patients exposed to ^{224}Ra. Two-step models have been invoked to explain the observations. However, the available data do not appear to have been fully exploited, and further epidemiological work, including more refined dosimetry and statistical analysis, would be informative.

324. The only other cancers that are clearly attributable to radium are those of the paranasal sinuses and mastoid air cells, which occurred 20 and more years after commencement of exposure in persons with body burdens of ^{226}Ra and ^{228}Ra in the United States. Further dosimetry and analyses of the available data would allow the existing risk to be better described. There is some evidence of an increase of leukaemia and of breast and liver cancer in patients treated with ^{224}Ra. Increases of some cancers in dial painters in the United Kingdom and in the United States may have resulted from external gamma-radiation. The further study of these increases and the associated dosimetry, including the role of external gamma-radiation, would be informative.

325. There is overwhelming evidence that injected Thorotrast causes liver cancer and leukaemia, and it seems likely that it also causes a variety of other cancers. For liver cancer the minimum latency period is about 15 years since first injection, but for leukaemia it is shorter, no more than about 5 years. Both the excess absolute risk and the excess relative risk of liver cancer increase as the amount of Thorotrast injected increases. The excess relative risk of liver cancer is higher in those injected at younger ages, but the excess absolute risk does not appear to depend strongly on age at exposure. For a given administered amount, males and females have similar risks. Based on the German, Japanese and Portuguese studies, the cumulative risk of liver cancer is estimated to be about 300 10^{-4} Gy^{-1}. This implies a low-LET risk of 15 10^{-4} Gy^{-1} (w_R = 20), or 15 10^{-4} Sv^{-1} for all radiations, which is broadly comparable with risk estimates from low-LET exposures determined from the life span study (Table 5). It should be noted, however, that although these estimates of liver cancer risk are attributed to the alpha particles of thorium, the chemical effects of thorium may also enhance the radiation risk. On the other hand, the doses are high, and with cell-killing and other factors potentially involved, the risk estimates may be lower than those for lower doses. The excess of leukaemia among patients treated with Thorotrast appears to be concentrated among tumours of the acute myeloid type (including a relatively large number of the rare erythroleukaemias); it is, however, small compared with the low-LET risks derived from the survivors of the atomic bombings, i.e. it is compatible with an RBE closer to 1 than to 20 for this end-point. The risk of osteosarcoma is similar to risks derived from ^{224}Ra exposures.

326. The information contained in the studies of the effects of Thorotrast has not yet been fully exploited, and additional analyses would be helpful. For example, dose-response analyses based on the number of liver

tumours per person-year at risk, classified by total dose accumulated up to 10 years previously and other relevant factors, would allow an examination of the shape of the dose-response curve in the German study, as well as a comparison with the results of other studies. In the German study, relative risks and significance levels for cancers other than liver cancer have not been calculated and would be informative, as would explicit risk estimates in the Danish, Portuguese and Japanese studies.

327. The studies of men occupationally exposed to other radionuclides, e.g. polonium and uranium, suggest that these exposures may be associated with increased risks of specific types of cancer, but the studies are small in size and at present have only a limited length of follow-up. Further information is necessary before definite conclusions can be drawn.

328. The question of whether high-LET alpha radiation induces tumours other than those caused by low-LET radiation or more frequently than would be expected based on values of RBE or of the radiation weighting factor is not completely answered by the information available so far. Liver cancer risks from high-LET and low-LET radiation appear to be broadly similar in spite of age and other differences in the populations exposed. The increases in deaths from myeloid leukaemia in the Thorotrast studies are clear-cut, but risk estimates are less than expected from low-LET risk estimates unless an RBE closer to 1 than to 20 is used.

329. Risk estimates for bone cancer induction ($5\ 10^{-4}\ Sv^{-1}$) are derived from studies of ^{224}Ra but supported also by the Thorotrast studies. Risk estimates for liver cancer ($15\ 10^{-4}\ Sv^{-1}$) from the Thorotrast studies are broadly similar to risk estimates derived from studies of low-LET radiation.

C. RADON

330. Numerous studies of underground miners exposed to radon and its decay products in the air of uranium and other hard rock mines have shown an increased risk of lung cancer in comparison with non-exposed populations. The subject was reviewed in the UNSCEAR 1977 Report [U5], in which it was estimated that the number of lung cancers caused by exposing 1 million people to 1 WLM (equivalent to 5 mSv [I18]) would be 200-450, compared with a natural rate of lung cancer of 10,000-40,000 per million people in a lifetime. In the UNSCEAR 1988 Report [U2], the range estimated was similar. Radon is classified as a Group I carcinogen by the International Agency for Research on Cancer [I19]. The recognition that hard rock miners face an occupational risk has resulted in reduced exposure levels in many such mines. However, in recent years other important sources of exposure to radon have been recognized; for example, it is now established that in places where air exchange with outdoor air is restricted, especially in homes, offices and other buildings, radon may accumulate in appreciable concentrations. In response to the need to conserve energy used for heating, construction methods that improve insulation and sharply restrict ventilation have been introduced in some countries (for example, Sweden), increasing the radon exposure of the general population. There are now a large number of investigations of the risks associated with non-occupational exposure to radon, as it is a subject of considerable public concern in many countries.

1. Occupational exposures

331. The principal sources of quantitative information on the effects of exposure to radon remain the occupational studies of miners of uranium and other hard rock. For some studies (e.g. [T3, T10]) detailed exposure histories are available, but for many others the exposure history of individual miners is not known precisely, so the estimates of individual exposure are subject to considerable error. Another difficulty arises when attempts are made to generalize to the population from the results of these studies: in all of them the exposures are expressed in terms of working level months (WLM). The dose to the bronchial epithelium varies somewhat with the fraction of unattached radon progeny in the air and with the breathing rate. Mine exposure levels have been reduced over time and cumulative exposures in most countries no longer exceed 100 WLM. Interest now focuses on the possible effects of lower exposures of more recent workers, but further results are still expected from study of the higher exposures received in the past.

332. Since the last review of the subject by the Committee in the UNSCEAR 1977 Report [U5], a number of other groups have used data from the studies of exposed miners to estimate the lifetime risk of lung cancer mortality following lifetime exposure to radon progeny. In one study, the BEIR IV Committee [C6] obtained detailed tabulations of data from four of the principal cohort studies and carried out extensive statistical analyses. Data on individual smoking habits were not available from all four studies, and the Committee made the implicit assumption that the joint effect of cigarette smoking and radon was multiplicative. Their findings indicated that the risk of lung cancer mortality was best explained by a model in which the excess relative risk was directly proportional to the cumulative exposure, but modified by the age at which the risk occurred and by the length of time since exposure. The excess relative risk decreased with attained age and declined with increasing time since exposure.

333. The BEIR IV Committee [C6] also considered the distribution of lung cancers by specific histological type and concluded that the greatest excess was found for

small-cell carcinomas but that other histological types were also increased. More recent work includes a study in China [Y6], in which a significant dose response for adenocarcinoma was found. However, a different result has been found in a recent comparison of lung cancer induced in miners and in survivors of the atomic bombings [L23]. This study, conducted by a team of Japanese and American pathologists, found that radiation-induced cancers were more likely to be small-cell cancers than adenocarcinomas. A significant dose response was found for small-cell cancers for both the miners and the survivors of the atomic bombings.

334. Since the publication of the report of the BEIR IV Committee [C6], a number of new studies of lung cancer in miners exposed to elevated levels of radon have been carried out [C31, H16, H37, L17, M18, R9, S5, S24, S57, S58, T3, T9, T10, W1, X1]. Two of these studies, one on Chinese tin miners [X1], the other on West Bohemian uranium miners [S57, T3], each included a larger number of lung cancers than had been included in all four datasets available to the BEIR IV Committee for their review. They have thus enabled the appropriateness of the model used by the BEIR IV Committee [C6] to be examined. Both new studies included a substantial amount of information on the effects of exposures more than 15 years previously, and the findings with respect to variation in the excess relative risk of lung cancer per unit exposure with time since exposure are summarized in Table 52. Both find clear evidence that the effect of exposures in the distant past is less than the effect of more recent exposures. In both studies, the excess relative risk per unit exposure for exposures 15-24 years previously is about half that of exposures 5-14 years previously. In both studies, the effect of exposures received 25-34 years previously is less than that of exposures 15-24 years previously, and exposures more than 35 years previously have little effect.

335. These two studies also find strong evidence that the excess relative risk per unit exposure varies with attained age, as can be seen from Table 53. In the Chinese study, estimates of the excess relative risk per unit exposure, including the entire follow-up period, indicate a 25-fold variation in the excess relative risk per unit exposure with attained age. In the West Bohemian study, estimates specific to the period 5-14 years since exposure indicate that the excess relative risk per unit exposure at ages 35-44 years is twice the value for attained ages of 55 years and above, and at ages <35 years it is eight times as great. The two studies, therefore, independently confirm two of the conclusions of the BEIR IV Committee [C6], namely that the effect of radon exposure is concentrated 5-14 years after exposure and that the excess relative risk of lung cancer per unit exposure decreases with attained age.

336. Recently a joint analysis of data from 11 studies of male underground miners was conducted [L21]. Over 2,700 lung cancer deaths occurred among 68,000 miners who provided nearly 1.2 million person-years for observation (Table 54). The mines were in Australia, Canada, China, the former Czechoslovakia, France, Sweden and United States. The excess relative risk of lung cancer was linearly related to cumulative exposure to radon progeny, estimated in WLM. The excess relative risk per unit exposure in the combined data was 0.49% WLM^{-1} (Table 54). As in the analysis carried out by the BEIR IV Committee [C6], the excess relative risk per WLM was strongly modified by various factors. It decreased with attained age, with time since exposure and with time after cessation of exposure to radon. Among miners first exposed to radon in the mines when they were less than 20 years old and less than 10 years old, 861 and 110 lung cancer deaths were observed. However, the excess relative risk per WLM was not related to age at first exposure. Over a broad range of total cumulative dose, higher lung cancer risk was associated with underground exposures received at low rates, suggesting that either low exposure rates or long duration of exposure, or both, may be especially hazardous (Figure XV).

337. It is difficult to interpret the data on lung cancer in miners because of the confounding effects of cigarette smoking and exposure to other carcinogens in mine air, such as arsenic. The joint effects of exposures to radon and other agents are considered in Section III.C.3. Two studies, both of uranium miners in the Colorado Plateau, have demonstrated that exposure to increased levels of radon gives rise to an elevated risk of lung cancer, even when few or no cigarettes are smoked.

338. The first of these studies used data from the New Mexico tumour registry in a population-based case-control study of lung cancer in Navajo men [S25]. The 32 cases of lung cancer included all those occurring among men between 1969 and 1982. For each case, two deceased controls with non-respiratory cancer were selected. Of the 32 cases, 72% had been employed as uranium miners, whereas no controls had documented experience in this industry. The lower 95% confidence limit for the relative risk of lung cancer associated with uranium mining was 14.4. Information on cigarette smoking was available for 21 of the 23 affected uranium miners: 8 were non-smokers, and the median consumption by the remainder was in the range 1-3 cigarettes per day.

339. The second study of lung cancer among non-smoking uranium miners was carried out by the United States Public Health Service on 516 white men in the cohort of Colorado Plateau miners who were recorded as never having smoked cigarettes, pipes or cigars [R9]. Follow-up was carried out from 1950 through 1984. Age-specific mortality rates for non-smokers from a United States study of veterans were used for comparison. Fourteen deaths from lung cancer were observed among the non-smoking miners, while 1.1 deaths had been

expected, yielding a relative risk of 12.7 (95% CI: 8.0-20.1). The median exposure in this group was 296 WLM (mean 720 WLM).

340. Data on tobacco use were available for 6 of the 11 cohorts analysed by Lubin et al. [L21]. No clear pattern of risk for the joint association of smoking and exposure to radon emerged, although the data were most consistent with a relationship that was intermediate between multiplicative and additive. In cohorts with information on the use of tobacco, 64 lung cancers occurred among nearly 3,000 non-smoking miners. Analyses revealed a linear exposure-response trend that was about three times higher in never-smokers than in smokers. Data on exposure to arsenic was available for two cohorts. Exposure to arsenic increased the risk of lung cancer, and adjusting for it reduced the estimated excess relative risk per WLM.

2. Effect of exposure rate

341. Substantial evidence has emerged in recent years that among miners exposed to high concentrations of radon, higher excess relative risks per unit dose are associated with a lower average exposure rate and/or a longer duration of exposure. The effects of these two factors, exposure rate and exposure duration are often difficult to disentangle and express separately. The reasons for a higher risk at lower exposure rate or longer duration are unclear. Possible explanations include random or systematic errors in estimates of exposure rate, a non-linear relation between exposure and dose to the target cells at high exposure rates, cell-killing at high exposure rates, and effects related to the multistage nature of carcinogenesis.

342. The initial evidence of the exposure rate effect was put forward by Howe et al. [H15, H16] in studies of the workforces of two Canadian uranium mines, both operated by the former Eldorado Resources Ltd. In the first study, of workers at the Beaverlodge mine in Saskatchewan, the excess relative risk per unit exposure was initially given as 3.28% WLM^{-1} (95% CI: 2.08-4.48). A re-evaluation of a sample of the dosimetry data at the Beaverlodge mine has indicated, however, that there was considerable underestimation of the exposures and that the initial risk estimates should be multiplied by a factor of 0.65 [H23]. A further re-evaluation of exposure histories in this study is under way. Using the currently available factor of 0.65 gives an excess relative risk per unit exposure of 2.13% WLM^{-1} (95% CI: 1.35-2.91). In the second study, of workers at the Port Radium mine in the Northwest Territories, the excess relative risk per unit exposure was about one eighth of this value, or 0.27% WLM^{-1} (95% CI: 0.11-0.43). When expressed in terms of excess absolute risks, the contrast between the findings for the two mines is also striking: the risk was 13.5 10^{-6} PY^{-1} WLM^{-1} at Beaverlodge (95% CI: 7.67-19.4) after allowing for the revised dosimetry, while it was 3.10 10^{-6} PY^{-1} WLM^{-1} at Port Radium (95% CI: 1.89-4.32). The chief difference between the two mines seems to be in the average exposure rates, which were estimated to be 8 WLM a^{-1} at Beaverlodge after allowing for the revised dosimetry and 109 WLM a^{-1} at Port Radium.

343. The question of an increased carcinogenic effect per unit exposure for lower exposure rates or longer duration times has arisen in connection with other high-LET exposures and is widely known as the "inverse dose-rate effect". The word "inverse" is used because the effect of dose rate is inverse to that usually found with low-LET radiations, where higher dose rates are almost always more effective than lower dose rates. The phenomenon in cellular transformation systems [H14] and a possible biological basis for it involving the initiation of carcinogenic steps at only one brief portion of the cell cycle [R3] have been widely discussed (see, for example [B13, S30]). The effect is not seen in all circumstances. The possible biological basis for the effect has been elaborated in a model that appears to account for most features of the effect [B10, U1 (Annex F)].

344. Evidence that higher excess relative risks per unit exposure are associated with lower average exposure rates in miners has also been reported in three other cohorts. First, Hornung and Meinhardt [H18], in an analysis of data from the study of uranium miners in the Colorado Plateau, found statistically significant evidence of an exposure rate or duration effect, implying that among groups of miners receiving equivalent cumulative exposures, those exposed to lower levels for longer periods of time were at greater risk of lung cancer. They also found that this effect varied with cumulative exposure level. Thus a miner with total exposure in the range 0-830 WLM at a rate one tenth as great as another miner of the same age, smoking status and cumulative exposure would have a 58% greater risk of lung cancer. Secondly, in the cohort of Chinese tin miners the excess relative risk per WLM declined smoothly with increasing exposure rate from 1.74% among miners exposed at less than 10 WLM a^{-1} to 0.48% among those exposed at more than 30 WLM a^{-1} [X1], as is summarized in Table 55. When these risks are adjusted for arsenic exposure, which may be correlated with radon exposure, there is still a small but consistent decrease in excess relative risk with increasing exposure rate. Thirdly, Ševc et al. [S5, S57] reported a similar effect within the cohort of miners in the former Czechoslovakia. However, it should also be noted that some studies [T10] do not find that risk increases with a lower exposure rate.

345. Some additional evidence comes from a comparison of estimates of risk derived from the four cohorts of miners studied by the BEIR IV Committee [C6]. These and other estimates are presented in Table 56.

Beaverlodge was among the cohorts studied. Results for workers at the Port Radium mine were not available at that time, but they are shown in the Table for the sake of comparison, as are the recent results for Newfoundland, Cornwall, China, former Czechoslovakia, France and New Mexico. The estimates of risk calculated by the original authors are given in Table 56 along with the estimates calculated by the BEIR IV Committee [C6] in one of their joint analyses. Values given by the original authors may not be comparable as the different studies covered different ages and follow-up periods. The estimates of the BEIR IV Committee [C6] are specific to age 55-64 years and to exposure in the preceding 5-14 years and thus take account of the different age-structures and follow-up periods in the different studies. Miners at Malmberget had exposure rates comparable with those at Beaverlodge and also experienced high excess relative risks per WLM. Miners in the Colorado Plateau had exposure rates similar to those at Port Radium and experienced lower risks per WLM.

346. The above evidence suggests that lower exposure rates, i.e. longer durations, are associated with higher risk per unit dose. This issue has been investigated further in a special study of the effect of exposure rate in the cohort of West Bohemian uranium miners [T3]. Using detailed information on the dates during which each man in the cohort worked in the 19 mine shafts involved, together with 39,000 shaft- and calendar-year-specific radon measurements, the exposure rate experienced by each man in each month of his employment was calculated. The effect of variations in exposure rate was then studied using power transformations of the monthly exposure rates. When data from the entire cohort were considered, models constructed in this way provided a substantially better fit to the data than models based on cumulative exposure in the usual way, and they confirmed that exposures delivered at higher exposure rates (i.e. over shorter times) were associated with lower excess relative risks per unit exposure. Among the class of models studied, the best-fitting model was obtained by taking a square-root transformation of the monthly exposure rates. In this analysis, miners were enrolled in the study five years after starting underground work. During those first five years on the job, some of the men had worked in shafts where concentrations exceeded 10 WL. However, the high radon concentrations did not persist more than five years after the men had started underground work, and they now occur only very exceptionally, either occupationally or environmentally. The analysis was therefore repeated, this time including only men who had never worked in shafts where the radon concentration exceeded 10 WL. When this was done, the excess relative risk per WLM did not vary with exposure rate or with calendar year, age at start of underground work or duration of exposure. In addition, there was no evidence of non-linearity in the excess relative risk with time-weighted cumulative exposure. The excess relative risk per WLM estimated by considering only men who worked in shafts with radon concentrations below 10 WL was just over twice that estimated from the entire cohort [T3]. If this is confirmed in other studies, it would be more appropriate to base risk predictions for domestic or modern occupational exposure on the experience of men exposed only at lower rates of exposure (much less than about 10 WL). The variations in excess relative risk with time since exposure and attained age for the West Bohemian miners exposed to less than 10 WL are given in Table 57.

3. Other issues: cigarette smoking, arsenic and cancer at other sites

347. The joint effects of exposure to radon and tobacco smoke in uranium miners were discussed briefly in the UNSCEAR 1982 Report [U4] in a general review of the effects of exposure to radiation in conjunction with other carcinogenic agents. Since then a number of additional reports have become available, and the subject has been reviewed by Lubin [L5]. The available data, however, have many limitations. Not only are the histories of exposure to radon not well known for individuals, but obtaining accurate cigarette smoking histories is often difficult. In addition, the resulting risk may well be strongly dependent on factors other than the total exposure to radon or cigarette smoke. For example, laboratory experiments with rats have shown that radon followed by smoke is four times more hazardous than the same exposures given in reverse sequence [G13].

348. The BEIR IV Committee [C6] analysed the combined effects of exposure to radon and cigarette consumption in two populations of uranium miners. For the first of these the Committee obtained access to an on-going case-control study of lung cancer in a cohort of New Mexico uranium miners. The cohort included 4,051 subjects with at least one year of documented underground employment. The cases included all Hispanic and non-Hispanic white males diagnosed with lung cancer. For each of the 69 cases, four controls were selected. The number of years of employment was used as the measure of exposure to radon, and cigarette smoking histories for both cases and controls were available from pre-employment or annual physical examinations. The data were fitted with models in which the relative risk of lung cancer was either multiplicative in the two exposures (relative risk = (1 + ar) (1 + bs), where r is years employed as a miner and s is number of cigarettes per day) or additive (relative risk = 1 + ar + bs). The results suggested that the multiplicative relative risk model gave a better fit for the data, although there was not a statistically significant difference between the fit of the additive and multiplicative models. After the publication of the report of the BEIR IV Committee [C6] the original investigators carried out a full analysis of the New Mexico data using individual estimates of exposure to radon progeny rather than years

of employment [S24, S58]. They did not find evidence of a departure from a multiplicative interaction and concluded that the combined risk of the two agents was adequately described by the product of the individual relative risks.

349. The joint effects of cigarette smoking and radon in uranium miners in the Colorado Plateau have also been studied in detail. Whittemore and McMillan [W9], using data up to 1977, showed that an additive relative risk model of the form described in the preceding paragraph gave a poor fit to the data, while a multiplicative model could not be rejected. The BEIR IV Committee [C6] extended this analysis and applied it to data up to 1982. Their findings supported the conclusions of Whittemore and McMillan. The Committee also fitted the data with some general risk models and found that they were consistent with a wide range of sub-multiplicative to supra-multiplicative models. They found no clear reason except parsimony to accept precisely the multiplicative model. In another analysis of the Colorado data through 1982, Hornung et al. [H18] concluded that the effects of the two exposures were better modelled using power functions (using the notation of the previous paragraph, relative risk = $r^a s^b$ and relative risk = $r^a + s^b$ for multiplicative and additive forms, respectively) rather than linear functions. When this was done, the joint effect of exposure to radon progeny and cigarette smoking was greater than additive but slightly less than multiplicative. The departure from multiplicativity was not quite significant ($p = 0.06$), and the authors showed that the evidence of sub-multiplicativity came chiefly from data in the most recent follow-up period, 1977-1982.

350. Seven further studies of miners in Sweden (three studies), former Czechoslovakia (one study), China (one study) and Canada (two studies) present data on the joint effects of cigarette smoking and exposure to radon. Of the Swedish studies, the first, which presents data from Hammar, suggested a protective effect of smoking that might be attributable to smokers having a thickened mucous layer in critical bronchial regions [A7]. The study, however, was small (29 cases), and the results may have been biased, as mine foremen were the source of smoking status for miners. Moreover, controls may have included persons who died from tobacco-related causes [C6, L5]. The second Swedish study presented data from Malmberget and included data on 51 lung cancer deaths [R8]. After a complex series of assumptions it was estimated that the relative risk for non-smokers was 10.0, while that for smokers was 2.9. The authors concluded that mining- and smoking-related risks combined additively. However, as noted by both the BEIR IV Committee [C6] and by Lubin [L5], this conclusion seems to go beyond the evidence as presented, and the estimated relative risks appear compatible with a joint model that is greater than additive but less than multiplicative. The third Swedish study, from Kiruna and Gallivare, was a case-control study and included 60 lung cancer deaths [D13]. Among non-smokers and smokers with lifetime cigarette consumptions of less than and more than 150,000 cigarettes, the estimated relative risks for those with no underground mining experience were 1.0, 2.4, and 8.4, respectively, while in the same smoking categories the relative risks for underground miners were 5.4, 21.7, and 69.7, respectively. Thus the results appear consistent with a multiplicative relative risk model, although no formal testing was carried out.

351. A nested case-control study has been carried out within the cohort of Beaverlodge uranium miners to assess any possible contribution of confounding by smoking and other mining experience to the risk estimate derived from the original cohort study [L15]. Next of kin were interviewed for 46 lung cancer cases and 95 controls enrolled in the original study who had died between 1950 and 1980. Controls were matched to cases by province of death, year of birth and year of death. The data were analysed using conditional likelihood methods and linear logistic models. An initial model included terms representing cumulative radon exposure lagged by five years and number of cigarettes smoked per day. When a third term representing the interaction between these two effects was included, its estimated coefficient was small and negative and its 95% confidence interval included zero. Thus the data showed no significant departure from a multiplicative effect of radon exposure and cigarette smoking, although the best-fitting model indicated that the combined risk of the two agents was slightly less than the product of the individual relative risks.

352. In the study in former Czechoslovakia [S5], smoking data were available only for clay shale miners. Three deaths occurred among non-smoking miners compared with 0.3 expected according to national mortality rates; for miners who smoked, 19 deaths were observed compared with 11.7 expected. On that basis the relative risks among non-smokers and smokers were 10.0 and 1.6, respectively, while the attributable risks (excess deaths) were 16.6 and 26.1 10^{-6} PY^{-1} WLM^{-1}, respectively. In the Chinese study [X1], the smoking information was limited to method of use and there was a substantial amount of missing data. Nevertheless, the analyses rejected both additive and multiplicative relationships but were consistent with a range of models intermediate between additive and multiplicative. Finally, in the Canadian study of fluorspar miners [M18], the excess absolute risk coefficients among those who never smoked (based on 7 deaths) and among current smokers (based on 57 deaths) were very similar at 7.6 and 6.3 10^{-6} PY^{-1} WLM^{-1}, respectively. A few of the studies of non-occupational exposure to radon have presented data separately for smokers and non-smokers [E3, L1, S20]. These suggest that the joint effects of cigarette smoking and radon are more than additive. However, they do not yet provide enough evidence to allow a detailed examination of the alternatives.

353. When all the data on the joint effects of cigarette smoking and radon are considered, there is ample confirmation of the conclusion of the BEIR IV Committee [C6], namely, that the joint effect of the two agents is more than additive. However, there is a mounting body of evidence that the effect is somewhat less than multiplicative. The strongest evidence for this conclusion comes from the two studies of non-smoking miners [R9, S25]. The relative risks reported in those studies are substantially higher than the risks reported in studies that included smokers. However, some corroborative evidence of submultiplicativity is also available from the most recent analysis of the Colorado miners [H18], from the Newfoundland fluorspar miners [M18], from the clay shale miners in former Czechoslovakia [S5] and from the Malmberget study [R8]. It should also be noted that for low-LET radiation the joint effects of smoking and exposure to radiation from the atomic bombings of Hiroshima and Nagasaki appear to be less than multiplicative for lung cancer. Recent work *in vitro* on the interaction between high-LET radiation and tobacco smoke condensate also suggests an interaction for the end-point of oncogenic transformation that is less than multiplicative [P26]. The implications of this conclusion are far-reaching, in so far as predictions based on models in which multiplicativity between radon and cigarette smoking is assumed, either implicitly or explicitly, can generally be assumed to be too low for populations in which there are substantial numbers of non-smokers, including many populations of women, and too high for populations with high proportions of smokers.

354. In the study of Chinese tin miners [X1], the arsenic content of the ore averaged about 1.34% by weight. Thus, many of the miners were exposed to dusts containing arsenic, which is a known carcinogen [E5, I19]. As can be seen from Table 55, adjustment for arsenic exposure reduced the estimated relative risk of lung cancer within each category of cumulative radon exposure in these data, and the estimate of the excess relative risk per unit exposure, averaged over the entire follow-up period, was reduced from 0.6% WLM^{-1} (95% CI: 0.4-0.8) to 0.2% WLM^{-1} (95% CI: 0.1-0.2). In Ontario gold miners [K10] an excess lung cancer risk could not be attributed to radon exposure alone and was associated with exposure to arsenic in the mines.

355. The miners of the West Bohemian cohort were employed in two separate mines: Jáchymov, with arsenic levels of 0.5% in the dust on average (7.1% maximum), and Horní Shavkov, in which arsenic was negligible, i.e. average concentrations of 0.01% in the dust (0.05% maximum). While no comparison was possible between miners employed exclusively at one or the other mine, in miners who spent less than 20% of their time at Jáchymov and who were 55 years old or more the excess relative risk of lung cancer was 1.36% WLM^{-1} 5-14 years after exposure (Table 57), similar to the value derived by the BEIR IV Committee [C6] of 1.5% WLM^{-1}. However, for similar miners with more than 20% of their work experience at Jáchymov the excess relative risk was 1.8 times greater, indicating a clear effect of arsenic exposure (Table 57).

356. The effects of exposure to other agents in miners exposed to radon has received little attention in the literature up to now, but these results suggest that exposure to arsenic-containing dusts may have an important influence on the subsequent risk of lung cancer in some groups of miners exposed to radon, and further evaluation of this possibility is required. Mine exposures other than arsenic might also influence the risk of lung cancer either directly or indirectly, possibly via mechanisms related to cellular irritation and increased tissue proliferation [C21]. The role of dust (silica), diesel and blasting fumes and other heavy metals has also not been clearly elucidated and requires additional study.

357. Significant increases have been noted for certain cancer sites other than the lung in miners exposed to radon. These include salivary gland cancer in Newfoundland fluorspar miners [M18], laryngeal cancer in French uranium miners [T10], multiple myeloma, cancer of the stomach and cancers of the liver and gall-bladder in Swedish iron-ore miners [R8], basal cell skin carcinoma in mid-Bohemian uranium miners [S64] and cancers of the liver and gall-bladder and extrahepatic bile ducts in West Bohemian uranium miners [T9]. However, no consistency exists across studies, and many of the reported increases may not be related to radon exposure. In the West Bohemian miners [T9] mortality was related to cumulative radon exposure only for cancers of the gall-bladder and extrahepatic bile ducts and multiple myeloma, while in the Chinese tin miners a significant but weak relationship between mortality and a joint association of exposure to radon and arsenic was found only for lymphoma [X1]. Neither of these two studies of miners nor a small case-control study of domestic radon and childhood cancer in Sweden [S27] found evidence for an increased risk of leukaemia following radon exposure.

358. It has been suggested from geographical correlation studies that radon in the ambient air may cause cancers at sites other than the lung [E1, E9, H38, L8, W6]. It has also been suggested that the dose received from radon by the red marrow is higher than that usually estimated because radon is very soluble in the fat cells scattered throughout the marrow [R17]. Other information on the dose to organs other than the lung is also available [H47]. These correlation studies are subject to numerous biases, however. Interpretation is restricted by the fact that exposure to individuals, much less dose to tissues, is not known. Furthermore, attempts to reproduce findings (on leukaemia, for example) using more refined statistical methods and smaller geographical areas have not been productive [M13]. Thus, there is no reproducible evidence that radon causes any increase in fatal cancers other than lung cancer.

4. Residential exposures

359. Although the studies of occupational exposure to radon provide a guide to the likely effect of exposures in homes and other buildings, there is inevitably a substantial degree of uncertainty in extrapolating between the two environments. One reason for this is simply that the occupational studies provide data about the effects of exposure primarily in adult males and give no information about effects in women, who in many populations have substantially different smoking habits from men, apart from intrinsic differences in, for example, breathing rates, or about the effects of exposure in early childhood. Another source of uncertainty arises from differences between mines and the indoor environment regarding several factors that affect the dosimetry of radon decay products, including ventilation patterns; the fraction of unattached radon decay products and the particle size distribution in inhaled air; the equilibrium between radon and its decay products; and differences in the breathing rates and breathing route (nose versus mouth) of men at work and at home. As already noted, the effects of mine exposures are subject to confounding or enhancing from mine contaminants such as silica, arsenic, diesel exhaust and blasting fumes [A17].

360. It is therefore necessary, despite the technical and methodological difficulties involved, to directly examine the effects of exposure to radon and its decay products in the indoor environment. The published results of studies addressing this issue are summarized in the following paragraphs, based on a review by Samet [S28]. Many more studies are currently under way, and much additional information on this topic should soon become available. In most studies of indoor radon, the concentrations are measured in becquerel per cubic meter. The conversion factor between radon gas concentration in dwellings and effective dose was considered in Annex A of the UNSCEAR 1993 Report [U1]. It is noted, however, that a committee of the National Academy of Sciences of the United States [C21] concluded that the dose per unit exposure tends to be 20%-30% lower in the home environment than in the mines. Thus, risk estimates based on the experience of miners will tend to overestimate risks from radon in homes [C21].

361. Several studies of non-occupational exposure to radon have been carried out in which lung cancer incidence or mortality rates in different geographical areas have been correlated with measures of exposure of the inhabitants. These are summarized in Table 58. In 8 of the 19 studies there was a significant positive association between radon exposure and lung cancer incidence or mortality, in 1 study there was a positive correlation that did not reach statistical significance, in 4 there was no relation between lung cancer and radon concentration, and in 6 there was an inverse relation. As discussed in Section II.B.2, such geographical studies are always difficult to interpret. This is especially true in the study of diseases such as lung cancer, for which the major determinant of risk is smoking habits and the incidence of which varies geographically. Another recent paper [S26] examined and reviewed features of 15 of these geographical studies, all of which are represented in Table 58, and concluded that their shortcomings rendered them uninformative on the lung cancer risk associated with indoor radon.

362. The association between radon exposure and the development of lung cancer has been tested much more directly in case-control and cohort studies, and the results of studies available to date are summarized in Table 59. Early studies, e.g. that of Axelson et al. [A3], relied on surrogate measures of exposure, such as whether houses were constructed of wood or stone and whether the residents lived on the ground or upper floors. In 10 studies, however, measurements have been made in some or all of the houses. In six of these studies (five in Sweden and one in the United States), a significant increase in lung cancer was associated with increasing exposure after adjusting for cigarette smoking. In two of them (in Canada and the United States), an increase in lung cancer was observed with increasing exposure, but the result was not statistically significant. In one (in Sweden) increased exposure was associated with increased risk for rural dwellers but not for urban, and in two (in China and Canada) there was no increase in lung cancer risk with exposure to radon in any category of smokers, despite the fact that the exposure categories varied from less than 70 Bq m^{-3} to 300 Bq m^{-3} or more. The difficulties of evaluating studies of domestic radon exposure are clearly exemplified by the recent publication of two of the largest and more carefully conducted studies. One, in Sweden [P36], shows a clear-cut association between radon exposure and lung cancer incidence, but the other, in Canada [L10], finds no such association.

363. Each of the case-control studies has limitations that hinder interpretation of the results, in addition to the problems previously mentioned of indirect indications of exposure and variability in sampling with small numbers. It might be illustrative to mention a few of the problems. The study in China was the only one that could reject statistically a risk estimated by extrapolating results from studies of miners; however, there was concern that the high indoor air pollution from coal combustion products might have hindered the ability to detect an effect or measure the exposure. The study in New Jersey had mainly extremely low exposures, and the suggested excess was based on a comparison of only six cases and two controls who received exposures greater than 150 Bq m^{-3}; thus the numbers were too small to provide useful exposure-response information. The study of women in Stockholm was difficult to interpret because adjustment for subject occupancy (that is, the time a woman actually spent in her home) removed the significance of the association. The time response, moreover, was inconsistent

with the miner studies in so far as more recent exposures had little effect, and it was the exposures many years in the past, for which the measurements were most uncertain, that had the greatest influence on risk. Furthermore, the level of significance of the trend tests was markedly different depending on whether continuous (actual) or categorical (grouped) radon concentrations were analysed. This influence on the choice of trend statistics and delineation points for grouped analyses points to a lack of robustness in the data being analysed. The national Swedish study is important, but it is somewhat peculiar in that risk seemed concentrated in men and not in women and few details were given regarding occupancy and time response [P36]. The Canadian study had generally higher exposures than the Swedish national survey, but no risk was evident [L10]. The new study in Missouri was methodologically sound: exposure measurements were made close in time to the diagnosis of lung cancer in the over 500 cases of lung cancer in non-smoking women and the spouses served as controls [A18]. Moreover, the fact that the women were not smokers enhanced the probability of detecting a radon effect by removing the confusing influence of cigarette smoking. Despite the many strengths of this study, an increased risk could not be associated with residential radon exposure. On the other hand, and despite the large number of cases, the data were not strong enough to preclude the possibility that the risk might be similar to that predicted by extrapolating from high doses the risks obtained in studies of miners.

364. The difference seen in the various studies with regard to the radon effect over categories of cigarette smoking also point to a lack of robustness in the data being analysed. Clearly, cigarette smoking was the most striking cause of lung cancer in these series, with relative risks almost 14 times as great for persons smoking more than one pack per day in Stockholm, New Jersey and Canada, and 3 times as great among heavy smokers in China. In the study in Stockholm, the steepest gradient in risk occurred among those smoking more than 20 cigarettes per day, but there was no comparable dose-response trend for non-smokers [P30]. In the national Swedish study, radon-associated risks were not significantly different for different smoking status. In New Jersey, an increasing dose-response trend was observed in smokers of less than 25 cigarettes per day, no dose-response was seen in non-smokers and there was a negative dose-response trend in the heaviest smokers [S51].

365. It is planned to address the inconsistencies seen in indoor studies and to provide more precise information on the level of risk possibly attributable to residential exposures by conducting pooled or combined analyses of the many series that are being conducted throughout the world [S73]. The first attempt to pool similar studies was recently made [L3]. A combined analysis of the studies in China, Stockholm and New Jersey revealed no evidence for a radon-related risk, despite the availability of nearly 1,000 lung cancer cases. It was concluded that the lack of consistency within and across studies for a radon effect overall and within smoking categories severely limited the inferences that could be made regarding the risk related to residential radon. Interestingly, the combined data, which showed no association between lung cancer and radon exposure, were not inconsistent in a statistical sense with extrapolations based on investigations of underground miners. Because of the wide confidence limits about the risk estimates for low exposure categories in indoor studies, it cannot be excluded that the level of lung cancer risk from indoor radon may be less than the level predicted from studies of miners. Thus, the combining of the more than 15 studies conducted with generally similar methodologies will help to resolve the inconsistencies that currently exist.

5. Summary

366. Since the publication of the report of the BEIR IV Committee [C6], two major studies of radon-exposed miners have been carried out, providing much additional data [T3, X1]. In addition, a detailed analysis of 11 miner studies involving almost 2,700 lung cancer cases has been published [L21]. These studies and the comprehensive analysis strongly support the BEIR IV Committee's conclusions that the excess risk of lung cancer following radon exposure is concentrated 5-14 years following exposure and that the excess relative risk of lung cancer per unit exposure decreases with attained age.

367. There is now considerable evidence that risk of lung cancer from radon exposure depends inversely on exposure rate (or on duration of exposure) at high exposure levels. In any case, to make estimates of the risk of radon-induced lung cancer applicable to modern occupational or domestic exposures, they should be based on those studies in which exposures and work history data are as complete as possible and the exposures are at lower rates.

368. In two recent large studies of miners [T3, X1] estimates of the risk of lung cancer following radon exposure were substantially reduced when adjustment was made for the effect of exposure to arsenic-containing dusts. The possible role of arsenic in causing lung cancer in radon-exposed miners has received little attention up to now and requires further evaluation, as does concomitant exposure to other mine contaminants that might influence lung cancer induction.

369. There is now strong evidence to suggest that the joint effect of exposure to radon and tobacco smoke on the subsequent relative risk of lung cancer is less than multiplicative, as was assumed by the BEIR IV Committee [C6], although it seems likely that it is more than additive. These findings imply that estimates based on the BEIR IV Committee's model will underestimate the risks for

non-smokers and overestimate the risks for smokers. It is recommended that all the available data relating to these two issues should be further analysed so that firm conclusions can be drawn.

370. There is little direct evidence on the risks of lung cancer resulting from residential exposure to radon. Statistically weak studies are generating conflicting results. For the time being, risk estimates should continue to be based on the studies of occupational exposures, although without adjustment these are likely to overestimate risks in the domestic environment. They cannot be based on geographical studies, which are always difficult to interpret and which in this case give inconsistent results. However, the substantial number of case-control and cohort studies that are under way might provide useful information. There is no consistent evidence that radon causes cancer in tissues other than the lung.

IV. OTHER RELEVANT STUDIES OF CARCINOGENIC EFFECTS

371. In the last 15 years a number of studies have been carried out of individuals living near nuclear installations and of participants in atmospheric nuclear tests. Some of these have recorded unexpected associations between exposure to radiation and a subsequent increase in the incidence of or mortality from cancer in either the exposed persons or their offspring. In addition, persons suffering from certain types of hereditary disease have been identified as being highly sensitive to radiation, and there is some evidence that there may be other subgroups of the population who are not overtly affected by such diseases but who may also be especially sensitive to the effects of radiation.

A. CANCER NEAR NUCLEAR INSTALLATIONS

372. Several studies have reported increased incidence or mortality rates for leukaemia and non-Hodgkin's lymphoma in young people living near nuclear installations. The studies caused public concern and led to much additional research. The principal findings to date are summarized below.

373. First, an excess of leukaemia was reported in young people in the village of Seascale, four miles south of Sellafield, which is the site of the principal nuclear fuel reprocessing plant in the United Kingdom. Operations involving radioactive materials began at Sellafield in 1950. The excess was reported in a television programme in November 1983 and was later confirmed by an official enquiry [B7]. The Seascale cluster might be considered, from one point of view, to be suspect: it was the occurrence of the cases that determined both the geographic boundary and the age definition of the cluster. Television reporters first went to Sellafield, not Seascale, seeking excesses of cancer among adult workers, not leukaemia among young people in the general population [M6]. One of the investigations that had been recommended in the report of the enquiry subsequently demonstrated that by July 1986 there had been five deaths from leukaemia in the slightly more than 1,000 children born to mothers resident in Seascale in the years 1950-1983 compared with an expected number, based on national rates over the same period, of 0.53 [G1]. The results are summarized in Table 60. Four of the five deaths were in children of age 0-4 years. In contrast, there were no deaths from leukaemia and lymphoma compared with 0.83 expected over the same period in some 1,500 children who moved into the area at some time after their birth [G2]. Although less marked, the contrast extended to other fatal cancers and also to non-fatal cancers for the limited time-period 1971-1984, when expected numbers of non-fatal cancers could be derived from the national cancer register, as is shown in Table 60. Two of the six other cancers were non-Hodgkin's lymphoma. At young ages non-Hodgkin's lymphomas consist of a variety of conditions, one of the most common of which is distinguished from acute lymphocytic leukaemia only arbitrarily by the number of lymphoblasts in the blood, and for the study of aetiology such cases may better be classed with leukaemia than treated as a separate disease. When all cases of leukaemia and non-Hodgkin's lymphoma were studied that occurred in young people resident in Seascale in the slightly longer period 1950-1988, the cluster in Seascale was found not to be limited to children who had been born there [K27]. Eleven cases were diagnosed in this period, and statistically significant excesses were found in young people not born in Seascale as well as in those who were born there, the risks being increased by a factor of approximately 10 in each case (9.9 for those not born in Seascale and 10.2 for those who were).

374. Estimates of the average annual dose in the red marrow that one-year-old children living in Seascale are thought likely to have received from Sellafield discharges are given in Table 61. Before 1960, the estimated doses from Sellafield exceeded those estimated from natural radiation by about 20%, but afterwards they were much lower. Doses of this magnitude would not be expected to cause the observed ninefold increase in fatal leukaemia in the children of mothers resident in Seascale.

375. Further study of the affected young people resident in Seascale and of those born and resident in West Cumbria Health District, in which Sellafield is situated, has given potentially conflicting results. In one study [G21] comparisons were made between the characteristics of young people born in the West Cumbria Health District who developed leukaemia, Hodgkin's disease or non-Hodgkin's lymphoma while still resident in the district and the characteristics of the up to 16 times more numerous young people born in the district, resident in it and registered in the same birth registers as the affected young people, half of whom were also matched for parental residence in the same civil parish. The results gave no reason to think that the affected young people had had any more exposure to radioactive contaminants in the general environment than their controls, as evidenced by such factors as the frequency of playing on the beach or eating fish. However, they did suggest that the fathers were more likely to have been employed at Sellafield (RR = 2.82; 95% CI: 1.07-7.40) and to have been exposed to relatively large doses of external radiation, particularly in the six months prior to the conception of the children, as is shown in Table 62. Other findings (some of which were statistically significant) include two that would have been expected from previous observations — namely, an increased relative risk with a history of maternal exposure to abdominal radiography during pregnancy and with maternal age — and some evidence that several other types of paternal employment might also be associated with an increase of risk. None, however, suggested as large an increase as relatively heavy paternal irradiation, and no other factor seemed capable of accounting for the excess leukaemias. In a later study [K27] it was shown that parental exposure before the child's conception could not alone account for the Seascale cluster. The results are summarized in Table 63. They show that the cluster was not limited to young people whose fathers received substantial doses of radiation at work. The one boy with acute lymphocytic leukaemia was not born in Seascale and his father fell into the second lowest dose category in the Table (0.01-49.9 mSv), having received only 5.5 mSv. This boy and the four young people whose fathers had not been exposed would alone have caused a significant excess in Seascale (RR = 4.5; $p < 0.01$). It may also be noted that most of the paternal doses were received outside Sellafield [P37]. Another study of the influence of parental exposure examined the data already collected in the Oxford survey and found no association between leukaemia incidence or the incidence of all cancers and external exposure of fathers [S34]. In a recent commentary [D28] it is concluded that the hypothesis that preconception paternal radiation exposure is associated with leukaemia in children is not supported by known radiation genetics, or by the body of knowledge about the heritability of childhood leukaemia or by what is known about radiation and leukaemia risks in other studies. The authors conclude that the association between paternal irradiation and leukaemia is likely to be a chance finding.

376. Shortly after the Seascale cluster was discovered, another cluster was reported near Dounreay [H4]. The results are less impressive, as a highly significant excess is obtained only by concentrating on an arbitrarily defined period and a geographic area that bisects the local urban centre of Thurso (5 cases against 0.5 expected within 12.5 km of Dounreay in the period 1979-1984). Extending the time back 10 years (to 1968) and the distance to 25 kilometres, increases the expected number of cases by 2.5 and the observed number by only 1. Nevertheless, Dounreay has the only other plant in the United Kingdom where nuclear fuel is reprocessed, making the excess difficult to dismiss. The excess, however, is equally, if not more difficult, to attribute to nuclear waste from Dounreay itself and to waste arriving near Dounreay by coastal current from Sellafield, as the estimated annual radiation dose to the bone marrow from low-LET radiation from these sources is substantially less than the maximum annual dose received from fallout nationally following the atmospheric testing of nuclear weapons that took place in the early 1960s, with the estimated annual bone marrow dose from high-LET radiation being no more than the maximum received from fallout. Further studies in the Dounreay area included a case-control study [U13] and a follow-up study of two cohorts of children, one of which included those born to local mothers (birth cohort) and the other of which included those who moved to the area after birth and attended school in the area (schools cohort) [B40]. The case-control study found that the higher incidence of childhood leukaemia could not be explained by paternal occupation at Dounreay or by paternal exposure to external ionizing radiation before conception. There was, however, an association with the use of beaches within 25 km of Dounreay. The follow-up study found a non-significantly higher incidence of childhood leukaemia in the birth cohort and a significantly higher incidence in the schools cohort, suggesting that place of birth was not a more important factor than place of residence in the aetiology of childhood leukaemia in the Dounreay area. One possible explanation of this excess is that it is caused by population mixing linked to the oil industry in this otherwise rural area of northern Scotland [K17]. Another report [K2] did not find any association between the preconception exposure of fathers and leukaemia clusters in Scotland or North Cumbria.

377. The next finding to be reported was an excess of childhood leukaemia near one of three nuclear sites that are close to one another in the south of England [R18]. Exposures due to radioactive discharges from the sites are estimated to be minimal. The observations are summarized in Table 64. Made over a period of 14 years, they suggest that there might have been an excess of some 37% in children under 15 years of age living within 15 km of one of three neighbouring sites, and because the population of the area was substantial, this amounts to about 20 cases in all. The greatest excess occurred in children under five years of age living within 5 km of one or other of the

sites, of which there were 9 cases, giving an excess of 134%. The excess was not limited to leukaemia but was also observed with other types of childhood cancer, with 30 cases of cancer other than leukaemia occurring in children under 5 years of age who were resident in wards less than 10 km from the sites, compared with 19.4 expected [C30]. A further case-control study of children aged 0-4 years with leukaemia or non-Hodgkin's lymphoma established that the above-average rates of leukaemia in the area could not be accounted for by parental employment in the nuclear industry [R24].

378. Several other smaller clusters of leukaemia have been reported near other installations in the United Kingdom, and another cluster of four cases against 0.5 expected has been observed in the north ward of Egremont, a small town 7 km north of Sellafield [C11]. However, it is unclear to what extent some of these clusters represent a selection of high rates that occur by chance and a neglect of low rates. To assess the significance of these excesses it is necessary to have a detailed and uniform account of the distribution of all cases throughout the country. So far, this can be provided only for mortality in the 402 county districts in England and Wales for 1969-1978 [C7]. After adjustment for urban-rural status, population size, social class and health region and compared with all the other county districts, there was a 15% increase in mortality from leukaemia of all types ($p = 0.01$), a 21% increase in mortality from lymphoid leukaemia ($p = 0.01$) and a 25% increase in the mortality from Hodgkin's disease ($p = 0.05$) in young people under 25 years of age in county districts with more than 0.1% of their population living within 16 km of one of the 15 principal nuclear installations in the country. No excess of any other cancer, nor of all malignancies combined, nor of seven groups of malignancies in adults was found.

379. An excess of lymphoid leukaemia of about 21% in these specified districts corresponds to about eight excess deaths per year. The extent to which the risk of leukaemia is related to paternal employment in the nuclear industry is unknown. It did not, however, account for the excess found in the small town north of Sellafield, as none of the fathers of the four young people who developed leukaemia while resident in the area had been irradiated occupationally before the child's conception, despite the fact that many Sellafield workers lived in the north Egremont ward and had received a collective dose that was greater than that received by the workers who lived in Seascale. The overall results of the study do not support the idea that the excess is due to environmental contamination by radiation [C7]. No trend in relative risk was observed with increasing proximity to an installation, as measured by the trend from districts with a low proportion of their population within 16 km of the installation (RR = 1.18) through districts with 10%-65% of their population within 16 km (RR = 1.29) to districts with 66% or more of their population within 16 km (RR = 1.16). The excess relative risks between the districts around Sellafield and those around the other installations differed by a factor of 5 (RR = 1.94 and 1.20), many times more than the difference between the estimated annual doses from Sellafield received by children living in the vicinity and the doses estimated for children living near the other nuclear installations [S16].

380. Further support for the idea that the excess lymphoid leukaemia seen near nuclear installations other than Sellafield is not caused by environmental radiation comes from a second study using the same data that examined mortality around "potential" nuclear installations [C18]. Six sites that had been seriously considered for nuclear installations but where no installation was built were included, as were two further sites where nuclear power stations were built but that did not start operating until after the period for which mortality data were available. After adjustment for the same factors as in the original study, mortality patterns were very similar to those found near the existing installations. The relative risks of mortality from leukaemia of all types and from Hodgkin's disease at ages 0-24 years were 1.14 and 1.50, respectively. The total population near the 8 potential installations was considerably smaller than that near the 15 existing installations; as a result, the number of leukaemia deaths was much smaller and the increase in mortality did not reach statistical significance. However, the increase in Hodgkin's disease did ($p < 0.05$).

381. Reports of small clusters of childhood leukaemia around nuclear installations in the United Kingdom in the 1980s prompted several large-scale systematic surveys. In the United Kingdom, lymphoid leukaemia among persons under 25 years old was found to be generally increased in populations living near nuclear fuel reprocessing or weapons production facilities but not in those near plants that generated electrical energy [C19, F10]. Mortality from Hodgkin's disease at ages 0-24 years was also increased, whereas mortality from lymphoid leukaemia at ages 25-64 years was significantly reduced. Overall there was no general increase in cancer deaths in the vicinity of these nuclear facilities.

382. Leukaemia in children has also been studied in the vicinity of nuclear installations in France, the United States (studies near the Three Mile Island plant and of the country as a whole), Canada and the former Federal Republic of Germany [C26, H33, H39, H42, J4, M35, M44, M46, V10]. The results from France and the United States provide no evidence that any excesses similar to those in the United Kingdom have occurred. The negative evidence from France is impressive, as the communes near nuclear installations experienced slight (statistically non-significant) deficiencies of leukaemia in comparison with both national rates and the rates observed in control communes in the same departments, at all ages under 25

years and at ages 0-4 years, and at distances up to 10 km and 21 km from the installations [H39, V10]. The United States study [J4] examined cancer rates in counties and could, therefore, have missed very localized excesses such as that in Seascale. However, this study was otherwise large and very complete. Over 900,000 cancer deaths in 113 counties in the United States containing or adjacent to 62 nuclear facilities were compared to 1,800,000 cancer deaths in control counties with similar population and socioeconomic characteristics [J4]. Overall, and for specific groups of nuclear installations, there was no evidence that mortality for any cancer, including childhood leukaemia, was higher in counties with nuclear reactors than in control counties. For childhood leukaemia, the relative risk in the counties with nuclear reactors, compared with their controls was 1.03 after plant start-up and 1.08 before start-up. For all leukaemia, the relative risks were 0.98 after start-up and 1.02 before.

383. The first Canadian study [C26, M35] examined cancer incidence and mortality rates in Ontario in children born within a 25 km radius of a nuclear installation and compared them with the values expected on the basis of provincial rates. No significant increases were found. The second study [M44] used a case-control design. The case series consisted of 112 children aged 0-14 years who died from or who were diagnosed with leukaemia in 1950-1988 and were born to mothers who, at the time of their child's birth, lived near an operating nuclear facility in Ontario. Eight controls per case, matched on date of birth and mother's residence at the time of birth, were identified from birth certificates. Data on occupational radiation exposure were obtained by a computerized record linkage with the Canadian National Dose Registry and subsequent examination of employers' records. It was concluded that there was no association between childhood leukaemia and the occupational exposure of fathers to ionizing radiation before conception as a consequence of employment in the nuclear industry in Ontario. No association was detected for external whole-body dose, tritium dose or radon exposures. Odds ratios were close to 1.0 for all radiation dose categories and occupations except for uranium mining, which had a larger, but not statistically significant, odds ratio.

384. The study based on the registry of childhood malignancies in the former Federal Republic of Germany included 1,610 cases diagnosed at ages under 15 from 1980 to 1990. The design was based on that of the first systematic British study [F10] and included all communities at least one third of whose areas fell within 15 km of the 18 nuclear power plants and 2 research reactors that started operation between 1960 and 1988. For each nuclear installation a nearby control region was selected with similar regional structure and population density and whose cancer patients would have attended the same paediatric oncology centre. The relative risk compared with the control areas was 0.97 for all malignancies (95% CI: 0.81-1.60) and 1.06 for acute leukaemia (95% CI: 0.88-1.28) in all regions within a 15 km radius of a nuclear installation. Increased relative risks were, however, seen in the subgroup of acute leukaemia before 5 years of age, especially in areas less than 5 km from the installations (RR = 3.01; 95% CI: 1.25-10.31) and for lymphomas (RR = 1.67; 95% CI: 1.07-2.76) [M46]. However, leukaemia incidence in the control areas was unusually low compared with national rates. As in the United Kingdom, a comparable and even more pronounced increase was observed in regions where nuclear power plants had been projected.

385. None of the increases in leukaemia and non-Hodgkin's lymphoma reported in the vicinity of nuclear installations seems likely to have been caused by environmental exposure to discharged radioactive materials, nor can any of them be accounted for by paternal exposure to radiation at work before the affected young people were conceived. While Gardner et al. [G21] thought that the association they observed between parental irradiation and the subsequent development of leukaemia and non-Hodgkin's lymphoma in offspring born and living in West Cumbria suggested causation, the increases seen in a segment of the West Cumbrian population are implausible quantitatively when compared with existing experimental evidence, as is shown in the UNSCEAR 1993 Report, Annex G [U1]. No similar associations were observed in the children of the survivors of the atomic bombings in Japan [Y3, Y4, Y5] or in any of the other studies cited above, such as in Scotland. One possibility is that the observations relating to parental irradiation in West Cumbria are attributable to chance [D28]. Another possible explanation is that the excesses are due to the spread of infection resulting from the mixing of populations from urban and rural areas [K4, K14, K17].

B. PARTICIPANTS IN ATMOSPHERIC NUCLEAR WEAPONS TESTS

386. The issue of cancer incidence in participants in nuclear tests came to public attention when, at the instigation of one of the participants who had developed leukaemia, the United States Centre for Disease Control undertook a study of the mortality of men who had participated in the test Smoky in Nevada in 1957. The results showed that in 1957-1979, 10 men had developed leukaemia (including the index case that prompted the investigation) in comparison with 4.0 that would have been expected on the basis of the United States national rates over the same period of observation, a difference that was statistically significant (p = 0.008, two-sided test) [C1, C2]. The data from the Smoky test were analysed in more detail as part of a later study of participants in a series of five weapons tests [R2]. In the study it was found that while the mean recorded doses for all participants was 0.009 Sv,

leukaemia deaths among Smoky participants tended to occur among those with higher doses or those exposed in unusual ways that might have led to additional unrecorded internal doses from the ingestion or inhalation of radionuclides. No excess of leukaemia deaths (46 observed versus 52 expected) was seen in the complete analysis nor in any group other than the Smoky participants. The study also found no evidence of either a general excess of cancer or excess risks for any specific sites in the full study population. The only other significant finding occurred in the analysis of genital cancers among participants in the Redwing series of tests carried out at Bikini and Enewitok atolls in 1956 (16 cases observed compared with 8.35 expected; p = 0.02, two-sided). This increase was largely made up of prostate cancers (11 cases). It could be the kind of chance finding that must be expected when many different rates are compared in a single cohort (see paragraph 31). Prostate cancer has not been associated with exposure to high-dose-rate, low-LET radiation in the studies of survivors of the atomic bombings or studies of external medical radiation. Although significant increases of prostate cancer have been reported in employees of the United Kingdom Atomic Energy Authority [B4] who were monitored for various radionuclides including tritium, it was not found in employees of the United Kingdom Atomic Weapons Establishment [B5] nor has it been seen in radiation workers in general. Further follow-up of the Atomic Energy Authority employees [F11] revealed that the elevated mortality was limited to employees in a single plant. A case-control study found the increase was limited to men who worked in environments contaminated by tritium and other radionuclides [R25]. Whether this is a chance finding from the examination of multiple subgroups or the effect of a particular radionuclide cannot be determined from the present data.

387. The United States data have been criticized on two grounds. First, the comparison with national rates is inappropriate [C1, C2, R2]. Secondly, the comparison within the series of tests that included the test Smoky shows that mortality from all cancers increased with the size of the recorded dose [B14], whereas when all five series are combined, the difference in rates between the high- and low-dose groups is trivial. This is to be expected, as the recorded doses were much too small for any detectable effects to be seen on the basis of the recent results from the survivors of the atomic bombings in Japan. The mean recorded gamma dose was 9 mSv, and only 1.7% of the participants had doses in excess of 50 mSv. Among the Smoky participants the deaths from leukaemia tended to occur in persons with higher recorded doses: 6 of the 10 leukaemia deaths occurred in participants with at least 0.03 Sv recorded dose, compared with only 1.34 expected (p = 0.03). It is, however, notable that 4 of the 10 leukaemia deaths in the Smoky participants occurred in a group of 730 men in the task forces Warrior and Big Bang (constituting 21% of the total number of Smoky participants), who were exposed in unusual ways, including witnessing a shot from trenches within 3 km of ground zero and carrying out exercises in areas contaminated by recent previous tests. It is not impossible that these men received substantial additional unrecorded doses internally from the ingestion and inhalation of radionuclides. One of the leukaemia cases may have resulted from radiotherapy for another malignancy.

388. As a result of the findings in participants in the Smoky test and of the concern expressed by participants in the British tests, the United Kingdom Ministry of Defence commissioned an investigation of the health of participants in the programme associated with the British tests that were carried out in Australia and on islands in the Pacific between 1952 and 1958 [D3, D4]. The study included 21,000 men who were identified in the records of the Ministry of Defence as having participated in the programme and also a control group of similar size, made up of men of the same ages, same services, same proportion of officers and same date of entry to the study who did not participate in the programme but who had served in tropical or subtropical areas while the tests were being conducted. The results showed that not only were the total mortality and the mortality from all neoplasms combined almost identical in both groups but they were also substantially less than the rates expected on the basis of the corresponding mortality rates in England and Wales. In a later study of the same populations with longer follow-up [D24], statistically significant differences between the participant and control group were recorded for 4 of the 27 cancers examined. In two instances the mortality was higher in the participants (leukaemia, 1.75 times higher; bladder cancer, 2.69 times higher) and in two instances the mortality was lower in the participants (cancer of the mouth, tongue and oesophagus, 0.45 times that in the controls; lung cancer, 0.85 times that in the controls). In no case was the mortality in the participants significantly greater than would have been expected from the national mortality rates (standardized mortality ratio for leukaemia of 1.00 and for bladder cancer of 1.04), the significant excess in the participants being due in each case to an unusually low mortality in the controls. With the statistical test used, some statistically significant differences were to be expected by chance alone when 27 different diseases were examined, and it was concluded that all of the findings are likely to have been due to chance. A small, statistically non-significant excess of leukaemia in the first 25 years after participation could not be ruled out (standardized mortality ratio of 1.23, based on 20 deaths), but it could not be attributed to any known cause, as it was not concentrated in men known to have been exposed to radiation or men involved in any particular operation or employed in any particular job.

389. In a separate study of New Zealand participants in the British atmospheric nuclear weapons tests in 1957-1958, 528 men known to have participated in the

tests and a control group of 1,504 men who were in the Royal New Zealand Navy during the same period but who did not participate in the tests were followed until 1987 for both mortality and cancer incidence [P27]. Mortality from causes of death other than cancer and mortality and cancer incidence from cancers other than haematological malignancies were similar among test participants and controls. There were seven deaths from haematological malignancies among test participants (RR = 3.25; 90% CI: 1.12-9.64; p = 0.02), including four from leukaemia (RR = 5.58; 90% CI: 1.04-41.6; p = 0.03). For haematological cancers the relative risk was 1.94 (90% CI: 0.74-4.84; p = 0.10) and for leukaemia it was 5.51 (90% CI: 1.03-41.1; p = 0.03). However, as noted by the authors, one of the four leukaemia cases was diagnosed as chronic lymphocytic leukaemia, and the temporal distribution of the cases, one within 15 years of exposure and three more than 25 years after exposure, makes it unlikely that the cases resulted from radiation exposure from the tests.

390. Interpretation of these three sets of results from of participants in nuclear weapons tests is a matter of some controversy (see, e.g. [B14, J7]). The results are not clear-cut; however, taken together, they provide no evidence for a greater risk of leukaemia in exposed participants than in unexposed individuals or for a greater risk than expected from other sources, such as the atomic bombings.

C. SENSITIVE SUBGROUPS OF THE POPULATION

391. It has long been known that patients with certain hereditary diseases have an increased sensitivity to radiation. As an example, patients with ataxia-telangiectasia are at increased risk of developing certain cancers and often develop devastating necrosis of normal tissue when given therapeutic irradiation [S43]. Retinoblastoma patients with the *rb* gene are at increased risk of subsequently developing a wide variety of other tumours, and when treated with radiation they appear to have a higher excess absolute risk of second cancers than patients irradiated for the non-genetic form of the disease [D12, E4].

392. The increased radiosensitivity associated with hereditary diseases such as those mentioned above is not necessarily confined to persons in whom the genetic disease is manifest. Again, it has been suggested by way of example that ataxia-telangiectasia heterozygotes may have an increased risk of developing cancer. Swift et al. [S43] ascertained retrospective cancer incidence rates in 110 families of ataxia-telangiectasia patients in the United States. Among the adult blood relatives of the affected patients, a total of 138 cancer cases were ascertained compared with 103 expected based on cancer incidence rates for Connecticut; among spouse controls who were not consanguineous with the affected patients, 32 cancers were found compared with 40.9 expected. Using a maximum-likelihood method that accounts for each relative's prior probability of heterozygosity, it was estimated that for men who are heterozygous for ataxia-telangiectasia the relative risk of all sites and types of cancer combined was 2.3 (p = 0.014), while for women it was 3.1 (p = 0.004). Among the blood relatives of affected patients, the commonest cancer was breast cancer, with 27 cases observed compared with 20.5 expected based on Connecticut cancer incidence rates, while for spouse controls there were 3 observed compared with 7.2 expected. These data led to an estimated heterozygote relative risk for breast cancer of 6.8 (p = 0.006).

393. In a further prospective study of 161 families affected by ataxia-telangiectasia, Swift et al. [S59] compared cancer incidence and mortality in 1,599 adult blood relatives of patients with ataxia-telangiectasia and 821 of their spouses, who served as controls. This study included many of the same families as the previous study, but the periods of follow-up in the two studies did not overlap; in the later study the individuals were followed prospectively for a mean of 6.4 years. Cancer rates were significantly higher in the group of blood relatives than in the spouses, specifically in the subgroup of 294 blood relatives who were known to be heterozygous for the ataxia-telangiectasia gene. The estimated relative risk of cancer of all types among obligate heterozygotes as compared with non-carriers was 3.9 (p < 0.005) in men and 2.7 (p < 0.005) in women, and that of breast cancer in women was 3.8 (no significance level given).

394. In a case-control substudy, documented occupational and fluoroscopic diagnostic exposures to radiation in the 19 female blood relatives in whom breast cancer was diagnosed during the period of prospective observation were compared with exposures for 57 matched blood relatives who did not have breast cancer. Among the blood relatives, women with breast cancer were more likely than controls without cancer to have been exposed to selected sources of ionizing radiation (RR = 5.8; p = 0.005), suggesting that at least part of the observed excess of breast cancer may be due to an increased susceptibility to ionizing radiation. These findings, however, have not been confirmed and were the subject of considerable debate (see [S59]). It appears that fluoroscopic procedures that contributed negligible breast exposures, such as barium enemas, were considered meaningful; the age distributions of breast cancer cases and their controls were significantly different, with the controls being much younger and less likely to develop breast cancer; and when general population rates were used to compute expected number of breast cancers among the ataxia-telangiectasia heterozygous women, the excess risk and the level of significance were no longer striking.

395. The studies of Swift et al. are the largest to date of cancer among the relatives of ataxia-telangiectasia

patients. A smaller study of the families of 67 patients in the United Kingdom found that three parents of affected patients had died of cancer [P2]. Two of the deaths were from breast cancer compared with only 0.17 expected based on mortality rates in England and Wales, and the excess was statistically significant ($p < 0.05$). However, no excess mortality was found among the grandparents of the affected patients. The interpretation of these studies must remain at present uncertain. More studies of the relatives of affected patients comparable in size to that of Swift et al. and a method of identifying ataxia-telangiectasia heterozygotes in the general population would be very useful to clarify the situation. The frequency of ataxia-telangiectasia heterozygotes in the population is estimated to be between 1% and 8% [S43]. If the finding of an increased tendency for them to develop cancer is confirmed it would be of major significance, whether or not it is caused by an increased radiosensitivity. For further discussion of this topic, see Annex E in the UNSCEAR 1993 Report [U1].

396. There are a number of genetic conditions for which radiation has enhanced the development of secondary cancers. Children with familiar retinoblastoma have a deletion in chromosome 13 that predisposes to osteosarcoma, and radiation appears to cause a second mutation in an osteoblast that leads to a high rate of osteosarcoma development. Further, these children are at a significantly increased risk for developing cancers of the connective tissue and brain and skin melanoma, because radiotherapy appears to enhance the inborn susceptibility to cancer development [E4]. Children treated for medulloblastoma who have basal cell nevus syndrome develop multiple basal cell carcinomas in irradiated skin at an exceptionally high rate [S71]. Patients with a rare familial cancer syndrome called the Li Fraumeni syndrome are at increased risk for cancer development largely because of an inherited mutation in the p53 tumour suppression gene. Radiation appears to enhance the development of secondary tumours among patients with this syndrome [L24].

397. In a new analysis of the breast cancer risk among atomic bomb survivors, a high excess relative risk associated with radiation dose was apparent for early-onset breast cancer, defined as cancer diagnosed before age 35 years [L20]. These early-onset cases occurred almost exclusively among women exposed before the age of 20 years. The authors suggested that this may be the consequence of a small radiation-susceptible subgroup of exposed women. The authors cautioned that the interpretation of these data needed to be confirmed in studies from other irradiated populations or in experimental investigations.

CONCLUSIONS

398. In recent years a substantial number of new epidemiological studies of the carcinogenic effects in man of exposure to external low-LET radiation at high dose rate have been undertaken. Many of these studies are of a high quality, and they contribute substantially to the knowledge of the consequences of human exposure to this type of radiation under these conditions. In addition, many of the studies that have been under way for several years have been extended. The new information is broadening the base of data on which to evaluate the association between radiation exposure and cancer incidence or mortality.

399. The Committee acknowledges that for the time being quantitative evaluations must continue to be based primarily on the findings of the life span study cohort of survivors of the atomic bombings. However, as the material in this Annex illustrates, some other studies are able to provide useful risk estimates for a substantial number of specific sites. It would be very useful if more parallel analyses of multiple studies could be made, as was done for the radon studies or the studies of cancer of the breast. The Committee recognizes that a limitation of the life span study is its inability to address directly the effects of low-dose-rate exposures. Other studies are needed to provide information on the appropriateness of risk estimates derived from high-dose-rate exposures.

400. In the UNSCEAR 1988 Report [U2], estimates of the risk of exposure-induced death and loss of life expectancy following exposure to 1 Gy of low-LET radiation delivered at high dose rate were made on the basis of estimates of the excess absolute and excess relative risks, averaged for age at exposure and sex, observed in the life span study cohort of survivors of the atomic bombings in Japan using data through 1985. These estimates were calculated on the basis of unweighted organ absorbed dose. Individuals whose estimated kerma doses were greater than 4 Gy were included in the calculation, and a linear dose-response relation was assumed both for leukaemia and for other sites of cancer. For leukaemia the risks were assumed to apply from the second year following exposure up to 40 years after exposure, while for other types of cancers they were assumed to apply from 10 years after exposure for the rest of life. It was assumed that there was no additional variation in the risks with time since exposure and that risks for men and women were the same. While age-specific coefficients were used to estimate the total risk of all cancer, for specific sites of cancer the only estimates that were presented were those based on risk coefficients that had been averaged over all ages at exposure.

401. Data on the incidence of cancer in the life span study have become available for the first time and may be compared with the mortality data, which have been extended through 1987 [R23]. Estimates of the fatal risks associated with exposure to high-dose-rate, low-LET radiation in this Annex have been based on this extended life span study mortality data. The analyses of the more recent data differ from those used to describe the risks in the earlier life span study data. In particular, excess risks in the life span study were described using general models that allowed for differences in age at exposure and sex in the excess risks. Survivors with shielded kerma estimates in excess of 4 Gy were excluded, and risks were computed in terms of weighted organ doses in which the neutron dose was given a weight of 10 and the gamma dose, a weight of 1. Because the time-constant additive (i.e. constant absolute risk) model no longer fits the life span study data, risk projections based on the constant absolute risk model are not included in this Annex. The Committee acknowledges that risk projections must take into account important variations in radiation-induced risks with sex, age at exposure and time since exposure and that more efforts are needed to distinguish significant aspects of the different patterns of risk seen for specific sites from the random variability inherent in comparisons of site-specific risks.

402. The life span study data for solid tumours from 1950 to 1987 are consistent with linearity between 0.2 Sv and 4 Sv (weighted dose), but they provide little direct evidence about the shape of the dose response at lower doses. Risk estimates are presented in this Annex for doses of both 1 Sv and 0.2 Sv and for a Japanese population with the demographic characteristics that prevailed in about 1950. The results for 0.2 Sv show a slightly greater risk proportionally than for 1.0 Sv because the risk of exposure-induced death is non-linear in dose. A linear dose-response model does not fit the life span study leukaemia data, but a linear-quadratic model does. The data from the life span study (1950-1987) lead to a lifetime risk estimate for solid tumours of 10.9% at 1 Sv; the estimate for 0.2 Sv is 2.4% (Table 31). For leukaemia the corresponding numbers are 1.1% at 1 Sv and 0.14% at 0.2 Sv. For all cancers, solid tumours plus leukaemia, at 1 Sv the risk is 12.0% and at 0.2 Sv it is 2.5%. The corresponding estimate of risk for all cancers in the UNSCEAR 1988 Report [U2] using a multiplicative projection was 10.7% at 1 Gy (organ-absorbed dose). Projections beyond the follow-up period for models in which risks were allowed to decrease at longer projection times lead to lifetime risk estimates 20%-40% lower than 12.0% (Table 31).

403. No attempt has been made in the above risk estimates to address the issue of dose rate for either solid tumours or leukaemia. The estimates are presented without any adjustment for such effects. The application of a small dose and dose-rate effectiveness factor was recommended in the UNSCEAR 1993 Report [U1], Annex F, "Influence of dose and dose rate on stochastic effects of radiation". If a factor such as 2 (as was used by ICRP [I10]) is applied, the above estimates for solid tumours for both 0.2 Sv and 1 Sv would be halved. Those for leukaemia would also be approximately halved for a chronic exposure of 1 Sv as compared with an acute exposure, while those at 0.2 Sv, having already been determined by a linear-quadratic response, would stay the same.

404. The publication of the life span study incidence data, available for 1958-1987, is an important addition to the body of knowledge on the late effects of radiation. The strengths of the incidence data include the high quality of the diagnostic information and the larger number of cases, especially for sites such as breast, thyroid and skin, with lower lethality. Their limitations include the absence of data on solid cancer incidence for the first 13 years after exposure in the Japanese population and the need to adjust for migration. It is noteworthy that in the incidence analyses two cancers, that of the oesophagus and multiple myeloma, for which statistically significant risks had been seen in the mortality data, did not exhibit statistically significant risks.

405. Additional information has become available on the effects of prenatal exposure to radiation since the subject was last reviewed in depth in the UNSCEAR 1986 Report [U3]. The evidence regarding the causal nature of the increase in childhood cancer following exposure of the mother's abdomen to diagnostic x rays in pregnancy is still equivocal. The best estimate of the excess absolute risk of developing cancer before age 15 years following prenatal exposure to x rays at high dose rate is about 5 10^{-2} Gy^{-1}, similar to the lifetime risk for adults. This estimate does not include any cancers induced by prenatal exposure to radiation that may arise in the age group of 15 years and above. Data from the survivors *in utero* at the time of the atomic bombings in Japan initially revealed an increase in cancers in the ages of 15-39 years, but no new cancers were found in the exposed group in the next four years of follow-up [Y1]. The excess is large but not, so far, statistically significant.

406. In the UNSCEAR 1988 Report [U2] it was noted that significant excess cancer mortality had been seen for the first time for some cancers at doses between 0.2 and 0.5 Gy. There is also some limited evidence that directly points towards the carcinogenicity of doses in the <0.2 Gy range, although each of these studies has weaknesses. The studies include childhood cancer among those exposed to doses of about 0.01 Gy *in utero*, for which there is an equivocal but possibly significant increase; both incidence and mortality for all cancers other than leukaemia in the life span study cohort, for which, although not significant, there are smooth increases in relative risk with increasing dose in dose categories in the <0.2 Gy range; and thyroid cancer in the Israeli tinea capitis study.

407. Since the publication of the UNSCEAR 1988 Report [U2], a substantial amount of new information has become available on workers at nuclear plants in the United Kingdom and in North America. The studies provide some quantitative information on the effects of protracted low-dose exposures to low-LET radiation. Although at present the confidence limits are wide, the significant value for leukaemia in the study in the United Kingdom, 0.8% Sv^{-1}, is similar to values derived from the life span study. Also similar to the life span study results is a non-significant value for all solid tumours of about 10% Sv^{-1}. Additional data acquired through extensions of the follow-up and the inclusion of extra cohorts should increase the information available from this study in future years. On the other hand, studies of workers in the United States, which are of somewhat lower power (smaller population), have found no evidence of an association between exposure and leukaemia or all cancer.

408. Data have recently become available on several populations in the southern Urals, including workers with substantial exposures in a nuclear plant and populations exposed as a result of environmental releases and accidents. In the study of workers, only limited information is available so far on the methods of data collection, and detailed internal analyses are only in the process of being carried out. The study could be very informative, however, since the doses were substantial and the workers had individual dose measurements, good medical supervision and a good control cohort. It is hoped that the analysis of these important worker experiences will be completed and published. For the studies of populations exposed to environmental releases, it would be desirable to improve the quality of the dose estimates and the follow-up data. Initial findings based on the follow-up of the Techa River cohort suggest elevated risks of leukaemia and solid cancers. While a better understanding of the dosimetry and the follow-up is needed, results are broadly similar to those derived from atomic bomb survivors for leukaemia.

409. No studies have yet provided evidence that an increase in thyroid cancer can be clearly attributed to ^{131}I. However, the relative importance of the various factors that might affect these results is not clear. These factors include the low dose rates involved in ^{131}I exposures, the beta-ray distribution in the gland, the rather limited follow-up in some studies and biological factors such as the cell-killing, which usually follows treatment with ^{131}I. In addition, most of the evidence concerns those exposed in adult life, and by analogy with recent results for thyroid cancer following external low-LET radiation, the possibility of substantial risks following childhood exposure to ^{131}I remains. Medical exposures to ^{131}I have been associated with increased risks for some other sites of cancer, but the available data do not provide strong evidence of a causal relationship. It should be noted that leukaemia has not been observed in many thousands of adult patients treated with ^{131}I. Further studies of the effects of internal exposures to low-LET radiation from ^{131}I and other radionuclides would be desirable, especially in children, for whom there is little information.

410. Studies of patients who have been injected with or workers who have ingested one or another of the various isotopes of radium provide conclusive evidence of the ability of alpha radiation in the skeleton to cause bone tumours. The excess absolute risk does not appear to depend strongly on sex or age at exposure and amounts to about 5 10^{-4} Sv^{-1} for lifetime. The minimum induction period appears to be short, no more than about 3 years, and the studies of patients injected with short-lived ^{224}Ra provide clear evidence that the period of expression of the radiation-induced tumours lasts no more than about 30 years. Further analyses of the data on the risk of bone tumours in persons exposed to the various isotopes of radium in terms of the risk per person-year at risk as related to dose, treatment period, time since exposure and other factors would be helpful in enabling comparisons with studies of populations exposed to other types of radiation. In addition to an increase in bone tumours, excesses of cancers of the breast, liver, paranasal sinuses and mastoid air cells, and also of multiple myeloma, have been reported in patients injected with or workers contaminated with ^{226}Ra. Further study of these increases and the associated dosimetry, including, where appropriate, the role of external gamma-radiation would be informative.

411. There is strong evidence that injected Thorotrast is a cause of liver cancer. The minimum induction period is about 10 years, and relative risks are higher among those injected at younger ages, although absolute excess risks do not depend strongly on age at exposure. Risks are similar in males and females, and the cumulative risk of malignant liver tumours, although subject to many uncertainties owing to possible confounding or enhancing factors, is estimated to be about 300 10^{-4} Gy^{-1}, or 15 10^{-4} Sv^{-1}, quite similar to results from low-LET exposure. There is also definite evidence that injected Thorotrast is a cause of leukaemia, especially myeloid leukaemia and erythroleukaemia, and it seems likely that it is also a cause of a variety of other cancers. Perhaps because the dose is highly localized at sites of microscopic dimensions, risk estimates for leukaemia based on patients exposed to Thorotrast are much less than would be expected based on low-LET risks (i.e. an RBE of 1 fits better than 20). Nevertheless, additional analyses of the risks of liver cancer and leukaemia in the German, Japanese and Portuguese studies based on the concept of person-years at risk would be helpful, as would calculation of the relative risk and corresponding significance levels for other types of cancer.

412. Additional studies and many more cases of lung cancer attributable to radon in mines have become available for analysis since the publication of the report of

the BEIR IV Committee [C6]. There is some recent evidence that the excess risk of lung cancer following occupational exposure to radon may be greater at low exposure rates than at high exposure rates, but not all studies show this. Studies of the joint effect of exposure to external, predominantly low-LET radiation from the atomic bombs and to tobacco smoke on the subsequent relative risk of lung cancer has shown that the two factors combine to increase the relative risk in a fashion that is more than additive but less than multiplicative. There is now strong suggestive evidence that the joint effect of exposure to radon and tobacco smoke on the subsequent relative risk of lung cancer may also be less than multiplicative. Although it may eventually need to be modified to take these two factors into account, the empirical model proposed by the BEIR IV Committee [C6] adequately summarizes the effects of occupational exposure to radon. Arsenic is a newly recognized confounding factor in some radon exposure circumstances, raising the question whether exposures to other substances in mine air, such as silica or diesel exhaust, might also modify risk. Studies of cancers other than those of the lung following radon exposure provide little evidence of a positive association.

413. At present there is little direct evidence on the risks of lung cancer resulting from residential exposure to radon. Although a substantial number of geographical studies have been carried out, difficulties in their interpretation render them unsuitable for use in risk estimation. Until results become available of case-control and cohort studies that are currently under way, estimates of the risks of residential exposure to radon must therefore continue to be based on the results of studies of the effects of occupational exposures.

414. Following reports of a leukaemia cluster near the Sellafield reprocessing plant in the United Kingdom, several other clusters were found in the United Kingdom. The subsequent study of the potential radiation doses that might have been received from radionuclide releases at or near the sites suggests that the clusters arose either by chance or for reasons other than radiation exposure. A tentative explanation based on an association of childhood leukaemia and paternal exposure has largely been discounted following extensive investigations of the Sellafield area and elsewhere and because there is no sound genetic basis for this effect.

415. In future research, several steps could be taken to facilitate the comparison of risk estimates from different epidemiological studies. First, risk estimates per unit dose should be published. Admittedly this is difficult for many studies because estimates of dose are unavailable or of poor quality. Nonetheless, crude estimates of dose for the exposed population, along with clear statements of the limitations of those estimates, are essential for a comparison of risk estimates from different studies. When estimates of doses to individuals or groups are used in an analysis, it is useful to describe their uncertainties and to attempt to evaluate the effect of these on risk estimates. Pierce et al. [P31] provide methods for this. Even if individual doses are not available, investigators should be encouraged to use modern regression and modelling methods [B28, B29, P32, V12] in their analyses of effect modification and temporal patterns of risk. Such methods are especially useful as an alternative to subset analyses.

416. A willingness on the part of investigators to share data that are more comprehensive than the data usually offered in most publications would be another step to facilitate comparative analyses of available data. For example, data on survivors of the atomic bombings in Japan are available from the Radiation Effects Research Foundation and can be used in comparative analyses of other radiation-exposed populations. Other groups should be encouraged to make their data available in a like manner so as to encourage parallel analyses. The Committee also notes the need for more careful, formal examination of the evidence for similarities or differences in risk between cancer types. The joint analysis methods suggested by Pierce and Preston [P20] should be useful in such investigations. There is, in summary, much scope for enhancing the value of epidemiological studies, making them better able to contribute to the quantification of risks from radiation exposures.

Table 1
Examples of high and low cancer rates in various populations [a]
[P10]

Site of cancer	Sex	High cancer incidence: Population	Rate	Low cancer incidence: Population	Rate
Nasopharynx	Males	Hong Kong Singapore (Chinese) United States, Hawaii (Chinese)	28.5 18.1 14.1	German Democratic Republic United Kingdom, Oxford Finland	0.3 0.3 0.3
	Females	Hong Kong Singapore (Chinese) United States, Los Angeles (Chinese)	11.2 7.4 3.0	United Kingdom, Birmingham Finland German Democratic Republic	0.1 0.1 0.2
Oesophagus	Males	France, Calvados Brazil, Porto Alegre Bermuda (black)	26.5 25.9 24.9	Algeria, Setif Israel Italy, Latina	1.4 1.4 1.8
	Females	India, Bangalore China, Tianjin Brazil, Porto Alegre	8.8 8.0 6.9	Romania, Cluj Spain, Zaragoza Poland, Warsaw (rural)	0.4 0.4 0.4
Stomach	Males	Japan, Yamagata Portugal, Villa Nova de Gaia Costa Rica	93.3 47.8 46.9	India, Ahmedabad Gambia United States, Los Angeles (Filipino)	2.1 3.9 4.0
	Females	Japan, Yamagata United States, Los Angeles (Korean) Ecuador, Quito	42.9 22.9 22.7	Gambia India, Ahmedabad Kuwait (Kuwaitis)	1.5 1.5 2.0
Liver	Males	Thailand, Khon Kach China, Qidong Mali, Bamako	90.0 89.9 47.9	Netherlands, Maastricht Ireland Paraguay, Asunciòn	0.8 1.1 1.1
	Females	Thailand, Khon Kach China, Qidong Mali, Bamako	38.3 24.5 21.4	France, Isère Netherlands, Eindhoven United Kingdom, Birmingham	0.4 0.4 0.6
Lung and bronchus	Males	New Zealand (Maori) United States, New Orleans (black) United Kingdom, Scotland	119.1 115.9 88.1	Mali, Bamako Ecuador, Quito India, Madras	4.8 8.3 8.5
	Females	Canada, Northwest Territories and Yukon United States, Hawaii (Hawaiian) China, Tianjin	51.8 39.5 33.2	India, Madras Algeria, Setif Spain, Granada	1.4 1.7 2.5
Breast	Females	United States, San Francisco Bay Area (white) United States, Hawaii (Hawaiian) Brazil, Porto Alegre	104.2 100.2 78.5	Gambia Algeria, Setif China, Qidong	3.4 6.4 9.5
Cervix	Females	Peru, Trujillo Brazil, Goiania India, Madras	54.6 48.9 47.2	Israel United States, Hawaii (Japanese) China, Qidong	2.6 3.6 3.7

[a] Numbers given are age-standardized (world) annual incidence rates per 100,000 population.

Table 2
Epidemiological studies of the effects of exposures to low-LET radiation

Study	*Type of study*	*Population studied*		*Follow-up (years)*	*Total person years* [a]	*Type of exposure*	*Type of dosimetry*	*Cancers studied* [b]
		Characteristics	*National origin*					
HIGH-DOSE-RATE EXPOSURES								
Exposure to atomic bombings								
Life span study [S7, R23]	Mortality	39,593 exposed persons 46,716 unexposed persons 55.5% females Age: 0->90 (29.0) [c]	Japan	47 (maximum)	2,185,335 (28.8) [d]	Gamma and neutron radiation from nuclear explosions	Individual estimates derived from detailed shielding histories	Leukaemia,* lymphoma, myeloma, oral cavity, oesophagus,* stomach,* colon,* rectum, liver,* gall-bladder, pancreas, lung,* female breast,* uterus, ovary,* prostate, bladder,* central nervous system
Life span study [T15, P33]	Incidence	37,270 exposed persons 42,702 unexposed persons 55.5% females Age: 0->90 (26.8)	Japan	42 (maximum)	1,950,567 (24.4) [e]	Gamma and neutron radiation from nuclear explosions	Individual estimates derived from detailed shielding histories	Leukaemia,* non-Hodgkins lymphoma,* myeloma, oral cavity, salivary gland,* oesophagus, stomach,* colon,* rectum, liver,* gall-bladder, pancreas, lung,* female breast,* non-melanoma skin,* uterus, ovary,* prostate, bladder,* central nervous system, thyroid*
Treatment of malignant disease								
Cervical cancer [B11]	Incidence	82,616 exposed women 99,424 unexposed women Age: <30->70 (26.8)	Canada, Denmark Finland, Norway, Sweden, United Kingdom, United States, Yugoslavia	0->30	1,278,950 (7.0)	Radiotherapy, including external beam and intra-cavity application and experimental reconstruction	Data on typical range of estimates for specific organs and phantom measurements	Oral cavity, salivary gland, oesophagus,* stomach, small intestine,* colon, rectum,* liver, gall-bladder, pancreas,* lung, breast, uterus, other genital,* kidney, bladder, melanoma, other skin tumours, bone, connective tissue, leukaemia (non-CLL),* myeloma, lymphoma
Cervical cancer [B21]	Case-control 4,188 cases 6,880 controls	10,286 exposed women 782 unexposed women Age: <30->70 (26.8)	Austria, Canada, Czechoslovakia, Denmark, Finland, Germany, Iceland, Italy, Norway, Sweden	0->30 (7.0 years per case)	n.a.	Radiotherapy, including external beam and intra-cavity application and experimental reconstruction	Individual doses from therapy records	Stomach,* pancreas, small intestine, colon, rectum,* breast, uterine corpus,* vagina,* ovary, vulva, bladder,* bone,* connective tissue, leukaemia (non-CLL),* myeloma, lymphoma, thyroid
Breast cancer [B18]	Case-control 655 cases 1,189 controls from a cohort of 41,109 women	449 exposed women 1,395 unexposed women Age: <45->60 (51)	United States	7-55 (~13 years per case)	n.a.	Radiotherapy	Individual doses from therapy records and experimental measurements	Contralateral breast among women less than 45 years old at exposure,* contralateral breast in older women

Table 2 (continued)

Study	*Type of study*	*Population studied*		*Follow-up (years)*	*Total person years* [a]	*Type of exposure*	*Type of dosimetry*	*Cancers studied* [b]
		Characteristics	*National origin*					
Breast cancer [S60]	Case-control 529 cases 529 controls from a cohort of 56,540 women	157 exposed women 901 unexposed women Age: <45->60 (51)	Denmark	12-47 (~16 years per case)	n.a.	Radiotherapy	Individual doses from therapy records and experimental measurements	Contralateral breast
Breast cancer [C24]	Case-control 90 cases 264 controls from a cohort of 82,700 women	110 exposed women 244 unexposed women Age: <50->70 (61)	United States	<12 (~5 years per case)	n.a.	Adjuvant radiotherapy	Individual doses from therapy records and experimental measurements	Acute non-lymphocytic leukaemia and myelodysplasia syndrome*
Childhood cancers [T5, T6, T7]	Case-control within 9,170-member cohort	112 exposed persons [f] 388 unexposed persons 45% females Age: 0-18 (7)	Canada, France, Netherlands, Italy, United Kingdom, United States	5-48	50,609 (5.5)	Adjuvant radiotherapy	Individual doses from therapy records and experimental measurements	Thyroid,* leukaemia, bone sarcomas*
Treatment of benign disease								
Childhood skin haemangioma [F12, F15]	Case-control within 14,467-member cohort	144 exposed persons [g] 314 unexposed persons 67% females Age: 0-1.5 (0.6)	Sweden	25-64	n.a.	Radiotherapy	Average doses based on experimental measurements with age adjustment	Thyroid,* brain,* bone and soft tissue, breast
Ankylosing spondylitis [D6]	Mortality	14,106 exposed persons [h] 16.5% females Age: 20-60	United Kingdom	<3->50	183,749 (13.0)	X-ray therapy	Average doses based on information from a 1 in 15 sample	Leukaemia,* other neoplasms (except colon)*
Tinea capitis [M20, R5, R10, R15]	Incidence / mortality	10,834 exposed persons 16,226 unexposed persons 50% females Age: <1-15 (7.1)	Israel	26-38	686,210 (25.3) [d]	X-ray-induced epilation	Individual doses from phantom measurements based on institution and age	Incidence: thyroid,* skin,* brain,* salivary gland,* breast Mortality: head and neck,* leukaemia
New York tinea capitis [A1, S8, S61]	Incidence	2,226 exposed persons 1,387 unexposed persons 16.1% females Age: 1-19 (7.7)	United States	20-39	98,881 (25.4)	X-ray-induced epilation	Representative doses based on standard treatment	Thyroid,* skin,* brain, leukaemia, salivary gland
New York acute post-partum mastitis [S9]	Incidence	571 exposed women 993 unexposed women Age: 14->40 (27.8)	United States	20-35	38,784 (25.1)	X-ray therapy	Individual doses from therapy records	Breast*

Table 2 (continued)

Study	*Type of study*	*Population studied*		*Follow-up (years)*	*Total person years* [a]	*Type of exposure*	*Type of dosimetry*	*Cancers studied* [b]
		Characteristics	*National origin*					
Rochester thymic irradiation [H22, S10]	Incidence	2,652 exposed persons 4,823 unexposed persons 42% females Age: 0-1	United States	23->50	220,777 (29.5)	X-ray therapy	Individual doses from therapy records	Thyroid,* breast,* skin
Swedish benign breast disease [M8]	Incidence	1,216 exposed women 1,874 unexposed women Age: 10->85	Sweden	5-60	56,900 (18)	X-ray therapy	Individual doses from therapy records and phantom measurements	Breast*
Metropathia haemorrhagica [D23]	Mortality	2,067 exposed women Age: 35-60	United Kingdom	5->30	53,144	X-ray therapy	Individual doses from therapy records and phantom measurements	Pelvic sites,* leukaemia, multiple myeloma,* lymphoma, all other sites
Benign gynaecological disease [I7, I8, I12]	Mortality	4,153 exposed women Age at exposure: 13-88 (46.6)	United States	0-60	109,910 (26.5)	Intra-uterine ^{226}Ra	Individual doses from therapy records and phantom measurements	Leukaemia,* other haematolymphopoietic cancers, uterus,* bladder,* rectum,* other genital,* colon, bone (in pelvis), liver and gall-bladder, stomach, kidney, pancreas*
Peptic ulcer [G11]	Mortality	1,831 exposed persons 1,778 unexposed persons 21.2% females Mean age: 49	United States	20-48	77,757 (21.5)	X-ray therapy	Individual doses from therapy records and experimental measurements	Stomach,* colon, pancreas,* lung,* leukaemia,* female breast,* oesophagus, rectum, liver, larynx, bone and connective tissue, bladder, kidney, brain, thyroid, non-Hodgkin's lymphoma,* myeloma
Diagnostic examinations								
Massachusetts TB fluoroscopy [B31]	Incidence	2,367 exposed women 2,427 unexposed women Age: 12-30 (26)	United States	0->50	54,609 (11.4)	Multiple chest fluoroscopies	Individual exposures from medical records and doses from phantom measurements and computer simulations	Breast*
Massachusetts TB fluoroscopy [D14]	Mortality	6,285 exposed persons 7,100 unexposed persons 49% females Age: 12-30 (26)	United States	0->50	331,206 (24.7)	Multiple chest fluoroscopies	Individual exposures from medical records and doses from phantom measurements and computer simulations	Breast,* oesophagus,* lung, leukaemia
Canadian TB fluoroscopy [M19]	Mortality	18,940 exposed women 12,800 unexposed women	Canada	28-50	773,400 (24.4)	Multiple chest fluoroscopies	Individual exposures from medical records and doses from phantom measurements	Breast*

Table 2 (continued)

Study	*Type of study*	*Population studied*		*Follow-up (years)*	*Total person years* [a]	*Type of exposure*	*Type of dosimetry*	*Cancers studied* [b]
		Characteristics	*National origin*					
Scoliosis [H24]	Incidence	973 exposed women Age: <5-20 (12.3)	United States	3->30	21,691 (25.3)	Diagnostic x rays	Average dose based on number of treatments and estimated doses from published literature	Breast*
LOW-DOSE OR LOW-DOSE-RATE EXPOSURES								
Prenatal exposures								
Oxford survey of childhood cancers [M22, B22]	Case-control 14,491 cases 14,491 controls	3,797 exposed persons 25,185 unexposed persons 56% females Exposure: *in utero*	United Kingdom	20 (maximum)	n.a.	Maternal x rays during pregnancy	Number of exposures with a model for dose per exposure [B23]	Leukaemia,* lymphoma,* Wilms' tumour,* tumours of the central nervous system,* neuroblastoma,* bone, other solid tumours,* all solid tumours*
Childhood cancers [M9]	Case-control 1,342 cases 14,294 controls	1,506 exposed persons 14,130 unexposed persons 49.2% females Exposure: *in utero*	United States	20 (maximum)	n.a.	Maternal x rays during pregnancy	Number of exposures	Leukaemia,* solid tumours
Survivors of atomic bombings [Y3]	Cohort	816 exposed persons 814 unexposed persons 53.1% females Exposure: *in utero*	Japan	39	50,811 (31.2) [d]	Maternal exposure to gamma and neutron radiation at high dose rate	Mother's estimated uterus dose	All cancer*
Occupational exposures								
United Kingdom workers registry [K20, K21]	Cohort	95,217 exposed persons 8.1% females	United Kingdom	4-40	1,218,000 (12.8)	Exposure in nuclear power plant and weapons production	Recorded exposures to external radiation	Leukaemia, all cancer
Sellafield [S14]	Cohort	14,000 subjects 19% females	United Kingdom	1-28 (22)	303,547	Fuel processing and reactor operation	Recorded exposures to external radiation	Leukaemia, all cancer
United Kingdom Atomic Energy Authority [F11]	Cohort	21,545 subjects 8% females	United Kingdom	1-42 (23)	473,990	Nuclear and reactor research and fuel processing	Recorded exposures to external radiation	Leukaemia, all cancer
United Kingdom Atomic Weapons Establishment [B5]	Cohort	9,389 subjects 9% females	United Kingdom	1-37 (18.3)	172,000	Weapons research	Recorded exposures to external radiation	Leukaemia, all cancer

Table 2 (continued)

Study	Type of study	Population studied		Follow-up (years)	Total person years [a]	Type of exposure	Type of dosimetry	Cancers studied [b]
		Characteristics	National origin					
Atomic Energy of Canada Ltd. [G3, H45]	Cohort	8,977 subjects 0% females	Canada	1-29	157,101	Nuclear and reactor research and related technologies	Recorded exposures to external radiation	Leukaemia, all cancer
Hanford [G16, G17]	Cohort	23,704 exposed men 0% females	United States (whites)	1-37	492,326 (20.8)	Exposures in nuclear fuel cycle and research	Recorded exposures to external radiation	Leukaemia, all cancer
Oak Ridge National Laboratory [G9, G17, W21]	Cohort	6,332 exposed men 0% females	United States (whites)	1-34	130,428 (20.6)	Exposures in nuclear fuel cycle and research	Recorded exposures to external radiation	Leukaemia, all cancer
Rocky Flats [G17, W7]	Cohort	5,897 exposed men 0% females	United States (whites)	1-28	82,721 (14.0)	Exposures in nuclear fuel cycle and research	Recorded exposures to external radiation	Leukaemia, all cancer
Chelyabinsk [B27]	Cohort	5,086 exposed men 0% females	Former USSR	35-40	167,790	Exposures in nuclear fuel cycle and research	Recorded exposures to external radiation	Leukaemia, all cancer
Environmental exposures								
Techa River [K18]	Incidence/ mortality	28,000 exposed persons 394,000 unexposed persons [i] 56% females 0-90 years	Former USSR (ethnic Russians and Tartar/Bashkirs	30	422,000 (15.7) [j]	Internal and external exposures to radioactive waste discharged by nuclear weapons production plant	Rough estimates of average doses based on preliminary dose reconstruction efforts and recent measurements of internal doses	Leukaemia,* other cancers
China background radiation, thyroid [W17]	Prevalence	1,001 high background 1,005 low background 100% females Age: 50-65	China	> 50	n.a.	Higher levels of natural background radiation	TLD, area and personal monitors	Thyroid (nodules)
International Chernobyl project [I3, M21]	Prevalence	853 contaminated areas 813 uncontaminated areas Age: fetal to >50	Former USSR	n.a.	n.a.	Releases and environmental contamination from power plant	Estimates of average dose by village	Thyroid (nodules)
Radionuclides exposures: iodine								
Swedish therapeutic ^{131}I [H12, H27]	Incidence	10,552 exposed persons 82% females Age: 13-70	Sweden	8-33	139,018 (13.6)	Treatment of hyperthyroidism	Average administered activity (multiple treatments)	Stomach,* kidney,* brain,* all other sites

Table 2 (continued)

Study	*Type of study*	*Population studied*		*Follow-up (years)*	*Total person years* [a]	*Type of exposure*	*Type of dosimetry*	*Cancers studied* [b]
		Characteristics	*National origin*					
United States thyrotoxicosis [D9, S2]	Incidence	21,740 exposed persons 12,148 unexposed persons 79% females Age: <10-80	United States	0-30	271,000 (8)	Treatment of hyperthyroidism	None	Leukaemia, thyroid
Marshall Islands fallout [H28, R21]	Prevalence	2,273 exposed persons 55% females Age: 5->60	Marshall Islands	29-31	n.a.	Short-lived radionuclides from nuclear explosion	Estimated average dose; distance was also used as surrogate	Thyroid
Utah ^{131}I fallout [R12, S48, S49]	Case-control	1,177 cases 5,330 controls	United States	up to 23	n.a.	Fallout from nuclear weapons tests	Based on residence histories and fallout deposition records	Leukaemia

[a] Mean per person in parentheses.
[b] An asterisk denotes sites for which statistically significant excesses are reported in the exposed group (cohort studies) or for which a higher proportion of the cases were exposed to radiation (case-control studies).
[c] Age at exposure, mean in parentheses.
[d] Excludes first five years.
[e] Excludes first 13 years.
[f] The distribution of patients is as follows: thyroid: 23 and 89; leukaemia: 25 and 90; bone sarcomas: 64 and 209 for exposed and unexposed persons, respectively.
[g] The distribution of patients is as follows: thyroid: 24 and 43; brain: 31 and 51; bone: 9 and 30; breast: 80 and 190 for exposed and unexposed persons, respectively.
[h] In the ankylosing spondylitis sample, 53% (7,431 persons) were censored 18 months after a second course of treatment. Persons in this group have an average of 3.5 years of follow-up. For the remaining 47% (6,675 persons), the average follow-up is 23.6 years.
[i] The control population in this study consists of the remaining population of unexposed areas near the contaminated zone. It is not a fixed cohort.
[j] Exposed group only.

Table 3
Strengths and limitations of major epidemiological studies of carcinogenic effects of radiation

Study	*Strengths*	*Weaknesses*
Exposures to atomic bombings		
Life span study [P33, R23, S7, T15]	Large population of all ages and both sexes not selected because of disease or occupation; Wide range of doses; Comprehensive individual dosimetry; Survivors followed prospectively for more than 40 years; Complete mortality ascertainment; Cancer incidence ascertainment	Acute, high-dose-rate exposure that provides no direct information on effects of gradual, low-dose-rate exposures; Restriction to 5-year survivors for mortality (13-year survivors for incidence); Possible contribution of neutrons somewhat uncertain; Possible effects of thermal or mechanical injury and conditions following the bombings uncertain
Treatment of malignant disease		
Cervical cancer cohort [B11]	Large-scale incidence study based on tumour registry records; Long-term follow-up; Relatively complete ascertainment of cancers; Non-exposed comparison patients	Very large doses to some organs result in cell-killing and tissue damage; Potential misclassification of metastatic disease for some organs; Potential misclassification of exposure; No individual dosimetry; Characteristics of patients with cervical cancer differ from general population
Cervical cancer case-control [B21]	Comprehensive individual dosimetry for many organs; Dose-response analyses; Other strengths as above [B8, B11]	As above [B11], except problem with individual dosimetry and comparisons with general population now removed; Small number of non-exposed cases; Partial-body and partial-organ dosimetry complex
Contralateral breast cancer [B18, S60]	Large numbers of incident cases within population-based tumour registries; Individual radiation dosimetry; Wide range of high doses	Limited number of young women; Possibility of overmatching resulting in some concordance of exposure between cases and controls; Potential misclassification of metastases or recurrence
Breast cancer, adjuvant radiotherapy [C24]	Comprehensive individual dosimetry for bone marrow compartments; Comprehensive ascertainment of treatment information to separate chemotherapy risk; Dose-response analyses	Very large high-dose partial-body exposure to chest wall probably resulting in cell-killing
Childhood cancers [T5, T6, T7]	Comprehensive individual dosimetry to estimate organ doses; Attempt to adjust for drug exposure; Dose-response analyses	Only high-dose exposures; Potential for some overmatching since hospital-based; Complete dosimetry not always available
Treatment of benign disease		
Childhood skin haemangioma [F12, F15]	Long-term and complete follow-up; Dosimetry for some cases and controls; Incidence ascertained; Protracted exposure to radium plaques	Relatively limited range of doses to most sites; Relatively small number of specific cancers
Ankylosing spondylitis [D6]	Large number of exposed; Long-term and complete mortality follow-up; Detailed dosimetry for leukaemia cases and sample of cohort; Small non-exposed groups evaluated for general assurance that leukaemia risk was unrelated to underlying disease	Comparisons with general population; Underlying disease related to colon cancer and possibly other conditions; Individual dose estimates available only for leukaemia cases and a 1 in 15 sample of the population
Israeli tinea capitis [M20, R5, R10, R15]	Large number exposed; Two control groups; Ascertainment of cancer from hospital records and tumour registry; Individual dosimetry for many organs	Dosimetry for some sites, e.g. thyroid, uncertain due to possible patient movement or uncertainty in tumour location; Limited dose range
New York tinea capitis [A1, S8, S61]	Relatively good dose assessment for skin and other cancers	Small number of cancers; No recent follow-up information; few females
New York acute post-partum mastitis [S9]	Individual estimates of breast dose from medical records; Incidence ascertained; Dose-response analyses	All exposed women were parous but comparison women were not (380 non-exposed and sisters of both exposed and non-exposed); Inflamed and lactating breast might modify radiation effect

Table 3 (continued)

Study	*Strengths*	*Weaknesses*
Rochester thymic irradiation [H22, S10]	Individual dosimetry for thyroid and some other sites; Sibling control group; Long follow-up; Fractionation effects could be evaluated; Dose-response analyses	Radiation treatment fields on newborns varied and dosimetry for some sites uncertain; Adjustment in analysis for sibship size uncertain; Questionnaire follow-up may have resulted in unascertained cases
Benign gynaecological disease [D23, I7, I8, I12]	Large number of exposed; Non-exposed women with benign gynaecological disease; Very long mortality follow-up; Individual dosimetry; Protracted exposure to radium implants (10-24 hours); Dose-response analyses	Uncertainty in proportion of active bone-marrow-exposed; Small numbers of certain cancers; Misclassification of certain cancers on death certificates, such as pancreas
Peptic ulcer [G11]	Individual dosimetry; Non-exposed patients with peptic ulcer; Exceptionally long follow-up (50 years); Some risk factor information available in records	Standardized radiotherapy precluded dose-response analyses; Non-homogeneous dose distribution within organs such that simple averaging might be misleading; Metastatic spread of stomach cancer was probably misclassified as liver and pancreatic cancer on death certificates; Possible selection of somewhat unfit patients for radiotherapy rather than surgery
Diagnostic examinations		
Massachusetts TB fluoroscopy [B31, B41, D14]	Incidence study with long-term follow-up (50 years); Individual dosimetry based on patient records and measurements; Non-exposed TB patients; Fractionated exposures occurred over many years; Dose-response analyses	Uncertainty in dose estimates related to fluoroscopic exposure time and patient orientation; Questionnaire response probably underascertained cancers; Debilitating effect of TB may have modified radiation effect for some sites, in particular the lung
Canadian TB fluoroscopy [M19]	Large numbers; Non-exposed TB comparison groups; Individual dosimetry for breast; Fractionated exposure over many years; Dose-response analyses	Mortality limits comparisons with incidence series, e.g. time response; Uncertainties in dosimetry limit precise quantification of risk; Different dose response for one sanatorium and rest of Canada may indicate errors in dosimetry, differential ascertainment of cancers, or biological response
Scoliosis [H24]	Adolescence possibly a vulnerable age for exposure; Dosimetry attempted based on number of films and breast exposure; Dose response attempted	Comparison with general population potentially misleading since disease associated with several breast cancer risk factors, such as nulliparity; Small numbers (11 cancers) compared with 6 expected; One cancer detected incidentally during special screening
Prenatal exposures		
Oxford survey of childhood cancers [B23, M22, S37, S52]	Very large numbers; Comprehensive evaluation of potential confounding; Early concerns over response bias and selection bias resolved	Uncertainty in fetal dose from obstetric x-ray examinations precludes precise estimate of radiation risk; Similar relative risks for leukaemia and other solid cancers may point to possible residual confounding
New England childhood cancer [M9]	Large numbers; Reliance on obstetric records to determine exposure	Uncertainty in fetal dose precludes precise estimate of radiation risk
Twin studies [H3, I15, M23, M32, R13, R16]	Higher proportion of twins than singletons received obstetric x-ray examinations for reasons unlikely to be due to maternal illness	Small number of cancers in general; Uncertainty in fetal dose precludes radiation risk estimate; For cohort studies, actual proportion exposed and comparisons with single births add uncertainty
Survivors of atomic bombings [Y1, Y2, Y3]	Not selected for exposure; Reasonably accurate estimate of dose; Mortality follow-up relatively complete; Follow-up until early adulthood	Small number of exposed individuals and small number of cases; Incidence determination may not be complete; Mechanical and thermal effects may have influenced results
Radionuclide exposures: iodine		
Diagnostic ^{131}I [H12, H27]	Large numbers; Nearly complete ascertainment of cancers through linkage with cancer registry; Administered activities of ^{131}I known for each patient; Low-dose-rate exposure	Comparison with general population; Reason for some examinations related to high detection of thyroid cancers, i.e. suspicion of thyroid tumour was often correct; Doses to organs other than thyroid very low; Population under surveillance

Table 3 (continued)

Study	*Strengths*	*Weaknesses*
United States thyrotoxicosis [D9, S2]	Large numbers of patients treated with ^{131}I; Large non-exposed comparison groups; Comprehensive follow-up effort	Individual radiation dose not computed; Short follow-up, not recent Only leukaemia and thyroid tumours evaluated
Swedish ^{131}I hyperthyroid [H40]	Large numbers; Nearly complete incidence ascertainment; Administered activities of ^{131}I known; Low-dose-rate exposure	Comparison with general population; Dose response not based on organ doses; High-dose cell-killing probably reduced possible thyroid effect; Patients selection for treatment
Marshall Islands fallout [H28, R21]	Population unselected for exposure; Comprehensive medical long-term follow-up; Individual dosimetry attempted	Mixture of radioiodines and gamma fields preclude accurate estimate of dose; Surgery and hormonal therapy probably influence subsequent occurrence of thyroid neoplasms; Small numbers
Utah ^{131}I fallout [R12, S48, S49]	Clinical examinations for thyroid cancer; Comprehensive dosimetry assessment attempted; Large number of leukaemia deaths; Protracted exposures at low rate	Great uncertainty in accurately estimating dose to thyroid and bone marrow; Estimated cumulative doses much lower than experienced from natural background radiation; Possible recall and surveillance bias
Environmental exposures		
Techa River [K18]	Large numbers exposed with relatively long follow-up; A wide range of estimated doses; Unselected population, attempted use of local population rates for comparison; Possibility to evaluate ethnic differences in cancer risk; Shows great potential for future studies	Dosimetry difficult and not individual; Quality of mixture of internal and external exposures complicates dosimetry Follow-up and cancer ascertainment uncertain; Contribution of chemical exposures unevaluated
China background radiation, thyroid [W17]	Blind clinical examinations of exposed and non-exposed; Lifetime exposure to low doses at low dose rate; Blood studies to confirm exposure; Extensive dosimetry for region; Stable population; Low prevalence of x-ray exposure; High participation rate; Assessment of possible confounders	Approximately one third of life-span exposed as children; Factors other than radiation levels existed that could not be evaluated; High prevalence of a mild goiter might have hindered detection of thyroid nodules; Doses very low
International Chernobyl project [I3, M21]	Dosimetry assessment for highly contaminated and control villages; Physical examination for thyroid nodules disease using ultrasound	Limited follow-up after accident (4.5 years) precluded detection of possible risk; Precise dosimetry uncertain Small sample size
Occupational exposures		
Nuclear workers [B27, D10, G16, G17, K20]	Often large numbers; Personnel dosimetry; Low-dose fractionated exposures; Could provide useful information in future	Doses so low that clear demonstration of radiation effect difficult; Possibly confounding influence of chemical, asbestos and other toxic exposures in workplace; Healthy worker effect; Mortality follow-up; Lifestyle factors, such as smoking histories, not available
Medical workers [A9, B39, E7, M24, S38, W14]	Often large numbers; Low-dose fractionated exposures over very long periods	Lack of information on individual doses precludes usefulness to date

Table 4
Comparison of incidence and mortality data for solid tumours in Life Span Study of survivors of the atomic bombings
[T15]

Cancer type	*Number of incident cases 1958-1987*	*Number of deaths 1950-1987* [a]
Oral cavity and pharynx	132	79
Digestive system		
Oesophagus	185	211
Stomach	2658	2365
Colon	457	277
Rectum	351	273
Liver	585	761
Gall-bladder	295	187
Pancreas	240	243
Other	26	91
Total	4797	4408
Respiratory system		
Nasal cavity	55	51
Larynx	80	52
Trachea, bronchus, lung	872	816
Other	20	22
Total	1027	941
Skin		
Melanoma	13	8
Other skin	168	28
Total	181	36
Female breast	529	186
Female genital		
Uterus, not otherwise specified	86	315
Cervix uteri	553	110
Uterine corpus	85	15
Ovary	133	104
Other	34	31
Total	891	575
Male genital		
Prostate	140	69
Other	20	6
Total	160	75
Urinary system		
Bladder	210	104
Kidney	73	45
Renal pelvis and ureter	28	4
Other	14	3
Total	325	156
Nervous system	125	75
Thyroid	225	49
Other and ill-defined sites	221	307
Total	8,613	6,887

[a] In some cases the number of deaths exceeds the number of incident cases. There are various reasons for this. First, the mortality follow-up is more than seven years longer than the follow-up for the incidence data. Secondly, for some sites, e.g. the liver and the pancreas, death certificate diagnosis is poor. For other, e.g. the uterus, no detail is provided.

Table 5
Summary risk estimates for solid tumours derived from data on survivors of the atomic bombings [a]

Cancer type	*Cancer mortality [S7] [b] 1950-1985*		*Cancer mortality [R23] [c] 1950-1987*		*Cancer incidence [T15] [c] 1958-1987*	
	Excess relative risk (Sv^{-1})	*Excess absolute risk* $(10^4\ PYSv)^{-1}$	Excess relative risk (Sv^{-1})	*Excess absolute risk* $(10^4\ PYSv)^{-1}$	*Excess relative risk* (Sv^{-1})	*Excess absolute risk* $(10^4\ PYSv)^{-1}$
Salivary gland	Not reported	Not reported	Not reported	Not reported	1.8 (0.2-6.0)	Not reported
Oral cavity	Not reported [d]	Not reported [d]	-0.2 (<0-0.3)	-0.1 (<0-0.1)	0.3 (<0-0.9)	0.2 (<0-0.7)
Digestive system	0.3 (0.2-0.4) [e]	4.0 (2.7-5.4) [e]	0.3 (0.2-0.5)	5.1 (3.0-7.3)	0.4 (0.3-0.5)	10.4 (7.0-14.0)
Oesophagus	0.6 (0.1-1.2)	0.5 (0.1-0.9)	0.6 (0.09-1.3)	0.5 (0.07-0.9)	0.3 (<0-1.0)	0.3 (<0-1.0)
Stomach	0.3 (0.1-0.4)	2.4 (0.5-3.5)	0.2 (0.1-0.4)	1.9 (0.5-3.5)	0.3 (0.2-0.5)	4.8 (2.5-7.4)
Colon	0.9 (0.4-1.5)	0.8 (0.4-1.3)	0.5 (0.1-1.2)	0.5 (0.06-1.1)	0.7 (0.3-1.3)	1.8 (0.7-3.0)
Rectum	-0.09 (<-0.2-0.4)	-0.1 (<-0.2-0.6)	0.1 (<-0.2-0.6)	0.1 (<-0.3-0.6)	0.2 (-0.2-0.8)	0.4 (-0.4-1.5)
Liver	0.1 (<-0.2-0.7) [f]	0.2 (<-0.2-0.5) [f]	0.5 (0.2-0.8) [h]	1.3 (0.5-2.2) [h]	0.5 (0.2-0.9)	1.6 (0.5-2.9)
Gall-bladder	0.5 (-0.1-1.6)	0.3 (-0.03-0.7)	0.3 (-0.1-1.0)	0.2 (-0.07-0.6)	0.1 (<0.2-0.7)	0.2 (<-0.4-1.1)
Pancreas	-0.2 (<-0.2-0.4)	-0.1 (<-0.2-0.3)	-0.2 (<-0.2-0.2)	-0.2 (<-0.3-0.2)	0.2 (<-0.2-0.8)	0.2 (<-0.4-1.1)
Respiratory system	1.9 (0.3-0.9) [e]	1.8 (1.0-2.7) [e]	0.6 (0.3-0.9)	2.0 (1.1-3.0)	0.8 (0.5-1.2)	4.4 (2.9-6.1)
Lung	0.6 (0.3-1.0)	1.7 (1.0-2.5)	0.7 (0.3-1.0)	1.9 (1.0-2.9)	1.0 (0.6-1.4)	4.4 (2.9-6.0)
Non-melanoma skin	0.2 (<-0.2-2.5)	0.0 (<-0.2-0.12)	0.31 (<-0.1-1.5)	0.03 (<0-0.1)	1.0 (0.4-1.9)	0.8 (0.4-1.4)
Female breast	1.2 (0.6-2.1) [e]	1.2 (0.6-1.9) [e]	1.3 (0.6-2.1)	1.3 (0.6-2.1)	1.6 (1.1-2.2)	6.7 (4.9-8.7)
Uterus	0.4 (0.02-0.8)	1.0 (0.07-2.1)	0.1 (<-0.2-0.6)	0.3 (<-0.5-1.4)	-0.2 (-0.3-0.1)	-1.1 (-1.9-0.7)
Ovary	1.3 (0.5-2.3) [e]	0.7 (0.2-1.3) [e]	1.2 (0.2-2.8)	0.7 (0.1-1.4)	1.0 (0.2-2.3)	1.1 (0.2-2.3)
Prostate	0.1 (<-0.2-0.9) [e]	0.0 (<-0.2-0.5) [e]	0.3 (-0.2-1.6)	0.2 (-0.1-0.9)	0.3 (-0.1-1.2)	0.6 (-0.3-2.2)
Urinary tract	1.3 (0.5-2.4)	0.7 (0.3-1.1)	1.3 (0.4-2.6)	0.7 (0.2-1.2)	1.2 (0.6-2.1)	2.1 (1.1-3.2)
Urinary bladder	1.4 (0.5-2.9) [e]	0.5 (0.2-0.9) [e]	1.5 (0.3-3.3)	0.5 (0.1-0.9)	1.0 (0.3-2.1)	1.2 (0.3-2.1)
Brain and CNS	Not reported [g]	Not reported [g]	0.4 (-0.1-2.0)	0.1 (-0.04-0.5)	0.3 (-0.1-1.3)	0.2 (-0.1-0.8)
Thyroid	Not reported	Not reported	0.094 (<-0.1-1.4)	0.016 (<-0.03-0.2)	1.2 (0.5-2.1)	1.6 (0.8-2.5)
All solid tumours	0.40 (0.3-0.5)	10.1 (7.8-12.4)	0.45 (0.3- 0.6)	11.1 (8.4-14.0)	0.63 (0.52-0.74)	29.7 (25-34)

[a] 90% CI given in parentheses.
[b] Based on organ-absorbed dose (neutron RBE = 1).
[c] Based on weighted dose in the organ (neutron RBE = 10).
[d] Shielded-kerma-based excess relative risk and excess absolute risk estimates for cancers of the nose (-0.2 and -0.03, respectively), larynx (0.5 and 0.1), and pharynx (-0.4 and -0.02), and tongue (-0.6 and -0.02) are presented in [S7].
[e] Values derived from shielded-kerma risk estimates in [S7].
[f] Based on cases explicitly coded as primary on the death certificates; excludes cases not explicitly coded as primary or secondary. Inclusion of such cases leads to an ERR estimate of 0.46 (0.2-0.8).
[g] Shielded kerma-based excess relative risk and excess absolute risk estimates for malignant brain tumours (0.03 and 0.01, respectively) and other CNS tumours (2.1 and 0.1) are presented in [S7].
[h] Includes liver cancer deaths specified as primary and also those specified as either primary or secondary.

Table 6
Summary risk estimates for lympho-haematopoietic tumours derived from data on survivors of the atomic bombings

Cancer type	*Cancer mortality 1950-1985 [S7] [a]*		*Cancer incidence 1950-1987 [P33] [b]*	
	Excess relative risk (Sv^{-1})	*Excess absolute risk* $(10^4\ PYSv)^{-1}$	*Excess relative risk* (Sv^{-1})	*Excess absolute risk* $(10^4\ PYSv)^{-1}$
Leukaemia	5.2 (3.8-7.1)	2.9 (2.4-3.5)	4.4 (3.2-5.5)	2.7 (2.2-3.2)
Lymphomas	-0.1 (<0-0.5) [c]	0.0 (<0-0.2) [c]	-0.1 (<0-0.7) [d]	-0.1 (<0-0.5) [d]
Multiple myeloma	2.3 (1.7-6.3)	0.3 (0.1-0.5)	0.4 (<0-1.7)	0.1 (<0-0.4)

[a] Based on organ-absorbed dose (neutron RBE = 1).
[b] Based on weighted dose in the organ (neutron RBE = 10). Estimates derived from summary tables in [P33].
[c] Values derived from shielded-kerma risk estimates in [S7]. Includes both Hodgkin's and non-Hodgkin's cases.
[d] Non-Hodgkin's cases only.

Table 7
Cancer incidence and mortality in survivors of the atomic bombings

Absorbed dose [a] *(Gy)*	*Mean weighted dose* [b] *(Sv)*	*Person-years*	*Number of subjects*	*Observed*	*Expected*
		Solid tumour incidence, 1958-1987 [T15]			
<0.01	0.00	893,767 [c]	42,702	4,286	4,281
0.01-0.1	0.04	454,734 [c]	21,479	2,223	2,174
0.1-0.2	0.14	113,414 [c]	5,307	599	553
0.2-0.5	0.33	124,131 [c]	5,858	759	637
0.5-1	0.74	60,096 [c]	2,882	418	290
1-2	1.42	30,031 [c]	1,444	273	146
>2	2.52	6,175 [c]	300	55	20
Total	0.11	1,682,349 [c]	79,972	8,613	8,103
		Solid tumour mortality, 1950-1987 [R23]			
<0.01	0.00	1,385,374	46,176	3,435	3,433
0.01-0.1	0.04	693,935	23,147	1,868	1,837
0.1-0.2	0.14	171,130	5,713	472	444
0.2-0.5	0.33	188,444	6,283	582	508
0.5-1	0.74	93,116	3,111	312	234
1-2	1.42	46,891	1,543	178	108
>2	2.52	9,984	336	40	18
Total	0.11	2,588,873	86,309	6,887	6,581
		Leukaemia incidence, 1950-1987 [P33]			
<0.01	0.00	1,166,110 [c]	45,159	90	81
0.01-0.1	0.04	609,964 [c]	23,304	38	42
0.1-0.2	0.15	155,982 [c]	5,880	8	11
0.2-0.5	0.34	167,675 [c]	6,387	27	12
0.5-1	0.76	85,711 [c]	3,304	24	6
1-1.5	1.35	31,295 [c]	1,209	19	2
1.5-2	1.92	15,082 [c]	584	8	1
2-4	2.83	11,101 [c]	444	17	1
Total	0.12	2,242,919 [c]	86,271	231	156

[a] Dose categories are defined in terms of unweighted intestinal doses for solid tumours and unweighted bone marrow doses for leukaemia.
[b] The mean doses are expressed as weighted intestinal doses for solid tumours and weighted bone marrow doses for leukaemia (RBE for neutrons = 10).
[c] Person-years adjusted for migration.

Table 8
Risk estimates for cancer incidence and mortality from studies of external low-LET exposures
The number of observed and expected cases as well as the mean dose and person-years for cohort studies are computed throughout this Table for exposed persons only. In the life span study the exposed group included survivors with organ doses of 0.1 Gy or more.

PART I: OESOPHAGUS

Study			*Observed cases*	*Expected cases*	*Mean dose (Sv)*	*Person-years*	*Average excess relative risk* [a] (Sv^{-1})	*Average excess absolute risk* [a] $(10^4\ PYSv)^{-1}$
Incidence								
Life span study [T15]								
	Sex	Male	68	65.5	0.24	298,700	0.16	0.35
		Female	17	13.7	0.23	493,900	1.06	0.29
	Age at exposure	<20 years	8	4.5	0.24	365,200	3.32	0.41
		>20 years	77	74.7	0.23	427,300	0.13	0.24
	All		84	78.9	0.23	788,580	0.29 (-0.2-1.0)	0.29 (-0.2-1.0)
Cervical cancer cohort [B11] [b]			12	11.0	0.35	178,243	0.26 (<-0.2-2.2)	0.16 (<-0.6-1.4)
Mortality								
Life span study [R23]								
	Sex	Male	81	74.7	0.23	471,800	0.37	0.58
		Female	22	16.1	0.23	731,300	1.61	0.36
	Age at exposure	<20 years	8	6.2	0.23	574,500	1.20	0.13
		>20 years	95	84.6	0.22	628,600	0.56	0.75
	All		103	90.8	0.23	1,203,100	0.59 (-0.0-1.4)	0.44 (<0-0.9)
Massachusetts TB fluoroscopy [D14]			14	13.8	1.06	169,425	0.01 (-0.2-0.6)	0.01 (-0.3-0.4)
Ankylosing spondylitis [D6, L6]			28	12.7	4.00	183,749	0.30 (0.1-0.5)	0.21 (0.1-0.3)

[a] 90% CI in parentheses derived from published data for the life span study and using exact Poisson methods for the other studies.
[b] Excludes cases during first 10 years of follow-up.

PART II: STOMACH

Study			*Observed cases*	*Expected cases*	*Mean dose (Sv)*	*Person-years*	*Average excess relative risk* [a] (Sv^{-1})	*Average excess absolute risk* [a] $(10^4\ PYSv)^{-1}$
Incidence								
Life span study [T15]								
	Sex	Male	679	660.4	0.24	298,700	0.12	2.61
		Female	628	561.3	0.23	493,900	0.52	5.86
	Age at exposure	<20 years	167	142.0	0.24	365,200	0.74	2.87
		>20 years	1140	1079.7	0.23	427,300	0.24	6.15
	All		1307	1221.7	0.23	792,500	0.30 (0.2-0.5)	4.68 (2.5-7.4)
Cervical cancer case-control [b] [B21]			348	117.3	2.00	n.a.	0.54 (0.1-1.5)	0.37 (0.03-1.0)
Mortality								
Life span study [R23]								
	Sex	Male	649	623.0	0.24	471,800	0.17	2.30
		Female	514	484.5	0.23	731,300	0.26	1.75
	Age at exposure	<20 years	96	91.2	0.24	574,500	0.22	0.35
		>20 years	1067	1016.3	0.22	628,600	0.22	3.51
	All		1163	1107.5	0.23	1,203,100	0.22 (0.1-0.4)	2.02 (0.5-3.5)
Ankylosing spondylitis [D6, L6]			55	54.3	1.65	130,616	0.01 (-0.1-0.2)	0.03 (-0.5-0.7)
Benign gynaecological disease [I8] [c]			23	21.8	0.20	77,878	0.28 (<-0.2-2.5)	0.78 (-2.6-4.6)
Peptic ulcer [G11]			40	17.4	14.80	35,815	0.09 (0.05-0.14)	0.43 (0.2-0.7)
Metropathia haemorrhagica [D23]			33	26.8	0.23	47,144	1.01 (<-0.2-2.8)	5.72 (<-2.4-16)

[a] 90% CI in parentheses derived from published data for the life span study and using exact Poisson methods for the other studies.
[b] The excess absolute risk estimate was computed using background incidence rates estimated using the cervical cancer cohort study [B11].
[c] The observed and expected number of cases are for 10-year survivors. The estimated number of expected cases incorporated an adjustment based upon the Poisson regression model given in [I8].

PART III: COLON

Study		Observed cases	Expected cases	Mean dose (Sv)	Person-years	Average excess relative risk (Sv^{-1}) [a]	Average excess absolute risk [a] $(10^4\ PYSv)^{-1}$
Incidence							
Life span study [T15]							
Sex	Male	109	90.7	0.23	297,500	0.87	2.66
	Female	114	103.0	0.22	491,100	0.48	1.01
Age at exposure	<20 years	32	28.0	0.23	363,300	0.62	0.48
	>20 years	191	165.7	0.22	425,300	0.70	2.71
All		223	193.7	0.23	788,600	0.67 (0.1-1.3)	1.65 (0.7-3.0)
Cervical cancer case-control [b] [B21]		409	409	24.0	n.a.	0.00 (0.00-0.01)	0.00 (0.00-0.01)
Mortality							
Life span study [R23]							
Sex	Male	63	56.8	0.23	471,800	0.47	0.57
	Female	66	59.7	0.23	731,300	0.47	0.38
Age at exposure	<20 years	11	9.6	0.23	574,500	0.60	0.10
	>20 years	118	106.9	0.22	628,600	0.47	0.80
All		129	116.5	0.23	1,203,100	0.47 (0.1-1.1)	0.45 (0.3-1.3)
Benign gynaecological disease [I8] [c]		75	46.6 [d]	1.30	77,878	0.47 (0.2-0.7)	2.81 (1.5-4.4)
Peptic ulcer [e] [G11]		25	20.6	6.00	118,300	0.04 (0.0-0.1)	0.07 (0.0-0.02)
Metropathia haemorrhagica [D23]		47	33.0	3.20	47,144	0.13 (0.03-0.3)	0.93 (0.2-1.8)

[a] 90% CI in parentheses derived from published data for the life span study and using exact Poisson methods for the other studies.
[b] The excess absolute risk estimate was computed using background incidence rates estimated using the cervical cancer cohort study [B12].
[c] The observed and expected number of cases are for 10-year survivors. The estimated number of expected cases incorporated an adjustment based upon the Poisson regression model given in [I8].
[d] Incorporates a standardized mortality ratio based upon internal comparison model in [I8].
[e] Since the original paper provides only a range of doses for this site, the value given here is a very crude estimate.

PART IV: LIVER

Study		Observed cases	Expected cases	Mean dose (Sv)	Person-years	Average excess relative risk (Sv^{-1}) [a]	Average excess absolute risk [a] $(10^4\ PYSv)^{-1}$
Incidence							
Life span study [T15]							
Sex	Male	174	151.5	0.24	299,600	0.61	3.09
	Female	110	107.2	0.23	496,600	0.11	0.24
Age at exposure	<20 years	63	47.2	0.24	367,000	1.39	1.78
	>20 years	221	211.4	0.23	429,200	0.19	0.95
All cases		284	258.6	0.24	796,300	0.41 (0.2-0.9)	1.33 (0.5-2.9)
Total (microscopically verified)		109	94.7	0.24	796,300	0.64	0.76
Cervical cancer cohort [b] [B11]		19	20	0.70	342,786	-0.13 (<0-0.9)	-0.03 (<0-0.2)
Mortality							
Life span study [R23]							
Sex	Male	202	178.6	0.24	471,800	0.54	2.04
	Female	150	141	0.23	731,300	0.27	0.53
Age at exposure	<20 years	56	43.4	0.24	574,500	1.26	0.90
	>20 years	296	276.2	0.22	628,600	0.31	1.35
All cases		352	319.6	0.23	1,203,100	0.44 (0.2-0.8)	1.18 (0.5-2.2)
Total [c] (primary on death certificate)		48	46.2	0.24	1,203,100	0.16	0.06
Ankylosing spondylitis [D6, L6]		6	5.6	1.60	130,616	0.05 (<0-0.7)	0.02 (<0-0.3)
Benign gynaecological disease [I8]		9	16.6	0.21	77,878	-0.5 (-0.5- -0.3)	<-1 (<-1- -0.5)
Peptic ulcer [G11]		9	9.0	4.60	35,815	0.00 (<0-0.2)	0.00 (<0-0.4)
Metropathis haemorrhagica [D23]		2	6.0	0.27	47,144	-0.5 (<-0.5-0.2)	<-0.2 (<-0.6-0.2)

[a] 90% CI in parentheses derived from published data for the life span study and using exact Poisson methods for the other studies.
[b] Excludes cases during first 10 years of follow-up.
[c] Estimate is based on deaths coded as primary liver cancer or liver cancer not specified as secondary or primary.

PART V: LUNG

Study		*Observed cases*	*Expected cases*	*Mean dose (Sv)*	*Person-years*	*Average excess relative risk* (Sv^{-1}) [a]	*Average excess absolute risk* [a] $(10^4 PYSv)^{-1}$
			Incidence				
Life span study [T15]							
Sex	Male	245	224.7	0.25	302,000	0.36	2.67
	Female	211	140.1	0.24	500,700	2.08	5.81
Age at exposure	<20 years	30	26.2	0.25	370,000	0.57	0.41
	>20 years	426	338.5	0.24	432,700	1.06	8.27
Time since exposure	5-19 years	85	67.8	0.24	288,566	1.04	2.45
	20-29 years	146	116.3	0.24	317,535	1.05	3.85
	30-42 years	225	186.4	0.24	314,545	0.85	5.05
All		456	364.7	0.25	802,700	1.00 (0.6-1.4)	4.55 (2.4-6.0)
			Mortality				
Life span study [R23]							
Sex	Male	226	214.7	0.25	471,800	0.21	0.95
	Female	207	149.0	0.24	731,300	1.59	3.24
Age at exposure	<20 years	27	29.8	0.25	574,500	-0.38	-0.20
	>20 years	406	334.0	0.24	628,600	0.88	4.68
Time since exposure	5-19 years	71	63.5	0.24	566,780	0.48	0.54
	20-29 years	135	90.3	0.24	316,912	2.04	5.80
	30-42 years	227	194.4	0.24	319,418	0.69	4.19
All		433	363.8	0.25	1,203,100	0.76 (0.3-1.0)	2.30 (1.0-2.9)
Ankylosing spondylitis [D6, L6]		224	184.5	1.80	130,616	0.12 (0.0-0.2)	1.68 (0.7-2.8)
Massachusetts TB fluoroscopy [D14]		69	81.8	0.84	169,425	-0.19(<-0.2-0.04)	-0.90 (<-1.8-0.2)
Peptic ulcer [G11]		99	58.4	1.80	35,815	0.39 (0.2-0.6)	6.29 (3.8-9.1)

[a] 90% CI in parentheses derived from published data for the life span study and using exact Poisson methods for the other studies.

PART VI: BONE AND CONNECTIVE TISSUE

Study		*Observed cases*	*Expected cases*	*Mean dose (Sv)*	*Person-years*	*Average excess relative risk* (Sv^{-1}) [a]	*Average excess absolute risk* [a] $(10^4 PYSv)^{-1}$
			Incidence				
Life span study [T15]							
Sex	Male	9	6.4	0.23	297,500	1.78	0.38
	Female	7	5.7	0.22	491,100	0.99	0.12
Age at exposure	<20 years	4	1.1	0.23	363,300	11.0	0.34
	>20 years	12	11.0	0.22	425,300	0.42	0.11
All		16	12.1	0.23	788,600	1.42 (<-0.2-4.5)	0.22 (<-0.1-0.7)
Skin haemangioma [F12]		8	2.9	0.4	379,283	4.33 (0.9-9.8)	0.33 (0.1-0.8)
Childhood radiotherapy [T6]		54	20.0	27.0	n.a.	0.06 (0.01-0.2)	n.a.
			Mortality				
Life span study [R23]							
Sex	Male	14	10.8	0.23	471,800	1.26	0.29
	Female	10	8.5	0.23	731,300	0.81	0.09
Age at exposure	<20 years	3	1.9	0.23	574,500	2.58	0.08
	>20 years	21	17.4	0.22	628,600	0.92	0.26
All		24	19.3	0.23	1,203,100	1.07 (<-0.2-3.3)	0.17 (<-0.1-0.5)
Ankylosing spondylitis [D6, L6] (bone only)		4	1.4	3.00	130,616	0.65 (0.00-1.9)	0.07 (0.00-0.2)

[a] 90% CI in parentheses derived from published data for the life span study and using exact Poisson methods for the other studies.

PART VII: SKIN (NON-MELANOMA)

Study		*Observed cases*	*Expected cases*	*Mean dose (Sv)*	*Person-years*	*Average excess relative risk* (Sv^{-1}) [a]	*Average excess absolute risk* [a] $(10^4 PYSv)^{-1}$
Incidence							
Life span study [T15]							
Sex	Male	41	31.4	0.33	324,100	0.92	0.89
	Female	57	44.4	0.32	538,900	0.88	0.72
Age at exposure	<20 years	21	7.7	0.32	399,300	5.37	1.04
	>20 years	77	68.2	0.33	463,700	0.39	0.58
	All	98	75.9	0.33	863,000	0.88 (0.4-1.9)	0.78 (0.4-1.4)
Israeli tinea capitis [R10, R19]		42	10.0	6.8	265,070	0.47 (0.3-0.7)	0.18 (0.1-0.25)
Cervical cancer case-control [B21]		80	85.7	9.6	342,786	-0.01 (-0.02-0.01)	-0.02 (-0.06-0.03)
New York tinea capitis (whites) [b] [S62]		83	24.0	4.5	52,000 [c]	0.55 (-0.02-0.01)	2.50 (1.9-3.2)
Rochester thymic irradiation [b] [S62]		14	4.2	2.3	87,000 [c]	1.04 (0.5-1.9)	0.50 (0.2-0.9)
Massachusetts TB fluoroscopy [b] [S62]		80	75.3	9.6	122,000 [c]	0.01 (-0.01-0.03)	0.04 (-0.08-0.2)
New York mastitis [b] [S62]		14	10.7	2.6	14,000 [c]	0.12 (-0.08-0.4)	0.90 (-0.6-3.1)
Thyroid irradiation [b] [S62, S63]		63	45.0	3.8	96,000 [c]	0.11 (0.03-0.2)	0.50 (0.2-0.9)

[a] 90% CI in parentheses derived from published data for the life span study and using exact Poisson methods for the other studies.
[b] From data presented in a review paper by Shore [S62].
[c] Person-years estimated from data in the review paper by Shore [S62].

PART VIII: FEMALE BREAST

Study		Observed cases	Expected cases	Mean dose (Sv)	Person-years	Average excess relative risk (Sv^{-1}) [a]	Average excess absolute risk [a] $(10^4\ PYSv)^{-1}$
Incidence							
Life span study [T15]							
Age at exposure	<20 years	122	62.8	0.28	202,600	3.32 (2.3-4.4)	10.3 (7.2-14)
	>20 years	173	137.1	0.27	308,000	0.98 (0.4-1.6)	4.36 (1.8-7.2)
Time since exposure	5-19 years	49	36.9	0.28	161,400	1.19	2.72
	20-29 years	87	63.5	0.27	175,800	1.34	4.86
	30-42 years	159	99.5	0.27	173,400	2.21	12.68
All		295	199.9	0.27	510,500	1.74 (1.1-2.2)	6.80 (4.9-8.7)
Massachusetts TB fluoroscopy [B31]		142	107.6	0.79	54.600	0.40 (0.2-0.7)	7.98 (3.6-13)
New York acute post-partum mastitis [S9]		54	20.8	3.70	9,800	0.43 (0.3-0.6)	9.14 (6.0-13)
Ankylosing spondylitis [D6, L6]		26	16.1	0.50	22,033	1.24 (0.3-2.5)	9.01 (2.0-18)
Swedish breast irradiation [B1, M8]		115	28.8	8.46	37.400 [b]	0.35 (0.3-0.4)	2.72 (2.2-3.3)
Cervical cancer case-control [c] [B21]		953	1083.0	0.31	n.a.	-0.2 (<-0.2-0.3)	<-0.3 (<-0.3-0.2)
Without ovaries		91	82.6	0.31	n.a.	0.33 (<-0.2-5.8)	n.a.
Contralateral breast							
Denmark [S60]		529	508.7	2.51	n.a.	0.02 (<-0.1-0.2)	n.a.
United States [B18]		655	550.4	2.82	n.a.	0.07 (<-0.1-0.2)	n.a.
Rochester thymic irradiation [d] [H22]		22	7.8	0.76	38,200	2.39 (1.2-4.0)	4.89 (2.4-8.1)
Skin haemangioma [F12]		56	36.4	0.20	379,283	4.2 (1.8-7.2)	4.1 (1.8-6.9)
Scoliosis patients [d] [H24]		11	6.1	0.13	21,691	6.37 (0.9-15)	17.8 (2.4-43)
Hodgkin's disease (Stanford) [H2]		25	6.1	~44.0	100,057	0.07 (0.04-0.11)	0.04 (0.03-0.07)
Mortality							
Life span study [R23]							
Age at exposure	<20 years	34	13.0	0.28	303,200	5.69 (3.2-9)	2.44 (1.4-3.7)
	>20 years	70	61.1	0.27	428,100	0.55 (<-0.2-1.5)	0.78 (<-0.3-2.1)
Time since exposure	5-19 years	35	28.7	0.28	338,258	0.80	0.68
	20-29 years	26	20.7	0.27	194,149	0.92	0.99
	30-42 years	43	24.2	0.27	198,925	2.87	3.49
All		104	74.1	0.27	731,300	1.79 (0.6-2.1)	1.81 (0.6-2.1)
Canada TB fluoroscopy [M19] [e]		212	157.5	0.43	282,300	0.81 (0.5-1.2)	4.52 (2.6-6.6)

[a] 90% CI in parentheses derived from published data for the life span study and using exact Poisson methods for the other studies.
[b] For cohort studies the numbers of observed and expected cases as well as the mean dose and person-years are computed for exposed persons only.
[c] Excess absolute risk of cervical cancer is computed using baseline incidence data derived from the cohort study [B12].
[d] Populations exposed as children.
[e] Original model included a factor to allow for differences between Nova Scotia and other Canadian provinces. The computation of the expected number of cases includes an allowance for this effect.

PART IX: BLADDER

Study		*Observed cases*	*Expected cases*	*Mean dose (Sv)*	*Person-years*	*Average excess relative risk* (Sv^{-1}) [a]	*Average excess absolute risk* [a] $(10^4\ PYSv)^{-1}$
Incidence							
Life span study [T15]							
Sex	Male	76	70.3	0.23	297,500	0.35	0.84
	Female	39	27.9	0.22	491,200	1.80	1.02
Age at exposure	<20 years	12	10.3	0.23	363,300	0.71	0.20
	>20 years	103	87.8	0.22	425,300	0.79	1.62
All		115	98.1	0.23	788,600	0.76 (0.3-2.1)	0.95 (0.3-2.1)
Cervical cancer case-control [b] [B21]		273	65.8	45	n.a.	0.07 (0.02-0.2)	0.12 (0.04-0.3)
Mortality							
Life span study [R23]							
Sex	Male	36	26.9	0.23	471,800	1.45	0.83
	Female	24	20.0	0.23	731,300	0.88	0.24
Age at exposure	<20 years	5	1.8	0.23	574,500	7.43	0.24
	>20 years	55	45.1	0.22	628,600	0.99	0.71
All		60	46.9	0.23	1,203,100	1.22 (0.3-3.3)	0.48 (0.1-0.9)
Ankylosing spondylitis [D6, L6]		20	16.7	1.45	130,616	0.14 (-0.1-0.5)	0.17 (-0.2-0.6)
Benign gynaecological disease [I8]		19	9.0	6.00	77,878	0.19 (0.1-0.3)	0.21 (0.1-0.4)
Metropathia haemorrhagica [D23]		20	6.7	5.20	47,144	0.39 (0.2-0.6)	0.54 (0.3-0.9)

[a] 90% CI in parentheses derived from published data for the life span study and using exact Poisson methods for the other studies.

[b] The excess absolute risk estimate was computed using background incidence rates estimated using the cervical cancer cohort study [B12].

PART X: BRAIN AND CENTRAL NERVOUS SYSTEM

Study		*Observed cases*	*Expected cases*	*Mean dose (Sv)*	*Person-years*	*Average excess relative risk* (Sv^{-1}) [a]	*Average excess absolute risk* [a] $(10^4\ PYSv)^{-1}$
Incidence							
Life span study [T15]							
Sex	Male	20	21.7	0.27	307,100	-0.30	-0.21
	Female	51	45.3	0.26	509,300	0.48	0.43
Age at exposure	<20 years	20	15.7	0.26	376,100	1.05	0.44
	>20 years	51	51.4	0.26	440,200	-0.03	-0.03
All		71	67.1	0.26	816,300	0.22 (<0-1.3)	0.18 (<0-0.8)
Israeli tinea capitis [R10]		60	8.4	1.5	283,930	4.08 (3.1-5.2)	1.21 (0.9-1.5)
New York tinea capitis [A1]		8	1.4	1.4	48,115	3.37 (1.3-6.7)	0.98 (0.4-1.9)
Skin haemangioma [F12]		16	17.1	0.1	379,283	<-0.2 (<-0.2-4.2)	<-0.1 (<-2-1.5)
Mortality							
Life span study [R23]							
Sex	Male	18	17.9	0.27	487,200	0.03	0.01
	Female	28	16.6	0.26	757,900	3.05	0.67
Age at exposure	<20 years	14	12.1	0.26	601,000	0.67	0.13
	>20 years	32	22.4	0.26	757,900	1.95	0.68
All		45	39.7	0.26	1,203,100	0.52 (<0-2)	0.17 (<0-0.5)
Ankylosing spondylitis [b] [D6, L6]		22	14.0	0.14	130,616	4.06 (0.4-8.9)	4.36 (0.5-9.5)
Pituitary adenoma (UK) [B9]		5	0.5	45.0	3760	0.19 (0.1-0.4)	0.26 (0.1-0.6)

[a] 90% CI in parentheses derived from published data for the life span study and using exact Poisson methods for the other studies.

[b] Excludes tumours of the spinal cord.

PART XI: THYROID INCIDENCE

Study		*Observed cases*	*Expected cases*	*Mean dose (Sv)*	*Person-years*	*Average excess relative risk* (Sv^{-1}) [a]	*Average excess absolute risk* [a] $(10^4\ PYSv)^{-1}$
Life span study [T15]							
Sex	Male	22	14.9	0.27	307,167	1.80	0.87
	Female	110	79.4	0.26	510,388	1.49	2.32
Age at exposure	0-9 years	24	7.6	0.21	185,507	10.25	4.21
	10-19 years	35	14.6	0.31	190,087	4.50	3.46
	20-29 years	18	17.5	0.28	132,738	0.10	0.13
	>20 years	73	67.1	0.26	442,000	0.3 (<-0.2-1.2)	0.5 (-0.3-1.9)
	>30 years	55	54.5	0.25	309,224	0.04	0.06
All		132	94.3	0.26	817,600	1.5 (0.5-2.1)	1.8 (0.8-2.5)
Tuberculosis, adenitis screening							
Age at exposure	<20 years	6	0.0	8.20	950	36.5 (16-72)	7.7 (3.3-15)
[H1, S61]	>20 years	2	0.2	8.20	3,100	1.2 (0.1-3.7)	0.7 (0.1-2.4)
Cohort studies in children							
Life span study [T15]							
Age at exposure	0-19 years	59	22.2	0.26	375,600	6.3 (5.1-10.1)	3.8 (2.7-5.4)
Israeli tinea capitis [R15] [b]		43	10.7	0.1	274,180	34 (23-47)	13 (9.0-18)
New York tinea capitis [S61]		2	1.4 [c]	0.1	79,500	7.7 (<0-60)	1.3 (<0-10.3)
Rochester thymic irradiation [d] [S54]		37	2.7	1.4	85,204	9.5 (6.9-12.7)	3.0 (2.2-4.0)
Childhood cancer [e] [T5]		23	0.4	12.5	50,609	4.5 (3.1-6.4)	0.4 (0.2-0.5)
Skin haemangioma [F12, F15]		14	7.1	0.23	n.a.	4.2 (0.1-29)	n.a.
Screening studies in children							
Lymphoid hyperplasia screening [d f] [P8, S61]		13	5.4 [b]	0.24	34,700	5.9 (1.8-11.8)	9.1 (2.7-18.3)
Thymus adenitis screening [M2, S61]		16	1.1 [b]	2.90	44,310	4.5 (2.7-7.0)	1.2 (0.7-1.8)
Michael Reese, tonsils [g] [S63]		309	110.4	0.60	88,101	3.0 (2.6-3.5)	37.6 (32-43)
Tonsil/thymus/acne screening [D18, S61]		11	0.2 [b]	4.50	6,800	12.0 (6.6-20)	3.5 (2.0-5.9)
Studies in adults							
Cervical cancer case-control [c]	[B21]	43	18.8	0.11	n.a.	12.3 (<0-70.0)	5.8 (<0-32.9)
Cervical cancer cohort [c h]	[B11]	16	12.5	0.11	342,786	2.5 (<0-6.8)	0.9 (<0-2.5)
Stanford thyroid [H6]		6	0.4	45.0	17,700	0.3 (0.1-0.7)	0.07 (0.03-0.1)

[a] 90% CI in parentheses derived from published data for the life span study and using exact Poisson methods for the other studies.
[b] Doses to the thyroid in this study may be much more uncertain than doses to organs directly in the x-ray beam.
[c] Expected number of cases computed using excess relative risk estimates given in [S61].
[d] Known dose. Person-years and expected number of cases estimated from data given in [S61].
[e] Based on cohort members with 15 or more years of follow-up and population expected rates for non-melanoma skin tumours of the head and neck tumours.
[f] This was a study of nodular disease and cancer cases were not confirmed.
[g] Study includes no unexposed controls; estimates of the number of expected cases were computed using the fitted excess relative risk reported in [S63]. Results are based on the new dosimetry described in [S63]. The large excess absolute risk in this study illustrates the impact of screening on thyroid cancer risk estimates. As described in [S63], a special thyroid screening programme in this cohort was initiated in 1974. This screening led to a large increase in the number of incident cases detected among both cases and controls. The paper describes an analysis in which allowance was made for the effect of screening. The screening-adjusted excess absolute risk was estimated as 1.7 $(10^4\ PYGy)^{-1}$.
[h] Excludes cases diagnosed during first 10 years of follow-up.

PART XII: LEUKAEMIA

Study		*Observed cases*	*Expected cases*	*Mean dose (Sv)*	*Person-years*	*Average excess relative risk* (Sv^{-1}) [a]	*Average excess absolute risk* [a] $(10^4\ PYSv)^{-1}$
Incidence							
Life span study [P33]							
Sex	Male	71	35.3	0.26	412,300	3.91	3.35
	Female	70	32.1	0.25	664,500	4.75	2.29
Age at exposure	<20 years	46	17.9	0.26	478,100	6.11	2.28
	>20 years	95	49.5	0.25	598,700	3.70	3.06
Time since exposure	5-10 years	29	5.1	0.25	160,900	18.69	5.87
	11-20 years	45	40.3	0.25	367,200	0.46	0.50
	21-30 years	34	18.5	0.25	277,900	3.32	2.21
	31-42 years	33	28.1	0.25	270,800	0.70	0.72
All		141	67.4	0.25	1,076,800	4.37 (3.2-5.6)	2.73 (2.0-3.5)
Cervical cancer case-control [b] [B21]		141	n.a.	7.20	n.a.	0.74 (0.1-3.8)	0.50 (0.1-2.6)
Childhood cancer [c] [T7]		25	n.a.	10.00	n.a.	0.00 (0.000-0.004)	n.a.
Breast cancer therapy [d] [C24]		38	n.a.	7.50	n.a.	0.19 (0.00-0.6)	0.89 (0.00-3.0)
Techa River population [K18]		37	19.3	0.50	388,880	1.84 (0.9-3.1)	0.91 (0.4-1.5)
Mortality							
Ankylosing spondylitis [D6, L6]		36	11.9	3.83	165,758	0.53 (0.3-0.8)	0.38 (0.2-0.6)
Israeli tinea capitis [e] [R5]		14	6.0	0.30	279,901	4.44 (1.7-8.7)	0.95 (0.4-1.9)
Benign gynaecological disease [I12]		47	27.6	1.19	246,821	2.97 (2.2-4.0)	1.25 (0.9-1.7)
Massachusetts TB fluoroscopy [f] [D14]		17	18.0	0.09	157,578	<-0.2 (<-0.2-4.5)	<-0.2 (<-0.2-5.1)
Peptic ulcer [G11]		8	2.9	1.55	35,815	1.13 (0.4-2.5)	0.92 (0.3-2.0)
Metropathia haemorrhagica [D23]		12	5.6	1.30	53,144	0.88 (0.3-1.9)	0.93 (0.3-2.0)

[a] 90% CI in parentheses derived from published data for the life span study and using exact Poisson methods for the other studies.

[b] CIs for the cervical cancer data on excess relative risk were estimated from the confidence region curves in [B42]; the excess absolute risk estimate uses incidence estimates from the cohort study [B12].

[c] Risk estimates based on an unmatched analysis of data given in [T5].

[d] The excess absolute risk for this study is computed based on annual incidence estimates and average follow-up times reported in [C35].

[e] A re-estimate of the dose to bone marrow in this study indicates a mean dose of 0.60 rather than 0.30. Consequently the excess relative risk becomes 2.22 [R27].

[f] Excludes cases of chronic lymphocytic leukaemia.

PART XIII: NON-HODGKIN'S LYMPHOMA

Study		*Observed cases*	*Expected cases*	*Mean dose (Sv)*	*Person-years*	*Average excess relative risk* (Sv^{-1}) [a]	*Average excess absolute risk* [a] $(10^4\ PYSv)^{-1}$
Incidence							
Life span study [P33]							
Sex	Male	41	33.2	0.26	412,400	0.91	0.73
	Female	35	38.3	0.25	664,500	-0.34	-0.20
Age at exposure	<20 years	17	15.8	0.26	478,100	0.30	0.10
	>20 years	59	55.7	0.25	598,800	0.24	0.22
All		76	71.5	0.25	1,076,900	0.25 (<-0.2-1.1)	0.17 (<-0.3-0.8)
Cervical cancer case-control [b] [B21]		94	n.a.	7.10	n.a.	0.21 (0.1-1.1)	0.04 (0.02-0.2)
Mortality							
Ankylosing spondylitis [D6, L6]		16	7.1	3.83	130,616	0.32 (0.1-0.6)	0.18 (0.1-0.3)
Benign gynaecological disease [I12]		40	42.5	1.19	246,821	-0.05 (<-0.2-0.2)	-0.08 (<-0.3-0.3)
Massachusetts TB fluoroscopy [D14]		13	13.1	0.09	157,578	-0.05 (<-0.2-6.5)	-0.04 (<-0.2-5.4)
Peptic ulcer [G11]		12	6.4	1.55	35,815	0.57 (0.1-1.3)	1.01 (0.1-2.4)
Metropathia haemorrhagica [D23]		4	5.3	1.30	47,144	-0.19 (<-0.2-0.6)	-0.22 (<-0.2-0.6)

[a] 90% CI in parentheses derived from published data for the life span study and using exact Poisson methods for the other studies.

[b] CIs for the cervical cancer data on excess relative risk were estimated from the confidence region curves in [B42]; the excess absolute risk estimate uses incidence estimates from the cohort study [B11].

PART XIV: MULTIPLE MYELOMA

Study		*Observed cases*	*Expected cases*	*Mean dose (Sv)*	*Person-years*	*Average excess relative risk* (Sv^{-1}) [a]	*Average excess absolute risk* [a] $(10^4\ PYSv)^{-1}$
Incidence							
Life span study [P33]							
Sex	Male	12	9.2	0.26	412,400	1.17	0.26
	Female	18	19.3	0.25	664,500	-0.28	-0.08
Age at exposure	<20 years	4	3.1	0.26	478,100	1.07	0.07
	>20 years	26	25.4	0.25	598,800	0.09	0.04
All		30	28.6	0.25	1,076,900	0.20 (< 0.2 1.7)	0.05 (<-0.05-0.4)
Cervical cancer case-control [b] [B21]		56	n.a.	7.10	n.a.	-0.10 (<-0.1-0.1)	-0.01 (<-0.03-0.02)
Mortality							
Ankylosing spondylitis [D6, L6]		8	4.7	3.83	130,616	0.19 (<-0.04-0.5)	0.07 (<-0.01-0.2)
Benign gynaecological disease [I12]		14	12.4	1.19	246,821	0.11 (<-0.2-0.6)	0.05 (<-0.1-0.3)
Peptic ulcer [G11]		3	3.1	1.55	35,815	-0.02 (<-0.2-1.0)	-0.02 (<-0.2-0.8)
Metropathia haemorrhagica [D23]		9	3.5	1.30	47,144	1.23 (0.3-2.7)	0.90 (0.2-2.0)

[a] 90% CI in parentheses derived from published data for the life span study and using exact Poisson methods for the other studies.

[b] The excess absolute risk estimate was computed using background incidence rates estimated from the cervical cancer cohort study [B12].

Table 9
Sex-specific risk estimates for cancer incidence and mortality from the Life Span Study

Cancer type	*Average excess relative risk (Sv^{-1})*		*Average excess absolute risk ($(10^4\ PYSv)^{-1}$)*		*Sex ratio*		
	Males	*Females*	*Males*	*Females*	*Average excess relative risk (Sv^{-1}) (Female : male)*	*Average excess absolute risk ($(10^4\ PYSv)^{-1}$) (Female : male)*	*Normal cancer incidence (Male : female)*
Incidence 1958-1987 [T15]							
All solid tumours	0.33	0.81	18.96	34.44	2.45	1.82	1.36
Non-sex specific solid tumours	0.34	0.88	18.89	26.19	2.59	1.39	1.85
Oesophagus	0.16	1.06	0.41	0.24	6.63	0.59	6.64
Stomach	0.12	0.52	2.61	5.86	4.33	2.25	1.94
Colon	0.87	0.48	2.66	1.01	0.55	0.38	1.35
Liver	0.61	0.11	3.09	0.24	0.18	0.08	2.43
Lung	0.36	2.08	2.67	5.81	5.78	2.18	2.85
Skin	0.92	0.88	0.89	0.72	0.96	0.81	0.86
Bladder	0.35	1.80	0.84	1.02	5.14	1.21	3.60
Thyroid	2.21	1.94	0.99	2.78	0.88	2.81	0.29
Bone and connective tissue	1.78	0.99	0.38	0.12	0.56	0.32	1.25
Incidence 1950-1987 [P33]							
Leukaemia	3.91	4.75	3.35	2.29	1.21	0.68	1.76
Non-Hodgkin's lymphoma	0.91	-0.34	0.73	-0.20	-0.37	-0.27	
Myeloma	1.17	-0.28	0.26	-0.08	-0.24	-0.31	
Mortality 1950-1987 [R23]							
All solid tumours	0.26	0.72	8.74	15.81	2.77	1.81	1.48
Non-sex specific solid tumours	0.25	0.60	7.90	10.45	2.42	1.32	1.82
Oesophagus	0.45	0.94	0.71	0.22	2.09	0.31	6.39
Stomach	0.16	0.35	2.04	2.26	2.19	1.11	1.84
Colon	0.66	0.03	0.82	0.03	0.05	0.04	1.35
Liver	0.49	0.33	2.06	0.66	0.67	0.32	2.02
Lung	0.20	1.96	0.96	3.89	9.80	4.05	2.56
Bladder	1.59	0.41	0.93	0.12	0.26	0.13	1.69
Bone and connective tissue	1.26	0.81	0.29	0.09	0.64	0.31	1.82

Table 10
Risk estimates for cancers other than leukaemia in relation to age at exposure

Age at exposure (years)	*Life span study mortality [R23]*				*Ankylosing spondylitis [D6]*	
	Both sexes		*Males only*		*Mainly males*	
	Excess relative risk (Sv^{-1})	*Excess absolute risk ($(10^4\ PYSv)^{-1}$)*	*Excess relative risk (Sv^{-1})*	*Excess absolute risk ($(10^4\ PYSv)^{-1}$)*	*Excess relative risk (Sv^{-1})*	*Excess absolute risk ($(10^4\ PYSv)^{-1}$)*
0-15	0.95	3.72	0.66	2.72		
15-24	0.74	8.39	0.52	6.71	0.20	1.67
25-34	0.56	14.40	0.39	13.85	0.12	2.41
35-44	0.42	19.35	0.30	19.24	0.14	6.08
45-54	0.31	21.85	0.24	21.98	0.17	12.75
>55	0.23	22.84	0.18	23.72	0.08	8.21

Table 11
Excess relative risk estimates for breast cancer in various cohorts in relation to age at exposure

<table>
<tr><th rowspan="2">Study</th><th colspan="9">Excess relative risk (Sv⁻¹) for age at exposure [a]</th></tr>
<tr><th>0-9 years</th><th>10-19 years</th><th>20-24 years</th><th>25-29 years</th><th>30-34 years</th><th>35-39 years</th><th>40-44 years</th><th>45-49 years</th><th>>50 years</th></tr>
<tr><td>Life span study [T15]</td><td>3.4 (58)</td><td>2.5 (141)</td><td>1.3 (77)</td><td>1.4 (54)</td><td>1.2 (58)</td><td>1.4 (49)</td><td colspan="3">0.7 (92)</td></tr>
<tr><td>Massachusetts TB fluoroscopy [B31]</td><td></td><td>0.7 (71)</td><td colspan="2">0.5 (102)</td><td colspan="2">0.2 (31)</td><td colspan="3">0.2 (13)</td></tr>
<tr><td>Canada TB fluoroscopy [M19]
Nova Scotia
Other provinces</td><td></td><td>
1.9 (14)
0.4 (48)</td><td colspan="2">
1.6 (47)
0.4 (233)</td><td colspan="2">
0.8 (8)
0.2 (73)</td><td colspan="3">
0 (4)
0 (55)</td></tr>
<tr><td>New York acute post-partum mastitis [S9]</td><td></td><td colspan="2">0.4 (33)</td><td>0.4 (44)</td><td colspan="2">0.6 (43)</td><td></td><td></td><td></td></tr>
<tr><td>Swedish benign breast [B1]</td><td></td><td>2.5 (15)</td><td colspan="2">1.9 (18)</td><td colspan="2">0.4 (28)</td><td colspan="2">0.1 (20)</td><td>0.1 (7)</td></tr>
<tr><td>Hodgkin's disease [H2]</td><td></td><td colspan="2">0.9 (4)</td><td colspan="2">0.4 (9)</td><td colspan="4">0.1 (2)</td></tr>
<tr><td>Contralateral breast [B18]</td><td></td><td></td><td></td><td></td><td colspan="3">0.2 (78)</td><td colspan="2">0.0 (128)</td></tr>
</table>

[a] Numbers in parentheses are numbers of cases involved.

Table 12
Excess relative risk for thyroid cancer in various cohorts in relation to age at exposure

<table>
<tr><th rowspan="2">Study</th><th colspan="7">Excess relative risk (Sv⁻¹) for age at exposure</th></tr>
<tr><th><1 year</th><th>1-4 years</th><th>5-9 years</th><th>10-14 years</th><th>15-19 years</th><th>20-39 years</th><th>>40 years</th></tr>
<tr><td>Life span study [T15]</td><td colspan="3">9.5</td><td colspan="2">3.0</td><td>0.3</td><td>-0.2</td></tr>
<tr><td>Israeli tinea capitis [R15]
Israeli-born
Non-Israeli-born</td><td colspan="2">
20.0
67.0</td><td colspan="2">
7.0
27.0</td><td></td><td></td><td></td></tr>
<tr><td>Michael Reese tonsil irradiation [S63]</td><td>3.6</td><td>2.8</td><td>1.4</td><td></td><td></td><td></td><td></td></tr>
</table>

Table 13
Estimated relative risk of cancer mortality per sievert weighted dose (RBE for neutrons = 10) in the life span study cohort [a]
[S7]

Cancer type	*Sex*	*Age at exposure*				
		<10 years	*10-19 years*	*20-29 years*	*30-39 years*	*≥40 years*
Leukaemia	Male Female	17.9 18.8	4.8 5.0	5.9 6.2	4.5 4.7	3.9 4.0
All except leukaemia	Male Female	1.96 2.92	1.60 2.20	1.52 2.04	1.23 1.45	1.16 1.32
Stomach	Male Female	1.40 1.83	1.71 2.47	1.65 2.36	1.09 1.19	1.09 1.20
Female breast	Female	2.54	2.89	1.96	2.09	1.03
Lung	Male Female	1.78 3.93		1.08 1.31	1.22 1.84	1.32 2.19
Colon	Male Female	3.82 8.09		1.39 1.97	1.67 2.67	1.17 1.42

[a] DS86 dosimetry, both cities, 1950-1985.

Table 14
Studies of obstetric x-ray examinations in relation to childhood cancer (including leukaemia) [a]
[B22]

Study	*Units of information*	*Per cent of total*	*Relative risk estimate [b]*	*Comments*
Oxford survey of childhood cancers [c]				
1975 [B23]	434.2	39.1	1.41	
1987 [K6, M22]	812.0	73.0	1.40	RR: 1.94 [K6], ~1.4 [M22]
Prenatal x-ray examinations, United States [M9]	114.7	10.3	1.47	
Inter regional epidemiological study of childhood cancer [H48] [d]	39.0	3.5	1.23	Weighting of matched estimates
Los Angeles [P16]	23.9	2.2	1.34	Leukaemia only
Louisiana [F13]	18.3	1.6	1.70	
Helsinki [S36]	17.9	1.6	1.18	
California [K13]	16.6	1.5	1.48	Leukaemia only
Tri-state [G14]	16.6	1.5	1.40	
Minnesota [A8]	10.2	0.92	1.28	Leukaemia only
Edinburgh/London [C15]	9.0	0.81	0.86	Leukaemia only; cohort study
Baltimore [D16]	8.3	0.75	1.02	Cohort study
Connecticut twins [H3]	5.4	0.49	1.83	RR adjusted for birth weight = 2.4
New Zealand [G15]	5.4	0.49	1.13	Leukaemia only
All others	14.3	1.3	-	
Total	1,111.6	100	1.39 [e]	

[a] The measure of information is the reciprocal of the variance of the log relative risk.
[b] Unadjusted unless otherwise stated.
[c] Based on deaths in England, Wales and Scotland, but not including cases in the Interregional epidemiological study of childhood cancer.
[d] A study of incident cases in north and central England.
[e] 95% CI: (1.31-1.47).

Table 15
Estimated excess absolute risk of incidence of all cancers over ages 0-14 years associated with prenatal x-ray examinations in the Oxford survey of childhood cancers
[M22]

Year of birth	*Excess absolute risk (10^{-4} Gy^{-1}) based on*	
	Deaths during 1964-1979 with adjustment for pregnancy illnesses and drugs	*Deaths during 1953-1979 with no adjustment for pregnancy illnesses and drugs* [a]
1946	203	185 (98-304)
1952	100	96 (50-152)
1957	49	56 (21-98)
1962	27	36 (6-73)

[a] Approximate 95% CI given in parentheses.

Table 16
Estimated excess absolute risk for all cancers except leukaemia in the life span study cohort
[P23]

Age at exposure	*Excess absolute risk* [a] *(10^4 PYSv)$^{-1}$*		
	Follow-up time 1956-1965	*Follow-up time 1966-1975*	*Follow-up time 1976-1985*
0-19 years	0.083	0.31	0.73
20-34 years	0.49	0.79	1.6
≥35 years	0.35	1.4	3.5

[a] Estimates are for males. The corresponding values for females are obtained by multiplying by a factor 1.56.

Table 17
Estimated relative risk of mortality for ranges of absorbed dose above and below 2.0 Gy in the life span study cohort [a]
[S7]

Site of cancer	*Organ absorbed dose (Gy)* [b]						
	0.01-0.05	*0.06-0.09*	*0.10-0.19*	*0.20-0.49*	*0.50-0.99*	*1.00-1.99*	*≥ 2.00*
All cancers except leukaemia	1.06 (1.0-1.12)	1.08 (0.98-1.19)	1.06 (0.97-1.16)	1.12 (1.03-1.21)	1.36 (1.23-1.51)	1.66 (1.45-1.90)	2.05 (1.66-2.50)
Leukaemia	0.99 (0.68-1.40)	0.61 (0.25-1.22)	1.08 (0.61-1.82)	1.79 (1.18-2.68)	4.15 (2.76-6.19)	8.01 (5.34-11.9)	18.6 (12.1-28.2)
Stomach	1.06 (0.96-1.16)	0.93 (0.78-1.10)	1.05 (0.90-1.22)	1.16 (1.02-1.32)	1.28 (1.07-1.53)	1.29 (0.99-1.64)	1.73 (1.19-2.43)
Lung	1.30 (1.11-1.54)	1.21 (0.89-1.62)	1.02 (0.75-1.36)	1.54 (1.22-1.93)	1.63 (1.19-2.19)	2.45 (1.73-3.38)	2.14 (1.16-3.59)
Female breast	1.12 (0.80-1.59)	1.02 (0.49-1.89)	1.10 (0.60-1.89)	1.39 (0.85-2.21)	2.67 (1.62-4.27)	2.39 (1.09-4.59)	4.22 (1.77-8.54)
Colon	1.04 (0.80-1.37)	1.01 (0.61-1.59)	0.53 (0.29-0.91)	0.93 (0.60-1.38)	1.04 (0.56-1.78)	2.23 (1.20-3.81)	5.87 (3.02-10.3)

[a] DS86 dosimetry, both cities, both sexes, all ages at exposure, 1950-1985; comparison is with the control (0 Gy) group.
[b] 90% CI given in parentheses.

Table 18
Adjusted and fitted relative risk of thyroid tumours among Israeli children irradiated for tinea capitis [a]
[R15]

Absorbed dose in thyroid (Gy)	*Malignant tumours*				*Benign tumours*			
	Observed cases	*Person-years*	*Relative risk*		*Observed cases*	*Person-years*	*Relative risk*	
			Adjusted [b]	*Fitted [c]*			*Adjusted [b]*	*Fitted [c]*
0	16	412,030	1.0	1.0	41	316,840	1.0	1.0
0.04-0.07 (mean dose 0.062)	15	106,690	3.3	3.0	17	82,350	1.1	1.9
0.08-0.14 (mean dose 0.102)	24	149.720	4.2	4.3	32	114,930	3.0	2.4
0.15-0.50 (mean dose 0.214)	4	17,770	6.1	8.0	6	12,840	4.5	4.0

[a] Considerable uncertainty may exist in the estimates of dose to the thyroid gland.
[b] Adjusted for sex, ethnic origin and attained age.
[c] Assuming a linear increase in relative risk with increasing dose.

Table 19
Cancer incidence and risks of breast cancer among women with tuberculosis given chest x-ray fluoroscopy
[B31]

Absorbed dose in breast (Gy)	*Number of women*	*Person-years*	*Mean absorbed dose (Gy)*	*Breast cancer cases*			*Relative risk [a b]*	*Excess risk [a b] $(10^4\ PY)^{-1}$*
				Observed	*Expected*	*Observed/ expected*		
0	2,367	48,919	0	87	100.9	0.86	1.00 (1.00)	0.00 (0.00)
0.01-0.99	1,675	33,724	0.36	75	70.6	1.06	1.21 (1.18)	3.89 (3.36)
1.0-1.99	553	15,453	1.36	44	28.0	1.57 [c]	1.66 [c] (1.76)	11.3 [c] (12.9)
2.0-2.99	135	3,757	2.42	14	6.62	2.11 [c]	2.22 [c] (2.46)	20.5 [c] (24.3)
>3	64	1,675	3.75	9	2.44	3.68 [c]	3.83 [c] (3.60)	39.7 [c] (36.2)
Unknown	146	2,356	-	5	5.98	0.84	1.02	0.45
Total [d]	2427	54,609	0.79	142	107.6	1.32 [c]	1.44 [c] (1.48)	8.00 [c] (8.50)

[a] Based on maximum likelihood methods to compute adjusted risk estimates of the ratio observed/expected values between exposed and non-exposed, adjusting for differences in background rates between subcohorts. Excess risk also derived from this approach.
[b] Fitted relative risk and fitted excess risk in parentheses. Based on best-fitting linear dose-response model accounting for age, calendar year, follow-up and subcohort.
[c] $p < 0.05$ (one-sided).
[d] For exposed individuals with known dose (excludes 0 and unknown dose).

Table 20
Evaluation of various dose-response models for breast cancer incidence among women with tuberculosis given chest x-ray fluoroscopy
[B31]

Model	*Mathematical expression [a]*	*Parameter*	*Estimate [b]*	*95% CI*	*Deviance [c] difference*
Linear	$I = a_{age,time}(1 + a_1D)$	a_1	0.71	(0.40-1.08)	0.26
Quadratic	$I = a_{age,time}(1 + a_2D^2)$	a_2	0.22	(0.10-0.41)	6.29
Linear-quadratic	$I = a_{age,time}(1 + a_1D + a_2D^2)$	a_1 a_2	0.85 -0.07	(0.19-1.73) (-0.28-0.21)	0.00

[a] I denotes breast cancer incidence. D denotes breast dose (cGy). $a_{age,time}$ denotes background parameter adjusted for age and time. The model also includes a multiplicative age-at-exposure effect on the excess relative risk (not shown).
[b] Based on these models, the estimated RRs for a 20-year-old exposed to 1 Gy are 1.70, 1.22 and 1.81 for the linear, quadratic and linear-quadratic models, respectively.
[c] The deviance is a measure of unexplained variance used to assess goodness of fit of different nested models. The deviance differences are between the indicated model and the linear-quadratic model. This difference can be interpreted as a single degree-of-freedom χ^2 statistic of the adequacy of the simpler method. The results indicate that the linear model describes these data as well as the linear-quadratic model ($p > 0.5$), but that the pure quadratic model does not fit well ($p = 0.01$).

Table 21
Dose fractionation effects in thyroid cancers among individuals in Rochester, New York, given x-ray treatment in infancy for supposed enlarged thymuses [a]
[S10]

Radiation characteristics	*Mean total dose (Gy)*	*Person-years*	*Number of cancers observed*	*Excess absolute risk* $(10^4\ PYGy)^{-1}$ [b]
Dose per fraction (Gy)				
0.01-0.49	0.18	34,268	4	6.1
0.50-1.99	2.22	8,622	6	2.2
2.0-5.99	3.14	14,340	12	2.3
Number of dose fractions				
1	0.74	29,414	7	2.9
2	1.55	22,417	6	1.5
3	2.54	5,445	9	3.8
Average interval between treatments				
0 days	0.74	29,334	7	2.9
1-4 days	1.41	13,896	5	2.4
>4 days	2.07	13,980	10	1.9

[a] Data shown are for individuals with total thyroid doses of less than 6 Gy.

[b] Excess in comparison to control group rates, with adjustment by sex, ethnicity and interval since irradiation to the combined distribution of these groups plus the control group.

Table 22
Summary of studies of medical personnel exposed to radiation showing cancer types with significantly (p < 0.005) increased risks
[C25]

Study	*Number of persons*	*Study period*	*Period of subgroup entry*	*Cancer types in excess*	*Relative risk* [a]	*Number of cases*
Mortality in male radiologists (United Kingdom) [S38]	1,338	1897-1954	Pre-1921	All [b]	1.41	64
				Pancreas	3.23	6
				Lung	2.18	8
				Skin	7.79	6
				Leukaemia	6.15	4
			1920-1954	No cancers differed significantly from expectation in this subgroup but cancer mortality increased significantly with time since entry to the study		
Mortality in male radiologists (United States) [M24]	6,524	1920-1969	1920-1939	All	1.38	
				Mouth and pharynx	2.17	
				Liver (secondary)	3.22	
				Skin	3.38	
				Unspecified sites	1.69	
				Acute leukaemia	2.23	
				Myeloid leukaemia	2.39	
				Lymphosarcoma	2.73	
			1940-1969	All	1.15	
				Mouth and pharynx	2.88	
				Lung	1.22	
				Multiple myeloma	2.05	
Cancer incidence in Chinese x-ray workers [W14]	27,011	1950-1985	Pre-1960	All	1.3	193
				Oesophagus	5.5	13
				Liver	1.8	37
				Skin	2.9	7
				Leukaemia	2.6	17
			1960-1969	All	1.3	90
				Liver	2.2	20
				Skin	1.6	1
				Leukaemia	3.0	13
			1970-1985	All	0.8	49
				Oesophagus	3.8	2
				Liver	1.2	8
				Skin	4.8	1
				Leukaemia	1.3	4
Mortality in Japanese radiological technologists [A9]	9,179	1969-1986	1969-1982 [c]	All	1.32	137
				Large intestine	2.22	20
				Neurocerebral	8.16	4
			1983-1986 [c]	Leukaemia	5.83	7
				Other [d]	1.92	27
Cancer incidence in Danish radiotherapists [E7]	4,281	1954-1982 [e]	≥10 years [c]	All	1.13	103
				Prostate	6.25	4

[a] Adjusted for age and period.
[b] Expected deaths calculated using rates in social class I males; RR = 1.72, using rates in male medical practitioners.
[c] Period of observation.
[d] All malignant neoplasms other than oesophagus, stomach, liver, pancreas, lung and leukaemia.
[e] Provisional results available only to 1984.

Table 23
Cancer incidence in diagnostic x-ray workers in China
[W14]

Age at first employment	*All cancers except leukaemia*		*Leukaemia*		*Breast cancer*		*Thyroid cancer*	
	Observed number	*Rate ratio*	*Observed number*	*Rate ratio*	*Observed number*	*Rate ratio*	*Observed number*	*Rate ratio*
<20 years	24	1.5	5	5.9	0	0.0	2	5.4
20-24 years	70	1.1	15	4.1	6	1.4	3	2.0
25-29 years	91	1.3	8	2.2	10	3.3	2	1.6
30-34 years	60	1.3	5	1.9	4	1.6	1	1.2
35-39 years	28	0.9	1	0.6	0	0.0	0	0.0
≥40 years	25	0.8	0	0.0	0	0.0	0	0.0
All ages	298	1.1	34	2.4	20	1.5	8	1.7

Table 24
Parameter estimates from the life span study used in lifetime risk computations [a]

Site	*Males*		*Females*	
	Excess relative risk *(Sv^{-1})*	*Age-at-exposure effect*	*Excess relative risk* *(Sv^{-1})*	*Age-at-exposure effect*
Oesophagus	0.23	0.015	1.59	0.015
Stomach	0.16	-0.035	0.62	-0.035
Colon	0.54	-0.033	1.00	-0.033
Liver	0.97	-0.027	0.32	-0.027
Lung	0.37	0.021	1.06	0.021
Bladder	1.00	0.012	1.19	0.012
Breast	-	-	1.95	-0.079
Ovary	-	-	1.42	-0.042
Other	0.59	-0.059	0.39	-0.059
All solid tumours	0.45	-0.026	0.77	-0.026

[a] The parameters are a (excess relative risk per sievert) and b (age-at-exposure effect) in the model ERR(D,e) = aD exp [b(e - 25)], where D is the weighted dose in intestines (RBE for neutrons = 10) and e is the age at exposure.

Table 25
Estimates of excess relative risk per sievert for solid tumours based on fits to data of the life span study [a]

Site	*Excess relative risk (Sv^{-1})*					
	For males at age of exposure			*For females at age of exposure*		
	10 years	*30 years*	*50 years*	*10 years*	*30 years*	*50 years*
Oesophagus	0.18	0.25	0.34	1.26	1.72	2.34
Stomach	0.27	0.13	0.07	1.05	0.52	0.26
Colon	0.90	0.46	0.23	1.64	0.84	0.43
Liver	1.45	0.85	0.49	0.48	0.28	0.16
Lung	0.27	0.41	0.63	0.78	1.17	1.77
Bladder	0.83	1.06	1.36	0.99	1.27	1.62
Breast	-	-	-	6.38	1.32	0.27
Ovary	-	-	-	2.64	1.15	0.50
Other	1.43	0.44	0.13	0.95	0.29	0.09
All solid tumours	0.66	0.40	0.23	1.14	0.68	0.40

[a] The parameters and model are presented in Table 24 and its footnote.

Table 26
Estimates of excess absolute risk for leukaemia for acute exposures of 0.2 Sv and 1 Sv of low-LET radiation based on fits to data of the life span study [a]
[P33]

Age at exposure (years)	*Excess absolute risk for males $(10^4\ PYSv)^{-1}$* [b]						*Excess absolute risk for females $(10^4\ PYGy)^{-1}$* [b]					
	0.2 Sv			*1 Sv*			*0.2 Sv*			*1 Sv*		
	5 years after exposure	*20 years after exposure*	*40 years after exposure*	*5 years after exposure*	*20 years after exposure*	*40 years after exposure*	*5 years after exposure*	*20 years after exposure*	*40 years after exposure*	*5 years after exposure*	*20 years after exposure*	*40 years after exposure*
0-19	2.3	0.18	0.01	17.7	1.4	0.05	0.62	0.22	0.05	4.8	1.7	0.4
20-39	1.5	0.21	0.02	11.6	1.6	0.12	0.41	0.26	0.14	3.2	2.0	1.1
≥40	1.2	0.43	0.11	9.5	3.3	0.82	0.34	0.53	0.96	2.3	3.9	8.2

[a] Using an excess absolute risk model with time-, sex- and age-at-exposure dependence (see equation 14).
[b] Dose is weighted dose to bone marrow (RBE for neutrons = 10).

Table 27
Death rates for selected causes in the population of Japan, 1985
[J5]

Age category (years)	Males				Females				Both sexes			
	Population (thousands)	Death rates (10^{-5})			Population (thousands)	Death rates (10^{-5})			Population (thousands)	Death rates (10^{-5})		
		All causes	*Leukaemia*	*Solid tumours*		*All causes*	*Leukaemia*	*Solid tumours*		*All causes*	*Leukaemia*	*Solid tumours*
0-4	3,792	158.8	1.3	2.9	3,615	131.2	1.4	2.4	7,408	145.3	1.4	2.6
5-9	4,345	26.6	2.6	2.1	4,131	15.3	1.7	1.5	8,476	21.1	2.2	1.8
10-14	5,115	19.9	2.2	2.0	4,865	13.1	1.6	1.9	9,980	16.5	1.9	2.0
15-19	4,571	69.8	2.7	3.5	4,351	23.7	1.7	2.1	8,922	47.2	2.2	2.8
20-24	4,134	81.4	2.2	3.6	4,000	31.8	1.3	3.2	8,133	57.1	1.8	3.3
25-29	3,915	80.7	2.1	6.9	2,840	40.7	1.5	8.3	7,754	60.9	1.8	7.6
30-34	4,524	93.3	2.8	12.3	4,461	55.6	1.9	16.9	8,985	74.5	2.3	14.6
35-39	5,365	131.9	3.4	25.5	5,309	76.0	2.3	29.5	10,675	104.2	2.8	27.5
40-44	4,527	227.7	3.8	50.4	4,559	124.1	3.0	52.4	9,086	175.6	3.4	51.4
45-49	4,072	371.7	4.8	98.4	4,125	184.6	3.6	79.9	8,197	277.1	4.2	89.0
50-54	3,911	624.6	5.8	207.7	3,991	289.7	4.7	124.2	7,902	455.6	5.2	165.6
55-59	3,395	906.7	7.5	356.4	3,577	414.9	5.4	180.3	6,972	654.3	6.5	266.0
60-64	2,365	1,314.9	10.2	536.5	3,013	663.0	6.4	269.2	5,377	948.7	8.1	386.3
65-69	1,770	2,159.4	14.6	822.3	2,404	1,106.4	7.6	382.1	4,174	1,554.0	10.6	569.2
70-74	1,497	3,707.7	20.4	1,204.2	2,053	1,998.4	9.9	555.4	3,549	2,717.5	14.3	828.4
75-79	1,014	6,581.0	25.8	1,705.0	1,472	3,871.3	12.9	794.3	2,485	4,980.5	18.2	1,167.1
80-84	541	10,799.1	25.6	2,019.0	889	7,165.7	12.5	980.2	1,429	8,540.5	17.4	1,373.3
85-89	203	18,136.2	24.5	2,257.3	399	13,067.1	12.7	1,142.8	602	14,725.6	16.5	1,510.4
>90	53	25,429.3	21.7	1,754.5	129	22,490.8	10.0	1,018.4	181	23,364.8	13.5	1,240.3

Table 28
Comparison of measures of lifetime risk of mortality from solid tumours following acute whole-body exposure of 1 Sv [a b]

Age at exposure (years)	*Risk of exposure-induced death (REID) (%)*	*Excess lifetime risk (ELR) (%)*
10	17.0	14.5
30	9.3	8.1
50	7.0	6.4
70	2.8	2.6

[a] Estimates were computed using sex- and age-at-exposure-specific relative risks estimated for the life span study cancer mortality data for the period 1950-1987. Risks were applied to 1985 death rates in Japan [J5]. Since sex-specific estimates of the excess risk were virtually identical, the estimates in this Table are averages of both the sexes. Projection beyond the follow-up period assumes constant relative risks.

[b] Dose is weighted dose with neutron RBE = 10.

Table 29
Estimates of lifetime risk for leukaemia following acute whole-body exposures of 0.2 Sv or 1 Sv from low-LET radiation

Age at exposure (years)	*Risk of exposure-induced death (REID) (%) [a]*			*Years of life lost per case (YLC) [b]*		
	Males	*Females*	*Both*	*Males*	*Females*	*Both*
Exposure of 0.2 Sv [c]						
Newborn	0.22	0.11	0.16	67	64	66
5	0.22	0.11	0.16	62	59	60
10	0.22	0.10	0.16	58	54	56
15	0.22	0.10	0.16	53	50	52
20	0.16	0.12	0.14	46	38	42
25	0.16	0.11	0.14	42	34	38
30	0.16	0.11	0.14	37	30	34
35	0.16	0.10	0.13	33	27	30
40	0.18	0.24	0.21	25	16	20
45	0.17	0.19	0.18	21	14	18
50	0.16	0.15	0.16	18	13	15
55	0.15	0.12	0.13	15	11	13
60	0.13	0.09	0.11	12	9	11
65	0.11	0.06	0.09	9	8	8
70	0.09	0.04	0.07	7	6	6
Exposure of 1 Sv [c]						
Newborn	1.66	0.81	1.23	67	64	66
5	1.67	0.81	1.24	63	59	61
10	1.67	0.81	1.24	58	54	56
15	1.66	0.80	1.23	53	50	51
20	1.25	0.91	1.08	46	38	42
25	1.25	0.88	1.06	42	34	38
30	1.24	0.84	1.04	37	31	34
35	1.23	0.80	1.01	33	27	30
40	1.40	1.80	1.60	25	16	20
45	1.34	1.45	1.39	21	14	18
50	1.25	1.15	1.20	18	13	15
55	1.15	0.90	1.03	15	11	13
60	1.03	0.68	0.86	12	9	10
65	0.88	0.50	0.69	9	8	8
70	0.71	0.34	0.53	7	6	6

[a] Computed using time-, sex- and age-at-exposure-specific excess absolute risk model for the life span study leukaemia incidence data for 1950-1987 [P33]. Risks were applied to Japanese death rates of 1985 [J5].

[b] Average years of life lost per exposure-induced death. The average years of life lost per exposed person is equal to years of life lost per case multiplied by risk of exposure-induced death divided by 100.

[c] Dose is weighted dose with neutron RBE = 10.

Table 30
Estimates of projected lifetime risk of solid tumours following acute whole-body exposure of 0.2 Sv or 1 Sv from low-LET radiation

Age at exposure (years)	*Risk of exposure-induced death (REID) (%)* [a] *for projection method with a 10-year latent period and relative risk*			*Years of life lost per case (YLC)* [b] *for projection method with a 10-year latent period and relative risk*		
	Assumed constant from 10 years after exposure	*Declining to risk for age 50 years at exposure* [c]	*Declining to zero risk at age 90 years* [d]	*Assumed constant from 10 years after exposure*	*Declining to risk for age 50 years at exposure* [c]	*Declining to zero risk at age 90 years* [d]
			Exposure of 0.2 Sv [e]			
Newborn	5.8	3.5	2.6	14.3	17.4	20.2
5	4.6	3.1	2.2	14.0	16.6	19.7
10	3.8	2.8	2.0	13.8	15.9	18.9
15	3.1	2.5	1.8	13.5	15.1	18.1
20	2.6	2.2	1.6	13.2	14.3	17.1
25	2.2	2.0	1.5	12.8	13.6	16.0
30	2.0	1.9	1.5	12.4	12.8	14.7
35	1.8	1.7	1.5	11.9	12.1	13.5
40	1.6	1.6	1.5	11.4	11.5	12.3
45	1.5	1.5	1.5	10.7	10.7	11.2
50	1.5	1.5	1.5	10.0	10.0	10.0
55	1.4	1.4	1.4	9.1	9.1	9.1
60	1.3	1.3	1.3	8.3	8.3	8.3
65	0.8	0.8	0.8	5.9	5.9	5.9
70	0.6	0.6	0.6	4.2	4.2	4.2
			Exposure of 1 Sv [e]			
Newborn	24.9	16.1	12.1	15.3	17.9	20.6
5	20.6	14.4	10.6	14.8	17.1	20.0
10	17.0	12.9	9.3	14.4	16.3	19.2
15	14.2	11.6	8.4	14.0	15.5	18.3
20	12.1	10.6	7.8	13.6	14.7	17.3
25	10.5	9.7	7.5	13.1	13.9	16.1
30	9.3	8.9	7.3	12.7	13.1	14.9
35	8.4	8.3	7.2	12.1	12.3	13.6
40	7.8	7.8	7.2	11.6	11.6	12.4
45	7.4	7.4	7.1	10.9	10.9	11.3
50	7.0	7.0	7.0	10.1	10.1	10.1
55	6.7	6.7	6.7	9.2	9.2	9.2
60	6.4	6.4	6.4	8.4	8.4	8.4
65	4.1	4.1	4.1	6.0	6.0	6.0
70	2.8	2.8	2.8	4.3	4.3	4.3

[a] Estimated percentage of population that would die due to radiation-induced cancer. Computed using sex- and age-at-exposure-specific relative risk estimated for the life span study cancer mortality data for 1950-1987. Risks were applied to Japanese death rates of 1985 [J5]. Since sex-specific estimates of the excess risk were virtually identical, the estimates are averaged over sex.

[b] Average years of life lost per exposure-induced death. The average years of life lost per exposed person is equal to years of life lost per case multiplied by risk of exposure-induced death divided by 100.

[c] Constant relative risk for first 45 years after exposure; risk then decreases linearly with age. At attained age 90 years, the risk is equal to that for a person aged 50 years at exposure.

[d] Constant risk for first 45 years after exposure; risk then decreases linearly with age. At attained age 90 years, the risk is zero.

[e] Dose is weighted dose with neutron RBE = 10.

Table 31
Comparison of estimates of lifetime risk of mortality from solid tumours and leukaemia following acute whole-body exposures of 0.2 and 1 Sv [a]

Projection method	*Risk of exposure-induced death (REID) (%)*		*Years of life lost per case (YLC)*	
	Following exposure of 0.2 Sv	*Following exposure of 1 Sv*	*Following exposure of 0.2 Sv*	*Following exposure of 1 Sv*
Solid tumours				
Constant relative risk [b]	2.4	10.9	11.2	11.6
Decline to risk for age at exposure 50 [b]	1.9	9.2	12.1	12.3
Decline to zero risk at age 90 years [b]	1.6	7.5	13.6	13.3
Constant relative risk (UNSCEAR 1988) [c]	-	9.7	-	11.4
Leukaemia				
Linear-quadratic dose-response model [d]	0.14	1.1	31	31
Constant relative risk (UNSCEAR 1988) [c]	-	1.0	-	26

[a] Dose is weighted dose with neutron RBE = 10. Values are projections based on data for 1950-1987 [R23]. Age distribution of population is that of Japan in 1985 [J5].
[b] Based on a linear dose-response model with sex- and age-at-exposure-specific risks. Projection methods are described in footnotes c and d of Table 30.
[c] Risks computed using age-specific risk coefficients and multiplicative risk projection [U2].
[d] Projection was made using the excess absolute risk model fit to the life span study pooled leukaemia data, using a non-linear dose-response model.

Table 32
Site-specific lifetime risks for solid tumours and leukaemia following a whole-body acute exposure of 1 Sv [a]

Site of cancer	*Risk of exposure-induced death (REID) (%)*		
	Males	*Females*	*Both*
Leukaemia [b]	1.3	0.9	1.1
Oesophagus	0.3	0.7	0.5
Stomach	0.9	2.0	1.4
Colon	0.5	0.6	0.6
Liver	2.2	0.3	1.2
Bladder	0.4	0.2	0.3
Lung	1.8	3.1	2.5
Breast	-	2.0	1.0
Ovary	-	0.5	0.3
Other	4.3	2.0	3.1
Total (except leukaemia) [c]	10.4	11.4	10.9
Total	11.7	12.3	12.0

[a] Projections are based on age-at-exposure-specific values computed using death rates for Japan in 1985 [J5]. Rates were averaged over age at exposure using the population of Japan in 1985.
[b] Leukaemia risks were computed using the excess absolute risk model presented in [P33]. This model has a non-linear dose response and the risk varies with time, sex and age at exposure. Projection beyond the current follow-up was based on this model.
[c] Solid tumour risks were computed using linear dose-response models with age-at-exposure and sex-specific relative risks and a 10-year latency period.

Table 33
Standardized mortality ratios for all causes of death and for malignant neoplasms for nuclear energy workers [D10]

Nuclear installation	*Standardized mortality ratios [a] for deaths due to*					
	All causes	*All malignant neoplasms*	*Cancer of digestive system*	*Lung cancer*	*Cancer of lymphatic and haematopoietic system*	*Leukaemia*
United Kingdom						
United Kingdom Atomic Energy Authority	0.76	0.78	0.79 (108)	0.71 (153)	0.98 (38)	1.2 (18)
Sellafield [b]	0.96	0.96	1.07 (132)	0.88 (147)	0.93 (27)	0.83 (10)
United Kingdom Atomic Weapons Establishment [c]	0.73	0.79	0.84 (90)	0.64 (85)	0.48 (11)	0.44 (4)
United States						
United Nuclear	0.83	0.97	0.84 (107)	1.02 (14)	0.88 (4)	1.19 (2)
Pantex	0.72	0.63	0.45 (8)	0.51 (12)	0.89 (7)	1.34 (4)
Rocky Flats [b]	0.63	0.71	0.73 (25)	0.65 (30)	0.64 (9)	0.75 (4)
Portsmouth	0.78	0.96	-	-	0.74 (15)	0.88 (7)
Oak Ridge National Laboratory	0.73	0.80	0.67 (41)	0.75 (59)	1.04 (28)	1.50 (16)
Oak Ridge, Y-12 [d]	1.00	0.96	0.73 (210)	1.23 (324)	0.83 (77)	1.04 (40)
Oak Ridge, Y-12 [e]	0.89	1.01	0.74 (33)	1.36 (89)	0.90 (19)	0.50 (4)
Oak Ridge federal nuclear plants	1.11	1.05	0.80 (490)	1.27 (850)	0.98 (195)	1.13 (92)
Hanford	0.78	0.85	0.86 (310)	0.84 (339)	0.82 (103)	0.71 (36)
Linde Air Products	1.18	1.12	1.38 (27)	1.02 (21)	0.93 (6)	-
Savannah River	0.75	0.74	0.70 (50)	0.83 (83)	0.95 (32)	1.50 (18)
Naval shipyards	0.76	0.99	0.94	1.07	0.82	0.91

[a] Observed deaths in parentheses.
[b] Substantial proportion of radiation workers monitored for plutonium.
[c] Substantial proportion of workers exposed to tritium.
[d] Workers employed 1943-1947.
[e] Workers employed only after 1947.

Table 34
Estimates of excess relative risk per sievert for all cancers and leukaemia from studies of workers in nuclear and related industries [a]
[C34]

Nuclear installation	*Subjects*	*All cancers*			*Leukaemia*			*Other observations*		*Ref.*
		Number of deaths	*Excess relative risk* (Sv^{-1})	*90% CI*	*Number of deaths*	*Excess relative risk* (Sv^{-1})	*90% CI*	*Cancer type*	*Number of deaths*	
Canada										
Atomic Energy of Canada Ltd.	8,977	219 [b]	0.05	-0.68-2.17	4 [c]	19.0	0.14-113			[G3]
United Kingdom										
Atomic Energy Authority	21,545	720 [b]	0.8	-1.0-3.1 [d]	31	-4.2	-5.7-2.6	Uterus Prostate	8 42	[F11]
Atomic Weapons Establishment	9,389	275	7.6	0.4-15.3 [d]	4	-	-	Lung Prostate [e]	85 20	[B5]
Sellafield	10,157	396	-	-	10	[f]	[f]	Bladder Multiple myeloma Lympho-haematopioetic	14 7 27	[S14]
National Registry for Radiation Workers	95,217	1,435	0.47	-0.12-1.2	47 [c]	4.28	0.40-13.6	Thyroid	9	[K20]
United States										
Hanford	36,971	1,413 [b]	-0.0	-0-1.9	44 [c]	-1.1	-0-1.9	Hodgkin's disease Pancreas	19 77	[G16]
Oak Ridge National Laboratory	8,313	346	3.27	1.26-5.3	28	6.38 [g]	-11.2-24	Lung	96	[W21]
Combined [h]	35,933	1,036	-1.0	<0-0.4	42 [c]	<0	<0-3.4	Multiple myeloma	12	[G19]

a Doses are lagged by 10 years for cancers and 2 years for leukaemia unless otherwise specified.
b Excluding leukaemia.
c Excluding chronic lymphocytic leukaemia.
d 95% confidence interval.
e Only among workers monitored for exposure to radionuclides.
f Significant trend.
g Doses are not lagged.
h Hanford, Oak Ridge National Laboratory and Rocky Flats combined.

Table 35
Risk estimates for all cancer and leukaemia (excluding chronic lymphocytic leukaemia) from occupational studies and survivors of the atomic bombings in Japan

Study	*Excess relative risk (Sv⁻¹)* [a]
All cancer	
United Kingdom [K20] National Registry for Radiation Workers [b]	0.47 (-0.12-1.20)
United States [G17] Hanford, Oak Ridge National Laboratory and Rocky Flats combined [c]	-1.0 (<0-0.4)
United Kingdom and United States combined [K21]	0.23 (<0-0.83)
Survivors of atomic bombings in Japan [d] Males only, exposed over age 20 years 1950-1970 (linear relative risk model, RBE = 10)	0.33 (0.11-0.60)
Leukaemia (excluding chronic lymphocytic leukaemia)	
United Kingdom [K20] National Registry for Radiation Workers [b]	4.3 (0.40-13.6)
United States [G17] Hanford, Oak Ridge National Laboratory and Rocky Flats combined [c]	<0 [e] (<0-3.4)
United Kingdom and United States combined [K21]	1.7 (<0-5.9)
Survivors of atomic bombings in Japan [d] [R23] Males only exposed over age 20 years 1950-1970 (linear-quadratic relative risk model, RBE = 10)	6.2 (2.7-13.8)

[a] Estimates for occupational studies based on a linear relative risk model. Doses lagged for 10 years (all cancer) or 2 years (leukaemia). 90% CI in parentheses.
[b] Based on monitored workers at sites operated by British Nuclear Fuels (including Sellafield), Ministry of Defence, United Kingdom Atomic Energy Authority and Nuclear Electric.
[c] Based on monitored white males employed at least six months at the Hanford site, Oak Ridge National Laboratory or at the Rocky Flats nuclear weapons plant.
[d] No low-dose-rate reduction factor has been included.
[e] Likelihood maximized at a value that would have led to negative excess relative risks.

Table 36
Comparison of risk estimates for mortality in survivors of atomic bombings in Japan and nuclear workers in the United Kingdom and in the United States

Group	*Size of cohort*	*Person-years*	*Collective dose (man Sv)*	*Average dose (mSv)*	*Excess relative risk (Sv⁻¹)* [a]		*Lifetime risks (% Sv⁻¹)* [a]	
					All cancer	*Leukaemia*	*All cancer*	*Leukaemia*
Survivors of atomic bombings [S7]	75,991	2,185,000	10,500	251	0.39 (0.32-0.46)	5.2 (3.8-7.1)	4 [b] (3-5)	0.4 [b] (0.3-0.55)
Nuclear workers in the United Kingdom [K20]	95,217	1,218,000	3,198	34	0.47 (-0.12-1.20)	4.3 (0.40-13.6)	10 (<0-26)	0.76 (0.07-2.4)
Nuclear workers in the United States [G17]	35,933	705,000	1,140	32	-0.99 (-1.6-0.38)	<-1.5 (<-1.5-3.4)	<0 (<0-8.2)	<0 (<0-0.60)

[a] 90% CI in parentheses.
[b] Based on ICRP [I10] with low-dose-rate reduction factor of 2.

Table 37
Cancer mortality among workers at the Mayak nuclear plant [a]
[N4]

Employees at	Percentage of deaths from cancer for total gamma dose	
	<1.0 Sv	>1.0 Sv
Facility A	5.7 ± 0.6	9.4 ± 1.2
Facility B	4.3 ± 0.4	8.1 ± 0.6
Both facilities	4.8 ± 0.4	8.4 ± 0.5

[a] Employees who started work before 1958.

Table 38
Cancer mortality among male workers at the Mayak reactors and reprocessing plant
[K23]

	Reactor workers		Reprocessing plant workers	
	1949-1953	1954-1958	1949-1953	1954-1958
Average cumulative gamma dose (Sv)	1.22	4.92	2.45	7.16
Number of workers with film badge record	1,286	509	1,812	1,479
Person-years (to 31 December 1989)	45,947	15,730	61,649	44,464
All cancer				
Observed	101	19	197	55
Expected	126.4	26.3	153.2	61.3
Observed/expected [a]	0.80 (0.7-0.9)	0.72 (0.4-1.1)	1.28 (1.1-1.4)	0.9 (0.7-1.1)
Leukaemia				
Observed	6	2	25	6
Expected	5.68	1.41	7.11	3.59
Observed/expected [a]	1.06 (0.5-2.1)	1.42 (0.3-4.5)	3.52 (2.4-4.4)	1.67 (0.7-3.3)

[a] 95% CI in parentheses.

Table 39
Lung cancer deaths in workers at the Mayak radiochemical plant
[H17]

Cumulative equivalent dose in lungs (Sv)	Number of workers	Observed cases	Expected cases
0-0.25	470	0	2.1
0.26-1.00	607	4	6.2
1.01-4.0	929	19	16.2
>4.00	340	22	8.1
All cases	2,346	45	32.6

Table 40
Mortality from breast cancer among female luminizing workers in the United Kingdom [B35]

	Absorbed dose from gamma rays	
	<0.2 Gy	*≥0.2 Gy*
	Women <30 years old at start of work	
Mean absorbed dose (Gy)	0.085	0.51
Number at risk	255	678
Person-years	8,569	27,299
Observed	5	16
Expected	3.00	10.62
Observed/expected	1.67 (p = 0.18) [a]	1.51 (p = 0.074) [a]
90% CI	(0.7-3.5)	(0.9-2.3)
	Women ≥30 years old at start of work	
Mean absorbed dose (Gy)	0.10	0.50
Number at risk	110	160
Person-years	3,200	5,816
Observed	5	2
Expected	2.39	4.49
Observed/expected	2.09 (p = 0.09) [a]	0.45 (p = 0.94) [a]
90% CI	(0.8-4.4)	(0.1-1.4)

[a] Single-sided significance value.

Table 41
Odds ratios for mortality from leukaemia among residents of Utah exposed to radioactive fallout from the Nevada test site [S48]

	Odds ratios [a]			*Test for trend*
	Low dose (0.0-2.9 mGy)	*Intermediate dose (3.0-5.9 mGy)*	*High dose (6.0-30 mGy)*	
All leukaemias	1.00	1.08 (0.91-1.27)	1.69 (1.01-2.84)	NS
All leukaemia except chronic lymphocytic leukaemia	1.00	1.08 (0.89-1.30)	1.72 (0.94-3.12)	NS
Age at exposure				
0-19 years	1.00	1.13 (0.74-1.72)	3.97 (1.18-13.3)	NS
20-49 years	1.00	0.99 (0.74-1.33)	1.39 (0.44-4.36)	NS
≥50 years	1.00	1.17 (0.86-1.58)	1.36 (0.56-3.30)	NS
Age at death				
0-19 years	1.00	1.20 (0.62-2.32)	5.82 (1.55-21.8)	p = 0.022
20-49 years	1.00	0.92 (0.59-1.42)	2.97 (0.56-15.6)	NS
50-64 years	1.00	0.82 (0.56-1.21)	1.48 (0.36-6.16)	NS
>65 years	1.00	1.26 (0.97-1.64)	1.06 (0.43-2.59)	NS
Year of death				
1952-1957	1.00	1.30 (0.43-3.88)	3.09 (0.76-12.5)	p = 0.041
1958-1963	1.00	0.89 (0.60-1.33)	2.40 (0.81-7.08)	NS
1964-1969	1.00	1.31 (0.89-1.91)	2.57 (0.76-8.67)	NS
1970-1975	1.00	0.94 (0.63-1.40)	0.31 (0.03-2.74)	NS
1976-1981	1.00	1.11 (0.78-1.57)	1.05 (0.25-4.41)	NS
Cell type				
Acute lymphocytic leukaemia	1.00	0.91 (0.58-1.43)	5.28 (1.66-16.7)	p = 0.009
Acute non-lymphocytic leukaemia	1.00	1.21 (0.94-1.55)	1.31 (0.50-3.42)	NS
Acute leukaemias [b]	1.00	1.13 (0.91-1.41)	2.12 (1.03-4.35)	p = 0.046
Chronic lymphocytic leukaemia	1.00	1.06 (0.76-1.50)	1.70 (0.61-4.73)	NS
Chronic non-lymphocytic leukaemia	1.00	0.94 (0.65-1.37)	0.68 (0.18-2.58)	NS
Other	1.00	0.95 (0.38-2.36)	3.66 (0.54-24.6)	NS

[a] 95% confidence interval in parentheses.
[b] Presence of acute lymphocytic leukaemia and acute non-lymphocytic leukaemia plus three other acute leukaemias not otherwise specified.

Table 42
Risk estimates for site-specific cancer mortality in the population living along of the Techa River, 1950-1982 [K18]

Ethnic group	*Leukaemia*	*Stomach*	*Oesophagus*	*Lung*
Excess absolute risk $(10^4\ PYGy)^{-1}$ [a b]				
Tartars and Bashkirs	0.93 (-0.097-1.96)	1.06 (-2.45-4.56)	0.55 (-0.91-2.01)	1.51 (-1.36-4.38)
Russians	0.80 (-0.06-1.54)	2.80 (-1.57-7.17)	4.35 (0.45-8.25)	-0.73 (-3.50-2.04)
Tartars, Bashkirs and Russians, assuming common effect	0.85 (0.24-1.45)	1.84 (-0.96-4.64)	1.51 (-0.23-3.24)	0.61 (-1.34-2.56)
Excess relative risk (Gy^{-1}) [a]				
Tartars and Bashkirs	2.55 (-0.48-5.76)	0.18 (-0.43-0.76)	1.00 (-0.69-3.68)	0.53 (-0.48-1.54)
Russians	4.97 (-1.83-11.8)	0.86 (-0.55-2.27)	2.31 (0.008-4.61)	-0.37 (-1.75-1.01)
Tartars, Bashkirs and Russians, assuming common effect	3.24 (0.48-6.00)	0.33 (-0.25-0.91)	1.89 (0.09-3.70)	0.31 (-0.50-1.11)

[a] Account taken of ethnic differences in background rates. 90% CI in parentheses.
[b] Since information on cause of death is available for only two thirds of deaths, excess absolute risk is likely to be underestimated.

Table 43
Incidence of leukaemia in individuals surveyed in the Techa River and Kyshtym accident studies [K18]

Dose in bone marrow (Gy)	*Person-years (thousands)*	*Cases of leukaemia*	*Incidence rate* [a] $(10^5\ PY)^{-1}$
Techa River study			
1.43 [b]	46.8	6	12.8 (5.6-25.3)
0.82 [b]	36.3	3	8.3 (2.3-21.3)
0.59 [b]	90.9	9	9.9 (5.2-17.3)
0.29 [b]	106.8	12	11.2 (6.5-18.2)
0.13 [b]	108.0	7	6.5 (3.0-12.2)
Control group 1	2,956	133	4.5 (3.9-5.2)
Control group 2	1,664	93	5.6 (4.6-6.6)
Kyshtym accident study			
0.2-0.25 [c]	22.6	3	13.3 (3.6-34.3)
0.03-0.12 [c]	112.9	9	8.0 (4.4-9.1)
0.003-0.03 [c]	357.9	23	6.4 (4.4-9.1)
Control	526.7	28	5.3 (5.3-7.3)

[a] 90% CI in parentheses.
[b] Mean dose.
[c] Range.

Table 44
Age-specific incidence of leukaemia, thyroid cancer and all other cancers in the Ukrainian districts Polesskoe, Narodichy and Ovruch, 1981-1990
[P34]

Year	*Incidence rate* (10^{-6}) [a]			
	0-14 years	*15-44 years*	*45-64 years*	*≥65 years*
Leukaemia				
1981	3 (1)	3 (2)	12 (5)	4 (1)
1982	0	7 (4)	5 (2)	4 (1)
1983	0	3 (2)	10 (4)	18 (4)
1984	9 (3)	5 (3)	12 (5)	9 (2)
1985	0	2 (1)	3 (1)	10 (2)
1986	3 (1)	2 (1)	12 (5)	8 (2)
1987	7 (2)	2 (1)	9 (4)	30 (7)
1988	4 (1)	2 (1)	17 (7)	26 (6)
1989	0	0	10 (4)	22 (5)
1990	8 (2)	2 (1)	13 (5)	33 (7)
Thyroid cancer				
1981	0	2 (1)	2 (1)	0
1982	0	3 (2)	5 (2)	0
1983	0	0	2 (1)	0
1984	0	2 (1)	5 (2)	0
1985	0	0	3 (1)	0
1986	0	0	2 (1)	0
1987	0	2 (1)	5 (2)	0
1988	0	0	7 (3)	0
1989	0	2 (1)	0	0
1990	11 (3)	4 (2)	3 (1)	0
All other cancers [b]				
1981	0	44 (26)	468 (199)	467 (106)
1982	3 (1)	53 (31)	473 (199)	565 (127)
1983	0	52 (30)	457 (189)	524 (116)
1984	9 (3)	45 (25)	483 (194)	586 (348)
1985	6 (2)	42 (23)	508 (202)	775 (154)
1986	10 (3)	46 (23)	453 (195)	577 (138)
1987	10 (3)	63 (31)	416 (177)	787 (186)
1988	7 (2)	65 (32)	479 (201)	711 (166)
1989	4 (1)	45 (22)	479 (200)	811 (188)
1990	0	49 (22)	444 (171)	893 (190)

[a] Number of cases in parentheses.
[b] Excluding leukaemia and thyroid cancer.

Table 45
Observed number of cancers and standardized incidence ratios for 1,762 women treated with ^{131}I for hyperthyroidism at Massachusetts General Hospital
[G7]

Cancer site	*Amount of ^{131}I received from all treatments combined*							
	None		*4-239 MBq*		*240-369 MBq*		*≥ 370 MBq*	
	Observed	*SIR* [a]	*Observed*	*SIR* [a]	*Observed*	*SIR* [a]	*Observed*	*SIR* [a]
All malignant neoplasms	32	1.0	41	0.9	53	0.9	48	1.0
All digestive organs	8	1.1	13	1.1	15	0.9	13	0.9
Pancreas	2	2.9	2	1.7	3	1.8	3	2.2
Breast	10	1.0	18	1.4	14	0.9	18	1.5
Brain	1	2.4	1	1.8	0 [c]	0	3	6.0 [d]
All other sites [b]	13	0.9	9	0.4	24	0.9	14	0.7
Number of women [e]	356		447		529		416	
Average age at entry (years)	36.1		46.3		48.6		53.5	
Number of person-years	7,414.1		7,250.7		8,699.7		6,146.0	

[a] SIR = Standardized incidence ratio (observed/expected). Expected numbers were based on age- and calendar-time-specific incidence rates for Connecticut white women. Unless indicated, standardized incidence ratios do not differ significantly from 1.
[b] Data not reported for any of these sites individually.
[c] Expected number is 0.7.
[d] 95% CI: 1.2-17.5.
[e] Excluded 14 women whose dose was unknown.

Table 46
Leukaemia incidence in patients in Sweden given ^{131}I for treatment or diagnostic purposes
[H26]

Group		Leukaemia other than CLL			Chronic lymphatic leukaemia (CLL)			All leukaemias		
		No.	SIR	95% CI	No.	SIR	95% CI	No.	SIR	95% CI
Total		130	1.09	0.91-1.29	65	1.08	0.84-1.38	195	1.09	0.94-1.25
Females		104	1.13	0.92-1.37	43	1.08	0.78-1.45	147	1.11	0.94-1.31
Males		26	0.94	0.62-1.38	22	1.09	0.68-1.65	48	1.01	0.74-1.33
Dose in bone marrow (mGy) [a]										
0.00-0.01	(0.009)	10	1.28	0.62-2.36	2	0.53	0.06-1.90	12	1.04	0.53-1.81
0.02-0.10	(0.058)	31	1.01	0.69-1.44	17	1.15	0.67-1.83	48	1.06	0.78-1.40
0.11-10	(3.5)	62	1.23	0.94-1.58	30	1.22	0.82-1.74	92	1.23	0.99-1.50
11-100	(39)	23	0.85	0.54-1.28	9	0.62	0.28-1.17	32	0.77	0.53-1.09
>100	(221)	4	1.04	0.28-2.67	7	3.17	1.27-6.53	11	1.82	0.91-3.26
Age at exposure										
<40 years		19	1.07	0.64-1.67	5	1.03	0.34-2.41	24	1.06	0.68-1.58
40-60 years		78	1.17	0.92-1.45	39	1.18	0.84-1.62	117	1.17	0.97-1.40
≥60 years		33	0.93	0.64-1.31	21	0.93	0.58-1.42	54	0.93	0.70-1.22
Time after exposure										
2-9 years		38	0.95	0.67-1.30	13	0.71	0.38-1.21	51	0.87	0.65-1.14
10-19 years		63	1.29	0.99-1.65	30	1.22	0.83-1.75	93	1.27	1.02-1.55
≥20 years		29	0.96	0.64-1.37	22	1.26	0.79-1.90	51	1.07	0.80-1.41

[a] Mean dose in parentheses.

Table 47
Radiation weighting factors recommended by ICRP
[I10]

Type of radiation	Energy range	Radiation weighting factor w_R
Photons, electrons, muons	All energies	1
Neutrons	<10 keV, >20 keV	5
Protons	>2 MeV	5
Neutrons	10-100 keV, 2-20 MeV	10
Neutrons	0.1-2 MeV	20
Alpha particles, fission fragments, heavy nuclei	All energies	20

Table 48
Causes of death in patients in the German Thorotrast study [V9]

Cause of death	*Number of deaths*	
	Exposed to Thorotrast	*Controls*
Diseases with high excesses		
Liver cancer	396	2
Liver cirrhosis	168	40
Myeloid leukaemia	35	3
Bone marrow failure	29	4
Diseases that probably have an excess		
Cancer of the extrahepatic bile ducts	26	6
Cancer of the pancreas	18	5
Non-Hodgkin's lymphoma	14	4
Plasmacytoma	7	1
Cancer of the oesophagus	7	1
Cancer of the larynx	5	1
Bone sarcoma	4	1
Malignant mesothelioma		
Pleural	3	0
Peritoneal	2	0
Diseases without apparent excess		
Lung cancer	47	44
Stomach cancer	30	43
Brain cancer	18	11
Prostate cancer	16	11
Colon cancer	10	16
Rectal cancer	8	12
Breast cancer	7	17
Kidney cancer	6	5
Cancer of the urinary bladder	5	5
Ovary cancer	4	4
Chronic lymphatic leukaemia	3	3
Hodgkin's disease	3	2
Adrenal cancer	2	1
Acute lymphatic leukaemia	1	0
Total number of deaths	2,110	1,463
Total number of patients with follow-up	2,334	1,912

Table 49
Cancer incidence in patients in the Danish Thorotrast study [A12]

Cancer site or type	*Observed*	*SIR*	*95% CI*
Cancer with statistically significant excess			
Liver	79	126	100-157
Liver, not primary	14	30	17-51
Gall-bladder	15	14	8.1-24
Leukaemia	23	10	6.5-15
Nasal cavity	2	10	1.2-36
Peritoneum	3	8.6	1.8-25
Multiple myeloma	4	4.6	1.2-12
Ovary	7	2.4	1.0-5.0
Lung, primary	23	2.3	1.4-3.4
Breast	18	1.8	1.1-2.8
Metastases	16	12	6.7-19
Other, unspecified	10	11	5.3-22
Cancer without statistically significant excess			
All other sites	93		
All sites (brain and CNS not included)	297	3.3	3.0-3.7

Table 50
Primary liver cancers and leukaemia in patients in the Danish Thorotrast study
[A12]

	Primary liver cancer		*Leukaemia*	
	Observed number	*SIR*	*Observed number*	*SIR*
Age at injection				
0-25 years	45	429	6	16
26-45 years	31	91	14	13
46-59 years	2	13	3	4.8
>59 years	1	30	0	1
Significance of trend		p < 0.00001		p = 0.011
Injected volume				
1-9 ml	2	57	0	0
10-19 ml	26	84	10	9.4
20-29 ml	31	151	9	12
>29 ml	20	244	4	13
Significance of trend		p < 0.0001		p = 0.31
Time since injection				
0-9 years	0	0	2	5.0
10-19 years	2	22	4	7.3
20-29 years	13	64	7	11
30-39 years	45	212	8	16
>39 years	19	255	2	12
Significance of trend		p < 0.00001		p = 0.07
Estimated mean cumulative dose [a]				
Liver				
0-2.9 Gy	7	23		
3.0-3.9 Gy	15	101		
4.0-4.9 Gy	13	176		
5.0-5.9 Gy	17	380		
>5.9 Gy	27	443		
Significance of trend		p < 0.00001		
Red bone marrow				
0-0.9 Gy			5	5.2
1.0-1.9 Gy			11	16
2.0-2.9 Gy			5	23
>9 Gy			2	26
Significance of trend				p = 0.0003

[a] Mean dose to liver and red bone marrow was estimated taking account of volume, latency period (lagged 10 years for liver) and radiophysical properties, including self-absorption.

Table 51
Deaths from cancer and other causes in 592 workers in a thorium processing plant [a]
[P28]

Cause of death	*Number of observed deaths*	*Number of expected deaths*	*Standardized mortality ratio*	*95% CI*
Cancer				
Digestive organs	13	6.82	1.91	(1.01-3.25)
Rectum	1	0.85		
Liver	2	0.62		
Pancreas	5	1.21	4.13	(1.34-9.63)
Lung	10	5.96	1.68	(0.81-3.09)
Skin	1	0.40		
Brain, central nervous system	3	0.63		
Lymphosarcoma	1	0.51		
Hodgkin's disease	2	0.34		
Leukaemia	2	0.91		
All cancers	38	21.69	1.75	(1.26-2.39)
Other diseases				
Circulatory system	59	65.76	0.90	(0.69-1.17)
Respiratory system	8	6.66	1.20	(0.52-2.37)
Digestive system	6	5.90	1.02	(0.37-2.22)
All other causes				
External causes	26	12.67	2.05	(1.33-3.01)
Motor vehicle accidents	13	4.31	3.02	(1.61-5.16)
Unknown causes	5			
All causes [b]	153	123.10	1.24	(1.06-1.45)

[a] Workers were employed for one year or longer in job categories with high exposures to thorium and thoron.
[b] Total person-years 10,621; mean age at entry 33.4 ± 13.1 SD; mean year at entry 1955.

Table 52
Variation with time since exposure in excess relative risk of lung cancer per unit exposure to radon
[T3, X1]

		Chinese tin miners [a]	*West Bohemian uranium miners* [b]
Number of lung cancers		981	503
Time since exposure (years)	5-14	1.0 [c]	1.0 [c]
	15-24	0.5	0.5
	25-34	0.4	0.2
	>35	0.2	0.0 [d]
p value for heterogeneity		0.05	<0.001

[a] Background risk adjusted for age and exposure to arsenic.
[b] Estimate of ERR/WLM adjusted for attained age and proportion of time spent in the Jáchymov mine. Men who worked in shafts with radon concentrations greater than 10 WL are excluded.
[c] Value fixed at 1.0. Other values are estimated ERR/WLM relative to this baseline category.
[d] Parameter constrained to be non-negative in model-fitting.

Table 53
Variation with attained age in excess relative risk of lung cancer per unit exposure to radon
[T3, X1]

Chinese tin miners [a]			*West Bohemian uranium miners* [b]		
Age attained (years)	*Number of lung cancers*	*Excess relative risk (% WLM^{-1})*	*Age attained (years)*	*Number of lung cancers*	*Excess relative riskRR (% WLM^{-1})*
<50	98	1.02	<35	7	11.08
50-54	158	0.18	35-44	35	2.72
55-59	219	0.10	45-54	151	2.15
60-64	245	0.23	≥55	310	1.36
≥65	261	0.04			
p value for heterogeneity		<0.001	p value for heterogeneity		<0.001

[a] Background risk adjusted for age and arsenic exposure, but estimates of ERR/WLM include entire follow-up period and are not specific to the 5-14 year period of maximum effect. Values are lower than in the West Bohemian cohort because risk decreases with time since exposure.
[b] Estimate of ERR/WLM adjusted for proportion of time spent in the Jáchymov mine and time since exposure. Values quoted are specific for men who had spent less than 20 per cent of their employment at the Jáchymov mine and for exposures in the preceding 5-14 years. They are therefore higher than in the Chinese tin miner cohort.

Table 54
Lung cancer deaths and estimated excess relative risk in cohorts of miners [a]
[L21]

Location	*Exposed miners*				*Unexposed miners*			*Excess relative risk per unit exposure [b] (% WLM⁻¹)*
	Number of miners	*Person-years*	*Average exposure (WLM)*	*Deaths from lung cancer*	*Number of miners*	*Person-years*	*Deaths from lung cancer*	
China	13,649	135,357	277.4	936	3,494	39,985	44	0.16 (0.1-0.2)
Czechoslovakia	4,284	103,652	198.7	656	0	4,216	5	0.34 (0.2-0.6)
Colorado	3,347	75,032	807.2	327	0	7,403	2	0.42 (0.3-0.7)
Ontario	21,346	319,701	30.8	282	0	61,017	9	0.89 (0.5-1.5)
Newfoundland	1,751	35,029	367.3	112	337	13,713	6	0.76 (0.4-1.3)
Sweden	1,294	32,452	80.6	79	0	841	0	0.95 (0.1-4.1)
New Mexico	3,457	46,797	110.3	68	12	12,152	1	1.72 (0.6-6.7)
Beaverlodge	6,895	68,040	17.2	56	1,591	50,345	9	2.21 (0.9-5.6)
Port Radium	1,420	31,454	242.8	39	683	22,222	18	0.19 (0.1-0.6)
Radium Hill	1,457	25,549	7.6	32	1,059	26,301	22	5.06 (1.0-12.2)
France	1,769	39,487	68.7	45	16	4,556	0	0.36 (0.0-1.3)
Total	60,570	908,983	161.6	2,620	7,176	242,332	116	0.49 [c] (0.2-1.0)

[a] Entries include 5-year lag interval for exposure to radon.
[b] 95% CI in parentheses.
[c] $p < 0.001$.

Table 55
Excess relative risk of lung cancer in relation to radon exposure rate for tin miners in China
[X1]

Exposure rate (WLM a^{-1})	*Number of lung cancer cases*	*Excess relative risk per unit exposure (% WLM⁻¹)*	
		Unadjusted for arsenic exposure	*Adjusted for arsenic exposure*
<10	92	1.74	0.49
10-14	115	1.15	0.37
15-19	200	1.09	0.37
20-24	211	0.83	0.23
25-29	234	0.70	0.19
≥30	129	0.48	0.16
p value for heterogeneity		<0.001	0.003

Table 56
Relative risks of lung cancer and average exposure rates in cohorts of miners exposed to radon

Location	*Average exposure rate ($WLM\ a^{-1}$)*	*Excess relative risk per unit exposure y (% WLM^{-1}) [a]*		*Ref.*
		Original [b]	*BEIR IV [c]*	
Beaverlodge [d]	8 [d]	2.1 [d]	3.31 [d]	[H15]
Malmberget [e]	5	3.6	3.6 [e]	[R8]
France	5	0.58	-	[T16]
Ontario	~10	1.3	1.8	[M14]
Cornwall	10	0.9	-	[H37]
New Mexico	~16	1.8	-	[S28]
China	22 ?	0.6	0.2 ?	[X1]
Czechoslovakia	28	1.7	-	[S57]
Newfoundland	~70	0.9	-	[M18]
Port Radium	109	0.27	-	[H16]
Colorado	124	- [f]	0.9 [g]	[H18]

[a] ERR/WLM depends on attained age (see text).
[b] Estimate given by original authors. Values may not be comparable, as studies cover different ages and follow-up periods.
[c] BEIR IV estimate, based on internal comparison but allowing the relative risk per WLM to vary between cohorts, for age 55-64, 5-14 years after exposure. When the relative risk is assumed to be the same for all cohorts, its estimated value is 2.5% per WLM.
[d] The average exposure rate for the Beaverlodge mine has been divided by 0.65, and the risk estimates multiplied by 0.65 to take into account the recent dosimetry revision.
[e] Estimates of exposure rate may be too low [C6].
[f] Original paper only gives estimates after allowing for exposure rate effect, so they are not directly comparable.
[g] Based on cumulative lagged exposures of less than 2,000 WLM only.

Table 57
Variation in excess relative risk of lung cancer mortality in men exposed to <10 WL in the West Bohemian mining cohort [a]
[T3]

Exposure category		*Fractional change in risk from baseline category [b]*
Employment at Jáchymov	<20% (baseline category)	1.00
	>20%	1.80 (1.27-2.97)
Time since exposure (years) [c]	5-14 (baseline category)	1.00
	15-24	0.47 (0.30-0.69)
	25-34	0.24 (0.07-0.41)
	≥35	0.00 [d]
Attained age (years)	>35	8.15 (2.70-21.1)
	35-44	2.00 (1.04-3.76)
	45-54	1.50 (1.04-2.46)
	>50 (baseline category)	1.00

[a] Excess relative risk per unit exposure for baseline category is 1.36% WLM^{-1} (95% CI: 0.52-3.54).
[b] 95% CI in parentheses.
[c] Ten-year window of exposure.
[d] Parameter constrained to be non-negative.

Table 58
Geographical studies of non-occupational exposure to radon and lung cancer
[S28]

Country/area	*Measure*		*Findings*	*Ref.*
	Outcome	*Exposure*		
Canada	Lung cancer mortality rates for 18 cities, 1966-1979	Geometric mean WL from survey of 14,000 homes done 1978-1980	No association of lung cancer mortality rates with radon daughter levels	[L9]
China	Lung cancer mortality rates for two areas of the Guangdong Province, 1970-1983	By area: "control" and "high background"	Similar lung cancer mortality rates in the two areas	[H10, H20]
Denmark and Sweden	Years of life lost from lung cancer	Surveys of residential radon exposures	Lung cancer rates in Denmark (lower radon) were significantly higher than in Sweden (higher radon)	[G23]
Finland	Comparison of lung cancer incidence with national mean	Regional survey in seven municipalities in high radon areas	Lung cancer rates did not differ from national mean	[C23]
France	Lung cancer mortality rates for the two regions Limousin and Poitou-Charentes, 1968-1982	By area: from the geology, indoor radon was estimated to be 3-4 times higher in Limousin region	Significantly lower lung cancer mortality rates in the high background region	[D11]
Italy	Lung cancer mortality rates for 31 towns in central Italy, 1969-1978	Soil geologic features	Non-significant increase for males and females in areas of higher exposure	[F9]
Japan	Lung cancer SMRs, 1952-1988,in Misasa spa area and neighbouring control area	Average outdoor radon concentrations	SMR for lung cancer lower in high radon area than in control area, but difference not statistically significant	[M45]
Norway	Lung cancer incidence for municipalities	By municipality: based on measurements in about 20 homes in each	Significant correlation between lung cancer incidence and mean radon concentration	[S29]
Sweden	Lung cancer mortality rates by county, 1969-1978	Estimated background gamma radiation, assumed to correlate with radon	Significant correlations for lung cancer rates in males and females with exposure	[E1]
United Kingdom	Lung cancer mortality rates in England and Wales, by county, 1980-1983	By county: based on national survey of 2,000 homes	For males and females, lung cancer rates inversely associated with county average radon levels	[H19]
United States				
Florida	Lung cancer in Florida compared with colorectal cancer, 1981-1983	Residence in central Florida phospate mining region	Twofold increase among male non-smokers. No significant increase for male smokers or women	[S17]
Florida	Lung cancer in Florida compared with colorectal cancer 1981-1983	Residence in three-county area with potential raised indoor radon level	25 per cent increase in lung cancer	[S31]
Florida	Database on episodes of serious illness by county	Statewide survey of residential radon levels	Lung cancer relatively less frequent in males with highest radon exposures and slightly more frequent in such females, compared to those with little exposure	[V13]
Iowa	Lung cancer incidence for municipalities of 1,000-10,000 residents, 1969-1979	Mean level of radium-226 in water supply	Significantly increasing lung cancer incidence for males with exposure; increase not significant for females	[B3]
Maine	Lung cancer mortality rates by county, 1950-1969	Estimated county average for radon concentration in water	Significant associations in males and females between lung cancer mortality and exposure	[H5]
Reading Prong	Lung cancer mortality rates by county, 1950-1969	By county: based on proportion within Reading Prong	For the three counties mostly within the Reading Prong area, lung cancer mortality for men was significantly elevated in all three and for women, in two	[F6]
Reading Prong	Lung cancer mortality rates by county, 1950-1979	By county: based on geology; three levels of exposure	For both sexes combined, lung cancer mortality followed a gradient consistent with exposure	[A5]
Various counties	Lung cancer mortality rates for 411 counties, 1950-1969	By county: geometric mean concentration measured in >10 homes	For males and females, lung cancer mortality rates inversely associated with county-average radon levels	[C13, C35]
Various counties	Lung cancer mortality rates, 1950-1969	Presence of phosphate deposit, mine or processing plant in county	Significant excess of high rates of lung cancer in counties with phosphate mills	[F5]

Table 59
Case-control and cohort studies of domestic exposure to radon and lung cancer
[S28]

Location	*Study design*	*Subjects*	*Exposure measure*	*Findings*	*Ref.*
Canada					
Ontario	Case-control	27 cases 49 controls	Exposures reconstructed on the basis of measurements	RR = 2.4 (95% CI: 0.8-7.1) with smoking adjustment for exposed vs. unexposed	[L1]
Winnipeg	Case-control	750 cases 750 controls	Year-long alpha track measurements	No association between relative risk and cumulative exposure	[L10]
China					
Shenyang	Case-control	308 cases 356 controls	Year-long alpha track measurements	No association between radon and lung cancer, with or without adjustment for cigarette smoking habits	[B37]
Sweden					
Stockholm	Case-control	292 female cases 584 controls	Geology and living near ground level	RR = 2.2 (95% CI: 1.2-4.0) for exposed vs. unexposed; exposure-response relationship not found	[S19]
Stockholm	Case-control	210 female cases 209 controls	Year-long alpha track measurements	Significant trend after adjusting for smoking, age and municipality. RR = 1.0, 1.4, 1.7, 2.3 in categories of increasing cumulative radon exposure, but not when adjusted for occupancy	[P30, S20]
South	Case-control	37 cases 178 controls	Residence type: wood, "mixed" or stone	RR = 1.8 ($p < 0.05$) for stone and mixed vs. wood	[A3]
South	Case-control	177 cases 677 controls	Residence type and geology, all homes; measurements lasting two months, some homes	Exposure associated with increased risk for rural, but not urban, dwellers	[A6]
North	Case-control	15 non-smoker 15 smoker case-control pairs	Construction characteristics	Estimated mean exposure significantly higher for smoking cases than controls; exposure not different for non-smokers	[P1]
North	Case-control	589 male cases 582 deceased controls 453 living controls	Residence type: wood or non-wood	RR not increased, with or without smoking adjustment; RR increased for those never employed in occupations associated with lung cancer	[D1]
Oeland	Case-control	23 cases 202 controls	Residence type and measurements lasting four months with alpha-sensitive film	RR = 4.3 (90% CI: 1.7-10.6) for low- vs. high-exposure home type; RR − 2.7 (90% CI: 1.4-18.5) for low- vs. high-exposure home type by measurement; multiplicative interaction with smoking	[E2, E3]
Stockholm	Case-control	11 non-smoker 12 smoker case-control pairs	Construction characteristics	Estimated mean exposures comparable for cases and controls regardless of smoking	[P1]
Nationwide	Case-control	1,360 cases 2,847 controls	Three months alpha track measurements	Compared with <140 Bq m^{-3}, RR = 1.3 (95% CI: 1.1-1.6) for 140-400 Bq m^{-3}, RR = 1.8 (95% CI: 1.1-29) for >400 Bq m^{-3}	[P7]
United States					
Maryland	Cohort	298 cases over a 12-year period	Housing characteristics	No associations of incidence rates with housing characteristics	[S11]
New Jersey	Cohort	752 persons who had lived in houses with documented high exposure	Measurements with carbon canister detectors	Excess lung cancer in high exposure homes, but not significant	[K11]
New Jersey	Case-control	433 cases 402 controls	Year-long alpha track measurements	Significant trend in risk after adjusting for smoking, age and occupation; few high-dose exposures	[S51]

Table 60
Cancer in two cohorts of young Seascale residents
[G1, G2, G22]

Disease	*Cohort born to mothers resident in Seascale*		*Cohort moved into Seascale after birth*	
	Observed	*Expected*	*Observed*	*Expected*
Fatal leukaemia	5 [a]	0.53	0	0.54
Other fatal cancer	4 [b]	1.07	1 [e]	1.50
Non-fatal leukaemia	1 [c]	0.08	0	0.03
Non-fatal other cancer	2 [d]	1.11	3 [f]	2.01

[a] Four with onset under age 5 years and one case of chronic myeloid leukaemia with onset at age 20 years.
[b] Non-Hodgkin's lymphoma at age 3 years, Wilms' tumour at age 4 years, retroperitoneal sarcoma at age 6 years and metastatic squamous carcinoma of tongue at age 28 years.
[c] Onset at age 5 years.
[d] Non-Hodgkin's lymphoma at age 1 year and melanoma at age 28 years.
[e] Melanoma at age 33 years.
[f] Hodgkin's lymphoma at age 26 years, non-Hodgkin's lymphoma at age 9 years, and carcinoma of lung at age 21 years.

Table 61
Average annual equivalent dose (RBE = 20) to red bone marrow of a one-year-old child in Seascale from exposure to effluents of the Sellafield plant
[S16]

Year	*Equivalent dose in Seascale from Sellafield plant* [a] *(μSv)*
1955	1200
1960	120
1965	58
1970	120
1975	330
1980	170

[a] The estimated average annual bone marrow dose from natural radiation is 990 μSv.

Table 62
Factors associated with increased risks of leukaemia in young people in West Cumbria
[G21]

Factor	*Number of cases*		*Relative risk* [a]	*95% CI*
	Positive	*Negative* [b]		
Mother's age at individual's birth ≥40 compared with <25 yrs	4	48	4.94	1.11-21.85
Father's age at individual's birth ≥40 compared with <25 yrs	8	40	1.87	0.59-5.91
Maternal abdominal radiation in pregnancy	4	31	1.74	0.44-6.82
Play on beach more often compared with less often than monthly	13	15	0.89	0.37-2.17
Eating fish more often compared with less often than weekly	16	13	1.26	0.50-3.21
Father's occupation				
Sellafield	9	37	2.82	1.07-7.40
Iron and steel industry	5	41	1.84	0.60-5.60
Farming	5	41	1.98	0.66-5.96
Chemical industry	5	41	1.39	0.49-3.97
Father's occupational dose of radiation before conception				
Previous 6 months				
1-4 mSv	3		1.30	0.32-5.34
5-9 mSv	1	38	3.54	0.32-38.88
≥ 10 mSv	4		7.17	1.69-30.44
Total				
1-49 mSv	3		1.12	0.31-4.05
50-99 mSv	1	38	0.69	0.08-5.73
≥ 100 mSv	4		6.24	1.51-25.76

[a] The total number of cases varied between questions because of variation in the source of the information.
[b] Compared with controls not matched for parish of residence.

Table 63
Leukaemia and non-Hodgkin's lymphoma in young people living in Seascale: extent of paternal irradiation before the child's conception
[K27]

Paternal external dose (mSv)	*Number of affected young people born*	
	in Seascale	*Elsewhere*
0	1 [a]	3
0.01-49.9	0	1
50.0-99.9	3	0
≥100	2 [a]	1

[a] Each of these categories included an additional individual who developed leukaemia within a few months of leaving Seascale for another address; the father of one had received a dose of more than 100 mSv and was included in Gardner's 1990 series [G21].

Table 64
Leukaemia in young people living near three nuclear sites in southern England
[R18]

Distance from site (km)	*Ratio of observed to expected cases [a] at various ages*		
	0-4 years	*5-14 years*	*0-14 years*
<5	9 : 3.85	1 : 3.83	10 : 7.68
5-9	20 : 10.56	11 : 10.33	31 : 20.89
10-14	14 : 12.96	21 : 14.11	35 : 27.07
≥15	10 : 6.63	3 : 7.06	13 : 13.69

[a] Cases included are from districts in which more than half of the population is within the defined area.

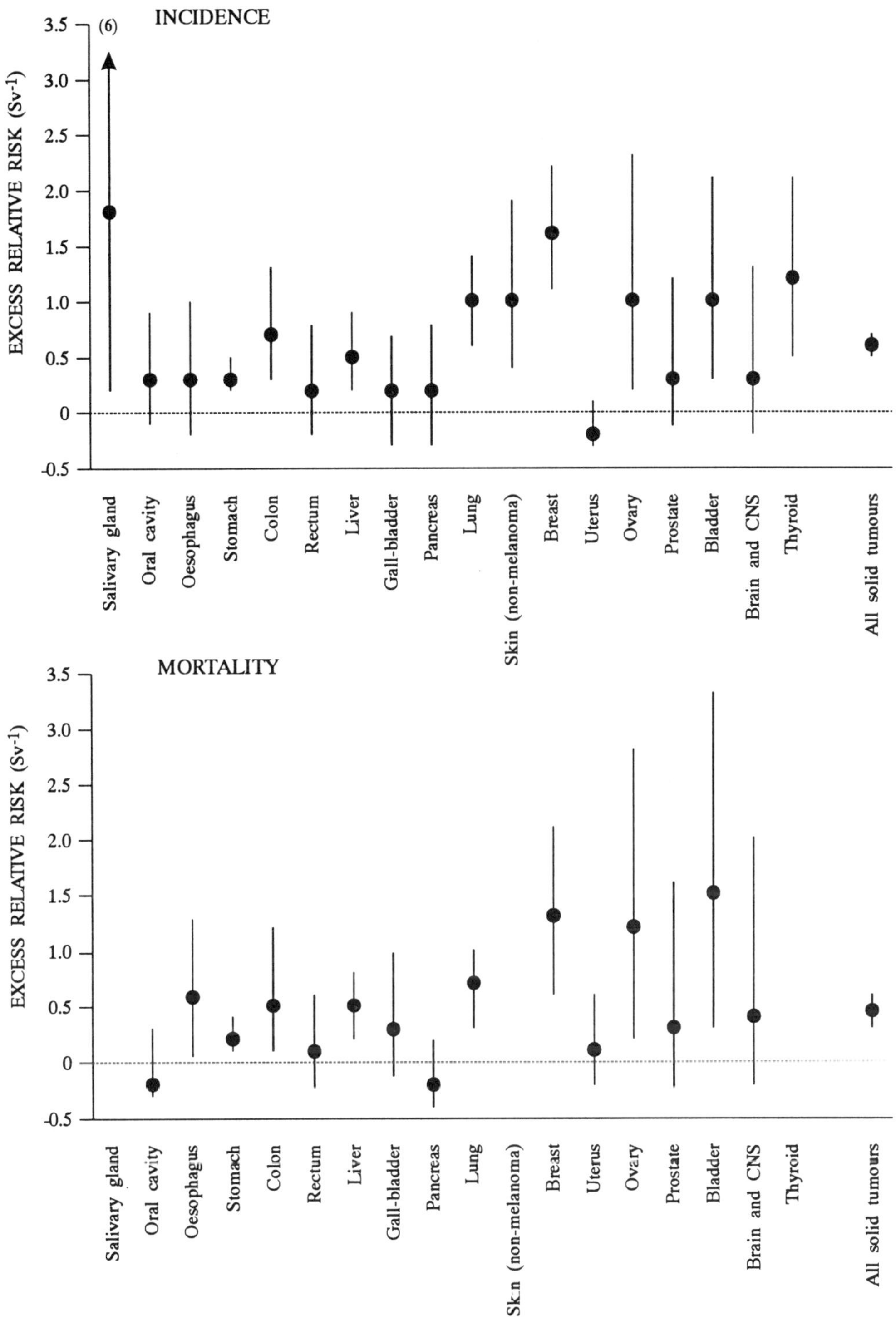

Figure I.
Excess relative risk per sievert (CI: 90%) for incidence (1958-1987) and mortality (1950-1987) from solid tumours, based on data of the life span study.

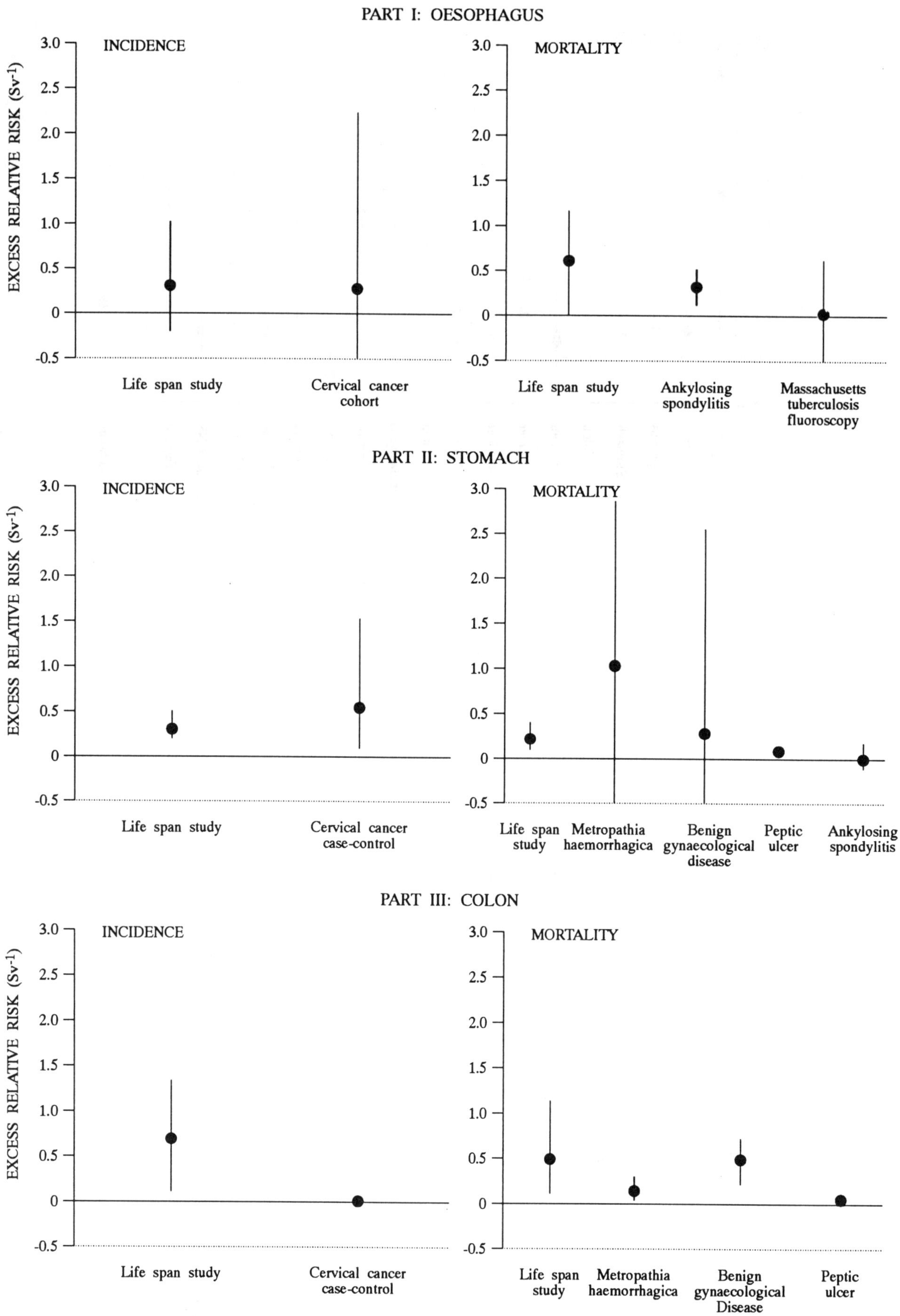

Figure II.
Comparative estimates of excess relative risk per sievert (CI: 90%) from epidemiological data on exposures under various conditions to low-LET radiation.

PART IV: LIVER

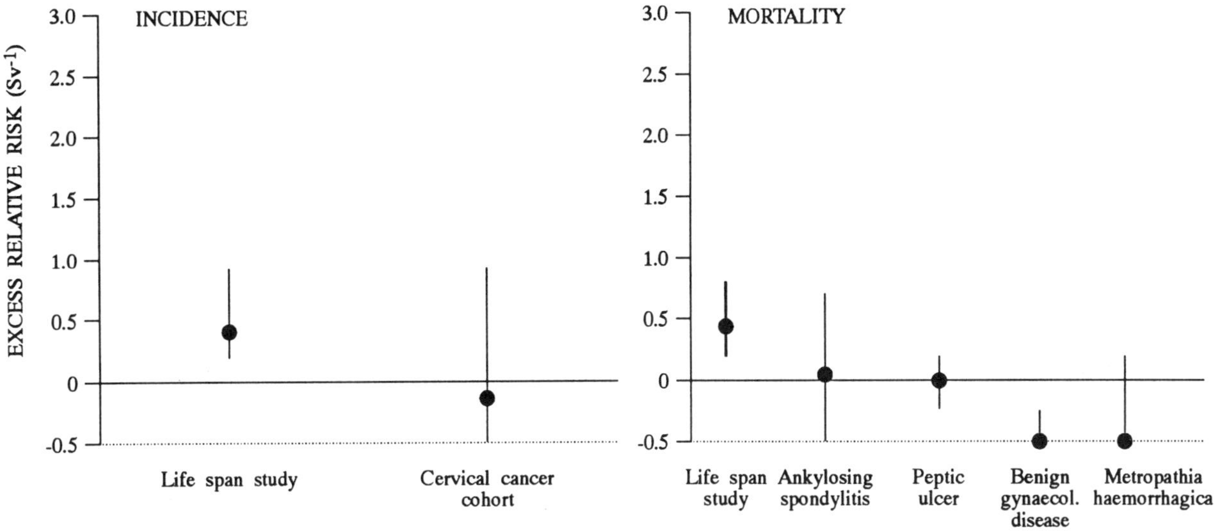

PART V: LUNG

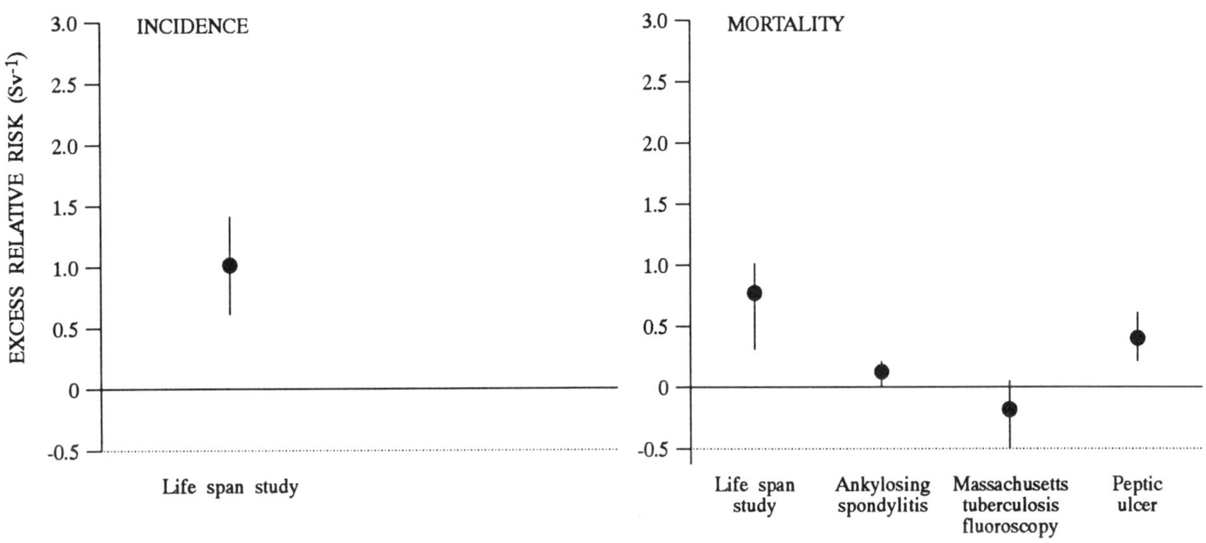

PART VI: BONE AND CONNECTIVE TISSUE

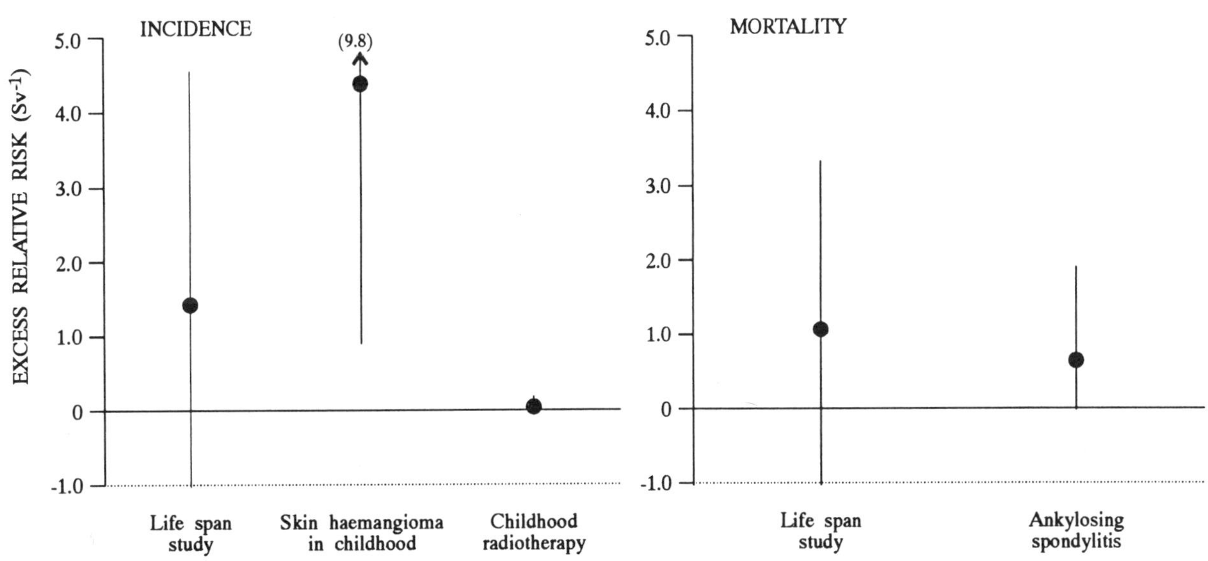

Figure II (continued)

PART VII: SKIN (NON-MELANOMA)

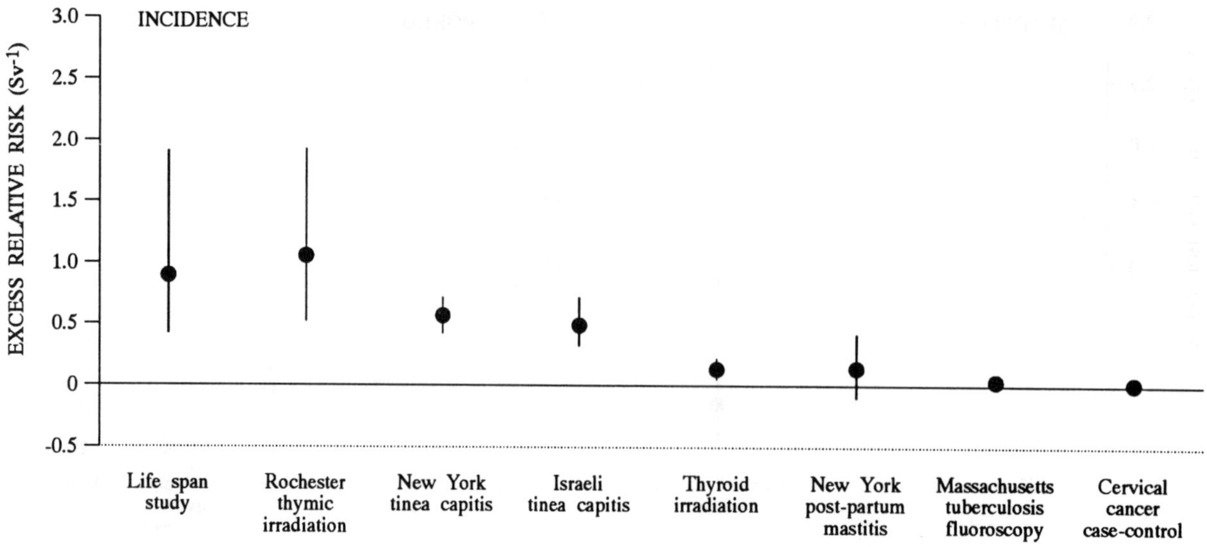

PART VIII: FEMALE BREAST

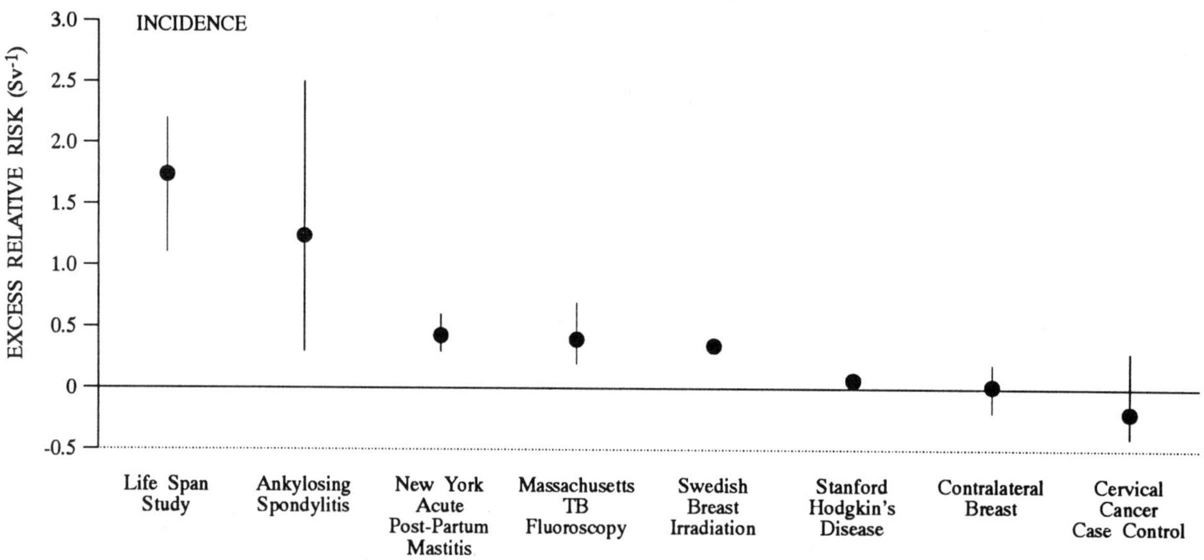

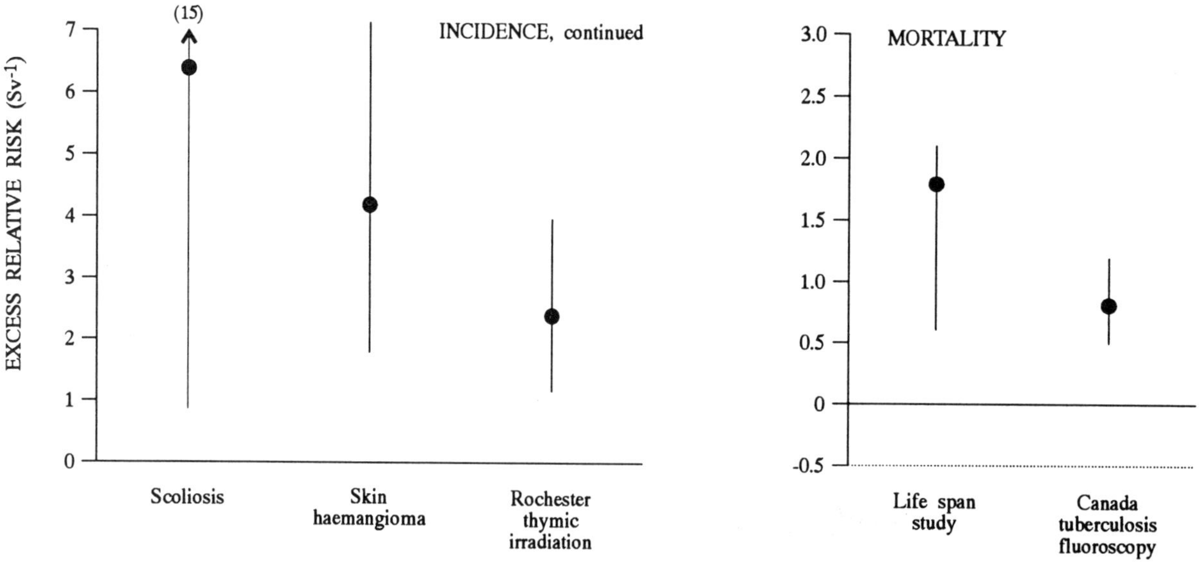

Figure II (continued)

PART IX: BLADDER

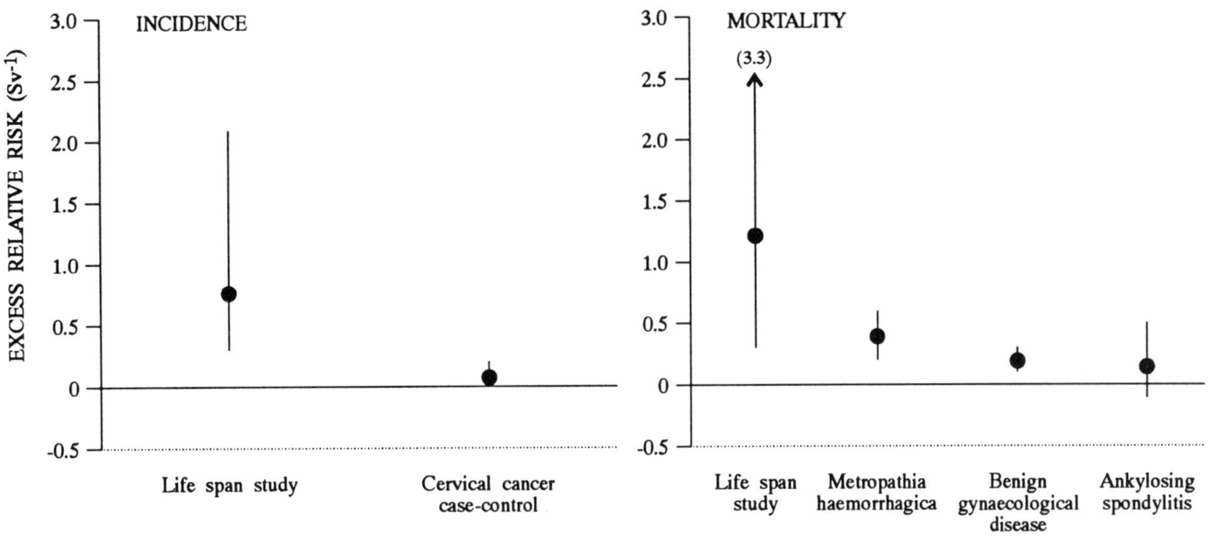

PART X: BRAIN AND CENTRAL NERVOUS SYSTEM

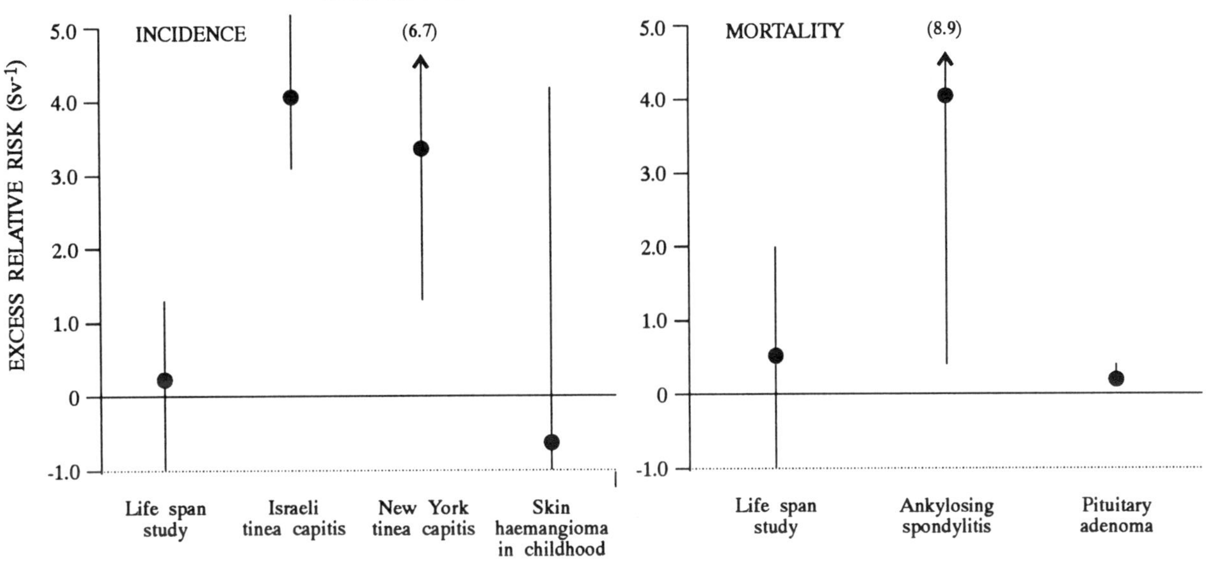

Figure II (continued)

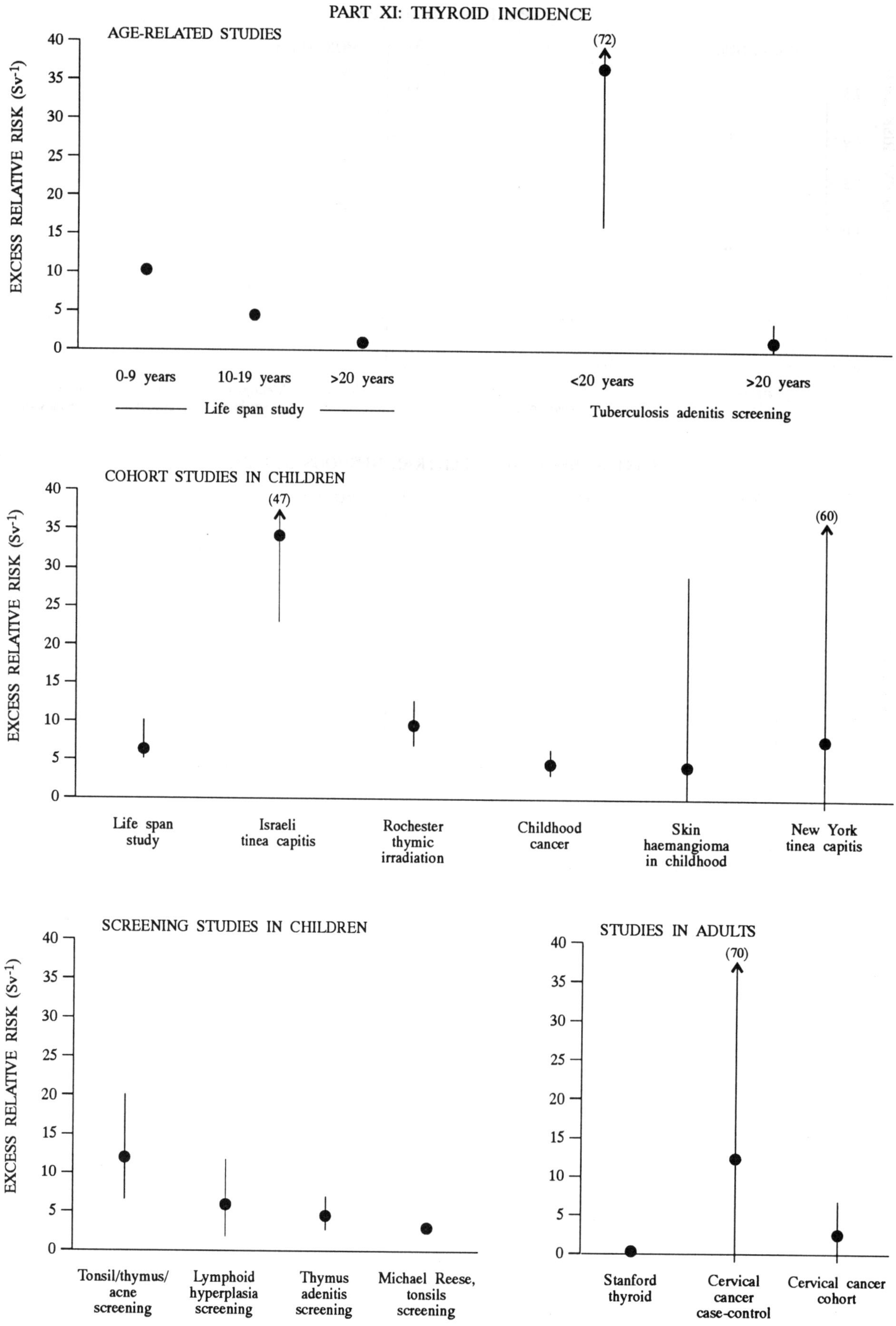

Figure II (continued)

PART XII: LEUKAEMIA

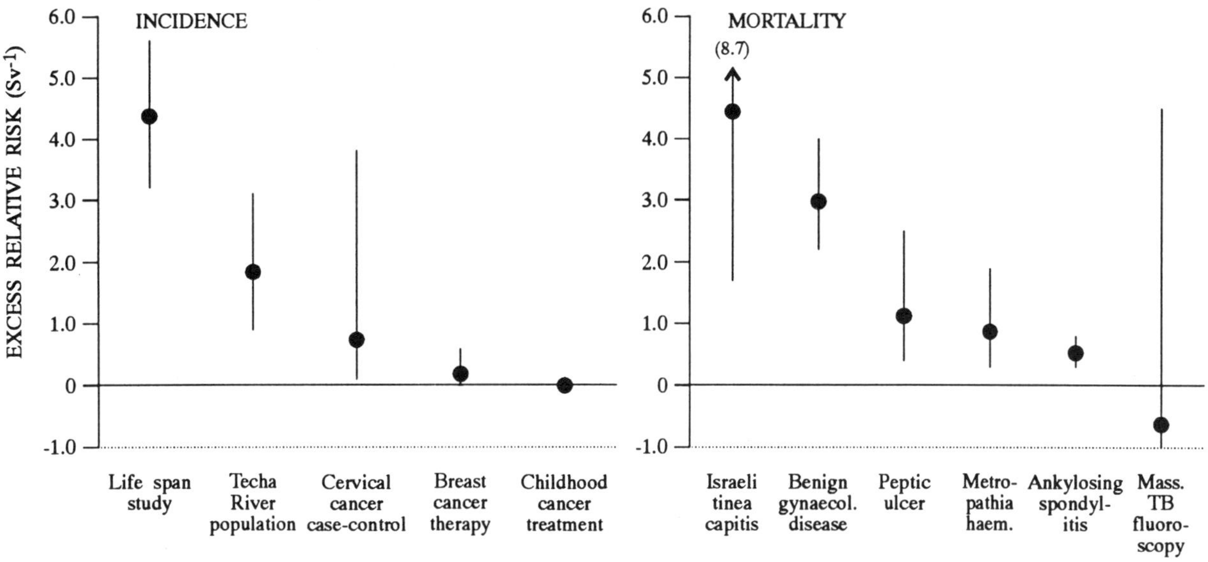

PART XIII: NON-HODGKIN'S LYMPHOMA

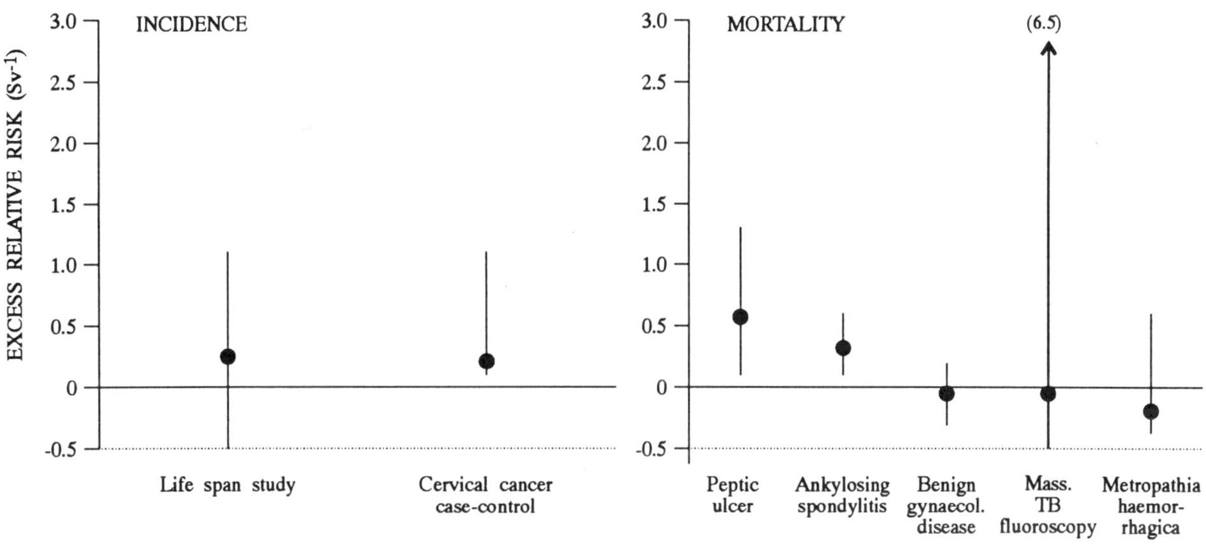

PART XIV: MULTIPLE MYELOMA

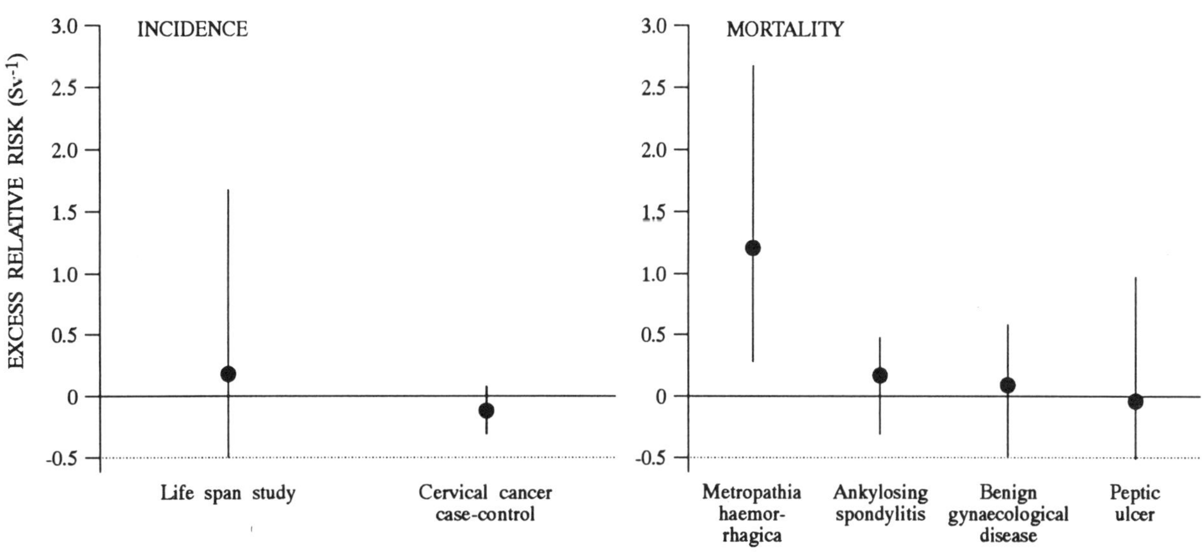

Figure II (continued)

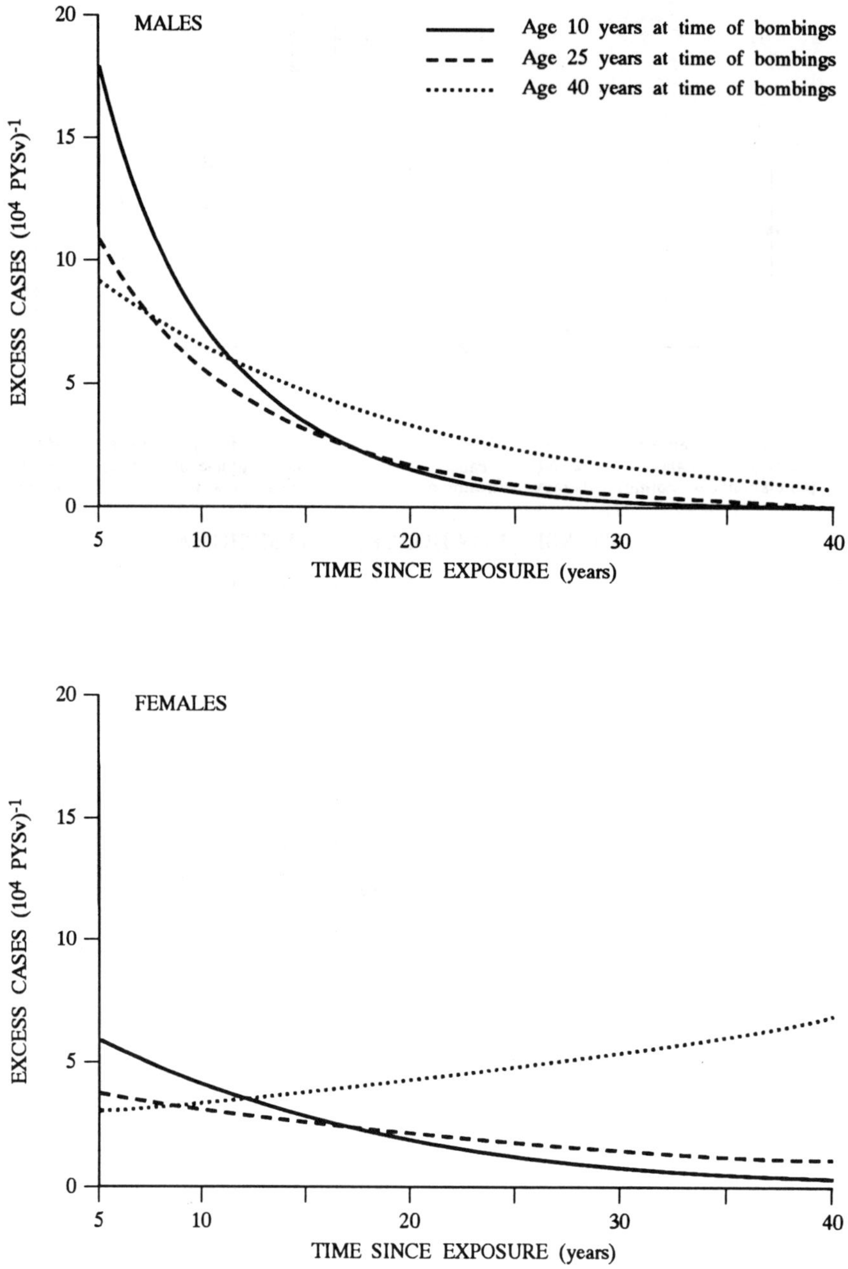

Figure III.
Temporal variation in excess absolute risk of leukaemia in the Life Span Study for a weighted dose in bone marrow of 1 Sv.
[P33]

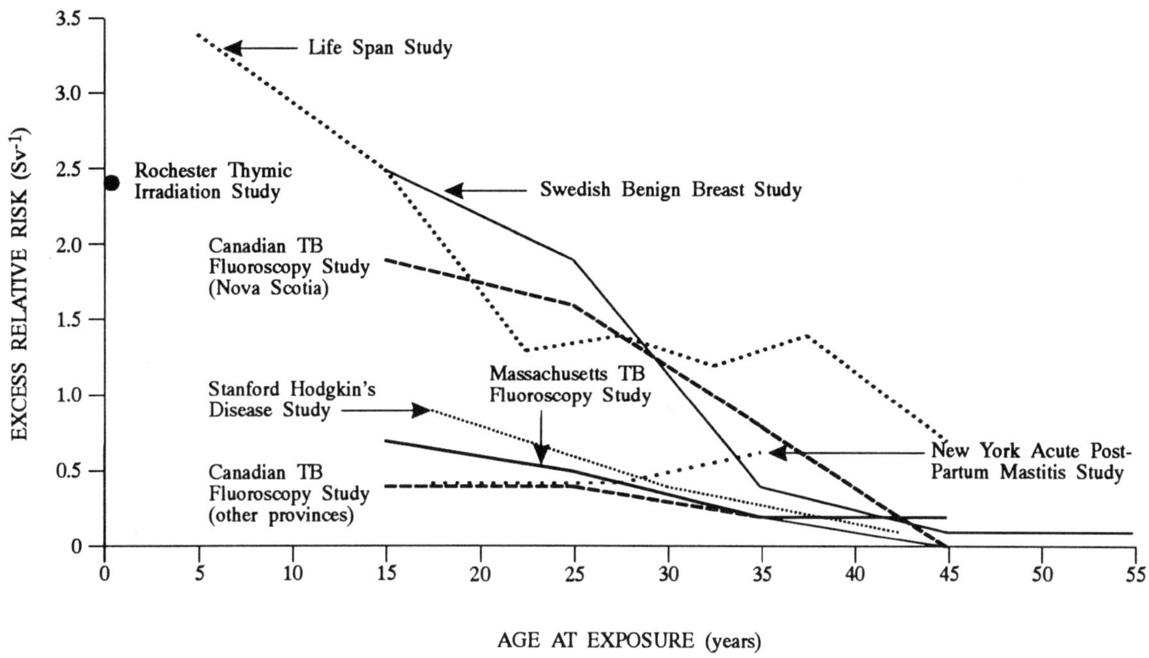

Figure IV.
Excess relative risk for the incidence of breast cancer in relation to age at exposure.

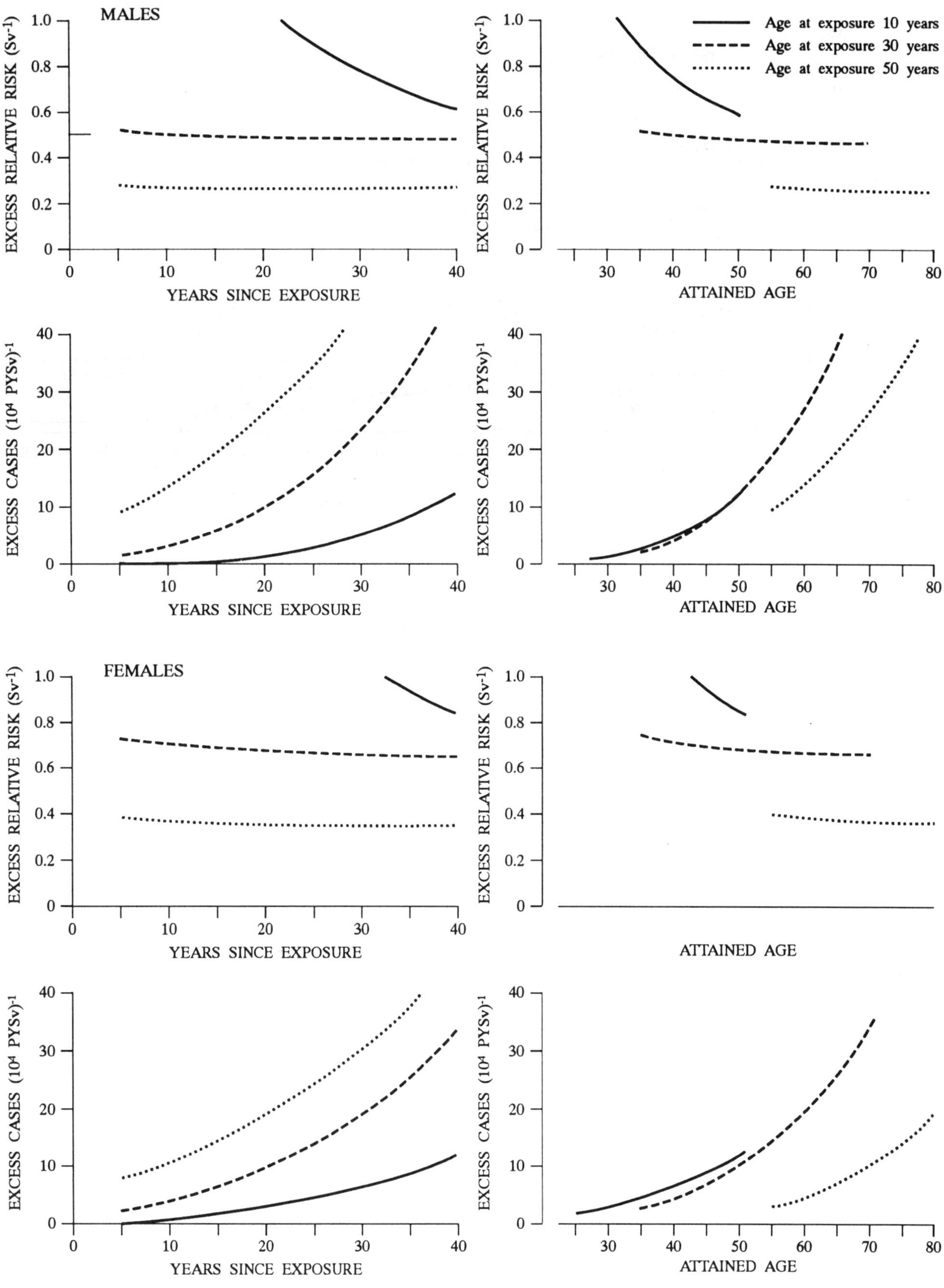

Figure V.
Excess relative and absolute risk of mortality from solid tumours for males (upper plots) and for females (lower plots) following an exposure of 1 Sv based on 1950-1987 mortality data of the life span study. [T15]

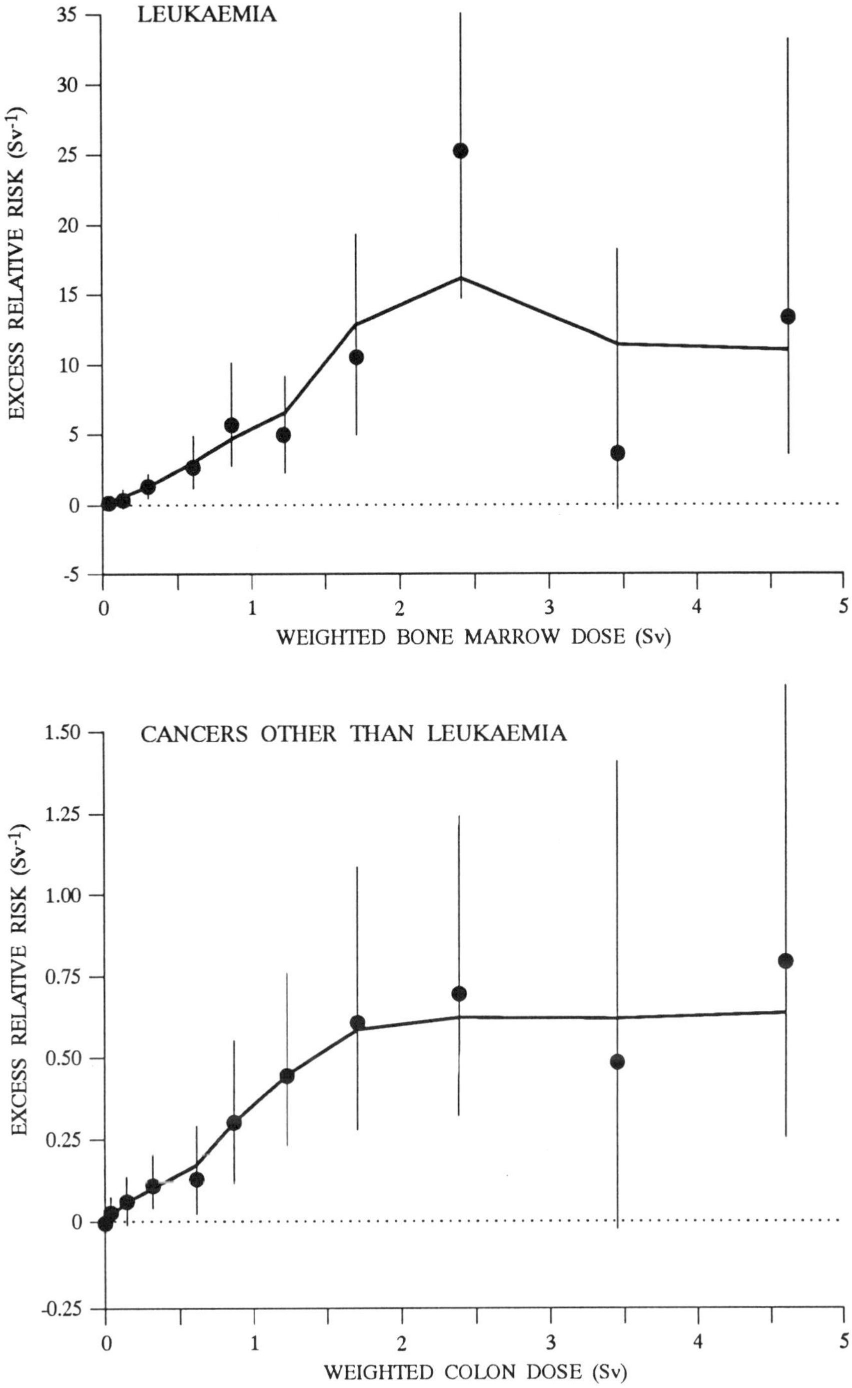

Figure VI.
Excess relative risk per Sv (and CI: 95%) for mortality from leukaemia and cancers other than leukaemia based on 1950-1987 data of the life span study.

The lines connecting smoothed points were obtained by weighting the estimated risk for the dose category (weight 0.5) and the risks in the two adjacent categories (weight 0.25 each). For the highest dose category weights of 0.25 and 0.75 were used for the next to last and last category risk estimates, respectively.

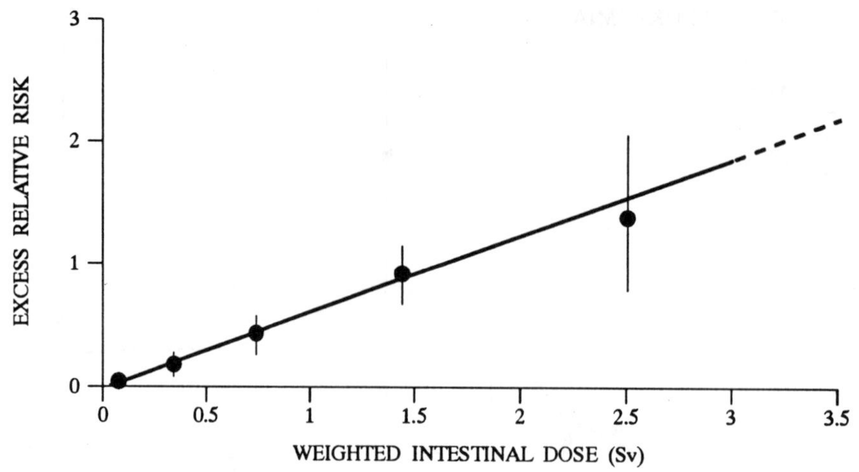

Figure VII.
Dose-response relationship for all ages and both sexes of solid tumour incidence based on 1958-1987 data of the Life Span Study.
The excess relative risk per Sv is 0.63 (95% CI: 0.52-0.74).
[T15]

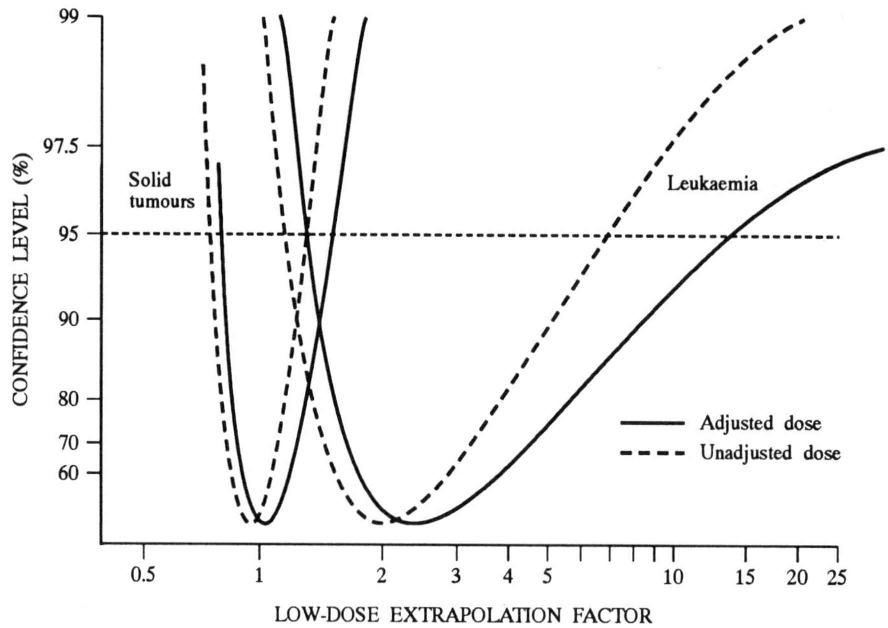

Figure VIII.
Dose-rate reduction factors for cancer incidence derived from data on survivors of the atomic bombings.

The minimum values correspond to the maximum likelihood estimates of the low-dose extrapolation factor (LDEF). The points where the curves cross the 95% likelihood level give the confidence bounds for the LDEF. Adjusted values account for bias introduced by random errors in individual dose estimates.

[V14]

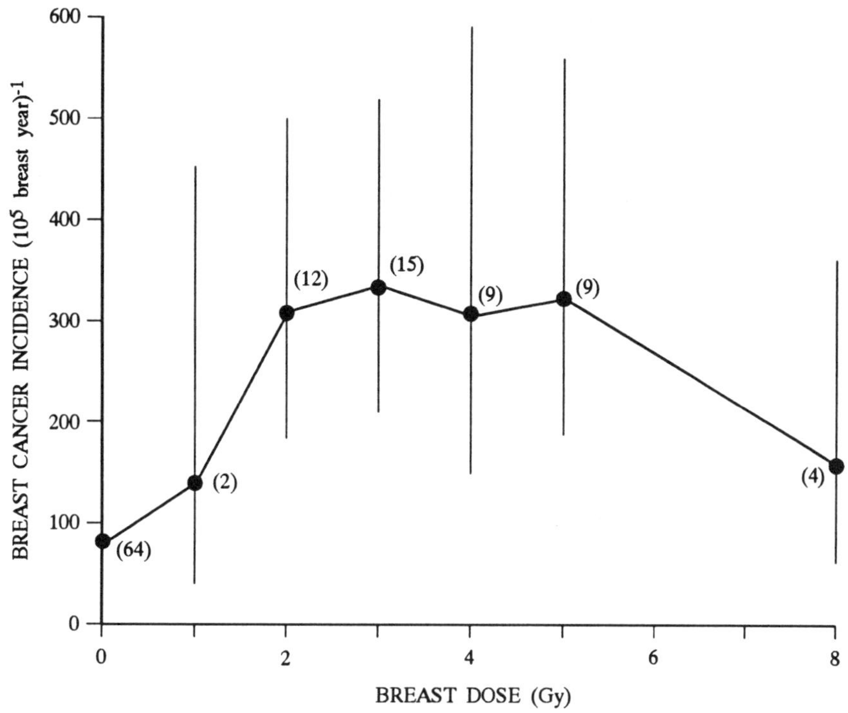

Figure IX.
Absolute risk of breast cancer incidence by single breast in a study of women in New York given radiotherapy for post-partum mastitis, adjusted for age at irradiation and interval since irradiation. Vertical bars are 90% CI and numbers in parentheses are number of cases.
[S9]

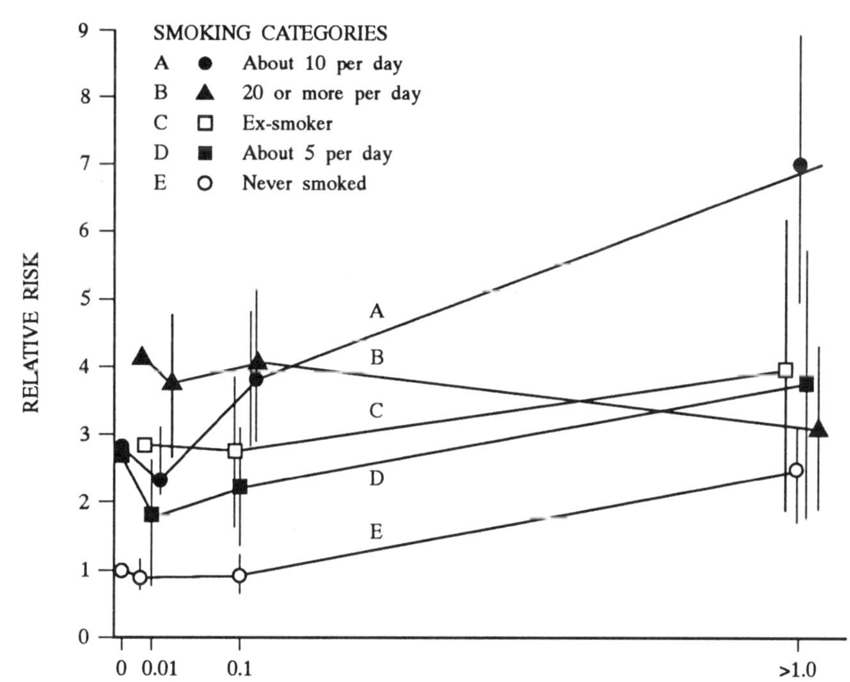

Figure X.
Relative risk for incidence of lung cancer in the Life Span Study cohort associated with cigarette smoking for T65 dose groups. Adjusted for city, sex, age at exposure and attained age.
Vertical bars indicate ± 1 SE.
[K8]

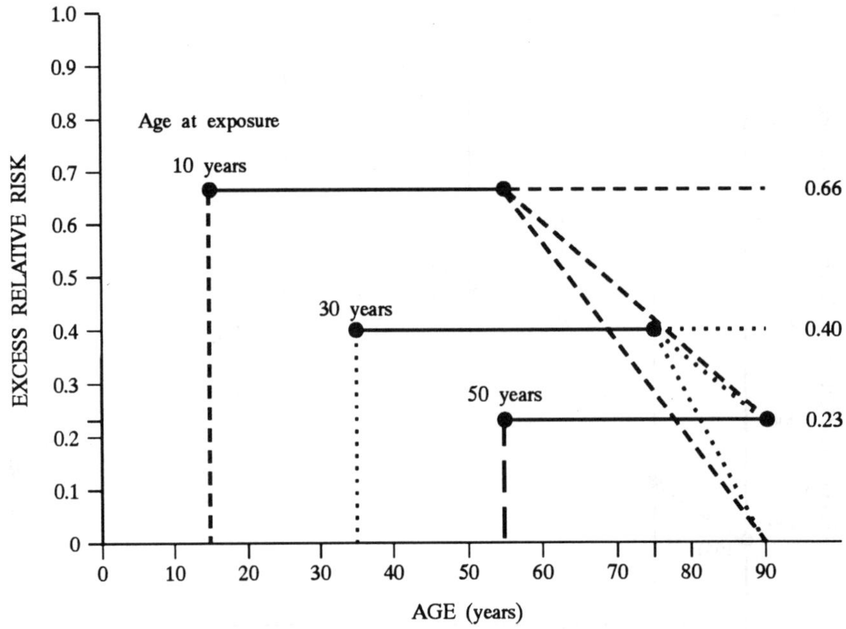

Figure XI.
Projection methods for lifetime risks of solid tumours in males.
The risk begins 5 years after exposure, is constant for 40 years and then either remains constant or declines to intermediate level or zero at age 90.

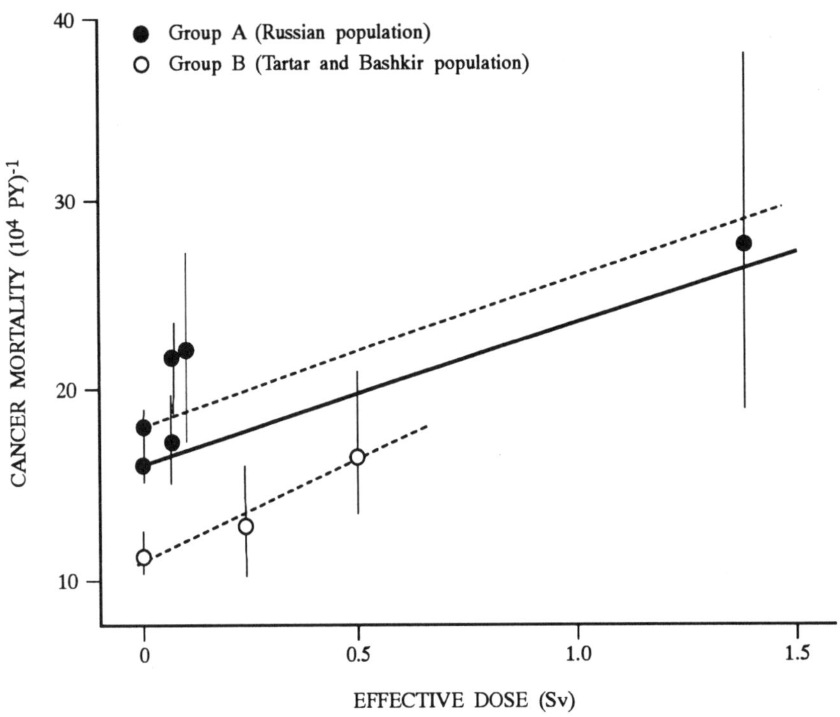

Figure XII.
Cancer mortality rates in populations living along the Techa River, USSR, 1950-1982.
[K18]

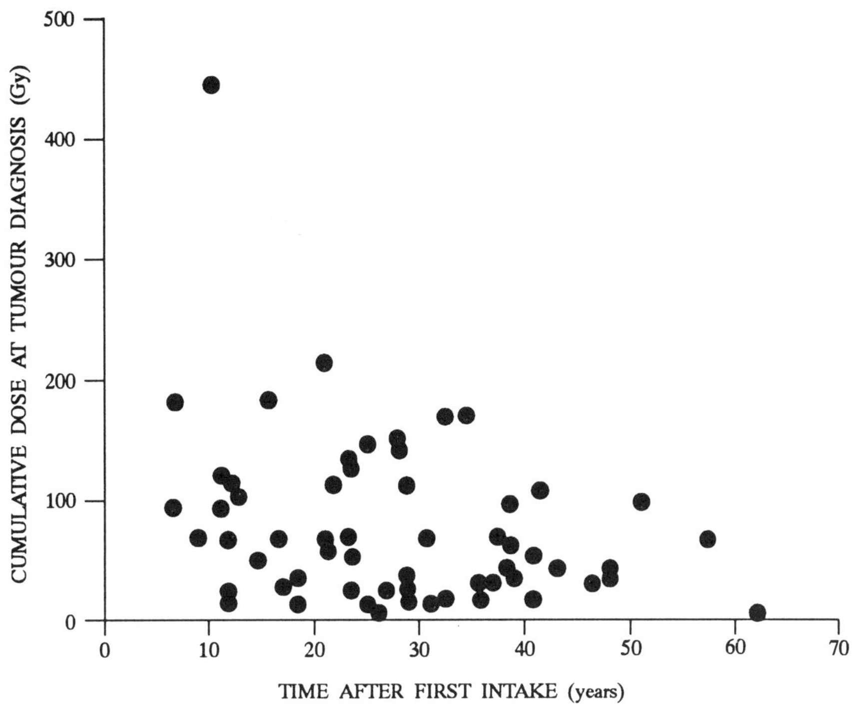

Figure XIII.
Bone sarcoma appearance times and average skeletal dose in persons in the United States exposed to ^{226}Ra.
[M4]

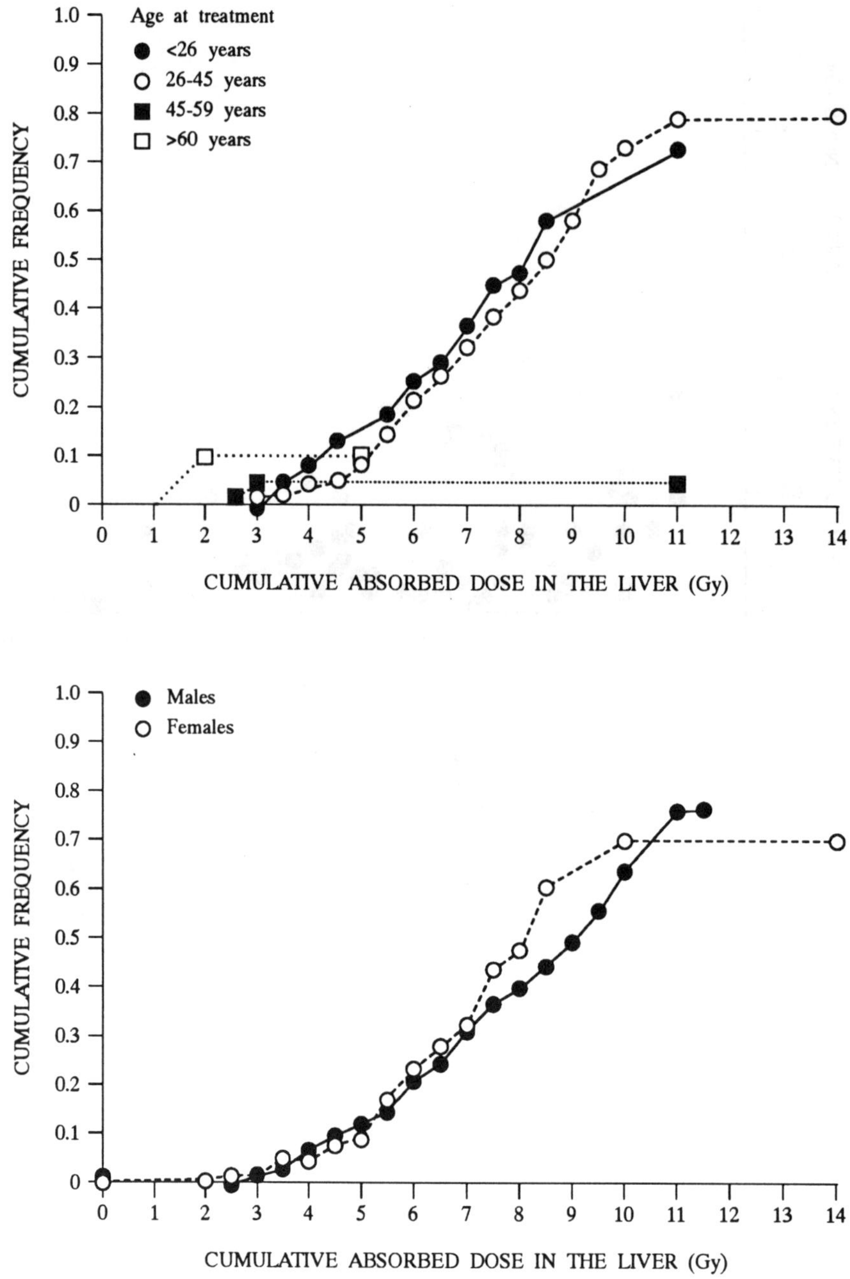

Figure XIV.
Cumulative incidence of primary liver tumours among Danish patients exposed to Thorotrast.
[A12]

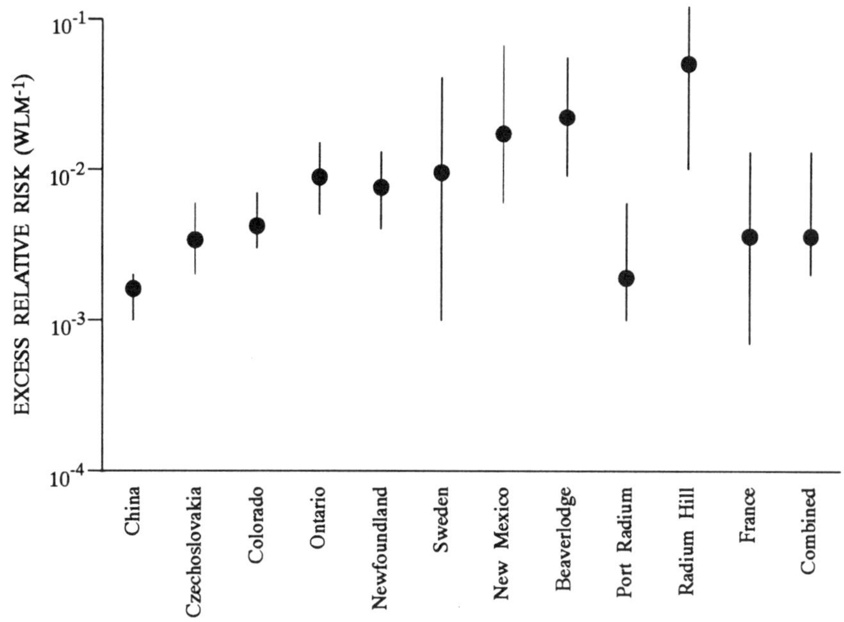

Figure XV.
Comparative estimates of excess relative risk per WLM (CI: 95%) from epidemiological studies of underground miners and from the combined data.
[L21]

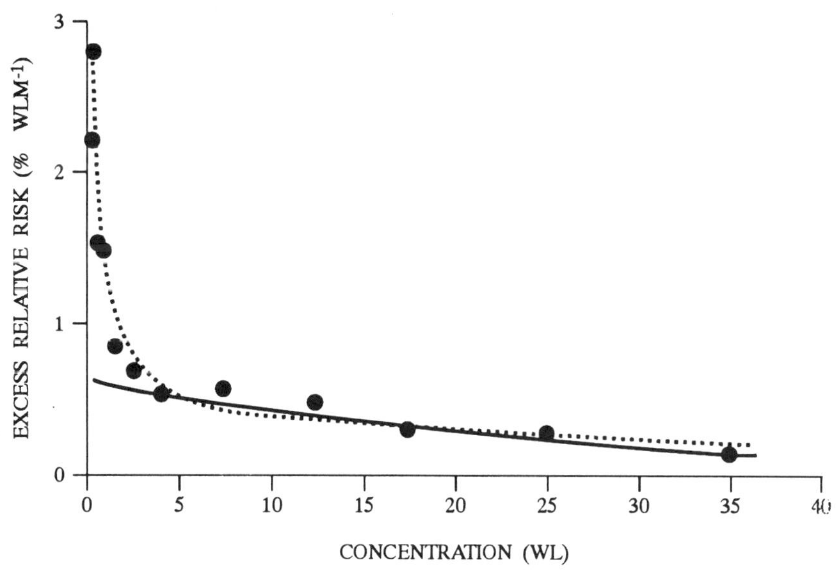

Figure XVI.
Estimated and fitted excess relative risk of lung cancer per cumulative exposure to radon and its decay products for the combined cohorts of miners.

The fitted curves are from models with an exponential function (solid line) and a power function (dotted line).

[L21]

Glossary

Dosimetric terms and abbreviations

absorbed dose	quantity of energy imparted by ionizing radiation to unit mass of matter, such as tissue. Unit: gray (Gy). 1 Gy equals 1 joule per kilogram.
DDREF	dose and dose-rate effectiveness factor. See also LDEF.
equivalent dose	quantity obtained by multiplying the average absorbed dose in a tissue or organ by a radiation weighting factor to allow for the different effectiveness of the various ionizing radiations in causing harm to tissue. Unit: sievert (Sv). The factor for gamma rays, x rays and beta particles is usually 1; for alpha particles and neutrons (100 keV-2 MeV) it is 20.
effective dose	quantity obtained by multiplying the equivalent dose in various specific tissues and organs by a tissue weighting factor appropriate to each and summing the products. Unit: sievert (Sv)
low dose	expression used in this document to refer to absorbed doses in the range 0-0.2 Gy or equivalent doses in the range 0-0.2 Sv. For further discussion see UNSCEAR 1993 Report, Annex F, "Influence of dose and dose rate on stochastic effects of radiation".
low dose rate	expression used in this document to refer to dose rates of below 0.1 Gy per day for all radiations. For further discussion see UNSCEAR 1993 Report, Annex F, "Influence of dose and dose rate on stochastic effects of radiation".
kerma	kinetic energy released per unit mass of material
LDEF	low-dose effectiveness factor. The factor by which the slope of a pure linear model fitted to the data should be divided to give the low-dose slope, i.e. the linear term in a linear-quadratic dose-response model. Also designated DDREF.
LET	linear energy transfer, indicating the amount of the initial energy of a particle given up per unit length of its path through material, as a result of ionizations and other interactions. High-LET radiations include alpha particles and neutrons. Beta and gamma radiations and x rays are low-LET radiations.
radiation weighting factor	factor representing the different effectiveness of different radiations in inducing stochastic effects
RBE	when low- and high-LET radiations cause an equal level of effect, the ratio of their absorbed doses is called the RBE (relative biological effectiveness) of the high-LET radiation relative to the low-LET radiation. For the purposes of routine radiation protection, RBE is replaced by a single quantity (the radiation weighting factor) that is judged to be reasonably representative of the range of RBEs for all relevant effects at low doses and low dose rates; the product of absorbed dose in gray (Gy) and radiation weighting factor is termed the equivalent dose and is measured in sievert (Sv). Low-LET radiations are assigned a radiation weighting factor of unity, so an absorbed dose of 1 Gy of low-LET radiation represents an equivalent dose of 1 Sv. An absorbed dose of 1 Gy of alpha particles represent an equivalent dose of 20 Sv.
shielded kerma	kinetic energy released per unit mass of material after the incident radiation has passed through intervening shielding material
tissue weighting factor	factor for a particular tissue representing the fractional risk of the detriment (cancer plus hereditary effects) attributed to that tissue when the whole body is irradiated uniformly
weighted dose	used in this Annex to represent approximate equivalent doses in the life span study of the survivors of the atomic bombings, obtained by applying a value of RBE for neutrons of 10.

Epidemiological terms and abbreviations

additive (absolute) risk model — a model in which a unit of exposure induces an increase in the basic non-exposed age-specific mortality rate. The increase is independent of and thus additive to the increases caused by other exposures.

multiplicative (relative) risk model — a model in which a unit of exposure induces an increase in the cancer mortality rate that is proportional to the underlying age-specific rate.

excess absolute risk (EAR) — the absolute difference between the instantaneous incidence or mortality rates between two groups of people, e.g. those exposed to radiation at a given level and those unexposed

excess relative risk (ERR) — relative risk minus one

person-years (PY) — a unit of measurement combining persons and time, used as denominator in instantaneous incidence and mortality rates. It is the sum of individual years that the persons in the study population have been at risk of developing or dying from the condition of interest.

relative risk (RR) — the ratio of the instantaneous incidence or mortality rates in two groups of people, e.g. those exposed to radiation at a given level and those unexposed

standardized mortality ratio (SMR) — the ratio of the number of deaths observed in the study population to the number of deaths that would be expected if it had the same age-, sex- and calendar-period-specific mortality rates as the standard population, which is often taken to be the relevant national population. The SMR is sometimes expressed as a percentage (i.e. multiplied by 100).

standardized incidence ratio (SIR) — analogous to the SMR but calculated for numbers of incident cases of disease

References

A1 Albert, R.E., R.E. Shore, N. Harley et al. Follow-up studies of patients treated by x-ray epilation for tinea capitis. p. 1-25 in: Radiation Carcinogenesis and DNA Alterations. (F.J. Burns, A.C. Upton and G. Silini, eds.) Plenum Press, New York and London, 1986.

A2 Albo, V., D. Miller, S. Leiken et al. Nine brain tumors (BT) as a late effect in children "cured" of acute lymphoblastic leukaemia (ALL) from a single protocol study. Proc. Am. Soc. Clin. Oncol. 4: 172 (1985).

A3 Axelson, O., C. Edling and H. Kling. Lung cancer and residency - a case referent study on the possible impact of exposure to radon and its daughters in dwellings. Scand. J. Work Environ. Health 5: 10-15 (1979).

A4 Armstrong, B.K., J.C. McNulty, L.J. Levitt et al. Mortality in gold and coal miners in Western Australia with special reference to lung cancer. Br. J. Ind. Med. 36: 199-205 (1979).

A5 Archer, V.E. Association of lung cancer mortality with precambrian granite. Environ. Health 42: 87-91 (1987).

A6 Axelson, O., K. Andersson, G. Desai et al. Indoor radon exposure and active and passive smoking in relation to the occurrence of lung cancer. Scand. J. Work Environ. Health 14: 286-292 (1988).

A7 Axelson, O. and L. Sundell. Mining, lung cancer and smoking. Scand. J. Work Environ. Health 4: 46-52 (1978).

A8 Ager, E.A., L.M. Schuman, H.M. Wallace et al. An epidemiological study of childhood leukaemia. J. Chronic Dis. 18: 113-132 (1965).

A9 Aoyama, T. Radiation risk of Japanese and Chinese low dose repeatedly irradiated population. J. Univ. Occup. Environ. Health. 11 (Suppl.): 432-442 (1989).

A10 Acquavella, J.F., L.D. Wiggs, R.J. Waxweiler et al. Mortality among workers at the Pantex weapons facility. Health Phys. 48: 735-746 (1985).

A11 Archer, V.E., J.K. Wagoner and F.E. Lundin. Cancer mortality among uranium mill workers. J. Occup. Med. 15: 11-14 (1973).

A12 Andersson, M. and H.H. Storm. Cancer incidence among Danish thorotrast-exposed patients. J. Natl. Cancer Inst. 84: 1318-1325 (1992).

A13 Akiba, S., J. Lubin, H. Ezaki et al. Thyroid cancer incidence among atomic bomb survivors, 1958-79. RERF TR/5-91 (1991).

A14 Abbatt, J.D. Human leukemic risk data derived from the Portuguese Thorotrast experience. p. 451-464 in: Radionuclide Carcinogenesis. (C.L. Sanders et al., eds.) CONF-720505 (1973).

A15 Andersson, M., B. Carstensen and J. Visfeld. Leukemia and other related hematological disorders among Danish patients exposed to Thorotrast. Radiat. Res. 134: 224-233 (1993).

A16 Andersson, M., M. Vyberg, J. Visfeldt et al. Primary liver tumors among Danish patients exposed to Thorotrast. Radiat. Res. 137: 262-273 (1994).

A17 Abelson, P.H. Mineral dusts and radon in uranium mines. Science 254: 777 (1991).

A18 Alavanja, M.C.R., R.C. Brownson, J.H. Lubin et al. Residential radon exposure and lung cancer among non-smoking women. J. Natl. Cancer Inst. (1994, in press).

B1 Baral, E., L.E. Larsson and B. Mattsson. Breast cancer following irradiation of the breast. Cancer 40: 2905-2910 (1977).

B2 Baverstock, K.F. and D.G. Papworth. The UK radium luminiser survey: significance of a lack of excess leukaemia. Strahlentherapie 80 (Suppl.): 22-26 (1986).

B3 Bean, J.A., P. Isacson, R.M.A. Hahne et al. Drinking water and cancer incidence in Iowa. II. Radioactivity in drinking water. Am. J. Epidemiol. 116: 924-932 (1982).

B4 Beral, V., H. Inskip, P. Fraser et al. Mortality of employees of the United Kingdom Atomic Energy Authority 1946-1979. Br. Med. J. 291: 440-447 (1985).

B5 Beral, V., P. Fraser, L. Carpenter et al. Mortality of employees of the Atomic Weapons Establishment. Br. Med. J. 297: 757-770 (1988).

B6 Bithell, J.F. and C.A. Stiller. A new calculation of the carcinogenic risk of obstetric x-raying. Stat. Med. 7: 857-864 (1988).

B7 Black, D. Investigation of the possible increased incidence of cancer in Western Cumbria. Her Majesty's Stationary Office, London (1984).

B8 Boice, J.D. Carcinogenesis - a synopsis of human experience with external exposure in medicine. Health Phys. 55: 621-630 (1988).

B9 Brada, M., D. Ford, S. Ashley et al. Risk of second brain tumour after conservative surgery and radiotherapy for pituitary adenoma. Br. Med. J. 304: 1343-1346 (1992).

B10 Brenner, D.J. and E.J. Hall. The interplay between dose, dose protection and radiation quality. p. 167-177 in: Advances in Radiation Biology. Effects of Low Dose and Low Dose Rate Irradiation. Volume 16. (O.F. Nygaard, W.K. Sinclair and J.T. Lett, eds.) Academic Press, San Diego, 1992.

B11 Boice, J.D., N.E. Day, A. Andersen et al. Second cancers following radiation treatment for cervical cancer. An international collaboration among cancer registries. J. Natl. Cancer Inst. 74: 955-975 (1985).

B12 Boice, J.D., M. Blettner, R.A. Kleinerman et al. Radiation dose and leukaemia risk in patients treated for cancer of the cervix. J. Natl. Cancer Inst. 79: 1295-1311 (1987).

B13 Baverstock, K.F. and J.W. Stather (eds.). Low Dose Radiation: Biological Bases of Risk Assessment. Taylor & Francis, London, 1989.

B14 Bross, I.D. and N.S. Bross. Do atomic veterans have excess cancer? New results correcting for the healthy soldier bias. Am. J. Epidemiol. 126: 1042-1050 (1987).

B15 Beral, V. and G. Reeves. Childhood thyroid cancer in Belarus. Nature 359: 680-681 (1992).

B16 Boivin, J.F., G.B. Hutchison, F.B. Evans et al. Leukaemia after radiotherapy for first primary cancers of various anatomic sites. Am. J. Epidemiol. 123: 993-1003 (1986).

B17 Blettner, M. and J.D. Boice Jr. Radiation dose and leukaemia risk: general relative risk techniques for dose-response models in a matched case-control study. Stat. Med. 10: 1511-1526 (1991).

B18 Boice, J.D., E.B. Harvey, M. Blettner et al. Cancer in the contralateral breast after radiotherapy for breast cancer. N. Engl. J. Med. 326: 781-785 (1992).

B19 Boice, J.D. Jr. Radiation and non-Hodgkin's lymphoma. Cancer Res. 52: 5489-5491 (1992).

B20 Boice, J.D., M. Blettner, R.A. Kleinerman et al. Radiation dose and breast cancer risk in patients treated for cancer of the cervix. Int. J. Cancer. 44: 7-16 (1989).

B21 Boice, J.D. Jr., G. Engholm, R.A. Kleinerman et al. Radiation dose and second cancer risk in patients treated for cancer of the cervix. Radiat. Res. 116: 3-55 (1988).

B22 Bithell, J.F. Epidemiological studies of children irradiated *in utero*. p. 77-87 in: Low Dose Radiation: Biological Bases of Risk Assessment. (K. Baverstock and J.W. Stather, eds.) Taylor & Francis, London, 1989.

B23 Bithell, J.F. and A.M. Stewart. Prenatal irradiation and childhood malignancy: a review of British data from the Oxford survey. Br. J. Cancer. 31: 271-287 (1975).

B24 Brincker, H., H.S. Hansen and A.P. Andersen. Induction of leukaemia by ^{131}I treatment of thyroid carcinoma. Br. J. Cancer. 28: 232-237 (1973).

B25 Bowlt, C. and P. Tiplady. Radioiodine in human thyroid glands and incidence of thyroid cancer in Cumbria. Br. Med. J. 299: 301-302 (1989).

B26 Baverstock, K.F., D. Papworth and J. Vennart. Risks of radiation at low dose rates. Lancet I: 430-433 (1981).

B27 Baisogolov, G.D., U.G. Bolotnikova, I.A. Golstian et al. Malignant neoplasms of haematopoietic and lymphatic systems in workers of the first nuclear facility (enterprise). Probl. Onkol. 5: 553-558 (1992). (in Russian)

B28 Breslow, N.E. and N.E. Day. Statistical methods in cancer research. The analysis of case-control studies. Volume I. IARC Publication No. 32 (1980).

B29 Breslow, N.E. and N.E. Day. Statistical methods in cancer research. The design and analysis of cohort studies. Volume II. IARC Publication No. 82 (1987).

B30 Buldakov, L.A. and A.K. Guskova. Effects of chronic exposure to radiation. J. Physician 5: 34-39 (1991).

B31 Boice, J.D. Jr., D.L. Preston, F.G. Davis et al. Frequent chest X-ray fluoroscopy and breast cancer incidence among tuberculosis patients in Massachusetts. Radiat. Res. 125: 214-222 (1990).

B32 Beck, W.L., D.L. Cookfair, C.L. Sowder et al. Radiation dosimetry for epidemiologic lung cancer study. p. 436-443 in: Epidemiology Applied to Health Physics. CONF-830101 (1983).

B33 Berk, P.D., J.D. Goldberg, N. Murray et al. Increased incidence of acute leukaemia in polycythemia vera associated with chlorambucil therapy. N. Engl. J. Med. 304: 441-447 (1981).

B34 Boice, J.D., C.E. Land and D.L. Preston. Ionizing radiation. in: Cancer Epidemiology and Prevention. (D. Schottenfeld and J.F. Fraumeni Jr., eds.) Oxford Press, 1994.

B35 Baverstock, K.F. and D.G. Papworth. The UK radium luminiser survey. p. 72-76 in: Risks from Radium and Thorotrast. BIR Report 21. (D.M. Taylor et al., eds.) British Institute of Radiology, London, 1989.

B36 Baverstock, K.F. and J. Vennart. A Note on radium body content and breast cancer in U.K. radium luminisers. Health Phys. 44: 575-577 (1983).

B37 Blot, W.J., Z.Y. Xu, J.D. Boice et al. Indoor radon and lung cancer in China. J. Natl. Cancer Inst. 82: 1025-1030 (1990).

B38 Boice, J.D., M.M. Morin, A.G. Glass et al. Diagnostic X-ray procedures and risk of leukaemia, lymphoma, and multiple myeloma. J. Am. Med. Assoc. 265: 1290-1294 (1991).

B39 Boice, J.D., J.S. Mandel, M.M. Doody et al. A health survey of radiologic technologists. Cancer 69: 586-598 (1992).

B40 Black, R.J., J.D. Urquhart, S.W. Kendrick et al. Incidence of leukaemia and other cancers in birth and schools cohorts in the Dounreay area. Br. Med. J. 304: 1401-1405 (1992).

B41 Boice, J.D., M. Rosenstein and E.D. Trout. Estimation of breast doses and breast cancer risk associated with repeated fluoroscopic chest examinations of women with tuberculosis. Radiat. Res. 73: 373-390 (1978).

B42 Baverstock, K., B. Egloff, A. Pinchera et al. Thyroid cancer after Chernobyl. Nature 359: 21-22 (1992).

C1 Caldwell, G.G., D.B. Kelley and C.W. Heath. Leukaemia among participants in military maneuvers at a nuclear bomb test. A preliminary report. J. Am. Med. Assoc. 244: 1575-1578 (1980).

C2 Caldwell, G.G., D.B. Kelley, M. Zack et al. Mortality and cancer frequency among military nuclear test (smoky) participants, 1957 through 1979. J. Am. Med. Assoc. 250: 620-624 (1983).

C3 Checkoway, H., R.M. Mathew, C.M. Shy et al. Radiation, work experience and cause specific mortality among workers at an energy research laboratory. Br. J. Ind. Med. 42: 525-533 (1985).

C4 Chmelevsky, D., A.M. Kellerer, H. Spiess et al. A proportional hazards analysis of bone sarcoma rates in German radium-224 patients. p. 32-37 in: The Radiobiology of Radium and Thorotrast. (W. Gössner et al., eds.) Urban and Schwarzenberg, München, 1986.

C5 Commission of the European Community. Thyroid cancer in children living near Chernobyl. EUR 15248 (1993).

C6 Committee on Biological Effects of Ionizing Radiation (BEIR IV). Health Risks of Radon and Other Internally Deposited Alpha-Emitters. United States National Academy of Sciences, National Research Council. National Academy Press, Washington, 1988.

C7 Cook-Mozaffari, P.J., S.C. Darby, R. Doll et al. Geographical variation in mortality from leukaemia and other cancers in England and Wales in relation to proximity to nuclear installations, 1969-1978. Br. J. Cancer 59: 476-485 (1989).

C8 Curtis, R.E., R.N. Hoover, R.A. Kleinerman et al. Second cancers following cancer of the female genital system in Connecticut, 1935-1982. Natl. Cancer Inst. Monogr. 68: 113-137 (1985).

C9 Cuzick, J. and B. de Stavola. Multiple myeloma: a case-control study. Br. J. Cancer. 57: 516-520 (1988).

C10 Curtis, R.E., J.D. Boice, R.A. Kleinerman et al. Summary: multiple primary cancers in Connecticut, 1935-1982. Natl. Cancer Inst. Monogr. 68: 219-242 (1985).

C11 Craft, A.W., L. Parker, S. Openshaw et al. Cancer in young people in the north of England, 1968-85. Analysis by census wards. J. Epidemiol. Community Health 47: 109-115 (1993).

C12 Committee on Biological Effects of Ionizing Radiation (BEIR V). Health Effects of Exposure to Low Levels of Ionizing Radiation. United States National Academy of Sciences, National Research Council. National Academy Press, Washington, 1990.

C13 Cohen, B.L. Experimental test of the linear no-threshold theory of radiation carcinogenesis. p. 319-330 in: Environmental Radon: Occurrence, Control and Health Hazards. (S.K. Majumdar et al., eds.) Pennsylvania Academy of Science, Pennsylvania, 1990.

C14 Curtis, R.E., B.F. Hankey, M.H. Myers et al. Risk of leukaemia associated with the first course of cancer treatment: an analysis of the surveillance, epidemiology and end results program experience. J. Natl. Cancer Inst. 72: 531-544 (1984).

C15 Court Brown, W.M., R. Doll and A. Bradford Hill. Incidence of leukaemia after exposure to diagnostic radiation in utero. Br. Med. J. 2: 1539-1545 (1960).

C16 Cragle, D.L., R.W. McLain, J.R. Qualters et al. Mortality among workers at a nuclear fuels production facility. Am. J. Ind. Med. 14: 379-401 (1988).

C17 Chmelevsky, D., H. Spiess, C.W. Mays et al. The reverse protraction factor in the induction of bone sarcomas in radium-224 patients. Radiat. Res. 124: S69-S79 (1990).

C18 Cook-Mozaffari, P., S.C. Darby and R. Doll. Cancer near potential sites of nuclear installations. Lancet II: 1145-1147 (1989).

C19 Cook-Mozaffari, P., F.L. Ashwood, T. Vincent et al. Cancer incidence and mortality in the vicinity of nuclear installations, England and Wales, 1950-1980. Studies on medical population subjects. Her Majesty's Stationary Office, London (1987).

C20 Cuzick, J. Radiation induced myelomatosis. N. Engl. J. Med. 304: 204-219 (1981).

C21 Committee on Comparative Dosimetry of Radon in Mines and Homes. United States National Academy of Sciences, National Research Council. National Academy Press, Washington, 1991.

C22 Checkoway, H., N. Pearce, D.J. Crawford-Brown et al. Radiation doses and cause-specific mortality among workers at a nuclear materials fabrication plant. Am. J. Epidemiol. 127: 255-266 (1988).

C23 Castren, O., A. Voutilainen, K. Winqvist et al. Studies of high indoor radon areas in Finland. Sci. Total Environ. 45: 311-318 (1985).

C24 Curtis, R.E., J.D. Boice, M. Stovall et al. Risk of leukemia after chemotherapy and radiation treatment for breast cancer. N. Engl. J. Med. 326: 1745-1751 (1992).

C25 Carpenter, L.M. Studies of cancer among medical personnel exposed to radiation. Radiat. Res. 124: 351-352 (1990).

C26 Clarke, E.A., J. McLaughlin and T.W. Anderson. Childhood leukaemia around Canadian nuclear facilities - Phase II. Final Report. Atomic Energy Control Board, Canada (1991).

C27 Court Brown, W.M., R. Doll, F.W. Spiers et al. Geographical variation in leukaemia mortality in relation to background radiation and other factors. Br. Med. J. 1: 1753-1759 (1960).

C28 Conard, R.A. Late radiation effects in Marshall islanders exposed to fallout 28 years ago. p. 57-71 in: Radiation Carcinogenesis: Epidemiology and Biological Significance. (J.D. Boice Jr. and J.F. Fraumeni Jr., eds.). Raven Press, New York, 1984.

C29 Cookfair, D.L., W.L. Beck, C. Shy et al. Lung cancer among workers at a uranium processing plant. p. 398-406 in: Epidemiology Applied to Health Physics. CONF-830101 (1983).

C30 Committee on Medical Aspects of Radiation in the Environment (COMARE). 3rd Report. (M. Bobrow, Chairman). Her Majesty's Stationary Office, London (1989).

C31 Chen, S.Y., R.B. Hayes, S.R. Liang et al. Mortality experience of haematite mine workers in China. Br. J. Ind. Med. 47: 175-181 (1990).

C32 Carpenter, L., P. Fraser, M. Booth et al. Smoking habits and radiation exposure. J. Radiol. Prot. 9: 286-287 (1989).

C33 Cardis, E. and J. Estève. International collaborative study of cancer risk among nuclear industry workers. I. Report of the feasibility study; II. Protocol. IARC 92/001 (1992).

C34 Cardis, E., E. Gilbert, L. Carpenter et al. Combined analyses of cancer mortality among nuclear industry workers in three countries. IARC 94/001 (1994).

C35 Cohen, B.L. Relationship between exposure to radon and various types of cancer. Health Phys. 65: 529-531 (1993).

C36 Curtis, R.E., J.D. Boice Jr., M. Stovall et al. The relation of leukaemia risk to radiation dose after cancer of the uterine corpus. J. Natl. Cancer Inst. (1994, in press).

D1 Damber, L.A. and L.G. Larsson. Lung cancer in males and type of dwelling. An epidemiological pilot study. Acta Oncol. 26: 211-215 (1987).

D2 Darby, S.C., J.H. Olsen, R. Doll et al. Trends in childhood leukaemia in the Nordic countries in relation to fallout from atmospheric nuclear weapons testing. Br. Med. J. 304: 1005-1009 (1992).

D3 Darby, S.C., G.M. Kendall, T.P. Fell et al. Mortality and cancer incidence in UK participants in UK atmospheric nuclear weapon tests and experimental programmes. NRPB-R214 (1988).

D4 Darby, S.C., G.M. Kendall, T.P. Fell et al. A summary of mortality and incidence of cancer in men from the United Kingdom who participated in the United Kingdom's atmospheric nuclear weapon tests and experimental programmes. Br. Med. J. 296: 332-338 (1988).

D5 Darby, S.C., E. Nakashima and H. Kato. A parallel analysis of cancer mortality among atomic bomb survivors and patients with ankylosing spondylitis given X-ray therapy. J. Natl. Cancer Inst. 75: 1-21 (1985).

D6 Darby, S.C., R. Doll, S.K. Gill et al. Long-term mortality after a single treatment course with X-rays in patients treated for ankylosing spondylitis. Br. J. Cancer 55: 179-190 (1987).

D7 Darby, S.C. The contribution of natural ionizing radiation to cancer mortality in the United States. p. 183-190 in: The Origins of Human Cancer. (J. Brugge et al., eds.) Cold Spring Harbor Laboratory Press, New York, 1991.

D8 Dionian, J., C.R. Muirhead, S.L. Wan et al. The risks of leukaemia and other cancers in Thurso from radiation exposure. NRPB-R196 (1986).

D9 Dobyns, B.M., G.E. Sheline, J.B. Workman et al. Malignant and benign neoplasms of the thyroid in patients treated for hyperthyroidism: a report of the Cooperative Thyrotoxicosis Therapy Follow-up Study. J. Clin. Endocrinol. Metab. 38: 976-998 (1974).

D10 Doll, R. and S.C. Darby. Occupational epidemiology: problems in reaching an overview. p. 105-112 in: Health Effects of Low Dose Ionizing Radiation. British Nuclear Energy Society, London, 1988.

D11 Dousset, M. Lung cancer deaths in two French regions with very different mean radon-222 concentration in dwellings. SCPRI, France, 1990.

D12 Draper, G.J., B.M. Sanders and J.E. Kingston. Second primary neoplasms in patients with retinoblastoma. Br. J. Cancer 53: 661-671 (1986).

D13 Damber, L. and L.G. Larsson. Underground mining, smoking and lung cancer: a case-control study in the iron ore municipalities in northern Sweden. J. Natl. Cancer Inst. 74: 1207-1213 (1985).

D14 Davis, F.G., J.D. Boice, Z. Hrubec et al. Cancer mortality in a radiation-exposed cohort of Massachusetts tuberculosis patients. Cancer Res. 49: 6130-6136 (1989).

D15 da Motta, L.C., J. da Silva Horta and M.H. Tavares. Prospective epidemiological study of Thorotrast exposed patients in Portugal. Environ. Res. 18: 162-172 (1979).

D16 Diamond, E.L., H. Schmerler and A.M. Lilienfeld. The relationship of intra-uterine radiation to subsequent mortality and development of leukaemia in children. A prospective study. Am. J. Epidemiol. 97: 283-313 (1973).

D17 Dalheimer, A.R., A. Kaul, M.D. Said et al. Investigation on the effect of incorporated radioactive and nonradioactive particles and their synergism by long-term animal studies - physicochemical and biokinetic properties of zirconium dioxide colloids. Strahlentherapie 80 (Suppl.): 167-171 (1986).

D18 DeGroot, L., M. Reilly, K. Pinnameneni et al. Retrospective and prospective study of radiation-induced thyroid disease. Am. J. Med. 74: 852-862 (1983).

D19 Doll, R. and R. Peto. The causes of cancer: Quantitative estimates of avoidable risks of cancer in the United States today. J. Natl. Cancer Inst. 66: 1191-1308 (1981).

D20 Dobson, R.L., T. Straume, A.V. Carrano et al. Biological effectiveness of neutrons from Hiroshima bomb replica: results of a collaborative cytogenetic study. Radiat. Res. 128: 143-149 (1991).

D21 Dickson, R.J. Late results of radium treatment of carcinoma of the cervix. Clin. Radiol. 23: 528-535 (1972).

D22 Dupree, E.A., D.L. Cragle, R.W. McLain et al. Mortality among workers at a uranium processing facility, the Linde Air Products Company Ceramics Plant, 1943-1949. Scand. J. Work Environ. Health 13: 100-107 (1987).

D23 Darby, S.C., G. Reeves, T. Key et al. Mortality in a cohort of women given X-ray therapy for metropathia haemorrhagica. Int. J. Cancer 86: 793-801 (1994).

D24 Darby, S.C., G.M. Kendall, T.P. Fell et al. Further follow-up of mortality and incidence of cancer in man from the United Kingdom who participated in the United Kingdom's atmospheric nuclear weapon tests and experimental programmes. Br. Med. J. 307: 1530-1535 (1993).

D25 Doll, R. Communication to UNSCEAR Secretariat (1990).

D26 Doshenko, V.N. Structure of death causes after considerable occupational chronic total γ-irradiation. J. Med. Radiol. 8: 40 (1991).

D27 Degteva, M.O., V.P. Kozheurov and M.I. Vorobiova. General approach to dose reconstruction in the population exposed as a result of the release of radioactive wastes into the Techa river. Sci. Total Environ. 142: 49-61 (1994).

D28 Doll, R., H.J. Evans and S.C. Darby. Paternal exposure not to blame. Nature 367: 678-680 (1994).

E1 Edling, C., P. Comba, O. Axelson et al. Effects of low-dose radiation - a correlation study. Scand. J. Work Environ. Health 8 (Suppl. 1): 59-64 (1982).

E2 Edling, C., H. Kling and O. Axelson. Radon in homes - a possible cause of lung cancer. Scand. J. Work Environ. Health 10: 25-34 (1984).

E3 Edling, C., G. Wingren and O. Axelson. Quantification of the lung cancer risk from radon daughter exposure in dwellings - an epidemiological approach. Environ. Int. 12: 55-60 (1986).

E4 Eng, C., F.P. Li, D.H. Abramson et al. Mortality from second tumors among long-term survivors of retinoblastoma. J. Natl. Cancer Inst. 85: 1121-1128 (1993).

E5 Enterline, P.E., G.M. Marsh, N.A. Esmen et al. Some effects of cigarette smoke, arsenic and SO_2 on mortality among US copper smelter workers. J. Occup. Med. 29: 831-838 (1987).

E6 Edmonds, C.J. and T. Smith. The long-term hazards of the treatment of thyroid cancer with radioiodine. Br. J. Radiol. 59: 45-51 (1986).

E7 Ennow, K., M. Andersson, G. Engholm et al. Epidemiological assessment of the cancer risk among the staff in radiotherapy departments in Denmark. p. 327-333 in: Low Dose Radiation: Biological Bases of Risk Assessment. (K.F. Baverstock and J.W. Stather, eds.) Taylor & Francis, London, 1989.

E8 Evans, R.D. The effect of skeletally deposited alpha-ray emitters in man. Br. J. Radiol. 39: 881-895 (1966).

E9 Eatough, J.P. and D.L. Henshaw. Radon and prostate cancer. Lancet 335: 1292 (1990).

F1 Faber, M. Current (1981) status of the Danish Thorotrast study. Health Phys. 44 (Suppl. 1): 259-260 (1983).

F2 Faber, M. Observations on the Danish Thorotrast patients. Strahlentherapie 80 (Suppl.): 140-142 (1986).

F3 Falk, H., N.C. Telles, K.G. Ishak et al. Epidemiology of Thorotrast-induced hepatic angiosarcoma in the United States. Environ. Res. 18: 65-73 (1979).

F4 Folley, J.N., W. Borges and T. Yamawaki. Incidence of leukaemia in survivors of the atomic bomb in Hiroshima and Nagasaki: Japan. Am. J. Med. 13: 311-321 (1952) and ABCC TR/30-59 (1959).

F5 Fleischer, R.L. A possible association between lung cancer and phosphate mining and processing. Health Phys. 41: 175-178 (1981).

F6 Fleischer, R.L. A possible association between lung cancer and a geological outcrop. Health Phys. 50: 823-827 (1986).

F7 Flodin, U., L. Andersson, C.G. Anjou et al. A case-referent study on acute myeloid leukaemia, background radiation and exposure to solvents and other agents. Scand. J. Work Environ. Health 7: 169-178 (1981).

F8 Flodin, U., M. Fredriksson, O. Axelson et al. Background radiation, electrical work and some other exposures associated with acute myeloid leukaemia in a case-referent study. Arch. Environ. Health 41: 77-84 (1986).

F9 Forastiere, F., S. Valesini, M. Arca et al. Lung cancer and natural radiation in an Italian province. Sci. Total Environ. 45: 519-526 (1985).

F10 Forman, D., P. Cook-Mozaffari, S.C. Darby et al. Cancer near nuclear installations. Nature 329: 499-505 (1987).

F11 Fraser, P., L. Carpenter, N. Maconochie et al. Cancer mortality and morbidity in employees of the United Kingdom Atomic Energy Authority, 1946-86. Br. J. Cancer 67: 615-624 (1993).

F12 Fürst, C.J., M. Lundell and L.E. Holm. Cancer incidence after radiotherapy for skin haemangioma: a retrospective cohort study in Sweden. J. Natl. Cancer Inst. 80: 1387-1392 (1988).

F13 Ford, D.D., J.C.S. Paterson and W.L. Treuting. Fetal exposure to diagnostic X rays and leukaemia and other malignant diseases in childhood. J. Natl. Cancer Inst. 22: 1093-1104 (1959).

F14 Friedman, G.D. Phenylbutazone, musculoskeletal disease and leukaemia. J. Chronic Dis. 35: 233-243 (1982).

F15 Fürst, C.J., M. Lundell and L.E. Holm. Tumors after radiotherapy for skin hemangioma in childhood. A case-control study. Acta Oncol. 29: 557-562 (1990).

F16 Frome, E.L., D.L. Cragle and R.W. McLain. Poisson regression analysis of the mortality among a cohort of World War II nuclear industry workers. Radiat. Res. 123: 138-152 (1990).

F17 Flodin, U., M. Fredriksson and B. Persson. Multiple myeloma and engine exhausts, fresh wood and creosote: a case-referent study. Am. J. Ind. Med. 12: 519-529 (1987).

F18 Flodin, U., M. Fredriksson, B. Persson et al. Acute myeloid leukaemia and background radiation in an expanded case-referent study. Arch. Environ. Health 45: 364-366 (1990).

F19 Frodkin, J.E., J.L. Mills, L.B. Schonberger et al. Risk of leukaemia after treatment with pituitary growth hormone. J. Am. Med. Assoc. 270: 2829-2832 (1993).

F20 Fisenne, I.M., P. Perry, N.Y. Chu et al. Measured ^{238}U and fallout ^{239}Pu in human bone ash from Nepal and Australia: skeletal alpha dose. Health Phys. 44: 457-467 (1983).

F21 Fisenne, I.M., P.N. Demoleas and N.H. Harley. Thorium excretion from a Thorotrast patient. The Radiobiology of Radium and Thorotrast. Strahlentherapie (Suppl.) 80: 151-156 (1986).

G1 Gardner, M.J., A.J. Hall, S. Downes et al. Follow-up study of children born to mothers resident in Seascale, West Cumbria (birth cohort). Br. Med. J. 295: 822-827 (1987).

G2 Gardner, M.J., A.J. Hall, S. Downes et al. Follow-up study of children born elsewhere but attending schools in Seascale, West Cumbria (schools cohort). Br. Med. J. 295: 819-822 (1987).

G3 Gribbin, M.A., J.L. Weeks and G.R. Howe. Cancer mortality (1956-1985) among male employees of Atomic Energy of Canada Limited with respect to occupational exposure to external low-linear-energy-transfer ionizing radiation. Radiat. Res. 133: 375-380 (1993).

G4 Gilbert, E.S. Some effects of random dose measurement errors on analyses of atomic bomb survivor data. RERF TR/12-82 (1982) and Radiat. Res. 98: 591-605 (1984).

G5 Gilbert, E.S. and S. Marks. An analysis of the mortality of workers in a nuclear facility. Radiat. Res. 79: 122-148 (1979).

G6 Greene, M.H., E.L. Harris, D.M. Gershenson et al. Melphalan may be a more potent leukemogen than cyclophosphamide. Ann. Intern. Med. 105: 360-367 (1986).

G7 Goldman, M.B., F. Maloof, R.R. Monson et al. Radioactive iodine therapy and breast cancer. A follow-up study of hyperthyroid women. Am. J. Epidemiol. 127: 969-980 (1988).

G8 Greene, M.H., R.C. Young, J.M. Merrill et al. Evidence of a treatment dose response in acute nonlymphocytic leukaemias which occur after therapy of non-Hodgkin's lymphoma. Cancer Res. 43: 1891-1898 (1983).

G9 Gilbert, E.S. Mortality of workers at the Oak Ridge National Laboratory. Health Phys. 62: 260-261 (1992).

G10 Gilbert, E.S. and J.L. Ohara. An analysis of various aspects of atomic bomb dose estimation at RERF using data on acute radiation symptoms. Radiat. Res. 100: 124-138 (1984) and RERF TR/9-83 (1983).

G11 Griem, M.L., R.A. Kleinerman, J.D. Boice Jr. et al. Cancer following radiotherapy for peptic ulcer. J. Natl. Cancer Inst. 86: 842-849 (1994).

G12 Goldman, K.P. Mortality of coal miners from carcinoma of the lung. Br. J. Ind. Med. 22: 72-77 (1965).

G13 Gray, R.G., J. Lafuma, S.E. Parish et al. Lung tumors and radon inhalation in over 2000 rats: approximate linearity across a wide range of doses and potentiation by tobacco smoke. p. 592-607 in: Life-Span Radiation Effects Studies in Animals: What Can They Tell Us? (R.C. Thompson and J.A. Mahaffey, eds.) Office of Scientific and Technical Information, United States Department of Energy, 1986.

G14 Graham, S., M.L. Levin, A.M. Lilienfeld et al. Preconception, intrauterine, and postnatal irradiation as related to leukemia. Natl. Cancer Inst. Monogr. 19: 347-371 (1966).

G15 Gunz, F.W. and H.R. Atkinson. Medical radiations and leukaemia: a retrospective survey. Br. Med. J. 1: 389-393 (1964).

G16 Gilbert, E.S., E. Omohundro, J.A. Buchanan et al. Mortality of workers at the Hanford site: 1945-1986. Health Phys. 64: 577-590 (1993).

G17 Gilbert, E.S., S.A. Fry, L.D. Wiggs et al. Analyses of combined mortality data on workers at the Hanford site, Oak Ridge National Laboratory and Rocky Flats nuclear weapons plant. Radiat. Res. 120: 19-35 (1989).

G18 Gibson, R., S. Graham, A. Lilienfeld et al. Irradiation in the epidemiology of leukemia among adults. J. Natl. Cancer Inst. 48: 301-311 (1972).

G19 Gilbert, E.S., D.L. Cragle and L.D. Wiggs. Updated analysis of combined mortality data on workers at the Hanford site Oak Ridge National Laboratory and Rocky Flats Weapons Plant. Radiat. Res. 136: 408-421 (1993).

G20 Gilbert, E.S., G.R. Petersen and J.A. Buchanan. Mortality of workers at the Hanford site: 1945-1981. Health Phys. 56: 11-25 (1989).

G21 Gardner, M.J., M.P. Snee, A.J. Hall et al. Results of case-control study of leukaemia and lymphoma among young people near Sellafield nuclear plant in West Cumbria. Br. Med. J. 300: 423-429 (1990).

G22 Gardner, M.J. Communication to the UNSCEAR Secretariat (1990).

G23 Gjørup, H.L. and H.J.M. Hansen. Lifetime loss through lung cancer in Denmark and Sweden in relation to radon levels. p. 165-169 in: Health Effects of Low Dose Ionising Radiation - Recent Advances and Their Implications. British Nuclear Energy Society, 1987.

G24 Goss, S.G. Sizes of population needed to detect an increase in disease risk when the levels of risk in the exposed and controls are specified: examples from cancer induced by ionizing radiation. Health Phys. 29: 715-721 (1975).

H1 Hanford, J.M., E. Quimby and V. Frantz. Cancer arising many years after radiation therapy. J. Am. Med. Assoc. 181: 132-138 (1962).

H2 Hancock, S.L., M.A. Tucker and R.T. Hoppe. Breast cancer after treatment of Hodgkin's disease. J. Natl. Cancer Inst. 85: 25-31 (1993).

H3 Harvey, E.B., J.D. Boice, M. Honeyman et al. Prenatal x-ray exposure and childhood cancer in twins. N. Engl. J. Med. 312: 541-545 (1985).

H4 Heasman, M.A., I.W. Kemp, J.D. Urquhart et al. Childhood leukaemia in northern Scotland. Lancet I: 266 and 385 (1986).

H5 Hess, C.T., C.V. Weiffenbach and S.A. Norton. Environmental radon and cancer correlations in Maine. Health Phys. 45: 339-348 (1983).

H6 Hancock, S.L., R.S. Cox and R. McDougall. Thyroid diseases after treatment of Hodgkin's disease. N. Engl. J. Med. 325: 599-605 (1991).

H7 Hoffman, D.A. Late effects of I-131 therapy in the United States. p. 273-280 in: Radiation Carcino-genesis: Epidemiology and Biological Significance. (J.D. Boice Jr. and J.F. Fraumeni Jr., eds.) Raven Press, New York, 1984.

H8 Hoffman, D.A., W.M. McConahey, E.L. Diamond et al. Mortality in women treated for hyperthyroidism. Am. J. Epidemiol. 115: 243-254 (1982).

H9 Hoffman, D.A., W.M. McConahey, J.F. Fraumeni Jr. et al. Cancer incidence following treatment of hyperthyroidism. Int. J. Epidemiol. 11: 218-224 (1982).

H10 Hofmann, W., R. Katz and C. Zhang. Lung cancer incidence in a Chinese high background area - epidemiological results and theoretical interpretation. Sci. Total Environ. 45: 527-534 (1985).

H11 Holm, L.E. Malignant disease following iodine-131 therapy in Sweden. p. 263-271 in: Radiation Carcinogenesis: Epidemiology and Biological Significance. (J.D. Boice Jr. and J.F. Fraumeni Jr., eds.) Raven Press, New York, 1984.

H12 Holm, L.E., K.E. Wiklund, G.E. Lundell et al. Thyroid cancer after diagnostic doses of iodine-131: a retrospective cohort study. J. Natl. Cancer Inst. 80: 1132-1138 (1988).

H13 Horta, J. da Silva, M.E. da Silva Horta, L. Cayolla da Motta et al. Malignancies in Portuguese thorotrast patients. Health Phys. 35: 137-151 (1978).

H14 Hill, C.K., A. Han and M.M. Elkind. Fission spectrum neutrons at a low dose rate enhance neoplastic transformation in the linear dose region (0-10 cGy). Int. J. Radiat. Biol. 46: 11-15 (1984).

H15 Howe, G.R., R.C. Nair, H.B. Newcombe et al. Lung cancer mortality (1950-80) in relation to radon daughter exposure in a cohort of workers at the Eldorado Beaverlodge uranium mine. J. Natl. Cancer Inst. 77: 357-362 (1986).

H16 Howe, G.R., R.C. Nair, H.B. Newcombe et al. Lung cancer mortality (1950-80) in relation to radon daughter exposure in a cohort of workers at the Eldorado Port radium uranium mine: possible modification of risk by exposure rate. J. Natl. Cancer Inst. 79: 1255-1260 (1987).

H17 Hohryakov, V.F. and S.A. Romanov. Lung cancer in radiochemical industry workers. Sci. Total Environ. 142: 25-28 (1994).

H18 Hornung, R.W. and T.J. Meinhardt. Quantitative risk assessment of lung cancer in U.S. uranium miners. Health Phys. 52: 417-430 (1987).

H19 Haynes, R.M. The distribution of domestic radon concentrations and lung cancer mortality in England and Wales. Radiat. Prot. Dosim. 25: 93-96 (1988).

H20 Hofmann, W., R. Katz and C. Zhang. Lung cancer risk at low doses of α particles. Health Phys. 51: 457-468 (1986).

H21 Hempelmann, L.H., W.J. Hall, M. Phillips et al. Neoplasms in persons treated with x-rays in infancy: fourth survey in 20 years. J. Natl. Cancer Inst. 55: 519-530 (1975).

H22 Hildreth, N.G., R.E. Shore and P.M. Dvoretsky. The risk of breast cancer after irradiation of the thymus in infancy. N. Engl. J. Med. 321: 1281-1284 (1989).

H23 Howe, G. Communication to UNSCEAR Secretariat (1992).

H24 Hoffman, D.A., J.E. Lonstein, M.M. Morin et al. Breast cancer in women with scoliosis exposed to multiple diagnostic X rays. J. Natl. Cancer Inst. 81: 1307-1312 (1989).

H25 Hagstrom, R.H., S.R. Glasser, A.B. Brill et al. Long term effects of radioactive iron administered during human pregnancy. Am. J. Epidemiol. 90: 1-10 (1969).

H26 Hall, P., J.D. Boice Jr., G. Berg et al. Leukaemia incidence after iodine-131 exposure. Lancet 340: 1-4 (1992).

H27 Holm, L.E., K.E. Wiklund, G.E. Lundell et al. Cancer risk in population examined with diagnostic doses of ^{131}I. J. Natl. Cancer Inst. 81: 302-306 (1989).

H28 Hamilton, T.E., G. van Belle and J.P. LoGerfo. Thyroid neoplasia in Marshall Islanders exposed to nuclear fallout. J. Am. Med. Assoc. 258: 629-636 (1987).

H29 Hoel, D.G. and G.E. Dinse. Using mortality data to estimate radiation effects on breast cancer incidence. Environ. Health Perspect. 87: 123-129 (1990).

H30 Hakulinen, T., M. Kenward, T. Luostarinen et al. Cancer in Finland in 1954-2008. Incidence, Mortality and Prevalence by Region. Finnish Foundation for Cancer Research, Helsinki, 1989.

H31 Hall, P., L.E. Holm, G. Lundell et al. Cancer risk in thyroid cancer patients. Br. J. Cancer 64: 159-163 (1991).

H32 Hadjimichael, O.C., A.M. Ostfeld, D.A. D'Atri et al. Mortality and cancer incidence experience of employees in a nuclear fuels fabrication plant. J. Occup. Med. 25: 48-61 (1983).

H33 Hatch, M.C., J. Beyea, J.W. Nieves et al. Cancer near the Three Mile Island nuclear plant: radiation emissions. Am. J. Epidemiol. 132: 397-412 (1990).

H34 Hatch, M. and M. Susser. Background gamma radiation and childhood cancers within ten miles of a US nuclear plant. Int. J. Epidemiol. 19: 546-552 (1990).

H35 High Background Radiation Research Group. Health survey in high background radiation areas in China. Science 209: 877-880 (1980).

H36 Hempelmann, L.H., C.C. Lushbaugh and G.L. Voelz. What happened to the survivors of the early Los Alamos nuclear accidents? p. 17-32 in: The Medical Basis for Radiation Accident Preparedness. (K.F. Hübner and S.A. Fry, eds.) Elsevier North Holland, Inc., 1980.

H37 Hodgson, J.T. and R.D. Jones. Mortality of a cohort of tin miners 1941-86. Br. J. Ind. Med. 47: 665-676 (1990).

H38 Henshaw, D.L., J.P. Eatough and R.B. Richardson. Radon as a causative factor in induction of myeloid leukaemia and other cancers. Lancet 335: 1008-1012 (1990).

H39 Hill, C. and A. Laplanche. Overall mortality and cancer mortality around French nuclear sites. A letter. Nature 347: 755-757 (1990).

H40 Holm, L.E., P. Hall, K.E. Wiklund et al. Cancer risk after iodine-131 therapy for hyperthyroidism. J. Natl. Cancer Inst. 83: 1072-1077 (1991).

H41 Holm, L.E. Cancer risks after diagnostic doses of ^{131}I with special reference to thyroid cancer. Cancer Detect. Prev. 15(1): 27-30 (1991).

H42 Hatch, M.C., S. Wallenstein, J. Beyea et al. Cancer rates after the Three Mile Island nuclear accident and proximity of residence to the plant. Am. J. Public Health 81: 719-724 (1991).

H43 Howe, G.R. Effects of dose, dose-rate and fraction on radiation-induced breast and lung cancers: the Canadian fluoroscopy study. p. 108-113 in: Proceedings of the International Conference on Radiation Effects and Protection, Mito, 1992. Japan Atomic Energy Research Institute, Tokyo, 1992.

H44 Howe, G.R., A.M. Chiarelli and J.P. Lindsay. Components and modifiers of the healthy worker effect: evidence from three occupational cohorts and implications for industrial compensation. Am J. Epidemiol. 128: 1364-1374 (1988).

H45 Howe, G.R., J.L. Weeks, A.M. Miller et al. A study of the health of the employees of Atomic Energy of Canada Limited. IV. Analysis of mortality during the period 1950-1981. AECL-9442 (1987).

H46 Harley, N.H., A.B. Kolber, R.E. Shore et al. The skin dose and response for the head and neck in patients irradiated with x-ray for tinea capitis: implications for environmental radioactivity. p. 125-142 in: Epidemiology Applied to Health Physics. CONF-830101 (1983).

H47 Harley, N.H. and E.S. Robbins. ^{222}Rn alpha dose to organs other than lung. Radiat. Prot. Dosim. 45: 619-622 (1992).

H48 Hopton, P.A., P.A. McKinney, R.A. Cartwright et al. X-rays in pregnancy and the risk of childhood cancer. Lancet ii(8458): 773 only (1985).

I1 Ishikawa, Y., T. Mori, Y. Kato et al. Lung cancers associated with Thorotrast exposure. High incidence of small cell carcinoma and implications for estimation of radon risk. Int. J. Cancer 52: 570-574 (1992).

I2 Iwasaki, T., M. Minowa, S. Hashimoto et al. Non-cancer mortality and life expectancy in different natural background radiation levels of Japan. p. 107-112 in: Low Dose Irradiation and Biological Defense Mechanisms. (T. Sugahara et al., eds.) Elsevier Science Publishers, Amsterdam, 1992.

I3 International Chernobyl Project. Assessment of radiological consequences and evaluation of protective measures. Technical Report. IAEA, Vienna (1991).

I4 International Commission on Radiation Units and Measurements. The quality factor in radiation protection. ICRU Report 40 (1986).

I5 Ichimaru, M. and T. Ishimaru. Atomic bomb radiation and adult T-cell leukemia. Acta Med. Nagasaki. 36(1): 170-175 (1991).

I6 International Agency for Research on Cancer. IARC Monographs on the Evaluation of the Carcinogenic Risk of Chemicals to Man: Some Miscellaneous Pharmaceutical Substances. Volume 13. IARC Lyon, 1976.

I7 Inskip, P.D., R.R. Monson, J.K. Wagoner et al. Leukemia following radiotherapy for uterine bleeding. Radiat. Res. 122: 107-119 (1990).

I8 Inskip, P.D., R.R. Monson, J.K. Wagoner et al. Cancer mortality following radium treatment for uterine bleeding. Radiat. Res. 123: 331-344 (1990).

I9 International Agency for Research on Cancer. IARC Monographs on the Evaluation of the Carcinogenic Risk of Chemicals to Humans: Tobacco Smoking. Volume 38. IARC, Lyon, 1985.

I10 International Commission on Radiological Protection. 1990 Recommendations of the International Commission on Radiological Protection. ICRP Publication 60. Pergamon Press, Oxford, 1990.

I11 Iwasaki, T., M. Minowa, S. Hashimoto et al. Cancer mortality rates by geographic distribution and levels of natural radiation dose in Japan. p. 503-514 in: Proceedings of the International Conference on High Levels of Natural Radiation, Ramsar, Iran, 1990. IAEA, Vienna, 1993.

I12 Inskip, P.D., R.A. Kleinerman, M. Stovall et al. Leukemia, lymphoma and multiple myeloma following pelvic radiotherapy for benign disease. Radiat. Res. 135: 108-125 (1993).

I13 Ichimaru, M. and K. Mabuchi. Multiple myeloma among atomic bomb survivors. J. Radiat. Res. 32 (Suppl.): 168-171 (1991).

I14 Ichimaru, M., T. Ishimaru, M. Mikami et al. Multiple myeloma among atomic bomb survivors in Hiroshima and Nagasaki, 1950-76: relationship to radiation dose absorbed by marrow. J. Natl. Cancer Inst. 69: 323-328 (1982).

I15 Inskip, P.D., E.B. Harvey and J.D. Boice Jr. Incidence of childhood cancer in twins. Cancer Causes Control 2: 315-324 (1991).

I16 Ishikawa, Y., T. Mori, Y. Kato et al. Systemic deposits of thorium in thorotrast patients with particular reference to sites of minor storage. Radiat. Res. 135: 244-248 (1993).

I17 Ishikawa, Y., Y. Kato, T. Mori et al. Alpha-particle dose to the liver and spleen tissues of Japanese thoro-trast patients. Health Phys. 63(5): 497-506 (1993).

I18 International Commission on Radiological Protection. Protection Against Radon-222 at Home and at Work. Annals of the ICRP 23(2). ICRP Publication 65. Pergamon Press, Oxford, 1993.

I19 International Agency for Research on Cancer. IARC Monographs on the Evaluation of Carcinogenic Risk to Humans: Man-made Fibres and Radon. Volume 43. IARC, Lyon, 1988.

I20 Inskip, P.D., M. Stovall and J.T. Flannery. Lung cancer risk and radiation dose among women treated for breast cancer. J. Natl. Cancer Inst. 86: 983-988 (1994).

J1 Jablon, S. Atomic bomb radiation dose estimation at ABCC. ABCC TR/23-71 (1971).

J2 Jablon, S. and H. Kato. Childhood cancer in relation to prenatal exposure to atomic-bomb radiation. Lancet I: 1000-1003 (1970).

J3 Jacobson, A.P., P.A. Plato and N.A. Frigerio. The role of natural radiations in human leukemogenesis. Am. J. Public Health 66: 31-37 (1976).

J4 Jablon, S., Z. Hrubec, J.D. Boice Jr. et al. Cancer in populations living near nuclear facilities. A survey of mortality nationwide and incidence in two states. J. Am. Med. Assoc. 265: 1403-1408 (1991).

J5 Japan, Ministry of Health and Welfare, Statistics and Information Department, Minister's Secretariat. Vital statistics 1985, Japan. Volume 1. Tokyo, 1985.

J6 Jensen, O.M., D.M. Parkin, R. Maclennan et al. Cancer registration, principles and methods. Int. Agency Res. Cancer, Monogr. 95: 159-169 (1991).

J7 Jablon, S. Do atomic veterans have excess cancer? New results correcting for the healthy soldier bias. Am. J. Epidemiol. 126: 1214 (1987).

K1 Kerber, R.A., J.E. Till, S.L. Simon et al. A cohort study of thyroid disease in relation to fallout from nuclear weapons testing. J. Am. Med. Assoc. 270: 2076-2082 (1993).

K2 Kinlen, L.J., K. Clarke and A. Balkwill. Paternal preconceptional radiation exposure in the nuclear industry and leukemia and non-Hodgkin's lymphoma in young people in Scotland. Br. Med. J. 306: 1153-1158 (1993).

K3 Kaul, A. and W. Noffz. Tissue dose in thorotrast patients. Health Phys. 35: 113-121 (1978).

K4 Kinlen, L.J. Evidence for an infective cause of childhood leukaemia: comparison of a Scottish New Town with nuclear reprocessing sites in Britain. Lancet II: 1323-1327 (1988).

K5 Kneale, G.W. and A.M. Stewart. Reanalysis of Hanford data: 1944-1986 deaths. Am. J. Ind. Med. 23: 371-389 (1993).

K6 Knox, E.G., A.M. Stewart, G.W. Kneale et al. Prenatal irradiation and childhood cancer. J. Soc. Radiol. Prot. 7: 177-189 (1987).

K7 Knox, E.G., A.M. Stewart, E.A. Gilman et al. Background radiation and childhood cancers. J. Radiol. Prot. 8: 9-18 (1988).

K8 Kopecky, K.J., E. Nakashima, T. Yamamoto et al. Lung cancer, radiation and smoking among A-bomb survivors, Hiroshima and Nagasaki. RERF TR/13-86 (1986).

K9 Kato, H. and W.J. Schull. Studies of the mortality of A-bomb survivors. 7. Mortality, 1950-1978: Part I. Cancer mortality. Radiat. Res. 90: 395-432 (1982).

K10 Kusiak, R.A., J. Springer, A.C. Ritchie et al. Carcinoma of the lung in Ontario gold miners: possible aetiological factors. Br. J. Ind. Med. 48: 808-817 (1991).

K11 Klotz, J.B., J.R. Petix and R.T. Zagraniski. Mortality of a residential cohort exposed to radon from industrially contaminated soil. Am. J. Epidemiol. 129: 1179-1186 (1989).

K12 Kellerer, A.M. and D. Barclay. Age dependences in the modelling of radiation carcinogenesis. Radiat. Prot. Dosim. 41: 273-281 (1992).

K13 Kaplan, H.S. An evaluation of the somatic and genetic hazards of the medical uses of radiation. Am. J. Roentgenol. 80: 696-706 (1958).

K14 Kinlen, L.J. The relevance of population mixing to the aetiology of childhood leukaemia. p. 272-278 in: Medical Response to Effects of Ionising Radiation. (W.A. Crosbie and J.H. Gittus, eds.) Elsevier Applied Science, London and New York, 1989.

K15 Kaldor, J.M., N.E. Day, F. Pettersson et al. Leukemia following chemotherapy for ovarian cancer. N. Engl. J. Med. 322: 1-6 (1990).

K16 Kaldor, J.M., N.E. Day, E.A. Clarke et al. Leukemia following Hodgkin's disease. N. Engl. J. Med. 322: 7-13 (1990).

K17 Kinlen, L.J., F. O'Brien, K. Clarke et al. Rural population mixing and childhood leukaemia: effects of the North Sea oil industry in Scotland, including the area near Dounreay nuclear site. Br. Med. J. 306: 743-748 (1993).

K18 Kossenko, M.M. and M.O. Degteva. Cancer mortality and radiation risk evaluation for the Techa river population. Sci. Total Environ. 142: 73-89 (1994).

K19 Kossenko, M.M., M.O. Degteva and N.A. Petrushova. Estimate of the risk of leukaemia to residents exposed to radiation as a result of a nuclear accident in the southern Urals. Phys. Soc. Resp. 2: 187-197 (1992).

K20 Kendall, G.M., C.R. Muirhead, B.H. MacGibbon et al. Mortality and occupational exposure to radiation: first analysis of the National Registry for radiation workers. Br. Med. J. 304: 220-225 (1992).

K21 Kendall, G.M., C.R. Muirhead, B.H. MacGibbon et al. First analysis of the National Registry for radiation workers. Occupational exposure to ionising radiation and mortality. NRPB-R251 (1992).

K22 Kazakov, V.S., E.P. Demidchik and L.N. Astakhova. Thyroid cancer after Chernobyl. Nature 359: 21 (1992).

K23 Koshurnikova, N.A., L.A. Buldakov, G.D. Bysogolov et al. Mortality from malignancies of the hemato-poietic and lymphatic tissues among personnel of the first nuclear plant in the USSR. Sci. Total Environ. 142: 19-23 (1994).

K24 Kato, H., M. Mayumi, K. Nishioka et al. The relationship of HB surface antigen and antibody to atomic bomb radiation in the Adult Health Study Sample. Am. J. Epidemiol. 117: 610-620 (1983).

K25 Kondo, S. Health Effects of Low-Level Radiation. Atomic Energy Research Institute, Kinki University. Kinki University Press, Osaka, Japan and Medical Physics Publishing, Madison, United States, 1994.

K26 Kozheurov, V.P. and M. Degteva. Dietary intake evaluation and dosimetric modelling for the Techa river residents based on in vivo measurements of strontium-90 in teeth and skeleton. Sci. Total Environ. 142: 63-72 (1994).

K27 Kinlen, L.J. Can paternal preconceptional radiation account for the increase of leukaemia and non-Hodgkin's lymphoma at Seascale. Br. Med. J. 306: 1718-1721 (1993).

L1 Lees, R.E.M., R. Steele and J.H. Roberts. A case-control study of lung cancer relative to domestic radon exposure. Int. J. Epidemiol. 16: 7-12 (1987).

L2 Land, C.E., J.D. Boice Jr., R.E. Shore et al. Breast cancer risk from low-dose exposures to ionizing radiation: results of parallel analysis of three exposed populations of women. J. Natl. Cancer Inst. 65: 353-376 (1980).

L3 Lubin, J.H., Z. Liang, Z. Hrubec et al. Radon exposure in residences and lung cancer among women: combined analyses of three studies. Cancer Causes Control 5: 114-128 (1994).

L4 Linos, A., J.E. Gray, A.L. Orvis et al. Low-dose radiation and leukaemia. N. Engl. J. Med. 302: 1101-1105 (1980).

L5 Lubin, J.H. Models for the analysis of radon-exposed populations. Yale J. Biol. Med. 61: 195-214 (1988).

L6 Lewis, C.A., P.G. Smith, I.M. Stratton et al. Estimated radiation doses to different organs among patients treated for ankylosing spondylitis with a single course of X rays. Br. J. Radiol. 61: 212-220 (1988).

L7 Land, C.E. Estimating cancer risks from low doses of ionizing radiation. Science 209: 1197-1203 (1980).

L8 Lucie, N.P. Radon exposure and leukaemia. Lancet II: 99-100 (1989).

L9 Létourneau, E.G., Y. Mao, R.G. McGregor et al. Lung cancer mortality and indoor radon concentrations in 18 Canadian cities. p. 470-483 in: Epidemiology Applied to Health Physics. CONF-830101 (1983).

L10 Létourneau, E.G., D. Krewski, J. Choi et al. Case-control study of residential radon and lung cancer in Winnipeg, Manitoba, Canada. Am. J. Epidemiol. 140/4: 310-322 (1994).

L11 Land, C.E., N. Hayakawa, S. Machado et al. A case control interview study of breast cancer among Japanese A-bomb survivors. I. Main effects. Cancer Causes Control 5: 157-165 (1994).

L12 Land, C.E. and W.K. Sinclair. The relative contributions of different organ sites to the total cancer mortality associated with low-dose radiation exposure. p. 31-57 in: Risks Associated with Ionising Radiation. Annals of the ICRP 22(1). Pergamon Press, Oxford, 1991.

L13 Lloyd, R.D., C.W. Mays, G.N. Taylor et al. Radium-224 retention, distribution, and dosimetry in beagles. Radiat. Res. 92: 280-295 (1982).

L14 Lyman, G.H., C.G. Lyman and W. Johnson. Association of leukemia with radium groundwater contamination. J. Am. Med. Assoc. 254: 621-626 (1985).

L15 L'Abbé, K.A., G.R. Howe, J.D. Burch et al. Radon exposure, cigarette smoking, and other mining experience in the Beaverlodge uranium miners cohort. Health Phys. 60: 489-495 (1991).

L16 Land, C.E. A nested case-control approach to interactions between radiation dose and other factors as causes of cancer. RERF CR/1-90 (1990).

L17 Lubin, J.H., Y. Qiao, P.R. Taylor et al. Quantitative evaluation of the radon and lung cancer association in a case control study of Chinese tin miners. Cancer Res. 50: 174-180 (1990).

L18 Little, M.P., M.M. Hawkins, R.E. Shore et al. Time variations in the risk of cancer following irradiation in childhood. Radiat. Res. 126: 304-316 (1991).

L19 Land, C.E., N. Hayakawa, S. Machado et al. A case control interview study of breast cancer among Japanese A-bomb survivors. II. Interactions between epidemiological factors and radiation dose. Cancer Causes Control 5: 167-176 (1994).

L20 Land, C.E., M. Tokunaga, S. Tokuoka et al. Early onset of breast cancer among atomic bomb survivors: evidence of a radiation-susceptible subgroup. Lancet 342: 237 (1993).

L21 Lubin, J.H., J.D. Boice, C. Edling et al. Radon and Lung Cancer Risk. A Joint Analysis of Eleven Underground Miner Studies. NIH Publication 94-3644. National Institute of Health, Bethesda, 1994.

L22 Linet, M.S. and J.D. Boice. Radiation from Chernobyl and risk of childhood leukaemia. Eur. J. Cancer 29A: 1-3 (1993).

L23 Land, C.E., Y. Shimosato, G. Saccomanno et al. A comparison of the histology of lung cancers in uranium miners and survivors of the atomic bombings of Hiroshima and Nagasaki. Radiat. Res. 134: 234-243 (1993).

L24 Li, F.P., J.F. Fraumeni Jr., J.J. Mulvihill et al. A cancer family syndrome in twenty-four kindreds. Cancer Res. 48: 5358-5362 (1988).

M1 Machado, S.G., C.E. Land and F.W. McKay. Cancer mortality and radioactive fallout in southwestern Utah. Am. J. Epidemiol. 125: 44-61 (1987).

M2 Maxon, H.R., E.L. Saenger, S.R. Thomas et al. Clinically important radiation-associated thyroid disease. A controlled study. J. Am. Med. Assoc. 244: 1802-1805 (1980).

M3 Mays, C.W. Cancer induction in man from internal radioactivity. Health Phys. 25: 585-592 (1973).

M4 Mays, C.W. Alpha-particle-induced cancer in humans. Health Phys. 55: 637-652 (1988).

M5 Mays, C.W. and H. Spiess. Bone sarcomas in patients given radium-224. p. 241-252 in: Radiation Carcinogenesis: Epidemiology and Biological Significance. (J.D. Boice Jr. and J.F. Fraumeni Jr., eds.) Raven Press, New York, 1984.

M6 MacMahon, B. Leukemia clusters around nuclear facilities in Britain. Cancer Causes Control 3: 283-288 (1992).

M7 Mays, C.W., H. Spiess, D. Chmelevsky et al. Bone sarcoma cumulative tumor rates in patients injected with radium-224. p. 27-31 in: The Radiobiology of Radium and Thorotrast. (W. Gössner et al., eds.) Urban and Schwarzenberg, München, 1986.

M8 Mattson, A., B. Ruden, R. Hall et al. Radiation-induced breast cancer and long term follow-up of radiotherapy for benign breast disease. J. Natl. Cancer Inst. 85: 1679-1685 (1993).

M9 Monson, R.R. and B. MacMahon. Prenatal X-ray exposure and cancer in children. p. 97-105 in: Radiation Carcinogenesis: Epidemiology and Biological Significance. (J.D. Boice Jr. and J.F. Fraumeni Jr., eds.) Raven Press, New York, 1984.

M10 Mori, T. and Y. Kato. Epidemiological, pathological and dosimetric status of Japanese Thorotrast patients. J. Radiat. Res. 32 (Suppl. 2): 34-45 (1991).

M11 Mori, T., Y. Kato, T. Kumatori et al. Epidemiological follow-up study of Japanese Thorotrast cases, 1980. Health Phys. 44 (Suppl. 1): 261-272 (1983).

M12 Mori, T., T. Kumatori, Y. Kato et al. Present status of medical study on Thorotrast-administered patients in Japan. Strahlentherapie (Suppl.) 80: 123-134 (1986).

M13 Muirhead, C.R., B.K. Butland, B.M.R. Green et al. Childhood leukaemia and natural radiation. Lancet 337: 503-504 (1991).

M14 Muller, J., W.C. Wheeler and J.F. Gentleman. Study of Mortality of Ontario Miners, 1955-1977. Part I. Report to Ontario Ministry of Labour, Ontario Workers' Compensation Board and Atomic Energy Control Board of Canada (1983).

M15 Mayya, Y.S., P.V. Joshi and K.S.V. Nambi. Radio-genic cancer risk attributes for the Indian population based on the revised ICRP risk coefficients. Radiat. Prot. Dosim. 46(2): 97-101 (1993).

M16 Miller, B.G. and M. Jacobsen. Dust exposure, pneumoconiosis, and mortality of coal miners. Br. J. Ind. Med. 42: 723-733 (1985).

M17 Muir, C., J. Waterhouse, T. Mack et al. Cancer Incidence in Five Continents. Volume V. IARC Publication No. 88. IARC, Lyon, 1987.

M18 Morrison, H.I., R.M. Semenciw, Y. Mao et al. Cancer mortality among a group of fluorspar miners exposed to radon progeny. Am. J. Epidemiol. 128: 1266-1275 (1988).

M19 Miller, A.B., G.R. Howe, G.J. Sherman et al. Mortality from breast cancer after irradiation during fluoroscopic examinations in patients being treated for tuberculosis. N. Engl. J. Med. 321: 1285-1289 (1989).

M20 Modan, B., A. Chetrit, E. Alfandary et al. Increased risk of breast cancer after low-dose irradiation. Lancet I: 629-631 and 1012-1013 (1989).

M21 Mettler, F.A., M.R. Williamson, H.D. Royal et al. Thyroid nodules in the population living around Chernobyl. J. Am. Med. Assoc. 268: 616-619 (1992).

M22 Muirhead, C.R. and G.W. Kneale. Prenatal irradiation and childhood cancer. J. Radiol. Prot. 9: 209-212 (1989).

M23 Mole, R.H. Antenatal irradiation and childhood cancer: causation or coincidence? Br. J. Cancer. 30: 199-208 (1974).

M24 Matanoski, G.M., P. Sartwell, E. Elliott et al. Cancer risks in radiologists and radiation workers. p. 83-96 in: Radiation Carcinogenesis: Epidemiology and Biological Significance. (J.D. Boice Jr. and J.F. Fraumeni Jr., eds.) Raven Press, New York, 1984.

M25 Miller, R.W. Effects of prenatal exposure to ionizing radiation. Health Phys. 59: 57-61 (1990).

M26 Mole, R.H. Carcinogenesis following medical uses of ionizing radiation. p. 100-113 in: Low Dose Radiation: Biological Bases of Risk Assessment. (K.F. Baverstock and J.W. Stather, eds.) Taylor & Francis, London, 1989.

M27 Mole, R.H. Alpha-particle irradiation and human leukaemia. p. 153-158 in: Risks from Radium and Thorotrast. BIR Report 21. (D.M. Taylor et al., eds.) British Institute of Radiology, London, 1989.

M28 Mole, R.H. Carcinogenesis by thorotrast and other sources of irradiation, especially other α-emitters. Environ. Res. 18: 192-215 (1979).

M29 Myers, M.H. and B.F. Hankey. Cancer survival in the United States. p. 168-178 in: Cancer Epidemiology and Prevention. (D. Schottenfeld and J.F. Fraumeni Jr., eds.) W.B. Saunders, Philadelphia, 1982.

M30 Mole, R.H. Fetal dosimetry by UNSCEAR and risk coefficients for childhood cancer following diagnostic radiology in pregnancy. J. Radiol. Prot. 10: 199-203 (1990).

M31 Mole, R.H. Ionizing radiations and human leukemia. in: Dameshek and Gunz's Leukaemia. 5th edition. (E.S. Henderson and T.A. Lister, eds.) Saunders, Philadelphia, 1990.

M32 Mole, R.H. Childhood cancer after prenatal exposure to diagnostic x-ray examinations in Britain: a review. Br. J. Cancer 62: 152-168 (1990).

M33 MacMahon, B. Some recent issues in low-exposure radiation epidemiology. Environ. Health Perspect. 81: 131-135 (1989).

M34 Modan, B. and A.M. Lilienfeld. Polycythemia vera and leukemia - the role of radiation treatment. Medicine 44: 305-344 (1965).

M35 McLaughlin, J.R., E.A. Clarke, D. Nishri et al. Childhood leukaemia in the vicinity of Canadian nuclear facilities. Cancer Causes Control 4: 51-58 (1993).

M36 Mori, T., T. Kumatori, S. Hatakeyama et al. Current (1986) status of the Japanese follow-up study of the thorotrast patients, and its relationships to the statistical analysis of the autopsy series. p. 119-124 in: Risks from Radium and Thorotrast. BIR Report 21. (D.M. Taylor et al., eds.) British Institute of Radiology, London, 1989.

M37 Mayer, K., K.S. Pentlow, R.C. Marcove et al. Sulfur-35 therapy for chondrosarcoma and chordoma. p. 185-192 in: Therapy in Nuclear Medicine. (R.D. Spencer, ed.) Grune Stratton, New York, 1978.

M38 Marshall, J.H. and P.G. Groer. A theory of the induction of bone cancer by alpha radiation. Radiat. Res. 71: 149-192 (1977).

M39 Marshall, J.H., P.G. Groer and R.A. Schlenker. Dose to endosteal cells and relative distribution factors for radium-224 and plutonium-239 compared to radium-226. Health Phys. 35: 91-101 (1978).

M40 Mole, R.H. Leukaemia induction in man by radionuclides and some relevant experimental and human observations. p. 1-13 in: The Radiobiology of Radium and Thorotrast. (W. Gössner et al., eds.) Urban & Schwarzenberg, München, 1986.

M41 Matanoski, G.M. Health effects of low-level radiation in shipyard workers. Report of US/DOE and St. Johns Hopkins School of Hygiene and Public Health DE-AC02-79EV10095 (1991).

M42 Muirhead, C.R. Projecting radiation-induced cancer risks across time and population. Soz. Präventivmed. 36: 249-254 (1991).

M43 Muirhead, C.R., R. Cox, J.W. Stather et al. Estimates of late radiation risks to the UK population. Documents of the NRPB 4(4): 15-157 (1993).

M44 McLaughlin, J.R., T.W. Anderson, E.A. Clarke et al. Occupational exposure of fathers to ionizing radiation and the risk of leukaemia in offspring - a case-control study. Research Report, INFO-0424. Atomic Energy Control Board, Canada (1992).

M45 Mifune, M., T. Sobue, H. Arimoto et al. Cancer mortality survey in a spa area (Misasa, Japan) with a high radon background. Jpn. J. Cancer Res. 83: 1-5 (1992).

M46 Michaelis, J., B. Keller, G. Haaf et al. Incidence of childhood malignancies in the vicinity of West German nuclear power plants. Cancer Causes Control 3: 255-263 (1992).

M47 McLaughlin, J.R., N. Krieger, M.P. Sloan et al. A historical cohort study of cardiac catheterization during childhood and the risk of cancer. Int. J. Epidemiol. 22: 584-591 (1993).

M48 Mays, C.W., D. Mays and R.A. Guilmette (eds.). Total-body evaluation of a Thorotrast patient. Workshop held in July 1990 at the National Cancer Institute, Bethesda. Health Phys. 63: 1-100 (1992).

N1 Nambi, K.S.V., Y.S. Mayya and P.V. Joshi. Database requirements for detecting radiogenic cancer risks from epidemiological surveys in exposed populations. p. 10-13 in: Proceedings of Second National Symposium on Environment, Jodhpur, India, 1993.

N2 Nambi, K.S.V., Y.S. Mayya, D.D. Rao et al. A study on cancer mortality in Tarapur-based atomic energy community. Arch. Environ. Health 47(2): 155-157 (1992).

N3 Nambi, K.S.V., Y.S. Mayya and S.D. Soman. A cancer mortality study in Bombay-based atomic energy community: 1975-1987. Indian J. Cancer 28: 61-69 (1991).

N4 Nikipelov, B.V., A.F. Lizlov and N.A. Koshurnikova. Experience with the first Soviet nuclear installation (irradiation doses and personnel health). Priroda 2 (February): 30-38 (1990).

N5 National Council on Radiation Protection and Measurements. Induction of thyroid cancer by ionizing radiation. NCRP Report No. 80 (1985)

N6 Nambi, K.S.V., Y.S. Mayya and S.D. Soman. An analysis of cancer mortality among atomic energy employees in Bombay and Tarapur. Curr. Sci. 59: 733-736 (1990).

N7 National Council on Radiation Protection and Measurements. The relative biological effectiveness of radiation of different quality. NCRP Report No. 104 (1990).

N8 Neuget, A.I., E. Robinson, W.C. Lee et al. Lung cancer after radiation therapy for breast cancer. Cancer 71: 3054-3057 (1993).

O1 Østerlind, A., J.H. Olsen, E. Lynge et al. Second cancer following cutaneous melanoma and cancers of the brain, thyroid, connective tissue, bone, and eye in Denmark, 1943-80. Natl. Cancer Inst. Monogr. 68: 361-388 (1985).

O2 O'Brien, T.R., P. Decouflé and P.H. Rhodes. Leukemia and groundwater contamination. J. Am. Med. Assoc. 257: 317 (1987).

O3 Olsen, J.H., M. Andersson and J.D. Boice. Thorotrast exposure and cancer risk. Health Phys. 58: 222-223 (1990).

O4 Olsen, J.H., M. Andersson and J.D. Boice. Danish epileptics given thorotrast. p. 116-118 in: Risks from Radium and Thorotrast. BIR Report 21. (D.M. Taylor et al., eds.) British Institute of Radiology, London, 1989.

O5 Oppenheim, B.E., M.L. Griem and P. Meier. Effects of low-dose prenatal irradiation in humans: analysis of Chicago lying-in data and comparison with other studies. Radiat. Res. 57: 508-544 (1974).

P1 Pershagen, G., L. Damber and R. Falk. Exposure to radon in dwellings and lung cancer: a pilot study. p. 73-78 in: Indoor Air. Volume 2. Radon, Passive Smoking, Particulates and Housing Epidemiology. (B. Berglund et al., eds.) Swedish Council for Building Research, Stockholm, 1984.

P2 Pippard, E.C., A.J. Hall, D.J.P. Barker et al. Cancer in homozygotes and heterozygotes of ataxia-telangiec-tasia and xeroderma pigmentosum in Britain. Cancer Res. 48: 2929-2932 (1988).

P3 Preston, D.L. and D.A. Pierce. The effect of changes in dosimetry on cancer mortality risk estimates in the atomic bomb survivors. RERF TR/9-87 (1987) and Radiat. Res. 114: 437-466 (1988).

P4 Preston, D.L., H. Kato, K.J. Kopecky et al. Life Span Study Report 10. Part I: Cancer mortality among A-bomb survivors in Hiroshima and Nagasaki, 1950-82. RERF TR/1-86 (1986) and Radiat. Res. 111: 151-178 (1987).

P5 Preston-Martin, S., A. Paganini-Hill, B.E. Henderson et al. Case-control study of intracranial meningiomas in women in Los Angeles County, California. J. Natl. Cancer Inst. 65: 67-73 (1980).

P6 Preston-Martin, S., M.C. Yu, B.E. Henderson et al. Risk factors for meningiomas in men in Los Angeles County. J. Natl. Cancer Inst. 70: 863-866 (1983).

P7 Pershagen, G., O. Axelson, B. Clavensjö et al. Radon in dwellings and cancer. A country-wide epidemiological investigation. Institute of Environmental Medicine, Karolinska Institute, Stockholm. IMM-rapport 2/93 (1993).

P8 Pottern, L.M., M. Kaplan, P. Larsen et al. Thyroid nodularity after childhood irradiation for lymphoid hyperplasia: a comparison of questionnaire and clinical findings. J. Clin. Epidemiol. 43: 449-460 (1990).

P9 Pierce, D.A., D.O. Stram and M. Vaeth. Allowing for random errors in radiation exposure estimates for the atomic bomb survivor data. Radiat. Res. 123: 275-284 (1990) and RERF TR/2-89 (1989).

P10 Parkin, D.M., C.S. Muir, S.L. Whelan et al. Cancer Incidence in Five Continents. Volume VI. IARC Publication No. 120. IARC, Lyon, 1992.

P11 Pierce, D.A. and M. Vaeth. Cancer risk estimation from the A-bomb survivors: extrapolation to low doses, use of relative risk models and other uncertainties. p. 54-69 in: Low Dose Radiation: Biological Bases of Risk Assessment. (K. Baverstock and J.W. Stather, eds.) Taylor & Francis, London, 1989 and RERF CR/2-89 (1989).

P12 Prentice, R.L., Y. Yoshimoto and M.W. Mason. Cigarette smoking and radiation exposure in relation to cancer mortality, Hiroshima and Nagasaki. RERF TR/1-82 (1982).

P13 Pierce, D.A. and M. Vaeth. The shape of the cancer mortality dose-response curve for atomic bomb survivors. RERF TR/7-89 (1989) and Radiat. Res. 126: 36-42 (1991).

P14 Preston-Martin, S., D.C. Thomas, S.C. White et al. Prior exposure to medical and dental X-rays related to tumors of the parotid gland. J. Natl. Cancer Inst. 80: 943-949 (1988).
P15 Preston, D.L., D.A. Pierce and M. Vaeth. Neutrons and radiation risk: a commentary. RERF Update 4 (1992-1993).
P16 Polhemus, D.W. and R. Koch. Leukaemia and medical radiation. Pediatrics 23: 453-461 (1959).
P17 Preston-Martin, S., D.C. Thomas, Y. & B.E. Henderson. Diagnostic radiography as a risk factor for chronic myeloid and monocytic leukaemia (CML). Br. J. Cancer 59: 639-644 (1989).
P18 Polednak, A.P., A.F. Stehney and R.E. Rowland. Mortality among women first employed before 1930 in the U.S. radium dial-painting industry. Am. J. Epidemiol. 107: 179-195 (1978).
P19 Petersen, N.J., L.D. Samuels, H.F. Lucas et al. An epidemiologic approach to low-level ^{226}Ra exposure. Public Health Rep. 81: 805-814 (1966).
P20 Pierce, D.A. and D.L. Preston. Joint analysis of site-specific cancer risks for the A-bomb survivors. Radiat. Res. 134: 134-142 (1993).
P21 Preston-Martin, S., W. Mack and B.E. Henderson. Risk factors for gliomas and meningiomas in males in Los Angeles county. Cancer Res. 49: 6137-6143 (1989).
P22 Preston-Martin, S., M.C. Yu, B. Benton et al. N-Nitroso compounds and childhood brain tumors: a case-control study. Cancer Res. 42: 5240-5345 (1982).
P23 Pierce, D.A., M. Vaeth and D.L. Preston. Variations with time and age of the excess cancer risk among A-bomb survivors. RERF TR/21-89 (1989) and Analysis of time and age patterns in cancer risk of A-bomb survivors. Radiat. Res. 126: 171-186 (1991).
P24 Polednak, A.P. and E.L. Frome. Mortality among men employed between 1943 and 1947 at a uranium-processing plant. J. Occup. Med. 23: 169-178 (1981).
P25 Petersen, G.R., E.S. Gilbert, J.A. Buchanan et al. A case-cohort study of lung cancer, ionizing radiation, and tobacco smoking among males at the Hanford site. Health Phys. 58: 3-11 (1990).
P26 Piao, C.O., S.A. Marino and T.K. Hei. Radiation, tobacco smoke condensate and oncogenic transformation. p. 185 in: 38th Annual Meeting of Radiation Research Society, New Orleans, Abstract. Radiation Research Society, Washington D.C., 1990.
P27 Pearce, N., I. Prior, D. Methven et al. Follow up of New Zealand participants in British atmospheric nuclear weapons tests in the Pacific. Br. Med. J. 300: 1161-1166 (1990).
P28 Polednak, A.P., A.F. Stehney and H.F. Lucas. Mortality among male workers at a thorium-processing plant. Health Phys. 44: 239-251 (1983).
P29 Polednak, A.P. Bone cancer among female radium dial workers. Latency periods and incidence rates by time after exposure: brief communication. J. Natl. Cancer Inst. 60: 77-82 (1978).
P30 Pershagen, G., Z.H. Liang, Z. Hrubec et al. Residential radon exposure and lung cancer in Swedish women. Health Phys. 63: 179-186 (1992).
P31 Pierce, D.A., D.O. Stram, M. Vaeth et al. The errors-in-variables problem: considerations provided by radiation dose-response analyses of the A-bomb survivor data. J. Am. Stat. Assoc. 87: 351-359 (1992).
P32 Preston, D.L. Statistical issues in modelling radiation effects on disease incidence. p. 141-146 in: The Future of Human Radiation Research. BIR Report 22. (G.B. Gerber, D.M. Taylor, E. Cardis et al., eds.) British Institute of Radiology, London, 1991.
P33 Preston, D.L., S. Kusumi, M. Tomonaga et al. Cancer incidence in atomic bomb survivors. Part III: Leukemia, lymphoma, and multiple myeloma, 1950-87. RERF TR/24-92 (1992) and Radiat. Res. 137: S68-S97 (1994).
P34 Prisyazhiuk, A., O.A. Pjatak, V.A. Buzanov et al. Cancer in the Ukraine, post-Chernobyl. Lancet 338: 1334-1335 (1991).
P35 Preston, D.L., J.H. Lubin and D.A. Pierce. "Epicure": Risk regression and data analysis software. Hirosoft International Corp., Seattle, Washington, 1992.
P36 Pershagen, G., G. Åkerblöm, O. Axelson et al. Residential radon exposure and lung cancer in Sweden. N. Engl. J. Med. 330: 159-164 (1994).
P37 Parker, L., A.W. Craft, J. Smith et al. Geographical distribution of preconceptional radiation doses to fathers employed at the Sellafield nuclear installation, West Cumbria. Br. Med. J. 307: 966-971 (1993).
P38 Parkin, D.M., E. Cardis, E. Masuyer et al. Childhood leukaemia following the Chernobyl accident: The European Childhood Leukaemia-Lymphoma Incidence Study (ECLIS). Eur. J. Cancer 29A: 87-95 (1993).
R1 Rinsky, R.A., R.D. Zumwalde, R.J. Waxweiler et al. Cancer mortality at a naval nuclear shipyard. Lancet I: 231-235 (1981).
R2 Robinette, C.D., S. Jablon and D.L. Preston. Mortality of studies in nuclear weapons tests participants. p. 1-47 in: Medical Follow-up Agency, National Research Council. National Academy Press, Washington, 1985.
R3 Rossi, H.H. and A.M. Kellerer. The dose rate dependence of oncogenic transformation by neutrons may be due to variation in response during the cell cycle. Int. J. Radiat. Biol. 50: 353-361 (1986).
R4 Ron, E. and B. Modan. Thyroid and other neoplasms following childhood scalp irradiation. p. 139-151 in: Radiation Carcinogenesis: Epidemiology and Biological Significance. (J.D. Boice Jr. and J.F. Fraumeni Jr., eds.) Raven Press, New York, 1984.
R5 Ron, E., B. Modan and J.D. Boice Jr. Mortality after radiotherapy for ringworm of the scalp. Am. J. Epidemiol. 127: 713-725 (1988).
R6 Rowland, R.E., A.F. Stehney and H.F. Lucas. Dose-response relationships for female radium dial workers. Radiat. Res. 76: 368-383 (1978).
R7 Rowland, R.E., A.F. Stehney and H.F. Lucas. Dose-response relationships for radium-induced bone sarcomas. Health Phys. 44 (Suppl. 1): 15-31 (1983).
R8 Radford, E.P. and K.G. St. Clair Renard. Lung cancer in Swedish iron miners exposed to low doses of radon daughters. N. Engl. J. Med. 310: 1485-1494 (1984).
R9 Roscoe, R.J., K. Steenland, W.E. Halperin et al. Lung cancer mortality among nonsmoking uranium miners exposed to radon daughters. J. Am. Med. Assoc. 262: 629-633 (1989).
R10 Ron, E., B. Modan, J.D. Boice Jr. et al. Tumors of the brain and nervous system after radiotherapy in childhood. N. Engl. J. Med. 319: 1033-1039 (1988).
R11 Ron, E., J. Lubin and A.B. Schneider. Thyroid cancer incidence. Nature 360: 113 only (1992).

R12 Rallison, M.L., B.M. Dobyns, F.R. Keating et al. Thyroid disease in children. A survey of subjects potentially exposed to fallout radiation. Am. J. Med. 56: 457-463 (1974).

R13 Rodvall, Y., Z. Hrubec, G. Pershagen et al. Childhood cancer among Swedish twins. Cancer Causes Control 3: 527-532 (1992).

R14 Rimm, I.J., F.C. Li, N.J. Tarbell et al. Brain tumors after cranial irradiation for childhood acute lymphoblastic leukemia. A 13-year experience from the Dana-Faber Cancer Institute and the Children's Hospital. Cancer 59: 1506-1508 (1987).

R15 Ron, E., B. Modan, D.L. Preston et al. Thyroid neoplasia following low-dose radiation in childhood. Radiat. Res. 120: 516-531 (1989).

R16 Rodvall, Y., G. Pershagen, Z. Hrubec et al. Prenatal x-ray exposure and childhood cancer in Swedish twins. Int. J. Cancer 46: 362-365 (1990).

R17 Richardson, R.B., J.P. Eatough and D.L. Henshaw. The contribution of radon exposure to the radiation dose received by red bone marrow. Int. J. Radiat. Biol. 57: 597 (1989).

R18 Roman, E., V. Beral, L. Carpenter et al. Childhood leukaemia in the West Berkshire and Basingstoke and North Hampshire District Health Authorities in relation to nuclear establishments in the vicinity. Br. Med. J. 294: 597-602 (1987).

R19 Ron, E., B. Modan, D.L. Preston et al. Radiation-induced skin carcinomas of the head and neck. Radiat. Res. 125: 318-325 (1991).

R20 Raabe, O.G., S.A. Book and N.J. Parks. Bone cancer from radium: canine dose response explains data for mice and humans. Science 208: 61-64 (1980).

R21 Robbins, J. and W. Adams. Radiation effects in the Marshall Islands. p. 11-24 in: Radiation and the Thyroid. (S. Nagataki, ed.) Excepta Medica, Tokyo, 1989.

R22 Roesch, W.C. (ed.). Final report on the reassessment of atomic bomb radiation dosimetry in Hiroshima and Nagasaki. RERF (1987).

R23 Ron, E., D.L. Preston, K. Mabuchi et al. Cancer incidence in atomic bomb survivors. Part IV: Comparison of cancer incidence and mortality. Radiat. Res. 137: S98-S112 (1994).

R24 Roman, E., A. Watson, B. Beral et al. Case-control study of leukaemia and non-Hodgkin's lymphoma among children aged 0-4 years living in West Berkshire and North Hampshire health districts. Br. Med. J. 306: 615-621 (1993).

R25 Rooney, C., V. Beral, N. Maconochie et al. Case control study of prostatic cancer in employees of the United Kingdom Atomic Energy Authority. Br. Med. J. 307: 1391-1397 (1993).

R26 Rossi, H.H. and M. Zaider. Neutron RBE for A-bomb survivors. RERF Update 3 (1 & 3-4) (1991).

R27 Ron, E. Communication to UNSCEAR Secretariat (1994).

S1 Sinclair, W.K. Science, Radiation Protection and the NCRP. p. 1-56 in: 17th L.S. Taylor Lecture. NCRP, Bethesda, 1993.

S2 Saenger, E.L., G.E. Thoma and E.A. Tompkins. Incidence of leukaemia following treatment of hyperthyroidism: Preliminary report of the Cooperative Thyrotoxicosis Therapy Follow-up Study. J. Am. Med. Assoc. 205: 147-154 (1968).

S3 Sandler, D.P., G.W. Comstock and G.M. Matanoski. Neoplasms following childhood radium irradiation of the nasopharynx. J. Natl. Cancer Inst. 68: 3-8 (1982).

S4 Shore, R.E., R.E. Albert and B.S. Pasternack. Follow-up study of patients treated by x-ray epilation for tinea capitis. Arch. Environ. Health 31: 21-28 (1976).

S5 Ševc, J., E. Kunz, L. Tomášek et al. Cancer in man after exposure to radon daughters. Health Phys. 54: 27-46 (1988).

S6 Shimizu, Y., H. Kato, W.J. Schull et al. Life Span Study Report 11. Part 1: Comparison of risk coefficients for site-specific cancer mortality based on the DS86 and T65DR shielded kerma and organ doses. RERF TR/12-87 (1987); *also published as*
Studies of the mortality of A-bomb survivors. 9. Mortality, 1950-1985: Part 1. Comparison of risk coefficients for site-specific cancer mortality based on the DS86 and T65DR shielded kerma and organ doses. Radiat. Res. 118: 502-524 (1989).

S7 Shimizu, Y., H. Kato and W.J. Schull. Life Span Study Report 11. Part 2: Cancer mortality in the years 1950-85 based on the recently revised doses (DS86). RERF TR/5-88 (1988); *also published as*
Studies of the mortality of A-bomb survivors. 9. Mortality, 1950-1985: Part 2: Cancer mortality based on the recently revised doses (DS86). Radiat. Res. 121: 120-141 (1990).

S8 Shore, R.E., R.E. Albert, M. Reed et al. Skin cancer incidence among children irradiated for ringworm of the scalp. Radiat. Res. 100: 192-204 (1984).

S9 Shore, R.E., N. Hildreth, E. Woodard et al. Breast cancer among women given X-ray therapy for acute postpartum mastitis. J. Natl. Cancer Inst. 77: 689-696 (1986).

S10 Shore, R.E., E. Woodard, N. Hildreth et al. Thyroid tumors following thymus irradiation. J. Natl. Cancer Inst. 74: 1177-1184 (1985).

S11 Simpson, S.G. and G.W. Comstock. Lung cancer and housing characteristics. Arch. Environ. Health 38: 248-251 (1983).

S12 Smith, P.G. and R. Doll. Late effects of x-irradiation in patients treated for metropathia haemorrhagica. Br. J. Radiol. 49: 224-232 (1976).

S13 Smith, P.G. and R. Doll. Mortality among patients with ankylosing spondylitis after a single treatment course with X-rays. Br. Med. J. 284: 449-460 (1982).

S14 Smith, P.G. and A.J. Douglas. Mortality of workers at the Sellafield plant of British Nuclear Fuels. Br. Med. J. 293: 845-854 (1986).

S15 Spiers, F.W., H.F. Lucas, J. Rundo et al. Leukaemia incidence in the US dial painters. Health Phys. 44 (Suppl. 1): 65-72 (1983).

S16 Stather, J.W., R.H. Clarke and K.P. Duncan. The risk of childhood leukaemia near nuclear establishments. NRPB-R215 (1988).

S17 Stockwell, H.G., G.H. Lyman, J. Waltz et al. Lung cancer in Florida. Risks associated with residence in the central Florida phosphate mining region. Am. J. Epidemiol. 128: 78-85 (1988).

S18 Stovall, M., S.A. Smith and M. Rosenstein. Tissue doses from radiotherapy of cancer of the uterine cervix. Med. Phys. 16: 726-733 (1989).

S19 Svensson, C., G. Eklund and G. Pershagen. Indoor exposure to radon from the ground and bronchial cancer in women. Int. Arch. Occup. Environ. Health 59: 123-131 (1987).

S20 Svensson, C., G. Pershagen and J. Klominek. Lung cancer in women and type of dwelling in relation to radon exposure. Cancer Res. 49: 1861-1865 (1989).

S21 Smith, P.G., R. Doll and E.P. Radford. Cancer mortality among patients with ankylosing spondylitis not given X-ray therapy. Br. J. Radiol. 50: 728-734 (1977).

S22 Stram, D.O. and S. Mizuno. Analysis of the DS86 atomic bomb radiation dosimetry methods using data on severe epilation. RERF TR/1-88 (1988) and Radiat. Res. 117: 93-113 (1989).

S23 Sposto, R. and D.L. Preston. Correcting for catchment area non-residency in studies based on tumor-registry data. RERF CR/1-92 (1992).

S24 Samet, J.M., D.R. Pathak, M.V. Morgan et al. Radon progeny exposure and lung cancer risk in New Mexico uranium miners: a case-control study. Health Phys. 56: 415-421 (1989).

S25 Samet, J.M., D.M. Kutvirt and R.J. Waxweiler. Uranium mining and lung cancer in Navajo men. N. Engl. J. Med. 310: 1481-1484 (1984).

S26 Stidley, C.A. and J.M. Samet. A review of ecologic studies of lung cancer and indoor radon. Health Phys. 65: 234-251 (1993).

S27 Stjernfeldt, M., L. Samuelsson and J. Ludvigsson. Radiation in dwellings and cancer in children. Pediatr. Hematol. Oncol. 4: 55-61 (1987).

S28 Samet, J.M. Radon and lung cancer. J. Natl. Cancer Inst. 81: 745-757 (1989).

S29 Stranden, E. Radon in Norwegian dwellings and the feasibility of epidemiological studies. Radiat. Environ. Biophys. 25: 37-42 (1986).

S30 Sinclair, W.K. Risk, Research and Radiation Protection. Radiat. Res. 112: 191-216 (1987).

S31 Stockwell, H.G., C.I. Noss, E.A. Ross et al. Lung cancer and indoor radon in Florida. Radiat. Prot. Dosim. 24: 475-477 (1988).

S32 Shore-Freedman, E., C. Abrahams, W. Recant et al. Neurolemmomas and salivary gland tumours of head and neck in childhood. Cancer 51: 2159-2163 (1993).

S33 Stram, D.O., R. Sposto, D. Preston et al. Stable chromosome aberrations among A-bomb survivors: an update. Radiat. Res. 136: 29-36 (1993).

S34 Sorahan, T. and P.J. Roberts. Chilhood cancer and paternal exposure to ionizing radiation: preliminary findings from the Oxford Survey of Childhood Cancers. Am. J. Ind. Med. 23: 343-354 (1993).

S35 Shore, R.E., E.D. Woodard, B.S. Pasternack et al. Radiation and host factors in human thyroid tumors following thymus irradiation. Health Phys. 38: 451-465 (1980).

S36 Salonen, T. Prenatal and perinatal factors in childhood cancer. Ann. Clin. Res. 7: 27-42 (1976).

S37 Stewart, A. and G.W. Kneale. Radiation dose effects in relation to obstetric X-rays and childhood cancers. Lancet I: 1185-1188 (1970).

S38 Smith, P.G. and R. Doll. Mortality from cancer and all causes among British radiologists. Br. J. Radiol. 54: 187-194 (1981).

S39 Spiess, H. and C.W. Mays. Bone cancers induced by ^{224}Ra (Th X) in children and adults. Health Phys. 19: 713-729 (1970).

S40 Spiess, H. and C.W. Mays. Protraction effect on bone-sarcoma induction of radium-224 in children and adults. p. 437-450 in: Radionuclide Carcinogenesis. (C.L. Sanders et al., eds.) CONF-720505 (1973).

S41 Spiess, H., C.W. Mays and D. Chmelevsky. Malignancies in patients injected with ^{224}Ra. p. 7-12 in: Risks from Radium and Thorotrast. BIR Report 21. (D.M. Taylor et al., eds.) British Institute of Radiology, London, 1989.

S42 Stebbings, J.H., H.F. Lucas and A.F. Stehney. Mortality from cancers of major sites in female radium dial workers. Am. J. Ind. Med. 5: 435-459 (1984).

S43 Swift, M., P.J. Reitnauer, D. Morrell et al. Breast and other cancers in families with ataxia-telangiectasia. N. Engl. J. Med. 316: 1289-1294 (1987).

S44 Stewart, A., W. Pennybacker and R. Barber. Adult leukaemias and diagnostic x rays. Br. Med. J. II: 882-890 (1962).

S45 Spengler, R.F., D.H. Cook, E.A. Clarke et al. Cancer mortality following cardiac catheterization: a preliminary follow-up study on 4,891 irradiated children. Pediatrics 71: 235-239 (1983).

S46 Shore, R.E. Radiation epidemiology: old and new challenges. Environ. Health Perspect. 81: 153-156 (1989).

S47 Shimizu, M., K. Noguchi, E. Sairenji et al. Correlation between natural radiation exposure and cancer mortality (5). J. Nihon Univ. Sch. Dent. 29: 314-320 (1987).

S48 Stevens, W., D.C. Thomas, J.L. Lyon et al. Leukemia in Utah and radioactive fallout from the Nevada test site. J. Am. Med. Assoc. 264: 585-591 (1990).

S49 Stevens, W., J.E. Till, J.L. Lyon et al. A case-control study of leukemia deaths in Utah (1952-81) and exposure to radioactive fallout from the Nevada test site (1952-58). Final Report. National Cancer Institute (1990).

S50 Stebbings, J.H. and W. Semkiw. Central nervous system tumours and related intracranial pathologies in radium dial workers. p. 63-67 in: Risks from Radium and Thorotrast. BIR Report 21. (D.M. Taylor et al., eds.) British Institute of Radiology, London, 1989.

S51 Schoenberg, J.B., J.B. Klotz, H.B. Wilcox et al. Case-control study of residential radon and lung cancer among New Jersey women. Cancer Res. 50: 6520-6524 (1990).

S52 Stewart, A., J. Webb and D. Hewitt. A survey of childhood malignancies. Br. Med. J. 1: 1495-1508 (1958).

S53 Sasaki, M.S., S. Saigusa, I. Kimura et al. Biological effectiveness of fission neutrons: energy dependency and its implication for the risk assessment. p. 31-35 in: Proceedings of the International Conference on Radiation Effects and Protection, Mito, 1992. Japan Atomic Energy Research Institute, Tokyo, 1992.

S54 Shore, R.E., N. Hildreth, P. Ovoretsky et al. Thyroid cancer among persons given x-ray treatment in infancy for an enlarged thymus gland. Am. J. Epidemiol. 137: 1068-1080 (1993).

S55 Sposto, R., D.O. Stram and A.A. Awa. An estimate of the magnitude of random errors in the DS86 dosimetry from data on chromosome aberrations and severe epilation. Radiat. Res. 128: 157-169 (1991).

S56 Straume, T., S.D. Egbert, W.A. Woolson et al. Neutron discrepancies in the DS86 Hiroshima dosimetry system. Health Phys. 63: 421-426 (1992).

S57 Ševc, J., L. Tomášek, E. Kunz et al. A survey of the Czechoslovak follow-up of lung cancer mortality in uranium miners. Health Phys. 64: 355-369 (1993).

S58 Samet, J.M., S.R. Pathak, M.V. Morgan et al. Lung cancer mortality and exposure to radon progeny in a cohort of New Mexico underground uranium miners. Health Phys. 61: 745-752 (1991).

S59 Swift, M., D. Morrell, R.B. Massey et al. Incidence of cancer in 161 families affected by ataxia-telangiec-tasia. N. Engl. J. Med. 325: 1831-1836 (1991) and correspondence regarding this subject by various authors and reply by M. Swift et al. in: N. Engl. J. Med. 326: 1357-1361 (1992).

S60 Storm, H.H., M. Andersson, J.D. Boice et al. Adjuvant radiotherapy and risk of contralateral breast cancer. J. Natl. Cancer Inst. 84: 1245-1250 (1992).

S61 Shore, R.E. Issues and epidemiological evidence regarding radiation-induced thyroid cancer. Radiat. Res. 131: 98-111 (1992).

S62 Shore, R.E. Overview of radiation-induced skin cancer in humans. Int. J. Radiat. Biol. 57: 809-827 (1990).

S63 Schneider, A.B., E. Ron, J. Lubin et al. Dose-response relationships for radiation-induced thyroid cancer and thyroid nodules: evidence for the prolonged effects of radiation on the thyroid. J. Clin. Endocrinol. Metab. 77: 362-364 (1993).

S64 Ševcová, M., J. Ševc and J. Thomas. Alpha irradiation of the skin and the possibility of late effects. Health Phys. 35: 803-806 (1978).

S65 Shimizu, Y., H. Kato and W.J. Schull. Dose-response analysis of atomic bomb survivors exposed to low-level radiation. p. 3-4 in: RERF Update, Autumn (1992).

S66 Sposto, R., D.L. Preston, Y. Shimizu et al. The effect of diagnostic misclassification on noncancer and cancer mortality dose response in the RERF Life Span Study. Biometrics 48: 605-617 (1992).

S67 Sinclair, W.K. Experimental RBE values of high-LET radiations at low doses and the implications for quality factor assignment. Radiat. Prot. Dosim. 13: 319-326 (1985).

S68 Sinclair, W.K., J.D. Abbatt, H.E. Farran et al. A quantitative autoradiographic study of radioactive distribution and dosage in human thyroid glands. Br. J. Radiol. 29: 36-41 (1956).

S69 Shimizu, Y., W.J. Schull and H. Kato. Cancer risk among atomic bomb survivors. J. Am. Med. Assoc. 264 (5): 601-604 (1990).

S70 Shigematsu, I. and J.W. Thiessen. Childhood thyroid cancer in Belarus. Nature 359: 681 (1992).

S71 Strong, L.C. Genetic and environmental interactions. Cancer 40: 1861-1866 (1977).

S72 Straume, T., L.J. Harris, A.A. Marchetti et al. Neutrons confirmed in Nagasaki and at the Army Pulsed Reactor Facility: implications for Hiroshima. Radiat. Res. 138: 193-200 (1994).

S73 Samet, J.M., J. Stolwijk and S.L. Rose. Summary: international workshop on residential Rn epidemiology. Health Phys. 60: 223-227 (1991).

T1 Taylor, D.M., C.W. Mays, G.B. Gerber et al. (eds.). Risks from Radium and Thorotrast. BIR Report 21. British Institute of Radiology, London, 1989.

T2 Tokunaga, M., C.E. Land, T. Yamamoto et al. Incidence of female breast cancer among atomic bomb survivors, Hiroshima and Nagasaki, 1950-1980. RERF TR/15-84 and Radiat. Res. 112: 243-272 (1987).

T3 Tomášek, L., S.C. Darby, T. Fearn et al. Patterns of lung cancer mortality among uranium miners in West Bohemia with varying rates of exposure to radon gas. Radiat. Res. 137: 251-261 (1994).

T4 Travis, L.B., C.E. Rochelle, J.D. Boice Jr. et al. Second cancers following non-Hodgkin's lymphoma. Cancer 67: 2002-2009 (1991).

T5 Tucker, M.A., A.T. Meadows, P. Morris Jones et al. Therapeutic radiation at young age linked to secondary thyroid cancer. Proc. Am. Soc. Clin. Oncol. 5: 827 (Abstract) (1986).

T6 Tucker, M.A., G.J. D'Angio, J.D. Boice Jr. et al. Bone sarcomas linked to radiotherapy and chemo-therapy in children. N. Engl. J. Med. 317: 588-593 (1987).

T7 Tucker, M.A., A.T. Meadows, J.D. Boice et al. Leukaemia after therapy with alkylating agents for childhood cancer. J. Natl. Cancer Inst. 78: 459-464 (1987).

T8 Tucker, M.A., C.N. Coleman, R.S. Cox et al. Risk of second cancers after treatment for Hodgkin's disease. N. Engl. J. Med. 318: 76-81 (1988).

T9 Tomášek, L., S.C. Darby, A.J. Swerdlow et al. Radon exposure and cancers other than lung cancer among uranium miners in West Bohemia. Lancet 341: 919-923 (1993).

T10 Tirmarche, M., A. Raphalan, F. Allin et al. Mortality of a cohort of French uranium miners exposed to relatively low radon concentrations. Br. J. Cancer 67: 1090-1097 (1993).

T11 Teppo, L., E. Pukkala and E. Saxén. Multiple cancer - an epidemiologic exercise in Finland. J. Natl. Cancer Inst. 75: 207-217 (1985).

T12 Tirmarche, M., A. Rannou, A. Mollie et al. Epidemiological study of regional cancer mortality in France and natural radiation. Radiat. Prot. Dosim. 24: 479-482 (1988).

T13 Tucker, M.A., J.D. Boice Jr. and D.A. Hoffman. Second cancer following cutaneous melanoma and cancers of the brain, thyroid, connective tissue, bone, and eye in Connecticut, 1935-82. Natl. Cancer Inst. Monogr. 68: 161-189 (1985).

T14 Thomas, D., S.C. Darby, F. Fagnani et al. Definition and estimation of lifetime detriment from radiation exposures: principles and methods. Health Phys. 63: 259-272 (1992).

T15 Thompson, D.E., K. Mabuchi, E. Ron et al. Cancer incidence in atomic bomb survivors. Part II: Solid tumors, 1958-87. RERF TR/5-92 (1992) and Radiat. Res. 137: S17-S67 (1994).

T16 Taylor, J.A., M.A. Watson, T.R. Devereux et al. p53 mutation hotspots in radon-associated lung cancer. Lancet 343: 86-87 (1994).

U1 United Nations. Sources and Effects of Ionizing Radiation. United Nations Scientific Committee on the Effects of Atomic Radiation, 1993 Report to the General Assembly, with scientific annexes. United Nations sales publication E.94.IX.2. United Nations, New York, 1993.

U2 United Nations. Sources, Effects and Risks of Ionizing Radiation. United Nations Scientific Committee on the Effects of Atomic Radiation, 1988 Report to the General Assembly, with annexes. United Nations sales publication E.88.IX.7. United Nations, New York, 1988.

U3 United Nations. Genetic and Somatic Effects of Ionizing Radiation. United Nations Scientific Committee on the Effects of Atomic Radiation, 1986 Report to the General Assembly, with annexes. United Nations sales publication E.86.IX.9. United Nations, New York, 1986.

U4 United Nations. Ionizing Radiation: Sources and Biological Effects. United Nations Scientific Committee on the Effects of Atomic Radiation, 1982 Report to the General Assembly, with annexes. United Nations sales publication E.82.IX.8. United Nations, New York, 1982.

U5 United Nations. Sources and Effects of Ionizing Radiation. United Nations Scientific Committee on the Effects of Atomic Radiation, 1977 Report to the General Assembly, with annexes. United Nations sales publication E.77.IX.1. United Nations, New York, 1977.

U6 United Nations. Ionizing Radiation: Levels and Effects. Volume I: Levels, Volume II: Effects. United Nations Scientific Committee on the Effects of Atomic Radiation, 1972 Report to the General Assembly, with annexes. United Nations sales publication E.72.IX.17 and 18. United Nations, New York, 1972.

U12 Upton, A.C. Risk estimates for carcinogenic effects of radiation. p. 1-29 in: Risks Associated with Ionising Radiation. Annals of the ICRP 22. Pergamon Press, Oxford, 1991.

U13 Urquhart, J.D., R.J. Black, M.J. Muirhead et al. Case-control study of leukaemia and non-Hodgkin's lymphoma in children in Caithness near the Dounreay nuclear installation. Br. Med. J. 302: 687-692 (1991).

U14 Ullrich, R.L. and R.J. Preston. Myeloid leukaemia in male RFM mice following irradiation with fission spectrum neutrons or γ rays. Radiat. Res. 109: 165-170 (1987).

V1 van Kaick, G., H. Muth, A. Kaul et al. Results of the German Thorotrast study. p. 253-262 in: Radiation Carcinogenesis: Epidemiology and Biological Significance. (J.D. Boice Jr. and J.F. Fraumeni Jr., eds.) Raven Press, New York, 1984.

V2 van Kaick, G., H. Muth, A. Kaul et al. Report on the German Thorotrast study. Strahlentherapie (Suppl.) 80: 114-118 (1986).

V3 Vaughan, J. Carcinogenic effects of radiation on the human skeleton and supporting tissues. p. 311-334 in: Radiation Carcinogenesis. (A.C. Upton et al., eds.) Elsevier Press, New York, 1986.

V4 Vaeth, M. and D.A. Pierce. Calculating excess lifetime risk in relative risk models. Environ. Health Perspect. 87: 83-94 (1990) and RERF CR/3-89 (1989).

V5 Voelz, G.L., R.S. Grier and L.H. Hempelmann. A 37-year medical follow-up of Manhattan project Pu workers. Health Phys. 48: 249-259 (1985).

V6 Voelz, G.L., G.S. Wilkinson, J.W. Healy et al. Mortality study of Los Alamos workers with higher exposures to plutonium. p. 318-327 in: Epidemiology Applied to Health Physics. CONF-830101 (1983).

V7 Voelz, G.L., A.R. Stevenson and C.C. Stewart. A 42-year medical follow-up of Manhattan project plutonium workers. Health Phys. 56 (Suppl. 1): S46 (1989).

V8 Valagussa, P., A. Santoro, F. Fossati-Bellani et al. Second acute leukemia and other malignancies following treatment for Hodgkin's disease. J. Clin. Oncol. 4: 830-837 (1986).

V9 van Kaick, G., H. Wesch, H. Lührs et al. The German thorotrast study - report on 20 years follow-up. p. 98-104 in: Risks from Radium and Thorotrast. BIR Report 21. (D.M. Taylor et al., eds.) British Institute of Radiology, London, 1989.

V10 Viel, J.F. and S.T. Richardson. Childhood leukaemia around the La Hague nuclear waste reprocessing plant. Br. Med. J. 300: 580-581 (1990).

V11 Voelz, G.L. and J.N.P. Lawrence. A 42-y medical follow-up of Manhattan project plutonium workers. Health Phys. 61: 181-190 (1991).

V12 Vaeth, M. Some statistical issues important to future developments in human radiation research. p. 161-164 in: The Future of Human Radiation Research. BIR Report 22. (G.B. Gerber, D.M. Taylor, E. Cardis et al., eds.) British Institute of Radiology, London, 1992.

V13 Vonstille, W.T. and H.L.A. Sacarello. Radon and cancer. J. Environ. Health 53: 25-28 (1990).

V14 Vaeth, M., D.L. Preston and K. Mabuchi. The shape of the cancer incidence dose-response curve for the A-bomb survivors. p. 75-79 in: Low Dose Irradiation and Biological Defense Mechanisms. Elsevier Science Publishers, Amsterdam, 1992.

W1 Woodward, A., D. Roder, A.J. McMichael et al. Radon exposures at the Radium Hill uranium mine and lung cancer rates among former workers, 1952-1987. Cancer Causes Control 2: 213-220 (1991).

W2 Walter, S.D., J.W. Meigs and J.F. Heston. The relationship of cancer incidence to terrestrial radiation and population density in Connecticut, 1935-1974. Am. J. Epidemiol. 123: 1-14 (1986).

W3 Wing, S., C.M. Shy, J.L. Wood et al. Job factors, radiation and cancer mortality at ORNL: follow-up through 1984. Am. J. Ind. Med. 23: 265-279 (1993).

W4 Wick, R.R. and W. Gössner. Follow-up study of late effects in Radium-224 treated ankylosing spondylitis patients. Health Phys. 44 (Suppl. 1): 187-195 (1983).

W5 Wick, R.R., D. Chmelevsky and W. Gössner. Radium-224: risk to bone and haematopoietic tissue in ankylosing spondylitis patients. p. 38-44 in: The Radiobiology of Radium and Thorotrast. (W. Gössner et al., eds.) Urban and Schwarzenberg, München, 1986.

W6 Wilkinson, G.S. Gastric cancer in New Mexico counties with significant deposits of uranium. Arch. Environ. Health 40: 307-312 (1985).

W7 Wilkinson, G.S., G.L. Tietjen, L.D. Wiggs et al. Mortality among plutonium and other radiation workers at a plutonium weapons facility. Am. J. Epidemiol. 125: 231-250 (1987).

W8 Wang, J. Statistical analysis of cancer mortality data of high background radiation areas in Yiangjiang. Chin. J. Radiol. Med. Prot. 13: 291-294 and 358 (1993).

W9 Whittemore, A.S. and A. McMillan. Lung cancer mortality among U.S. uranium miners: a reappraisal. J. Natl. Cancer Inst. 71: 489-499 (1983).

W10 Wang, J.X., J.D. Boice Jr., B.X. Li et al. Cancer among medical diagnostic X-ray workers in China. J. Natl. Cancer Inst. 80: 344-350 (1988).

W11 Wesch, H., W. Riedel, K. Hasenöhrl et al. German thorotrast study: results of the long-term animal studies on the effect of incorporated radioactive and nonradioactive particles. Strahlentherapie (Suppl.) 80: 186-188 (1986).

W12 Wesch, H., G. van Kaick, W. Riedel et al. Recent results of the German thorotrast study - statistical evaluation of animal experiments with regard to the nonradiation effects in human thorotrastosis. Health Phys. 44 (Suppl. 1): 317-321 (1983).

W13 Wiggs, L.D., C.A. Cox-DeVore and G.L. Voelz. Mortality among a cohort of workers monitored for ^{210}Po exposure: 1944-1972. Health Phys. 61: 71-76 (1991).

W14 Wang, J.X., P.D. Inskip, J.D. Boice Jr. et al. Cancer incidence among medical diagnostic x-ray workers in China, 1950 to 1985. Int. J. Cancer 45: 889-895 (1990).

W15 Wei, L., Y. Zha, Z. Tao et al. Epidemiological investigation of radiological effects in high background radiation areas of Yiangjiang, China. J. Radiat. Res. 31: 119-136 (1990).

W16 Wei, L., Y. Zha, Z. Tao et al. Cancer mortality study in high background radiation areas of Yiangjiang, China. Radiat. Biol. Res. Commun. 23: 209-220 (1988).

W17 Wang, Z., J.D. Boice Jr., L. Wei et al. Thyroid nodularity and chromosome aberrations among women in areas of high background radiation in China. J. Natl. Cancer Inst. 82: 478-485 (1990).

W18 Wick, R.R. and W. Gössner. Recent results of the follow-up of radium-224-treated ankylosing spondylitis patients. p. 25-28 in: Risks from Radium and Thorotrast. BIR Report 21. (D.M. Taylor et al., eds.) British Institute of Radiology, London, 1989.

W19 Waxweiler, R.J., V.E. Archer, R.J. Roscoe et al. Mortality patterns among a retrospective cohort of uranium mill workers. p. 428-435 in: Epidemiology Applied to Health Physics. CONF-830101 (1983).

W20 Wing, S., C.M. Shy, J.L. Wood et al. Reply to comments by Gilbert and Prichard. Health Phys. 62: 261-264 (1992).

W21 Wing, S., C.M. Shy, J.L. Wood et al. Mortality among workers at Oak Ridge national laboratory. J. Am. Med. Assoc. 265: 1397-1402 (1991).

W22 Werner, M.M. and D.K. Myers. Causes of death among long-term employees of Chalk River laboratories, 1966-1989. AECL-10293 (1990).

W23 Wiggs, L.D., C.A. Cox-DeVore and G.L. Voelz. Mortality among a cohort of workers monitored for exposure to the radiation isotope 210-polonium: 1944-1972. Health Phys. 61: 71-76 (1991).

W24 Wiggs, L.D., C.A. Cox-DeVore, G.L. Voelz et al. Mortality among workers exposed to external ionizing radiation at a nuclear facility in Ohio. J. of Occup. Med. 33: 632-637 (1991).

X1 Xuan, X.Z., J.H. Lubin, J.Y. Li et al. A cohort study in southern China of tin miners exposed to radon and radon decay products. Health Phys. 64: 120-131 (1993).

Y1 Yoshimoto, Y., M. Soda, W.J. Schull et al. Studies of children in utero during atomic bomb detonations. in: Proceedings of the 203rd National Meeting of the American Chemical Society, 1992, San Francisco. American Chemical Society, Washington D.C. (1994, in press).

Y2 Yoshimoto, Y., H. Kato and W.J. Schull. Risk of cancer among children exposed in utero to A-bomb radiations, 1950-84. RERF TR/4-88 (1988) and Lancet ii: 665-669 (1988).

Y3 Yoshimoto, Y., J.V. Neel, W.J. Schull et al. Malignant tumors during the first two decades of life in the offspring of atomic bomb survivors. RERF TR/4-90 (1990) and Am. J. Hum. Genet. 46: 1041-1052 (1990).

Y4 Yoshimoto, Y., W.J. Schull, H. Kato et al. Mortality among the offspring (F_1) of atomic bomb survivors, 1946-85. RERF TR/1-91 (1991) and J. Radiat. Res. 32: 327-351 (1991).

Y5 Yoshimoto, Y. Cancer risk among children of atomic bomb survivors. A review of RERF epidemiological studies. J. Am. Med. Assoc. 264(5): 596-600 (1990).

Y6 Yao, X.S., J.H. Lubin, Y.L. Qiao et al. Radon progeny exposure, tobacco use and lung cancer in Chinese tin miners. Radiat. Res. 138: 326-336 (1994).

ANNEX B

Adaptive responses to radiation in cells and organisms

CONTENTS

INTRODUCTION

1. The scientific community has been aware for many years of the possibility that low doses of radiation may result in changes in cells and organisms, which reflect an ability to adapt to the effects of radiation. In lower organisms, for example, enhanced proliferation in the presence of radiation at doses of a few microgray per day to a few milligray per day has been observed in experiments involving cultures of prokaryotes and eukaryotic cells [C1, C2, C3, C4, C5, C6, I1, I2, L4, P2, P3, T11].

2. The biological expression of adaptive and stimulatory responses in seeds and plants (e.g. [C14, H9, H10, R6, R7, S43]) has also been described. The extensive literature up to 1976 supporting radiation-associated adaptive effects was reviewed by Luckey [L1]. A more recent publication by the same author summarizes the literature between 1976 and 1991, involving about one thousand reports judged by him to demonstrate beneficial responses in animals and in human populations [L2].

3. It has been suggested in recent years that the conventional estimates of stochastic effects following exposure to low doses of ionizing radiation may have been overstated because no allowance was made for the possibility that small doses of radiation may condition cells so as to induce processes that reduce either the natural incidence of cancer in its various forms or the likelihood of excess cancers being caused by further radiation exposure.

4. An important observation from mammalian cell studies in support of these processes is that mitogen-stimulated human blood lymphocytes exposed *in vitro* appear to suffer less damage than would be expected following acute exposure to a few gray of low-LET radiation if they are first exposed to a dose of a few tens of milligray. This response to low-dose exposure, which remains effective for several hours, is referred to as an adaptive response.

5. Adaptive responses have been observed in other mammalian cell types, such as bone marrow cells and fibroblasts, but not consistently in spermatocytes or at all in embryo cells exposed in the pre-implantation stage. Changes in the composition of the culture medium can alter the adaptive response. In a wider context, it is known that changes similar to those observed in radiation-induced adaptive response can occur as a result of metabolic disturbances and after damage resulting from exposure to a variety of physical and chemical agents. The consequences of these cellular changes are referred to collectively in the literature as stress response or response to genotoxic stress.

6. Evidence of adaptive response has also been described in studies using laboratory animals, in which the animals were exposed either to single acute doses from a few tens of milligray to a few gray, or accumulated doses of up to a few gray over a lifetime. Reported manifestations of this form of adaptive response described in mammals after exposure to low doses of radiation include an accelerated growth rate in the young, an increase in reproductive ability, an extended life-span, stimulatory effects on the immune system and a lower-than-expected incidence of spontaneous tumours. A satisfactory explanation of mechanisms that might be responsible for such effects, which have not been consistently observed in different investigations, is not obvious. It may involve a DNA repair mechanism similar to that proposed for the cellular adaptive response, implying its immediate availability if cellular damage randomly occurs during the animal's lifetime. Involvement of the immunosurveillance system has also been proposed.

7. Four conferences (one on radiation hormesis, held at Oakland, California, in 1985 [S1]; one on low-dose radiation and the immune system, held at Frankfurt in 1987 [S21]; one on low-dose irradiation and biological defense mechanisms, held at Kyoto in 1992 [S2]; and one on low-level exposures to radiation and related agents, held at Changchun, China, in 1993 [I8]) have provided an opportunity for scientists working in the field of low-dose effects to present their experimental findings and debate the possible mechanisms involved in radiation-induced adaptive response.

8. This Annex, based mainly on recently published data, has been compiled by the Committee with a view to identifying the cellular mechanisms that may be involved in the adaptive response at low doses. It should be considered as a continuation of the discussions on radiation response contained in the UNSCEAR 1993 Report [U1]. One problem in identifying common mechanisms of response is that there are differences between the doses and dose rates used in cellular studies and those used in *in vivo* studies of laboratory animals. A further complication is that there are few studies available in which doses of a few milligray per year above the natural background radiation level have been used. Low doses were defined in the UNSCEAR 1993 Report [U1], Annex F, "Influence of dose and dose rate on stochastic effects of radiation", with the values depending on the level of investigation. At a microdosimetric level a low dose is defined as about 0.0002 Gy. In a similar context, the International Commission on Radiological Protection [I6] considered that low doses and low dose rates imply situations in which it is very unlikely that more than one event of energy deposition will occur in the critical parts of a cell within the time during which repair mechanisms in the cell can operate. For mammalian cells in culture, a low dose is defined as less than about 0.02 Gy. For the induction of human tumours, a low dose is defined as less than about 0.2 Gy. The same criteria can be applied in this Annex, although it should be recognized that much of the experimental data on effects in cells and animals is based on doses in excess of about 0.5 Gy.

9. Manifestations of adaptive responses in animals and in human populations are briefly addressed in this Annex. Evidence for the expression of an adaptive response in human populations exposed to low doses of radiation above the natural background level has not until now been clearly demonstrated nor has it been refuted. The possibility that exposure to low doses of radiation may affect the level of competence of immunosurveillance mechanisms in carcinogenesis is discussed. Details of the mechanisms of action of radiation in inducing cancers and serious hereditary effects in humans are not discussed. These aspects of the deleterious effects of radiation can be found in the UNSCEAR 1993 Report [U1], Annex E, "Mechanisms of radiation oncogenesis", Annex G, "Hereditary effects of radiation", and in this Report in Annex A, "Epidemiological studies of radiation carcinogenesis". These Annexes should be read in conjunction with this Annex to achieve a balanced view of the overall effects of low doses of radiation.

I. ADAPTIVE PROCESSES IN MAMMALIAN CELLS

10. Efforts have been made over the past decade to characterize the adaptive response induced by mutagens in mammalian cells. A response has been demonstrated in mitogen-stimulated, human blood T lymphocytes. Other cell types investigated for evidence of a response include proliferating lymphoblasts, bone marrow cells, spermatocytes, pre-implantation embryos and fibroblasts.

11. Cells respond to radiation-induced injury by the up-regulation of proteins involved in cell signalling and by the increased expression of genes involved in cell proliferation [A1, S3] and in the synthesis of DNA repair enzymes [L1, L5, M3, W2]. Qualitatively similar responses have been described as a result of cellular disturbance caused by temporary oxygen deprivation [B5, L7, M4, S9, S10] and glucose starvation [H1, S11].

12. The mechanisms involved in the adaptive response to low doses of radiation have been linked to a more general phenomenon in which the cells are able to respond to damage from a variety of physical and chemical agents. These agents include overheating [S8], UV-radiation [B3], trace amounts of mutagenic chemicals [B4, V1], local anaesthetics that alter membrane structure [L6] and heavy metals known to act as cellular poisons [C7].

A. EFFECTS IN HUMAN LYMPHOCYTES

1. Chromosome aberrations

13. It was reported in 1984 [O2] that when phytohaemagglutinin-stimulated human lymphocytes were grown in a culture medium containing tritiated thymidine and were exposed in the G_2 phase of the cell cycle to 1.5 Gy from x rays, the yield of chromatid aberrations was significantly less than the sum of yields of the aberrations induced by tritiated thymidine and x rays separately (Table 1). The response was observed to occur at a concentration of tritiated thymidine low enough to give an estimated one beta disintegration in each cell volume between exposure to the tritium and the x rays. The reduction in the expected number of aberrations was not considered to be attributable to a radiation-induced delay in cell cycle progression at this low concentration. Nor was it considered to be due to the selective killing of a radiosensitive population of lymphocytes that had incorporated tritiated thymidine [W3].

14. When the tritiated thymidine was present in the blood cultures throughout the entire culture period, the results were quite variable. This was shown, however, to be attributable to the fact that, in blood, the amount of tritiated thymidine incorporated into cells is highly dependent on the catabolism of thymidine to a degraded form which cannot be incorporated. Later experiments in which the tritiated thymidine was pulse-labelled while the cells were in the S phase markedly reduced this variability, and the uptake of tritiated thymidine was maximized.

15. A similar response was shown to be induced by exposing phytohaemagglutinin-stimulated lymphocytes in the S phase to a low dose from x rays (referred to variously in the literature as the conditioning, inducing, priming or adapting dose), followed by exposure of the cells in the G_2 phase to a high dose from x rays (called the challenge dose). This response was subsequently confirmed by some investigators [B6, B11, K4, L25, M30, O3, S13, S14, W4, W5], although not consistently by others who were using similar culture protocols [B7, H8, M21, S15, S16]. The lymphocyte cells from different donors show variable sensitivity, as is shown in Table 2.

16. It was postulated that the conditioning dose of radiation activated genes and that this was quickly followed by the synthesis of enzymes responsible for DNA repair. If these enzymes were available in adequate concentrations at the time the cells were exposed to a challenge dose, the extent of the repair of DNA damage was improved, so that fewer chromatid aberrations were observed than in cells receiving the challenge dose only. It was presumed that the repair enzymes were not immediately available to cells receiving the challenge dose only and that, in these circumstances, much of the damage was irreparable by the time a sufficient quantity of enzymes became available. This hypothesis was supported by the observation that the adaptive response could be blocked by the protein synthesis inhibitor cycloheximide (Table 3) and by 3-aminobenzamide (Table 4), an inhibitor of poly(ADP-ribose) polymerase, which is known to be induced during the repair of DNA strand breaks [A16].

17. Other characteristics of the *in vitro* lymphocyte adaptive response have been reported:

(a) the adaptive response to x rays requires a dose of at least 0.005 Gy delivered at a rate of more than 0.2 Gy min^{-1} [S14]. The implication of this finding is that a certain number of DNA lesions, perhaps of a specific type, need to occur within a fixed time in order to initiate the signal for expression of the adaptive response. In fact, there is a window of dose, 0.005 to 0.2 Gy, below which and above which the phenomenon was not observed. Such a narrow window has also been observed in experiments with radiomimetic compounds such as bleomycin, which, like x rays, induces double-strand breaks;

(b) the induction of the repair mechanism takes place between 4 and 6 hours after exposure to the conditioning dose and remains effective for three cell cycles [S13];

(c) when the cells were exposed to two conditioning doses delivered within a few hours of each other, the reduction in the amount of chromatid damage was found to be similar to that observed after a single dose. The second dose thus provided no additional protection against the damage caused by the first conditioning dose within this time (Table 5);

(d) it has recently been shown that a single dose of 0.005 Gy from ^{60}Co gamma rays did not create conditions for the adaptive response, but two doses, each of 0.005 Gy given in the same cell cycle, did do so [B22]. The effect of the two conditioning doses was optimum when they were given at 36 hours and 42 hours and the challenge dose of 1 Gy or 1.5 Gy was given at 48 hours after mitogen stimulation. It is implied from this study that the acute dose to induce maximum activity of the repair enzyme system is about 0.01 Gy.

18. The question why lymphocytes from some individuals do not respond remains unresolved. In fact, the lymphocytes in some cultures exposed to 0.02 Gy from x rays reacted synergistically to subsequent mutagenic treatment [O3]. The cells in this experiment were fixed 2-4 hours after the challenge dose, in contrast to those in other experiments, where fixation times were confined to 6 hours or more after the challenge dose. The unpredictable nature of the response has been confirmed using differences in micronuclei frequency as an end-point [P4]. Changes in the hydrogen ion concentration of the culture medium can affect the yield of induced chromatid aberrations (Table 6). This finding was confirmed when it was shown that adjusting the hydrogen ion concentration of the culture medium to pH 6.4 just before the challenge dose enhanced the effectiveness of the response [O3]. It was also shown that the response could be induced in cultured lymphocytes from donors who had not previously displayed the adaptive response, by adding compounds to the culture medium that could affect the metabolism of the phytohaemagglutinin-stimulated lymphocytes (e.g. interleukin-2, which stimulates proliferation). It is conceivable that the repair systems induced may react differently according to the culture conditions, a situation that should not be overlooked when considering the consequences of the conditioning dose. Thus, the composition of the culture medium may be crucial.

19. It has been pointed out that measurements of the frequencies of aberrations induced in asynchronous cell populations are likely to be misleading if expressed as a simple average from a single fixation time [S23]. The reason for this lies in the intercellular variability of the cell-cycle transition times. It is known that the intrinsic cellular radiosensitivity varies as the cells pass through the cell cycle, and the time of application of the conditioning or challenge doses of radiation may therefore be crucial. In these circumstances, the aberration score will always reflect the average of a mixture of cells having different radiosensitivities, and any shifts in the mixture ratios will influence the aberration yields. A double-labelling technique (Brd-U replication banding), which permits identification of the cell cycle position occupied by each scored metaphase at the time of the conditioning and challenge doses, has recently been described and may help to solve this problem [A10]. Using this technique, it was shown that a conditioning dose of 0.01 Gy from x rays delivered at a rate of 0.05 Gy min^{-1}, followed 6 hours later by a challenge dose of 1.5 Gy at a rate of 0.0044 Gy min^{-1}, resulted in a transient decrease in the frequency of total aberrations at 6 hours, but not at 9 hours, after challenge. Furthermore, when the cohorts at 3, 6 and 9 hours after the challenge dose were combined, there was no evidence of an adaptive response. This preliminary experiment serves to demonstrate the complex nature of the kinetics of cells in stimulated lymphocyte cultures and the possibility that an adaptive response may occur only in a narrow window of cell cycle when cells are particularly radiosensitive.

20. The experiments described above refer to the application of the conditioning dose in the S phase and a challenge dose from x rays in the G_2 phase of the cell cycle. Studies in which phytohaemagglutinin-stimulated lymphocytes were exposed to the conditioning dose at other stages of the cell cycle have been reported, but different laboratories have had different results, which has yet to be explained. An adaptive response was reported when the G_0 or G_1 phase cells were exposed to a conditioning dose and challenged in the late S or early G_2 phase [C8, K4, K5] and when cells were exposed to a conditioning dose in the G_1 phase and challenged in the G_1 phase [S17, W7]. No response was reported by other laboratories if the conditioning dose was given in the G_0 or the G_1 phase and the challenge dose in the G_1 phase [K4, S13]. These results are presented in Table 7.

21. An adaptive response, indicated by a reduced frequency of chromosome aberrations, has also been demonstrated with the conditioning dose given *in vivo*. Preliminary results of the cytogenetic monitoring of children living in a region of Ukraine contaminated after the Chernobyl accident indicate that the chromosome aberration yield in lymphocytes to a challenge dose *in vitro* is less than that in control lymphocytes from a challenge dose alone [P10]. This has to be confirmed by further studies, but there is supporting evidence from *in vivo* studies in the rabbit [L8]. The response of lymphocytes to both conditioning and challenge doses *in vitro* had been demonstrated earlier [C8]. In the *in vivo* study, four adult male rabbits were exposed to gamma-radiation at a dose rate of about 6 mGy h^{-1} for 9 hours each day for 36 days, giving a daily dose of 0.05 Gy. Blood samples were taken before the *in vivo* irradiations and further samples were collected at intervals of 6, 15, 18, 24, 30 and 36 days, the cumulative doses being 0.3, 0.75, 0.9, 1.2, 1.5

and 1.8 Gy, respectively. Six cultures were established from each blood sample. Two were analysed for baseline chromosome aberrations; two were exposed to a challenge dose of 1.5 Gy from x rays at a dose rate of 0.44 Gy min^{-1}, and phytohaemagglutinin was added to them immediately thereafter. The remaining two cultures were incubated with the addition of phytohaemagglutinin at the start of culture and exposed to a challenge dose of 1.5 Gy from x rays 48 hours later. The results from the analysis of blood lymphocytes from individual rabbits showed the same trend and are presented as average values in Table 8. They show that an adaptive response can be induced in lymphocytes exposed to 0.05 Gy d^{-1} *in vivo* when the challenge dose to the phytohaemagglutinin-stimulated cells is given *in vitro* in either the G_0 or the G_2 phase of the cell cycle. This could imply that chronic irradiation of circulating blood lymphocytes induces the synthesis of proteins in sufficient amounts to maintain a continuous and effective reservoir of repair enzymes.

2. Clone-forming ability and genomic stability

22. To study the effects of radiation on clone-forming ability and karyotypic abnormalities in human peripheral blood lymphocytes, cells were exposed to 3 Gy from x rays *in vitro* and either individual T-cell clones or long-term T-cell cultures were established [H16]. The karyotypes were analysed in G-banded chromosome preparations after proliferation for 9-34 days *in vitro.*

23. T-cell clonal karyotype abnormalities were found in 24 of 37 (65%) irradiated clones and in 2 of 43 (5%) control clones. Balanced reciprocal translocations and deletions were the predominating types of clonal aberrations. Complex aberrations and unstable karyotypes were found in about half of the irradiated clones. Some of the T-cell clones demonstrated sequential change from normal to aberrant karyotype. Other clones seemed to develop multiple, heterogeneous chromosomal aberrations during growth *in vitro.*

24. T cells irradiated with x rays and grown in long-term culture displayed karyotype abnormalities in 60%-80% of the cells, and the types of aberrations were similar to those found in the individual irradiated T-cell clones. An increasing number of cells with the same abnormal karyotype was observed when the cultivation time was extended, indicating preferential clonal proliferation.

25. These results demonstrate that a surprisingly high proportion of T cells with stable and often complex irradiation-induced chromosome aberrations are able to proliferate and form expanding cell clones *in vitro.* Furthermore, they indicate that x-irradiation induces latent chromosome damage and genomic instability in human T lymphocytes. What would be interesting would be to repeat this study by giving a conditioning dose of a few tens of milligray before the challenge dose of 3 Gy and observing if any reduction in proliferating T cells with stable aberration occurred.

3. Cell survival and mutation frequency

26. Cell survival and chromosome aberration yield have been measured in phytohaemagglutinin-stimulated lymphocytes exposed to 0.05 Gy from x rays followed by acute exposure to 2 or 4 Gy, both exposures being in the G_1 phase [S17]. In studies on six donors, the yields of chromosome exchanges and deletions were found to be less than in cells receiving the challenge dose only. Lymphocytes from only two of the six donors tested showed an adaptive response expressed as enhanced cell survival with a challenge dose of 2 Gy and none after a challenge dose of 4 Gy. A chromosome adaptive response, therefore, does not necessarily coincide with a cell survival adaptive response. Reductions in the number of cells with several aberrations (multiply aberrant cells) can be the result of a cytogenetic adaptive response, but if the proportion of non-aberrant cells is not increased, then a survival adaptive response will not be seen.

27. In a subsequent study, a lower challenge dose was used. Lymphocytes from the six donors were exposed to 0.05 Gy from x rays, followed by 1 Gy in the G_1 phase [S44]. Under these exposure conditions, most of the aberrant cells would be expected to contain only one chromosome aberration after the challenge dose. Cell survival adaptive responses were seen in four of the six donors, but the decrease in the numbers of singly aberrant cells was not in itself sufficient to account for the increase in cell survival. It was proposed, therefore, that some increase in cell survival could have been due to repair of lesions in cells that were at the level of the gene locus, which would not be recognized by the cytological techniques used to identify aberrations.

28. Cell survival has been measured concurrently with the yield of mutations using 6-thioguanine (TG) selection to detect clones mutated at the X-linked hypoxanthine phosphoribosyl transferase (*hprt*) locus [S18]. Tritiated thymidine was added during the G_0 phase, followed by exposure to 1.5 or 3.0 Gy from x rays in the G_1 phase. Cell survival was not affected (Table 9), but tritiated thymidine at concentrations of 3.7 and 37 kBq ml^{-1} in the culture medium produced a significant decrease in the number of mutations induced after the challenge dose from x rays compared with cells receiving the challenge dose only.

29. In support of this observation, the mutation frequency was reduced by 70%, while cell survival was not affected, when lymphoblastoid cells were exposed to 0.02 Gy from x rays, followed by a dose of 4 Gy in the G_1 phase [R8]. This decreased mutation frequency was

considered to be the result of an induced repair system, which was shown to be absent from mutant cells deficient in the *hprt* locus. Lymphocytes exposed in the G_1 phase to a conditioning dose of 0.01 Gy from x rays, followed by a challenge dose of 3 Gy from x rays in the G_2 phase, also showed a reduced mutation frequency compared with cells exposed to the challenge dose only (Table 10).

30. Reduced mutation frequency was demonstrated with human HL-60 cells exposed to a conditioning dose of 0.01 Gy from ^{60}Co gamma rays at a dose rate of 0.078 Gy min^{-1} and then exposed 18 hours later to a challenge dose of 2 Gy [Z6]. After irradiation, the cells were cultured for seven days in a non-selective RPM1-1640 medium to allow phenotypic expression of *hprt*$^-$ mutants. The frequency of *hprt* mutations resulting from the dose of 2 Gy was 26.9 10^{-6}. Treatment with the conditioning dose reduced the mutation frequency to 10.7 10^{-6}.

31. Mutant colonies exposed to the challenge dose showed gene deletions and rearrangements in 15 out of 32 colonies (46%). This compared to 12 out of 46 colonies (26%) first exposed to a conditioning dose. Since gene deletions and rearrangements are associated with unrepaired or error-prone DNA double-strand breaks, it could be concluded from this experiment that a DNA double-strand fidelity repair mechanism had been induced.

32. In contrast, the human lymphoblastoid cell line TK6, which is heterozygous for the thymidine kinase gene ($TK^{+/-}$), has been used in studies of mutation induction at two independent genetic loci. One of these is the *hprt* locus; the other is the autosomal thymidine kinase (*TK*) locus. The selective agents 6-thioguanine (TG) and trifluorothymidine (TFT) were used to measure mutation at the *hprt* and *TK* loci, respectively. Cell survival and mutation rate were measured after protracted exposure to tritiated water, followed by exposure to x rays at the rate of 0.8 Gy min^{-1} [T1]. The results of this experiment are illustrated in Figure I.

33. The cells were grown in a medium containing 0.74 MBq ml^{-1} of tritiated water, the tritium irradiating the cells at a dose rate of about 0.05 Gy d^{-1}. During the overall period of incubation in the presence of tritiated water, the cloning efficiency, determined after 10, 20 and 30 days of exposure, remained almost constant, and it was comparable to that found for unirradiated cells. After the challenge dose of up to 1.5 Gy from x rays, the survival curves for TK6 cells, pretreated or not with tritiated water for different lengths of time, were also similar, as shown in the upper plot of Figure I. These results showed that with a low-dose-rate, protracted conditioning exposure from incubation in tritiated water, no adaptive effect on cell survival was detectable. Furthermore, treatment with tritiated water had no significant effect on the induction of mutations. When the mutation frequency was plotted as a function of the accumulated dose, regardless of the radiation source and the modalities of treatment, a linear relationship was found, indicating that the mutagenic effects of protracted exposure to tritiated water and acute exposure to x rays were additive, as can be seen from the lower plot of Figure I.

4. Interaction with chemicals

34. A recent review of experiments involving the activation of bacterial oxidative stress genes provides a useful background for understanding adaptive mechanisms in eukaryotic cells [D5]. One of the mechanisms involved in DNA repair after exposure to low-LET radiation is thought to be similar to that operating after exposure to trace amounts of oxidizing radicals. In confirmation of this hypothesis, exposing lymphocytes to low concentrations of hydrogen peroxide, followed by a dose of 1.5 Gy from x rays, was found to induce the adaptive response (Table 11). Conditioning with a chemical and challenge with radiation is termed cross-adaptation.

35. Other studies (see also paragraph 74) have substantiated this finding, in which the adaptive response was shown to occur in donors whose lymphocytes were treated with 25-75 μM of hydrogen peroxide 24 hours before a challenge dose of 1.5 Gy from x rays [C13]. However, when the cells were repeatedly exposed to hydrogen peroxide at intervals of 24, 30 and 36 hours, the adaptive response was not observed. The authors did not give any explanation for this lack of response to repeated doses [W3].

36. A reduction in micronuclei frequency has been demonstrated in lymphocytes conditioned with hydrogen peroxide [D6]. Lymphocytes were exposed to a 30-minute pulse of hydrogen peroxide (25-250 μM) 24 hours after formation of the cultures and to a challenge dose of 1.5 Gy or 3 Gy from x rays 48 hours later.

37. An adaptive response can be induced in the presence of trace amounts of bleomycin, which is known to produce double-strand breaks during the G_2 and M phases of the cell cycle [V2, W8]. Thus, when lymphocytes cultured in the presence of low concentrations of bleomycin (0.01-0.1 μg ml^{-1}) for 48 hours were challenged with a high concentration (1.5 μg ml^{-1}) of bleomycin or with 1.5 Gy from x rays, lower than expected frequencies of chromatid and isochromatid breaks were found. This cross-adaptation was not observed if cells were exposed to methylating agents. In fact, radiation and methyl methane sulphonate act synergistically in the same way as a combination of methylating agents (Table 12). Conditioning with interferon (50 IF ml^{-1}) has been described [M30]. These experiments lend support to the view that cross-adaptation may operate in lymphocytes to reduce the damage caused by some, but not all, DNA-damaging agents.

5. Repair of specific DNA lesions

38. A study to identify the molecular lesions associated with conditioning doses of several mutagens has recently been reported [S19]. The end-points measured in the lymphocytes from two donors included chromatid and chromosome aberrations and sister chromatid exchanges.

39. To measure chromatid aberrations, the cells were exposed to 0.05 Gy from x rays 24 hours after phytohaemagglutinin stimulation (i.e. in the G_1 phase) and challenged with 2 Gy from x rays at 48 hours (i.e. in late S/early G_2 phase). To measure chromosome aberrations, the cells were exposed to 0.05 Gy from x rays 12 hours after phytohaemagglutinin stimulation and challenged with 2 Gy from x rays at 18 hours (i.e. both exposures in the G_1 phase). Cells from one donor showed an adaptive response when challenged in the late S or early G_2 phase, in contrast to cells from the other donor, which showed the adaptive response when challenged in the G_1 phase.

40. Three drugs were used to induce sister chromatid exchanges: etoposide (VP16), 1,3-bis(2-chloroethyl)-1-nitrosourea (BCNU) and cis-diamminedichloroplatinum(II) (cis-platin). Etoposide, a topoisomerase II inhibitor, prevents the creation and resealing of DNA strand breaks, as opposed to the base modifications caused by cis-platin and inter-strand cross-links by BCNU. The repair of DNA damage at specific sites from these drugs is the result of the synthesis of enzymes involved in excision and post-replication repair or the synthesis of damage-recognition proteins (esterases) that prevent cross-linking.

41. Cells were exposed to 0.05 Gy from x rays at 40 hours after phytohaemagglutinin stimulation and exposed to drugs (0.5 μM VP16, 10 μM BCNU or 0.67 μM cis-platin) 6 hours later for 2 hours. At 48 hours, the drugs were washed out and the cultures treated for 4 hours with 30 μM bromodeoxyuridine (BrdUrd). This technique ensured that only the cohort of cells that spent sufficient time in the S phase during the BrdUrd labelling would be scored. Small but statistically significant reductions in sister chromatid exchanges, consistent with an adaptive response, were observed (Table 13). Both donors responded similarly, showing reductions most often for VP16-induced sister chromatid exchanges. Although significant reductions were also observed for chromatid deletions, analysis of the data showed that they occurred independently of those for sister chromatid exchanges. These results are consistent with the view that damage to specific sites in DNA is repairable following a conditioning dose of x rays.

42. No adaptive response was obtained when cells from 10 donors were exposed to mitomycin C, with and without a prior conditioning dose of 0.01 Gy from x rays in the G_0 phase [M5]. This may be relevant and in contradiction to the observation that a statistically significant decreased number of sister chromatid exchanges was found in the lymphocytes of workers who had been occupationally exposed to low doses of radiation and whose blood lymphocytes were presumed to be in the G_0 phase while they were chronically irradiated [T2].

6. Summary

43. An adaptive response to low-LET radiation exposure has been demonstrated in mitogen-stimulated human lymphocytes when they are acutely exposed to a conditioning dose within the range 0.005-0.2 Gy prior to a challenge dose of a few gray. The response has been expressed as a reduction in the yield in chromatid or chromosome lesions, typically to about one half the yield expected. The adaptive response has been demonstrated when both the conditioning and challenge doses are applied at late stages (S/G_2 phases) of the cell cycle. However, there is disagreement as to whether or not the adaptive response occurs if the conditioning dose is applied in the resting or early stages (G_0/G_1) of the cell cycle. This important point needs to be clarified, since it has implications for the circumstances in which cells are chronically irradiated *in vivo.*

44. The cellular response is transient, lasting for about three cell cycles in culture. Since radiation-induced double-strand breaks are repaired, this could imply the production of specific repair enzymes in addition to those involved in the process of repair of damage in cells occurring during normal metabolism.

45. There appears to be individual donor variation, with no evidence of an adaptive response in the lymphocytes of some blood samples tested even though the culture procedures are identical to those producing a response. Why this is so is not known. Several explanations have been proposed. One possibility is that the adaptive response requires the maintenance of a narrow range of pH and the presence of specific growth-stimulating factors in the culture medium. Another possibility is that the adaptive response occurs at a precise time in the cell cycle, so that cells outside this phase do not respond. A third possibility is that *in vivo* factors such as the nutritional status or the immunocompetence of the living organism may influence the cellular response.

B. EFFECTS IN MOUSE CELLS

1. Splenic lymphocytes

46. The results of different studies with mouse lymphocytes have been contradictory. In one experiment, mice of the C57Bl/6 strain received a whole-body dose of 0.05 Gy from gamma rays, at the rate of 1.25 mGy min^{-1}

on four consecutive days [W9]. Groups of mice were killed at intervals up to 26 days thereafter. Lymphocytes isolated from the spleens of sham-irradiated and irradiated mice were exposed *in vitro* to UV-radiation to induce unscheduled DNA synthesis or to mitomycin-C to induce sister chromatid exchanges. The results showed a higher rate of unscheduled DNA synthesis and lower sister chromatid exchange frequencies in the irradiated mice than in the sham-irradiated controls. Irradiation *in vivo* with low doses of gamma rays was consistent with an increase in the rate of DNA repair, which is effective for approximately 12 days. The results support those published by Tuschl et al. [T2, T3] and Liu et al. [L9], who demonstrated that it is possible to induce the adaptive response *in vivo*.

47. However, the adaptive response was not observed when lymphocytes obtained from the spleens of female mice of the Heiligenberger strain were exposed *in vitro*, either to 0.05 Gy from x rays at 24 hours or 32 hours after phytohaemagglutinin-stimulation, followed by a challenge dose of 2 Gy at 40 hours; or to 0.1 Gy at 32 hours or 42 hours with the challenge dose at 48 hours [W5, W23]. A reduction in the number of chromosome aberrations as a result of exposure to conditioning doses was seen in the lymphocytes from only 1 of 14 mice tested. Because of the high variability of the radiation-induced break frequencies in the lymphocytes of the different donors, the authors concluded that this one positive result was due to chance and was not a genuine adaptive response.

48. To determine if this lack of an adaptive response was unique to the Heiligenberger strain, spleen lymphocytes were collected from C57Bl/6 female mice in which, as discussed above, adaptive response to UV-radiation-induced, unscheduled DNA synthesis and mitomycin-C-induced sister chromatid exchanges had been observed [W5]. A dose of 0.1 Gy from x rays was given after 32 hours, followed by a challenge dose of 1.5 Gy after 48 hours of culture. Initial results indicated the presence of an adaptive response in some of the C57Bl/6 mice. However, subsequent analysis of the aberration scores of parallel lymphocyte cultures revealed a high intra-individual variability. The authors concluded that the results were a reflection of this variability rather than of any induced adaptive response [W27].

49. Experiments have been reported in which colony-forming units (CFU-S) cells were exposed to low doses of radiation in the range 0.03-0.05 Gy [S20, S32]. The adaptive response was observed from 4 hours until 28 days after each challenge dose and was more pronounced after high-dose-rate exposure. The response could be potentiated by injecting the mice with 50 μg of polynucleotide Poly I-Poly C two days before the challenge dose.

2. Bone marrow cells

50. Male Kunming mice were exposed to a whole-body conditioning dose of 0.1 Gy from x rays, followed 2.5-3 hours later by a challenge dose of 0.75 Gy from x rays [C8, L10]. The combined exposure to the conditioning and challenge doses resulted in a smaller number of chromatid aberrations in bone marrow cells than in cells from animals receiving the challenge dose only. These results are given in Table 14. The whole-body exposure of female C57Bl/6 mice to a conditioning dose of 0.002-0.5 Gy from x rays, followed by a challenge dose of 0.65 Gy within 3 hours, also resulted in an adaptive response at all conditioning doses (Table 14). A similar adaptive response was observed when the animals were exposed to these low conditioning doses and then to a high dose of mitomycin C (0.5-50 mg kg^{-1}) instead of the challenge dose from x rays [Y2].

51. In a sequel to this experiment, mice were exposed to a range of whole-body doses from ^{60}Co gamma-radiation at a rate of 0.09 Gy min^{-1} and irradiated 3 hours later with a challenge dose of 1.5 Gy from x rays [J2]. Significantly lower chromosome aberration frequencies were observed in bone marrow cells after conditioning doses of 0.05, 0.10 and 0.20 Gy, but not 0.50 Gy, compared with animals receiving the challenge dose only. The protracted whole-body exposure of male mice to ^{60}Co gamma-radiation at the rate of 0.014, 0.025, 0.06 or 0.23 Gy d^{-1}, followed by a challenge dose of 0.9 Gy from x rays within 3 hours after protracted exposure had ceased, also resulted in adaptive responses [Y2].

52. Using a different end-point, male white SHK mice were given whole-body exposures to ^{137}Cs gamma-radiation at the rate of 1.3 mGy h^{-1} over periods up to 80 days and then exposed to a challenge dose of 1 Gy from x rays within a few hours after the protracted exposure ceased [G2]. The frequency of micronuclei in polychromatic erythrocytes in chronically irradiated mice exposed to the challenge dose was about one third of that observed in mice receiving the challenge dose only. It was also shown that chronic exposure before the challenge dose resulted in a marked decrease in single-strand breaks and an increase in DNA polymerase activity in splenic and liver cells, consistent with the availability of a reservoir of repair enzymes during chronic irradiation.

53. In another experiment, 9-12-week-old male mice of the Swiss albino strain were exposed *in vivo* to ^{60}Co gamma rays [F23]. The conditioning doses of either 0.025 or 0.05 Gy were given at a dose rate of 1.67 Gy min^{-1}. The challenge dose of 1 Gy was given at 2, 7.5, 13, 18.5 or 24 hours after the conditioning doses. At a time interval of 2 hours, both conditioning doses reduced the frequency of micronuclei in polychromatic erythrocytes and of chromosome aberrations in the bone marrow cells. After exposure to 0.025 Gy, the adaptive response remained for

24 hours. After exposure to 0.05 Gy, however, the adaptive response was not present when the challenge dose was given 13 or more hours later.

54. These experiments indicate that the adaptive response can be induced in bone marrow cells in some strains of mice after acute or chronic exposure to low-LET radiation *in vivo*, provided that the challenge dose is given within a few hours after the exposure to low doses has ceased. The results contrast, however, with those of Jacobsen-Kram and Williams [J1], who were unable to elicit an adaptive response in the bone marrow cells from their strain of mice irradiated *in vivo*. However, in the latter experiments, the challenge dose was given 24 hours after the conditioning dose, a time span possibly too long for the DNA repair enzymes to remain effective in rapidly dividing bone marrow cells.

3. Spermatocytes

55. Male Kunming mice were exposed to a whole-body dose of 0.01 Gy from x rays, followed 2.5-3 hours later with a challenge dose of 0.75 Gy from x rays [C8]. The number of chromatid aberrations in the spermatocytes of conditioned mice was less than in the spermatocytes of mice receiving the challenge dose only (Table 14).

56. In another experiment involving the whole-body irradiation of male Kunming mice, the adaptive response was shown by reduced chromosome damage and dominant lethal mutations [C17]. A conditioning dose of 0.05-0.2 Gy resulted in a statistically significant reduction ($p < 0.01$) of chromatid and isochromatid breaks in spermatocytes and in reciprocal translocations in spermatogonia, compared with cells from animals receiving only the 1.5-2 Gy challenge dose.

57. Cross-adaptation has been shown using x rays and low concentrations of mitomycin C, hydrogen peroxide and cyclophosphamide as the conditioning dose [M27]. Male Kunming mice were exposed to a conditioning dose of 0.05 Gy from x rays; 3 hours later, 0.1-0.5 mg ml^{-1} of mitomycin C or 0.1-1 M of hydrogen peroxide was injected intraperitoneally or directly into the testis. Twenty-four hours later, the mice were exposed to a challenge dose of 1.5 Gy from x rays. The frequency of aberrations in primary spermatocytes was markedly reduced with the use of mitomycin C or hydrogen peroxide. In contrast, cyclophosphamide in the range 0.05-0.5 mg ml^{-1} acted synergistically with the conditioning dose of x rays.

4. Mammary carcinoma cells

58. Exposing cultured mouse mammary carcinoma (SR-1) cells to a dose of 0.01 Gy from ^{60}Co gamma rays, followed by a dose of 3 Gy from gamma rays 18-24 hours later, resulted in a decreased frequency of induction of mutations at the *hprt* locus [Z5]. When cells were exposed to bleomycin (5-10 μg ml^{-1}) for 12 hours instead of 3 Gy from gamma rays, a similar reduction in mutagenic response was observed. Since bleomycin acts to produce double-strand breaks, it was presumed that the reduction in the frequency of radiation-induced mutations was also attributable to the repair of double-strand breaks.

5. Pre-implantation embryos

59. Mouse embryos of the Heiligenberger strain were exposed to a conditioning dose of 0.05 Gy from x rays at times corresponding to the late G_2/M phase of the four-cell stage or the G_1/S phase of the eight-cell stage embryos [M6, W5]. A challenge dose of 1.5 Gy from x rays was applied 6 hours later, and cells were arrested in metaphase immediately thereafter. The interval of 6 hours between the conditioning and challenge doses was chosen because it was found to be the appropriate time for the expression of the adaptive response in both human lymphocytes and in cultured Chinese hamster fibroblast cells [I5]. The results of these experiments are summarized in Table 15. The yields of chromosomal break frequencies and the percentages of aberrant cells give no indication of an adaptive response compared with cells receiving the challenge dose only.

60. It has been reported that rat mammary gland cells irradiated *in vivo* may have a higher repair capacity than cells irradiated *in vitro*, a phenomenon called *in situ* repair [G4]. To examine the possible influence of *in situ* repair on the adaptive response in embryos, the conditioning dose was applied *in vivo* and the embryos were left *in situ* until shortly before the challenge dose [W5]. One group of embryos given a conditioning dose of 0.05 Gy was irradiated *in vivo* with 2 Gy from x rays at 50 hours after conception and isolated 4 hours later. Another group was isolated at 48 hours after conception and irradiated with 2 Gy from x rays *in vitro* at 50 hours after conception. Colchicine was added at 55 hours and the metaphases of the embryos in the 8-16-cell stage were harvested at 62 hours after conception. The results of the *in situ* repair experiments, given in Table 15, indicate that *in situ* repair was not associated with an adaptive response. They suggest that the mouse embryos are either *in situ* repair-deficient or that the optimal conditions for the induction of an adaptive response have not been achieved.

61. Wojcik et al. [W5] pointed out that for an adaptive response to occur in pre-implantation mouse embryos, they must be able to perform DNA repair. Unscheduled DNA synthesis does occur in both

pronuclei of the one-cell embryo as well as in cells of the later developmental stages. Until the late two-cell stage, however, there is no gene expression in the embryo, and all proteins required are synthesized constitutively from mRNA inherited from the oocyte [J3]. At the late two-cell stage, the embryonic genes are switched on and much of the maternally inherited mRNA is destroyed. There is, however, evidence that despite the transcriptionally active genome, some genes inducible in somatic cells do not respond inductively to the changing environment in embryos of the preblastocyst stages. It is not clear whether induced expression of repair genes in the embryo is necessarily required for an adaptive response to radiation. However, the negative outcome of the above experiment could be due to a general inability of the preblastocyst embryo to adapt to changes in its environment.

62. To investigate the points further, pre-implantation embryos were exposed to a conditioning dose of 0.03-0.1 Gy, with a challenge dose 6-24 hours later [M6]. Table 16 gives the results of an experiment in which two-cell embryos were exposed to 0.05 Gy in the early G_2 phase and to 2 Gy when the embryos were in the late G_2 phase of the same cycle. None of the end-points measured indicates a statistically significant effect as a result of the conditioning dose.

63. It is recognized that up to the early two-cell stage, the absence of an adaptive response in early embryonic development could depend on specific traits of this system. Starting with the blastocyst, however, there is no reason why genes coding repair enzymes should not respond to signals calling for additional enzyme synthesis. A similar experiment to the one described for the embryo in the two-cell stage was therefore carried out using blastocysts in which the embryonic genome was active. No statistically significant difference was seen between the effects with and without the conditioning dose [M6].

64. It may be concluded that an adaptive response cannot be induced in pre-implantation embryos, at least with regard to the end-points measured, or that the conditions are entirely different from those determined in other systems, in particular in human lymphocytes.

65. The response of fetal tissue has also been examined. Pregnant Sprague-Dawley-derived rats were exposed to 0.02 Gy from ^{137}Cs gamma-radiation at a rate of 0.4 Gy min^{-1} at various times on day 15 of gestation, prior to receiving a challenge dose of 0.5 Gy. Fetuses were examined 6 hours and 24 hours after the challenge dose for changes in the developing cerebral cortex [H20]. There was no evidence of a cellular adaptive response under these conditions of exposure, but the authors pointed out that the different conditions of exposure need to be examined before the absence of an adaptive response can be conclusively stated.

C. EFFECTS IN FIBROBLASTS FROM VARIOUS SPECIES

1. Human embryonic and skin fibroblasts

(a) Life-span and mutation frequency

66. The effect on the life-span of human embryo fibroblast cells of chronic exposure to ^{60}Co gamma-radiation delivered at a dose rate of about 0.001 Gy h^{-1} for 10 hours per day has been investigated [S25]. The average life-span, which was measured by the number of mean population doublings, was 1.2-1.6 times longer in irradiated than in unirradiated cells. The number of chromosomes in the unirradiated cells remained constant throughout their life-span. Conversely, the irradiated cells showed numerical abnormalities with increasing time. These results indicated that the life-span of chronically irradiated cells at low dose rates was prolonged, but that the cells showed chromosomal changes consistent with abnormal phenotypes.

67. In another study [W10, W11] to determine the effect on the growth ability of human embryo fibroblast cells *in vitro*, the expression of abnormal phenotypes was measured after fractionated low-dose gamma-radiation. Cells were assayed for cell survival by their colony-forming ability, for mutation at the *hprt* locus and for transformation by foci formation. After a dose of about 2 Gy had been accumulated, the mean population doubling time was 1.3-1.6 times that of the controls. Although transformed foci were not observed until the cells had accumulated about 1 Gy, the numbers of cells with abnormal phenotypes increased thereafter with increasing dose. No cells, however, showed unlimited life-span *in vitro*.

68. The mechanism responsible for the increased growth potential of embryonic fibroblasts after fractionated low-dose gamma-radiation *in vitro* remains obscure. Although the data suggest that some damage is repaired during these exposures, it cannot be assumed that all of the damage resulting in the transformation of cells is repaired. The prolonged life-span may allow additional time for the expression of an otherwise unexpressed lesion, perhaps associated with the development of additional karyotypic changes and aneuploidy. This process might be very rare among human diploid fibroblasts grown *in vitro* [K6, K7] and might be specifically related to the immortalization of human cells. The mechanism leading to the prolongation of their life-span remains to be shown, but calcium ion may act as a signal transducer during cell cycling [I10].

(b) Cell survival and clone-forming ability

69. The human skin fibroblast cell line (AG1522) has been used to determine the effects of low-dose gamma-radiation, followed by a challenge dose of x rays [A2]. Cell survival, colony growth rate and micronuclei formation were measured to assess evidence of adaptive response. Protracted exposure of plateau-phase cells to gamma rays, delivered at a dose rate of 0.003 Gy min^{-1} over a period of 24 hours, reduced the effects of a challenge dose of 4.25 Gy from x rays given immediately after the chronic exposure. Figure II shows that up to a twofold improvement in cell survival and a twofold reduction in micronuclei formation were observed, compared with the results obtained when cells were exposed to the challenge dose only. Furthermore, after 7 days, the size of colonies from cells surviving the combined exposures was about four times as great as the size of colonies from cells given a challenge dose only.

70. The stimulation of clonogenicity at doses below 0.4 Gy from gamma rays has been observed in cultures of human skin fibroblasts (strain GM 2185). The results, illustrated in Figure III, are compatible with the hypothesis that cells that do not form clones in the absence of radiation are stimulated to do so by low doses of radiation; that is, additional colony-forming cells are recruited from formerly non-clonogenic cells. This enhanced clonogenicity was not observed in fibroblasts (A-T strain GM 2531) in which DNA repair was deficient, although the numbers of non-clonogenic cells are similar to those observed in the normal fibroblast strain. It can be implied from these results that enhanced clonogenicity is dependent on DNA repair competence, but other mechanisms could be involved [G5].

2. Chinese hamster cells

(a) Micronuclei formation frequency

71. Proliferating Chinese hamster cells, cloned from the V79-B310H cell line, were exposed to beta- or gamma-radiation from tritiated thymidine or tritiated water, followed by exposure to 1 Gy from ^{60}Co gamma rays [I5, I7, I11]. An adaptive response, expressed as a reduction in micronuclei frequency, was observed. The adaptive response was inhibited by 3-aminobenzamide (3AB) and was not observed after one cell division following the conditioning dose. The optimal range of the conditioning dose was estimated to be between 0.001 and 0.1 Gy on the basis of the amount of tritium incorporated into DNA. When tritiated thymidine was administered at lower or higher concentrations, a reduction in micronuclei induction was not observed.

72. Acute exposure to 0.01 or 0.05 Gy from gamma-radiation also induced an adaptive response to a challenge dose of 1 Gy, but exposure to high-LET radiation did not. This dependency on the type of radiation might reflect the quality and quantity of chromosomal lesions that trigger the adaptive response. The adaptive response did not fully develop until 4 hours after the challenge dose and was not observed if the time interval between the conditioning dose and the challenge dose was extended to 6 hours. The adaptive response can be attributed to the induction of a mechanism that repairs DNA damage.

(b) Cross-adaptation

73. Chinese hamster V79 cells exposed to conditioning doses (0.01-0.05 Gy) from gamma-radiation showed cross-adaptation to challenge doses of UV-B-radiation (97.5-195 J m^{-2}) and mitomycin C (25-50 $\mu g\ ml^{-1}$) but not to ethyl methane sulphonate (EMS) (100 $\mu g\ ml^{-1}$) or cis-platin (1 $\mu g\ ml^{-1}$), as evidenced by a reduction in the number of sister chromatid exchanges. This could imply that the adaptive response observed after radiation could be coupled to the repair network that copes with chromatin lesions induced by mitomycin C and UV-B [I7]. The results observed after exposure to cis-platin were contrary to those observed when human lymphocytes were exposed to this agent.

74. The effects of small amounts of hydrogen peroxide on the killing and mutation of Chinese hamster V79 cells by different agents is supportive of an adaptive response from damage due to oxidative free radicals [G11, S22]. It has been shown that low, non-toxic concentrations (e.g. 0.9 $\mu g\ ml^{-1}$) of hydrogen peroxide render V79 cells more resistant to subsequent killing by hydrogen peroxide (3-15 $\mu g\ ml^{-1}$), gamma rays (1-6 Gy) and N-methyl-N′-nitro-N-nitrosoguanidine (0.5-2.0 $\mu g\ ml^{-1}$). However, such pretreatment with hydrogen peroxide increased the mutation yield by N-methyl-N′-nitro-N-nitrosoguanidine or gamma rays, suggesting error proneness of the induced repair activity. Cycloheximide or benzamide prevented the induction of repair, and they also suppressed the increase in mutation yield.

75. The treatment of Chinese hamster V79 cells or H_4 rat hepatoma cells with low concentrations of hydrogen peroxide (1-5 μM) also resulted in an adaptive response, expressed as increased survival when the cells were exposed to high doses of hydrogen peroxide (0.1-1.5 mM) or to a challenge dose from gamma-radiation (up to 8 Gy) [L20]. This adaptive response was observed in both exponentially growing cells and plateau-phase cells, but there was a reduced *hprt* mutation frequency.

3. Mouse embryo cells

76. C3H10T½ plateau-phase mouse embryo cells were conditioned with doses of 0.1, 0.65 or 1.5 Gy from ^{60}Co gamma rays at a rate of 0.0025 Gy min^{-1}. Three and a

half hours later they were exposed to a dose of 4 Gy from gamma rays. The conditioning dose did not affect clonogenic survival, but it led to a reduction in micronucleus frequency in binucleate cells and to a twofold reduction in transformation frequency per viable cell when cells were subsequently exposed to 4 Gy from gamma rays. The data suggest that a conditioning dose of low-LET radiation induces an adaptive response in C3H10T½ cells, resulting in enhanced DNA double-strand break repair when the cells are exposed to the challenge dose. This enhanced repair appears to be error-free, since the cells are less susceptible to radiation-induced neoplastic transformation [A17].

4. Derived human epithelial cell line

77. Adaptive response has been described in experiments involving Hela cells [C21]. The cells were exposed to a conditioning dose of 0.03 Gy, followed by a challenge dose of 2 or 3 Gy. A decrease in the number of induced micronuclei occurred within 4 hours of the conditioning dose and lasted for three cell cycles. If the conditioning dose was increased to 0.4 Gy, the adaptive response disappeared and the cells subsequently showed increased radiosensitivity.

D. ADAPTIVE RESPONSE TO CHEMICAL MUTAGENS

78. An adaptive response in human keratinocytes exposed to low doses of the mutagen N-methyl-N′-nitro-N-nitrosoguanidine has been described [K8]. Growing and confluent human keratinocytes (Ha CaT cell line) were exposed to different concentrations of N-methyl-N′-nitro-N-nitrosoguanidine for one hour, and the number of single-strand DNA breaks was determined by measuring nucleoid sedimentation through neutral sucrose gradients. Strand breaks cause the supercoiled DNA structure of the nucleoids to relax, leading to a reduction in the sedimentation rate. When the growing cells were treated with low doses of N-methyl-N′-nitro-N-nitrosoguanidine, the nucleoids were found to sediment faster than in cells in the confluent phase. Similar shifts have been reported following mitogen activation of human lymphocytes [J4] and mouse splenic lymphocytes [G6]. This effect was attributed to the rejoining of DNA single-strand breaks present in confluent cells.

79. The ADP-ribosylation system of chromatin responds to radiation-induced damage by processing ADP-ribose residues through a complex series of synthetic and catabolic reactions. The key component of this multi-enzyme system is poly(ADP-ribose) polymerase, a zinc-containing protein that specifically binds to single- and double-stranded DNA breaks. Binding activates different catalytic reactions that lead to the synthesis of polymers covalently bound to the polymerase [N3]. These polymers then remove histones from the DNA, thereby allowing access to other proteins, e.g. DNA helicase A and topoisomerase I, to encourage DNA excision repair. Inhibitors of poly(ADP-ribose) polymerase suppress the adaptive response in mammalian cells.

80. Further evaluation [B9], however, has shown that differences in the nucleoid sedimentation rate might also be explained by changes in the amount of RNA and proteins, which affect the sedimentation velocity of the nucleoids. To test this hypothesis, keratinocytes were exposed to 0.005 μM of N-methyl-N′-nitro-N-nitrosoguanidine for 1 hour, followed by a challenge dose of 5 μM of N-methyl-N′-nitro-N-nitrosoguanidine 6 hours later. The results can be interpreted as reflecting fewer DNA breaks in the pretreated cells than in cells exposed to the challenge dose only. The presence of 2 mM of 3-aminobenzamide blocked this response. Repair in the presence of low doses of N-methyl-N′-nitro-N-nitrosoguanidine is consistent with an adaptive response of the cells to the mutagen. However, a synergistic rather than an adaptive response was observed in human lymphocytes pretreated with N-methyl-N′-nitro-N-nitrosoguanidine, followed by a challenge treatment with methyl methane sulphonate [W4].

81. Studies with hydrogen peroxide (0.1 μM conditioning dose, 100 μM challenge dose) and bleomycin (0.1 ng ml^{-1} conditioning dose, 100 ng ml^{-1} challenge dose) are also consistent with the view that exposure to a low dose of these mutagens results either in an overall decrease in the number of single-strand breaks or changes in the nucleoid cage of the DNA. They provide complementary evidence of an adaptive DNA repair process.

F. SUMMARY

82. The adaptive response has been demonstrated in proliferating cultured lymphocytes and fibroblasts. In addition to a reduction in chromosome aberrations, the response has been measured as a reduction in the expected number of sister chromatid exchanges, of induced micronuclei and of specific locus mutations. An increased survival rate and an increased proliferative capacity have been shown to be associated with increased mutation and transformation frequency in some experiments.

83. Bone marrow cells and spermatocytes from mice exposed *in vivo* to low doses (0.01-0.2 Gy) from x rays a few hours before challenge doses (0.75-2 Gy) to the cells showed reductions in the number of chromosome aberrations compared to cells exposed to the challenge dose alone. No adaptive response was observed in pre-implantation mouse embryo cells, even though these embryonic cells were tested at a stage of development in which they were considered to be capable of synthesizing their own DNA repair enzymes.

84. There is evidence of cross-adaptation between some toxic chemical agents and low-LET radiation. While it is reasonable to assume that some common repair pathways exist depending on the category of damage (for example, damage caused directly by the ionizing events or indirectly by induced hydroxyl radicals), the relation between random radiation-induced DNA damage and specific chemically induced DNA damage needs to be further resolved.

II. MECHANISMS OF ADAPTIVE RESPONSE

85. Studies of cultures of lymphocytes, bone marrow cells, melanoma cells and fibroblasts have provided insight into some aspects of the mechanisms involved in the adaptive response. These include:

(a) the effects of radiation on the up-regulation of genes and their influence on cell cycle kinetics;
(b) the identification of activated genes and their enzyme products specifically involved in radiation-induced DNA repair;
(c) the relationship between radiation-induced repair genes and those activated by other mutagens;
(d) the ability of cells to remove toxic radicals;
(e) the activation of membrane receptors and the release of growth factors;
(f) the effects of radiation on the proliferative response to mitogens.

Other mechanisms may be involved, such as enhanced immunosurveillance, which is discussed in Chapter III.

A. CELL CYCLE CONTROL

86. Research into the mechanisms involved in cell cycling is advancing rapidly [M23, N1, N2, N5, S28]. The division of a cell into a pair of genetically identical progeny depends on the precise timing of a sequence of events. To divide successfully, the cell must have completed DNA replication and repaired any DNA damage to the extent that allows the formation of chromosomes and their correct segregation.

87. Control of cell cycling is influenced by feedback mechanisms that can detect failure to complete the above processes and arrest progress at various stages in the cycle (e.g. progression from the G_1 to the S phase and from the G_2 to the M phase). Much of the basic knowledge of the mechanisms has been derived from studies with yeast cells and sea urchin eggs. Elucidating the mechanisms in mammalian cells is proving more complex than doing so in primitive cells. It is clear that the understanding is as yet incomplete, but there is sufficient information to allow speculating on the principles involved.

1. Protein synthesis

88. The key components in mammalian cell cycle control are two classes of protein, the kinases and the cyclins, which are synthesized in a well-conserved sequence. Cell-division-cycle (cdc) kinases are believed to act at several check-points, switching cell cycle progression on and off principally by interacting with cyclins. Activation of the kinase-cyclin complexes requires dephosphorylation of tyrosine phosphate, and possibly of threonine phosphate, on the kinase molecule.

89. Passage from the G_0 to the G_1 phase is thought to be triggered by low cell population density, cell size, the presence of mitogens and the activation of proto-oncogenes. Progression in the G_1 phase seems to be regulated by kinases similar to, but not identical with, cdc2 kinase encoded by the p34 ^{cdc2} gene [F20]. Cyclin D1 accumulates during the G_1 phase and associates with many cellular proteins, including cdk2, cdk4 and cdk5 kinases [M24, X1].

90. The G_1 to S phase transition appears to involve the complexing of cyclin E with cdk2 kinase and of cyclin D with cdk4 kinase. The cdc2-cyclin E complex may be particularly important for the transition from the G_1 to the S phase in human cells [K20].

91. A gene encoding a cyclin-like protein has recently been isolated from rat fibroblasts [T12]. It is referred to as cyclin G. Cyclin G mRNA is induced within 3 hours of growth stimulation, that is, during the transition from the G_1 to the S phase, and its level remains elevated with no apparent cell cycle dependency, indicating its close association with growth stimuli but not with the cell cycle. The kinase-cyclin G complex remains inactive until dephosphorylation of the kinase occurs.

92. A cdc2 kinase-cyclin A complex regulates S phase progression [G9, P11]. Essential accessory growth factors during the S phase include platelet-derived growth factor (PDGF) and insulin-like growth factor (IGF-1).

93. On passing through the G_2 phase, proteins that regulate the spindle assembly, chromosome condensation and nuclear envelope breakdown are synthesized. If the spindle assembly is not completed, then the cell is arrested in the M phase. A cdc2 kinase-cyclin B complex controls the transition from the G_2 to the M phase [G9, P11].

94. Cyclins A and B are rapidly degraded at the end of mitosis. The process induces the synthesis of enzymes that conjugate ubiquitin to the cyclins and thereby targets them for degradation by proteolytic enzymes. Degrading the cyclins negates the activity of the kinase-cyclin complexes, and the cells proceed to interphase.

95. The effect of radiation on the levels of cyclin present at different stages of the cell cycle has been studied. In normal cell cycling of unirradiated cells, the levels of cyclin B protein increase rapidly in the G_2 and M phases and decrease at the end of mitosis. If the cells are irradiated in the G_2 phase, cyclin B mRNA is readily detectable, although at slightly lower levels than in the unirradiated controls. However, cyclin B protein is markedly decreased in amount, as can be seen in Figure IV, and this is associated with a delay in the completion of the G_2 phase, which is associated with diminished levels of activated kinase-cyclin complex. This finding has been confirmed in experiments with Chinese hamster cells [L13].

2. Tumour suppressor genes

96. As the role of tumour suppressor genes was discussed in the UNSCEAR 1993 Report [U1] Annex E, only the points relevant to their role in cell cycling are referred to in this Annex. Current evidence indicates that the *Rb* tumour suppressor gene protein plays multiple roles in the control of the cell cycle, not only in regulating the response to early mitogenic signals to the cell but also in mediating the transitional phases of the cycle itself. The fundamental mechanism by which this is achieved is the repression of cell growth and division by the *Rb* binding of regulatory nuclear proteins, such as *E2F* and *Myc*, which drive proliferative responses. Mutational loss or inactivation of the *Rb* gene in an appropriate target cell may therefore be viewed as a principal means of relaxing these controls.

97. The phosphoprotein product of the *p53* tumour suppressor gene is also suspected of playing a role in cell cycle regulation [L22]. It is thought to function as an inhibitor of cell replication by delaying entry into the S phase of the cell cycle through influencing the assembly of the late G_1 protein complexes that initiate DNA replication. The inhibition of DNA synthesis is therefore an active physiological process, and loss of the *p53* gene results in loss of this control.

98. Another possible mechanism of action is that p53 protein, by virtue of its DNA-binding properties, may act as a transcriptional factor influencing critical gene expression controlling a cyclin-dependent protein kinase inhibitor (CKI). This is a p21 protein that can bind to and inhibit a wide variety of cyclin-dependent kinases [N5]. A simple hypothesis for cell cycle progression has been proposed: cyclin-dependent protein kinases build up at the G_1 phase and the G_2/M transition owing to the presence of cyclin-dependent protein kinase inhibitors. Surplus cyclins then trigger the inactivation of cyclin-dependent protein kinase inhibitors, and the cell proceeds through the cycle. Inducible DNA damage may also cause the build-up of cyclin-dependent protein kinase inhibitors, which may be reversible or irreversible.

99. Proto-oncogenes, originally isolated as functional genes supporting the proliferation of tumour cells, encode proteins that are involved in normal cellular proliferation. Some of these proto-oncogenes are therefore concerned with cell cycling. They are involved in signal transduction from the cell surface to the nucleus, thereby integrating growth signals so as to increase the biosynthesis of DNA. These oncoproteins include growth factors (e.g. *c-sis* encoding platelet growth factor), membrane binding receptors (e.g. *c-fms* encoding macrophage-colony stimulating factor receptor), signal mediators by subsequent phosphoryl reaction (e.g. *c-raf* encoding protein, which can be phosphorylated during signal transduction), transcriptional activators (e.g. *c-jun* and *c-fos* encoding AP-1 transcriptional activator protein) and replication-related proteins (e.g. *c-myc* encoding nuclear protein). Some oncoproteins possess DNA-binding activity after phosphorylation.

100. As examples, the transition from the G_0 to the G_1 phase has been associated with the increased expression of *c-fos*, *c-jun* and *c-myc* and *EGR-1* proto-oncogenes that become activated within minutes of a growth stimulus. All of the products are directly bound with DNA to activate transcription of the many genes necessary for entry into the growth cycle. The *c-fos* gene transiently expresses prior to differentiation in a wide variety of premature blood cells. After differentiation, several types of cell further express different oncogenes, such as *src* [G7], *c-sis* [P5] and *c-fms* [S29], depending on the type of cell. Progression from the G_1 to the S phase is thought to involve the up-regulation of the *ras* family genes.

B. GENE ACTIVATION

101. The disruption of DNA structure is a consequence of exposure to many physical, chemical and biological toxins. To a lesser degree, as described earlier, it is also a consequence of the changes that can take place during normal metabolism. It is not surprising, therefore, that cells have evolved a complex system of defence against circumstances that might irreversibly damage them. A major role is played by the activation of genes and gene products that initiate DNA repair processes.

102. Evidence for the activation of genes associated with growth control and DNA repair came initially from studies in prokaryotes. These studies provided an insight into the mechanisms operating in eukaryotic cells. For example, genotoxic stress in the bacterium *Escherichia coli* induces responses in which regulator genes (regulons) participate. These include the *lexA*/*recA*-mediated SOS response [L3, M1, P9, R1, S4, W2, W24], the adaptive response to alkylating agents [B1, K1, K2, K3, M2, O1, S4, S5, S6], the *oxyR*-mediated hydrogen peroxide response, the *soxRS*-mediated superoxide response, and the activation of heat shock protein (HSP) genes [D1, D5, F18, G1, S7, W8, W24, Z4].

103. The response of mammalian cells to mutagens, including radiation, is complex, but it is known that many of the genes involved in normal cell cycling are activated. These include genes responsible for growth stimulation, growth control and differentiation [R9]. Genes associated with growth control and activated by radiation exposure were discussed in a recent review article [F19]. Similar types of genes are activated by some alkylating agents and hydrogen peroxide.

1. Cell growth arrest

104. An immediate reaction in proliferating cells exposed to a mutagen is delayed progression through cell division, and a number of genes have been identified that inhibit cell cycle kinetics, among them the growth arrest and DNA-damage-inducible (DDI) genes [L3]. Delayed progression through cell division is accompanied by an increase in the rate of transcription of genes that encode for the production of enzymes to repair the DNA damage caused by the mutagen [W1]. The different types of repair enzymes produced in response to different types of genotoxic stress are probably interrelated in the sense that they are the products of similar regulons [S3].

105. Cell cycle delay is a primary response to DNA damage that represents active processes mediated by certain genes, such as those involved in the expression of the cyclins, *p53* tumour suppressor gene, *ras* oncogenes, and the *gadd* (growth arrest and DNA-damage), *gas* (growth-arrest-specific), *spr* (small proline-rich), *MyD* (myeloid differentiation) and *c/EBP* growth arrest genes [F19]. Genotoxic stress has the puzzling effect of conditioning genes associated with both growth stimulatory and inhibitory responses, but the main effect is inhibitory. Many transcription factors and genes activated soon after exposure to radiation are associated with both responses. Many of the genes involved in signal transduction have been implicated in both the initiation and progression stages of carcinogenesis, and the same genes are often induced by tumour promoters and DNA-damaging agents [C15, D5].

106. A summary of the DNA-damage-induced (DDI) genes found by various investigators to be induced within a few hours of exposure is given in Table 17. The complex multi-gene reaction after irradiation makes it difficult to characterize the molecular mechanisms of any particular group of genes. However, most of the DDI genes listed in Table 17 are probably involved immediately after DNA damage.

107. Experiments with cultured normal bone marrow progenitor cells and with myeloblastic (ML-1) leukaemic cells have shown that the levels of *p53* tumour suppressor gene protein transiently increase while the rate of DNA synthesis decreases after DNA damage, apparently occurring via a post-transcriptional mechanism [K21]. Since cells with wild-type (wt) *p53* genes exhibited transient arrest in both the G_1 and the G_2 phases after gamma-irradiation, while cells with absent or mutant *p53* genes arrested only in the G_2 phase, it was concluded that wt *p53* genes played a role in G_1 phase arrest.

108. This observation was further supported by experiments showing that the transfection of wt *p53* genes into malignant cells lacking endogenous *p53* genes partially restored the G_1 phase arrest after gamma-irradiation and that overexpression of a transfected mutant *p53* gene in tumour cells with wild-type endogenous *p53* genes abrogated the G_1 phase arrest after irradiation [K22].

109. Because the tumour cell lines used for the transfection experiments had multiple genetic abnormalities, the experiments were repeated in irradiated normal murine embryonic fibroblasts in which the *p53* genes had been disrupted by homologous recombination [K23]. Under these conditions, the loss of both *p53* alleles in otherwise normal fibroblasts led to the loss of G_1 phase arrest.

110. Further studies showed that irradiated cells from patients suffering from ataxia-telangiectasia were unable to induce the *gadd45* gene. Finally it was shown that wild-type but not mutant *p53* gene products bind strongly to a conserved element in the *gadd45* gene. It was concluded that in normal mammalian cells, *p53* and *gadd45* genes participate in a signal transduction pathway that controls cell cycle arrest in the G_1 phase following DNA damage and that this check-point pathway is defective in ataxia-telangiectasia patients.

111. The effect of radiation on the expression of two DNA-damage-induced genes, designated *gadd45* and *gadd153*, has been examined in cultured human lymphoblasts [P6]. These genes had previously been shown to be strongly induced by UV-radiation and alkylating agents in human and hamster cells. It was found that the *gadd45* gene, but not the *gadd153* gene, was also strongly induced by x rays. The level of *gadd45* mRNA increased rapidly after x-ray exposure at doses as low as 2 Gy (Figure V). After 20 Gy, *gadd45* induction, as measured by increased amounts of mRNA, was similar to that produced by the most effective dose of the alkylating agent methyl methane sulphonate. No induction was seen after treatment with 12-0-tetra-decanoylphorbol-13-acetate, a known activator of protein kinase C. Therefore, *gadd45* represents an x-ray-responsive gene whose induction is not mediated by protein kinase C. However, induction was blocked by the protein kinase inhibitor H7, so that induction is likely to be mediated by some other kinase.

2. Radiation-induced gene expression

112. The effects on gene expression of low doses of low- and high-LET radiations have been studied in cultures of

Syrian hamster embryo (SHE) cells [W16]. These fibroblasts are normal diploid cells that can be transformed into neoplastic cells by radiation. Genes coding virus-like 30 S elements, *c-fos* and β-protein kinase C have been shown to be activated by exposure to x rays (0.75 Gy) or gamma rays (0.9 Gy) but not by exposure to fission neutrons. Further studies [W17] have revealed that the induction of *c-fos* mRNA occurred within 3 hours of exposure, and a protein-binding site has been identified that mediates transcriptional response of the *c-fos* gene to serum factors [T5].

113. Of particular interest was the response to radiation of members of the protein kinase C (PKC) gene family, which has been shown to play an important role in tumour promotion and in the regulation of cell growth. The results of these experiments showed that exposure to gamma rays can induce increased expression of PKC mRNA within 1 hour of radiation exposure (Table 18). However, PKC inhibitors prevented the expression of PKC in Chinese hamster V79 cells [I11]. Dose effects were evident, with increased accumulation of PKC mRNA at higher doses. Levels of expression of PKC mRNA were increased sixfold over unirradiated controls after exposure to 0.75 Gy from x rays (Figure VI).

114. The induction of PKC mRNA occurred at a time when total cellular transcription was reduced following irradiation. The activation of PKC directly stimulates transcription of proto-oncogenes *c-fos* and *c-jun*, which are typical early immediate genes. Cellular levels of *c-fos* and *c-jun* mRNA also increase transiently after irradiation [I9, S48], and it has been reported that these oncogenes are induced via the activation of PKC after irradiation [H12]. These results suggest that supplementation of PKC after activation of the PKC gene is necessary for prolonged expression of *fos/jun* genes, since depletion of PKC and down-regulation of *fos/jun* mRNA occur after their activation. On the other hand, several cytokine genes such as interleukin (IL-Iβ) and tumour necrosis factor (TNF-α) are found to be continuously expressed by irradiation (Table 17). IL-Iβ is known as a radioprotector because the survival rate after a sublethal dose of radiation is increased by the administration of IL-Iβ [N6]. Since there is a potential AP-1 (transcriptional activated protein, complex of *fos* and *jun* protein) binding site at the 5′-upstream region of these genes, it is considered that the genes are continuously stimulated by regulators containing the products of the early immediate genes as a later response against radiation damage at a whole-body level [W25]. However, transient expression of the IL-Iβ gene is also observed together with transient expression of *fos/jun* early immediate genes within 1 hour of irradiation [I9]. This suggests the activation of early protective mechanisms as a response to whole-body irradiation.

115. The effects of neutrons and gamma rays on the expression of genes encoding the nucleus-associated H4-histone, c-jun, c-myc, Rb and p53 proteins have also been reported [W18]. Syrian hamster embryo cells were irradiated at various doses and dose rates. After incubation of the cell cultures for 1 hour following radiation exposure, the induction of transcripts for *c-jun* and H4-histone was shown to occur following gamma-ray exposure (Table 19) but not following neutron exposure. The expression of p53 protein was unaffected by either gamma-ray or neutron exposure. The increase in the relative amounts of *Rb* mRNA was marginal, and the expression of *c-myc* mRNA was repressed following exposure to gamma rays and was unaffected following exposure to neutrons.

116. These experiments provide support for the hypothesis that radiation induces different cellular responses to radiation-induced damage, be it DNA damage, oxidative damage, protein denaturation or some other intracellular event. Recent experiments implicating oxidative damage as the inducing agent for *c-fos*, *c-jun*, *c-myc* and other genes induced following DNA damage would suggest that gamma-ray induction of these genes may involve oxidative damage as the modulating agent.

117. The activation of oncogenes, including *c-raf*, *c-myc*, *v-src*, *Ki-ras*, *c-H-ras*, *v-H-ras*, *N-ras*, *v-k-ras* and *v-fms* genes, has been correlated with increased clonogenic survival over the dose range 1-6 Gy [M25, S45]. The adaptive response appeared to be a specific consequence of the *ras* gene mutation rather than transformation, since revertant cells that contained functional *ras* genes retained their clonogenic survival properties.

118. The effect of oncogene expression on the sensitivity to gamma-radiation of the haematopoietic (32DCl3) cell line has been measured [F21]. It was shown that these haematopoietic progenitor cells transfected with the oncogenes *v-erb-B*, *v-abl* or *v-src* also showed increased clonogenic survival when exposed at dose rates of 0.05 Gy min^{-1} over the dose range 1-10 Gy. Exposure of NIH3T3 fibroblast cells transfected with the oncogenes *v-abl*, *v-fms* or *H-ras* also showed increased clonogenic survival.

119. In a more recent experiment, rat embryo cells from the Fischer strain of rats and derived transfectants containing the *Ha-ras* oncogenes were irradiated with ^{60}Co gamma rays at dose rates between 0.018 and 0.72 Gy min^{-1} [O6]. The oncogene-containing cells exhibited higher survival levels at all doses compared with the non-transfected cells.

120. In contrast, the measurement of colony-forming ability following exposure to gamma-radiation was made on transformed human embryo retinoblast cell lines containing mutant *ras* genes [G10]. No correlation was found between transformation with activated *ras*, adenovirus or SV40 genes and increased radiation-

resistance. Nor was there any correlation between clonogenic survival and the level of expression of *ras*21, but two of the three *ras* transformants that were least sensitive to gamma radiation were from cell lines expressing the highest levels of *ras*21 polypeptide, which plays a pivotal role in signal transduction and possibly DNA repair. Notwithstanding the negative findings in some laboratories, it is generally accepted that radiation-induced oncogene activation can induce increased cell survival in some circumstances [W25].

121. A human *XRCC1* (x-ray repair cross complementing) gene has been isolated that affects the sensitivity of cells to radiation [C16, T13]. The Chinese hamster ovary cell mutant, EM9, exhibits extraordinarily high sister chromatid exchanges and is unable to effectively repair DNA breaks caused by radiation and certain alkylating agents. Introduction of the human *XRCC1* gene corrects the EM9 DNA-repair defect and is the first human gene to be cloned that has an established role in DNA strand-break repair. The gene is 33 kb in length and encodes a 2.2 kb transcript and a corresponding putative protein containing 633 amino acids. Constructs in which the open reading frame of the *XRCC1* gene were transcribed from the SV40 promoter, or the genomic promoter native to *XRCC1* gene, were compared with regard to their ability to correct the sister chromatid exchange defect in the EM9 mutant. These transfectants displayed significantly fewer sister chromatid exchanges than other transfectants. The results suggest that overexpression of the minigene from the SV40 promoter may increase the repair capacity of EM9 mutant cells relative to that of wild-type cells.

3. Induced protein products and DNA repair

(a) Human melanoma cells

122. Induced gene products synthesized in response to low doses of radiation in human melanoma (U1-Mel strain) cells and in a variety of other human normal and cancer-prone cells have been identified using two-dimensional gel electrophoresis [B12]. U1-Mel cells were chosen since they have a high capacity for potentially lethal damage repair. Eight proteins were induced by radiation, and two proteins were repressed. They were not found after heat shock treatment or exposure to UV-radiation or certain alkylating agents. The expression of one protein termed XIP269 (to indicate an x-ray-induced protein of approximately 269 kDa) at a dose of 0.05 Gy correlated very well with potentially lethal damage repair capacity. This protein was found to be down-regulated by exposure to caffeine or cycloheximide under conditions in which both potentially lethal damage repair and subsequent adaptive responses, expressed as cell survival, were prevented [H3].

123. In addition to characterizing x-ray induced proteins, the levels of x-ray-inducible genes have been measured in cDNA clones isolated by differential hybridization. Some of these genes were increased to over 20 times the background levels by as little as 0.05-0.2 Gy; four have been identified as T-diphorase, tissue-type plasminogen activator [B13, B14, H3], thymidine kinase and the proto-oncogene c-*fps*/*fes*.

124. The first phase of potentially lethal damage repair occurs very quickly (2-20 minutes), presumably to increase the chances of survival of irradiated cells. It is associated with a rapid resealing of single- and, at a later stage, double-stranded DNA lesions that are either created initially by x rays or produced as a result of the repair of various types of base damage. The second slower phase of potentially lethal damage repair proceeds over a period of a few hours following irradiation, during which attempts are made to repair the remaining double-stranded DNA breaks. This second phase of repair closely corresponds to the restructuring of gross chromosomal damage and can be partially blocked in some human cells by inhibiting protein synthesis [Y3]. The rapid repair of potentially lethal damage may be due to the immediate availability of constitutively synthesized repair enzymes such as DNA ligases, topoisomerases or polymerases [B15]. In contrast, it is proposed that the slow phase of potentially lethal damage repair requires the induction on demand of specific genes and gene products. These slow-phase, potentially lethal damage repair responses may be further enhanced if the genes are stimulated with low doses of radiation before a high challenge dose is given.

125. In a further experiment, confluence-arrested human normal (GM2936B and GM2907A) and neoplastic (U1-Mel, Hep-2 and HTB-152) cells were tested for evidence of adaptive survival recovery responses [B26, M9]. Cells were exposed to 0.05 Gy each day for four days at a rate of 1.13 Gy min^{-1} and then challenged with a dose of x rays giving a 20% cell survival. Only Hep-2 and U1-Mel cells pretreated with 0.05 Gy showed an improvement in survival after 4.5 Gy, compared with untreated cells. Two genes, *XIP5* (human growth hormone-related) and a gene transcript related to *XIP12* (human angiogenin-related), showed increased expression over time in these cells. Levels of cyclin A and, to a lesser extent, cyclin B increased in the pretreated cells only after the high challenge dose and were not expressed during exposure to the conditioning dose or in cells receiving only the challenge dose. A slight increase in glutathione S transferase mRNA was noted after the primary dose, but *p53* suppressor gene and 10 *XIP* genes were not activated.

126. Under the conditions of this experiment, U1-Mel cells did not progress into the S phase as measured by the uptake of tritiated thymidine into DNA. The induction of cyclin A under these conditions may thus indicate an involvement of cyclin A in specifically stimulating DNA

repair. Based upon these preliminary data, a model has been proposed in which an adaptive response to low doses of radiation may be achieved in mammalian cells [M9]. Initially cells are assumed to be in the G_0 phase or at some point in the G_1 phase. Upon repeated exposure to low doses of radiation, for example 0.05 Gy, cells progress to and pause at or near the G_1 phase or the beginning of the S phase of the cell cycle. Gene transcripts that build up slowly in response to conditioning doses then produce proteins (e.g. cyclin A) that regulate or control the transcripts that appear following a challenge dose. At that point, cells are poised to stimulate various DNA repair systems that are not inducible in the initial resting state.

127. In another experiment, human melanoma (G361 strain) cells were exposed to gamma radiation [O4]. Eleven induced proteins were extracted from cells with molecular weights between 43 and 98 kDa, while in P39 strain cells, 21 induced proteins were extracted after exposing the cells to 3 Gy. Their molecular weights ranged from 32 to 98 kDa, and four of these, with molecular weights of 57, 58, 77 and 88 kDa, were considered to be specific DNA repair enzymes. These proteins are of a lower molecular weight than those isolated and characterized by Boothman et al. [B12].

(b) Human lymphocytes

128. Using two-dimensional gel electrophoresis, a specific group of proteins was detected from human lymphocytes exposed to a conditioning dose of 0.01 Gy from x rays [W6, W8]. Cellular extracts from unirradiated lymphocytes and from other cell types were separated by electrophoresis and then exposed to a mixture of ^{32}P-labelled nick-translated and non-radioactive plasmid pCH110 on nitrocellulose membranes. Several bands that bind to the nick-translated DNA were detected, the protein binding occurring as early as 1 hour after irradiation and reaching its maximum by 6 hours. This binding was diminished by a prior proteinase K treatment of the extracts, indicating that the bands are related to proteins present in the extracts. The cellular extracts contained three proteins (of molecular weights 105 kDa, 35 kDa and 14-18 kDa) that reproducibly bound to the labelled DNA. The binding of the DNA probe to the 30-35 kDa and 14-18 kDa bands was twice as great as that found in unirradiated cells, and the 35 kDa band regularly separated into two bands.

129. The 30-35 kDa band proteins have been substantially purified by affinity chromatography. If enough of the proteins can be obtained, it should be possible to see if their introduction into cells will lead to a reduction in the yield of chromosomal aberrations induced by a dose of 1.5 Gy from x rays, even though the cells are not pre-exposed to the conditioning dose of 0.01 Gy. An alternative approach could be to introduce engineered genes into cells and look for the adaptive response. It is likely that the binding proteins are single-chain molecules and that the binding site does not require a conformation dictated by internal disulfides; and it is unlikely that the 14-18 kDa band protein is a subunit of the larger 34 kDa band protein. Further studies should confirm this.

130. The existence of a specific DNA-binding protein in the nuclei of human lymphoblast cells exposed to radiation, which was not detected in nuclear extracts from unperturbed cells, has been reported [S33]. The effects of this binding protein were shown to be dose-dependent and transient, reaching a maximum 1 hour after irradiation and disappearing from the nuclei by 9 hours. The protein was induced in cells by a mechanism not requiring *de novo* protein synthesis, and the response was specific for radiation and radiomimetic agents; neither UV-radiation nor heat shock invoked a response. The DNA-binding protein was present in the cytoplasm of unirradiated cells, apparently being translocated to the nucleus only after radiation exposure. Analysis demonstrated that the nuclear and cytoplasmic proteins were approximately the same size, that is, 43 kDa.

131. Similar experiments with irradiated lymphocytes from humans, mice and rabbits have been reported [L28]. Human lymphocytes *in vitro* and mice *in vivo* were irradiated with 200 kV x rays at a rate of 0.0125 Gy min^{-1}. Rabbits were exposed *in vivo* to ^{60}Co gamma-radiation at a dose rate of 0.0056 Gy h^{-1} for 9 hours, resulting in an accumulated dose of 0.05 Gy. Extracts of cells or separated cytosolic and nuclear fractions were subjected to two-dimensional electrophoresis. Four protein spots not present in the unirradiated cells appeared in the extracts of human lymphocytes 4 hours after *in vitro* exposure to 0.05 Gy, with molecular weights of 25, 167, 168 and 174 kDa. Nine spots were detected in the cytosolic extract from mouse lymphocytes 4 hours after *in vivo* exposure to 0.075 Gy with molecular weights of 51, 69-70, 145 and 160-179 kDa, and four spots were found in the nuclear extract with molecular weights of 70, 90, 230 and 247 kDa. Five spots were detected in the extract from rabbit lymphocytes with molecular weights of 105, 135, 138, 145 and 174 kDa. When compared to the protein spots identified after exposure of cells to mitomycin C and heat (41°C), it was found that the proteins induced by these treatments showed many aspects in common, although there were some differences in their electrophoretic mobility.

132. Crude extracts of splenic lymphocytes from irradiated and sham-irradiated mice were subjected to gel filtration with sephadex G-100 and aliquots tested for biological activity. Both stimulatory and suppressive effects were noted when separate fractions were added to normal splenocytes exposed to concanavalin A, the stimulatory effect of one protein fraction being more marked in the eluted fraction from the irradiated mice. These various proteins were found to have molecular

weights below 100 kDa. When the fraction with stimulatory effects was added in 10 ng amounts to lymphocyte cultures, it was found that the addition 42 hours after phytohaemagglutinin stimulation reduced the frequency of chromatid and isochromatid breaks produced by a dose of 1.5 Gy from x rays to a magnitude similar to that observed when a conditioning dose of 0.05 Gy from x rays was given 42 hours after phytohaemagglutinin stimulation.

4. Stress-response proteins

133. There are several examples of genotoxic stresses influencing the expression of proto-oncogenes and their products. Temporary hypoxia or glucose deprivation, for example, induces the expression of several proteins. The increased expression of intracellular proteins termed "oxygen regulated proteins" in Chinese hamster ovary cells has been described [W19]. These proteins are different from the heat shock proteins, but two with molecular weights of 80 kDa and 100 kDa appear to be identical to two proteins that are induced by glucose deficiency [S31].

134. Haemoxygenase is a protein associated with oxidative stress. Increased levels of this 33 kDa protein transcript have been observed in human skin fibroblasts and can be induced by treatment with UV-A-radiation, hydrogen peroxide and sodium arsenite [K9], heat shock [S8] and temporary hypoxia.

135. Changes in expression of protein kinase C can follow exposure to physical or chemical agents. Protein kinase C is often up-regulated in proliferative cells compared with quiescent cells in these circumstances. Moreover, the intracellular distribution of protein kinase C in cells is also affected [A3]. Consistent with this, low doses of UV-A-radiation have been shown to increase protein kinase C activity in cultured mammalian fibroblasts and also to inhibit EGF-binding [M10].

136. While some degree of homology between the effects of various stresses has been reported, there are clear differences. The expression of the *c-fos* and *c-jun* genes, both involved in the regulatory mechanisms of cell progression, are increased by heat shock [B16] and radiation [W16]. However, hypoxia followed by re-oxygenation has been shown to elevate only *c-fos* mRNA. Radiation causes up-regulation of the *gadd45* gene, while both UV-radiation and alkylating agents up-regulate mRNA for *gadd45* and *gadd153*. Up-regulation of the β-polymerase gene is an example of chemical agents that cause DNA damage affecting transcription, whereas x-radiation and UV, both potent agents for DNA damage, have no effect [F6].

137. The possibility that low doses of radiation induce an adaptive increase in the antioxidant defence mechanism has also been considered [N2]. The expression of stress proteins was given special attention, in view of the finding that these proteins are present in monocytes whose antioxidant activities have been elevated [C9, P7]. A similar response has been observed in mitogen-stimulated lymphocytes [F7, H4]. Male C57Bl/6 mice, six weeks of age, were exposed to a dose rate of 0.04 Gy d^{-1} from x rays on five consecutive days each week for four weeks [M19]. Three days after the last exposure, their spleens were assessed for the constitutive and mitogen-stimulated levels of heat shock protein (HSP) 70 mRNA and protein. Glyceraldehyde-3-phosphate dehydrogenase (GAPD), a housekeeping gene, was used as a reference. The results indicate that low doses augment the constitutive levels of HSP70 mRNA and HSP70 protein (Table 20). Thus, the magnitude of the proliferative response of splenocytes to T-cell mitogen stimulation can be directly related to the constitutive and mitogen-stimulated levels of HSP70 mRNA and protein. These results are consistent with the view that splenocytes need to accumulate some minimal constitutive level of HSP70 protein before they can undergo an augmented proliferative response to mitogenic stimulation and that T cells adapt to low doses by augmenting their HSP70 gene expression.

138. It was shown subsequently that the chronic irradiation of mice increases the expression of HSP70 genes in tissues [M11]. The mice were irradiated at a dose rate of 0.03 Gy d^{-1}, and the levels of HSP70 genes were analysed. Increased but transient expression of HSP70 in lung, spleen and intestinal cells was observed, commencing on day 5 of irradiation. In an extension of this experiment, the effect of dose rate was examined. The data indicated that chronic irradiation within the range 0.03-0.06 Gy d^{-1} can activate the transcription of HSP70 genes and their respective protein products. In this respect, the results were consistent with those of Nogami et al. [N4], who observed the induction of HSP70 protein in resting splenic T cells after chronic irradiation at a dose rate of 0.04 Gy d^{-1}, but not at rates above 0.1 Gy d^{-1} with a total dose not exceeding 0.2 Gy. However, increases in HSP70 protein were not seen in Chinese hamster ovary cells [A14] unless the dose was in the region of 400 Gy [S46].

139. In support of a common mechanism, the effects of viral and activated cellular oncogenes on the sensitivity of Syrian hamster (Osaka-Kanazawa) cells exposed to gamma-radiation, UV-radiation and heat shock have recently been described [S41]. Greater tolerance to gamma-radiation was conferred by the introduction of *v-mos* and *c-cot* genes, which encode for serine/threonine kinase. Cells transfected with *v-mos* and *c-cot* genes were also more tolerant to UV-radiation and heat shock. Of the activated *ras* genes, the *N-ras* gene developed a cell phenotype resistant to UV-radiation and gamma-radiation. The *Ha-ras* gene produced a cell type resistant to UV-radiation and heat shock, while introduction of the *Ki-ras* gene did not affect sensitivity. The *v-erb* B gene was

found to be involved in the development of resistance to heat shock. Transfection with *neo, c-myc and v-fgr* genes had little or no effect on survival. The karyotypes of the original cell type and the oncogene-containing cells were compared, and no alterations were seen in the cells carrying the foreign genes. These results suggest that activation of serine/threonine kinase may be involved in common processes occurring after gamma-radiation, UV-radiation and heat shock treatment, and that each oncogene may have a different effect on the development of a resistant phenotype. However, cDNA clones for a variety of DNA-damage-inducible (DDI) transcripts have been isolated that may represent more specific responses to DNA damage [F17].

C. OTHER MECHANISMS

1. Radical detoxification

140. In addition to cell cycle arrest and induced genes and gene products to initiate DNA repair, an alternative mechanism has been proposed to explain the adaptive response [F8, F9, F10, F11, F12, F13, F14, F15, F16, M12, Z1]. It relates to the ability of cells to remove toxic radicals. Radicals are known to be generated in small amounts and detoxified during normal metabolism. The process of detoxification involves the mobilization of enzymes, such as catalase, peroxidase, superoxide dismutase, from the cytosol [S34]. Membrane-bound vitamin E is thought to scavenge the radicals as they are formed; the latter are then detoxified by the cytosolic enzymes. The purpose of this radical detoxification system is to maintain the structural and functional integrity of the cell by preventing damage to cellular constituents such as membrane-bound DNA [S35].

141. Evidence in support of this hypothesis is based on the results of a series of experiments in which male Wistar rats were exposed to x rays within the dose range 0.05-0.5 Gy. Levels of superoxide dismutase were found to be increased in spleen, thymus and bone marrow 4 hours later, while lipid peroxides were decreased [Y4].

142. It has been reported that the concentration of serum thymidine kinase increases after acute exposure to radiation [H5]. This was associated with a decrease in the concentration of the enzyme in bone marrow cells and a delay in the incorporation of the radioactive thymidine analogue, 5-iodo-2-deoxyuridine (^{125}IUdr), into DNA in these cells. The function of thymidine kinase is to prepare extracellular thymidine for its incorporation into DNA. Measuring changes in thymidine kinase activity has proved to be a convenient probe for studying changes in radical detoxification.

143. In one procedure of this kind, female (NMRI strain) mice received whole-body ^{137}Cs gamma-irradiation, and the relative activity of thymidine kinase was measured in bone marrow cells after one or two exposures, each of 0.01 Gy. A dose of 0.01 Gy produced a reduction in enzyme activity that reached a minimum about 4 hours after the irradiation, with full recovery after 6-8 hours. If a second exposure to 0.01 Gy was given within 30 minutes of the first exposure, the decrease in enzyme activity was accelerated, as was the rate of recovery. If a second exposure to 0.01 Gy was given 4 hours after the first, there was no change in enzyme activity, and a second exposure 12 hours after the first resulted in a reduction of enzyme activity similar to that observed after a single dose of 0.01 Gy [F8].

144. These reductions in thymidine kinase activity closely followed the decreased rate of uptake of ^{125}IUdr into bone marrow DNA, and there was a simultaneous increase in the radical scavenger glutathione. These responses were absent in cells exposed to a strong magnetic field, which transiently alters lipid membrane structure. A deficiency of vitamin E, which is known to be a radical scavenger, also prevented the response.

145. The effect of low-dose radiation on ^{125}IUdr incorporation has also been measured in mouse intestines, spleen and thymus [M20]. Young female BALB/c mice were irradiated within the range 0.05-0.23 Gy with x rays at a dose rate of 0.05-0.2 Gy min^{-1}, followed at various times thereafter by an injection of ^{125}IUdr. The incorporation of ^{125}IUdr was decreased for several hours after a single exposure to x-radiation in a dose-dependent manner in spleen and thymus. At 4 hours after irradiation, for example, the decrease relative to controls was 79% in spleen and 86% in thymus. If the first irradiation was followed 4 hours later by a second irradiation (0.05 Gy or 0.1 Gy, for example) the second irradiation did not enhance the inhibitory effect of the first exposure, thus confirming the observations of Feinendegen et al. [F8].

146. To put these studies into perspective, it has been calculated that a radiation dose sufficient to create an average of one ionizing track per cell would produce enough radicals to delay the rate of DNA synthesis in metabolically active cells [F9]. This time delay was sufficient to ensure the availability of radical scavengers to cope with radicals produced by a second dose of radiation a few hours later.

147. Other calculations reveal that 0.01 Gy of low-LET radiation could produce about 6 nM of oxidative radicals in each cell [K10]. This concentration should be compared with the cellular steady-state concentration of radicals from normal metabolic processes involving oxygen, which is about 1 nM . Therefore, it has to be assumed that a small transient increase in radical concentration above that normally present in the cell is able to cause a measurable activation of normal detoxification mechanisms.

2. Activation of membrane receptors

148. The activity of adenylate cyclase increases markedly in isolated membranes from rat hepatocytes after acute exposure to between 1 and 2 Gy from gamma rays [K17]. This was confirmed by experiments with isolated plasma membranes from rat lung tissue, in which the increased adenylate cyclase activity, induced by a low concentration (0.02 mM) of isoproterenol, was further increased if the membranes were irradiated with gamma-radiation at the rate of 0.36 mGy d^{-1} [K18, R4]. When mitogen-stimulated human lymphocytes [K19] and Raji cells derived from human lymphoma cells [R5] were grown in the presence of serum growth factors, increased cellular proliferation was observed if the cells were irradiated at dose rates of 0.36 mGy d^{-1}. The conclusion from these studies was that low doses of radiation activated membrane-bound enzymes.

3. Stimulated proliferation of splenocytes

149. The effect of radiation on the *in vitro* proliferation of thymocytes and splenocytes was measured in rats previously irradiated *in vivo* [I3]. The animals received a whole-body dose of x-radiation in the range 0.01-2 Gy. Cells isolated from the spleen and thymus were then cultivated in the presence of various mitogens, and cell proliferation was evaluated by the rate at which tritiated thymidine was incorporated into DNA.

150. The results showed that the proliferation of splenocytes induced by concanavalin A (Con A) was enhanced by the use of x rays within the dose range 0.01-0.1 Gy, whereas that of thymocytes was not affected, as can be seen in Figure VII. Irradiation with 0.05 Gy also enhanced the proliferative rate of splenocytes stimulated by phytohaemagglutinin (PHA) or lipopolysaccharide (LPS), although their responses were less than that produced by Con A. This enhancement in the mitogen-induced proliferation of splenocytes was observed only within a few hours after irradiation, suggesting that low-dose, whole-body irradiation can induce an adapting effect in splenocytes.

151. This study was followed by a study of mouse splenocytes in which interleukin-1 (IL-1) production was measured [I4]. The bioavailability of intracellular IL-1 of splenocytes stimulated with LPS was enhanced following whole-body irradiation with 0.025 Gy from x rays. Furthermore, if splenocytes from an unirradiated mouse were exposed to Con A, the addition of a small amount of serum from an irradiated mouse augmented the proliferative effect of the Con A. This could imply the presence of growth-enhancing substances (e.g. cytokine) in the irradiated serum.

D. SUMMARY

152. Studies to characterize gene expression in relation to the radiation-induced adaptive response are continuing, and specific genes and their protein products induced after acute exposure to doses of radiation in the range of a few tens of milligray to a few gray have been identified. It has been shown that groups of genes coding for transcription factors, nuclear proteins, oncogenes, viruses, membrane receptors, functional proteins and enzymes can be activated during the few hours after acute exposure to radiation. Recent experiments have implicated oxidizing radicals as one activating mechanism.

153. While there is a general consensus that the adaptive response can be considered as a consequence of damage to the DNA molecule, other mechanisms have been proposed. A recent review of the mechanisms of induction of transcription factors by damaging environmental agents provides useful insight into these other mechanisms [H11]. As an example, reactive oxygen intermediates and free radicals may directly influence regulatory proteins, such as the transcription factors, which in turn may condition the cellular adaptive response by inducing the expression of genes. The presence of stress-induced proteins at dose rates of 0.04 Gy d^{-1} has also been demonstrated. Protein kinase C is a common factor in many of these responses, and since it plays a central role in cellular signal transduction, its activation could represent a general response to molecular damage.

III. EFFECTS ON THE IMMUNE SYSTEM

154. The observations of a radiation-induced adaptive response in mitogen-stimulated lymphocytes and of changes in the immune system after exposure to low doses of radiation could imply an essential role for immunocompetence in the living organism. Understanding the mechanisms of T-cell signalling and how they are affected by low doses of radiation may therefore explain some aspects of the adaptive response.

155. Steady progress is being made in identifying the receptors on T cells that permit specific recognition of the infinite variety of non-self molecules and in determining how the signal from the T-cell antigen surface receptor is transmitted to the cell interior. The functioning of the immunosurveillance system of the organism, its response to radiation and its possible involvement in adaptive response are considered in this Chapter.

A. CELLS OF THE IMMUNE SYSTEM

156. T cells are lymphocytes that develop in the thymus. They are primarily responsible for ensuring cellular immunity and delayed hypersensitivity response. They mediate their acquired immune responses first by activation of specific T cells, then by a phase of clonal expansion and finally by a phase in which some of the lymphocytes become effector cells.

157. There are at least three functionally distinct classes of T cells:

(a) cytotoxic T cells, which kill virus-infected cells and tumour cells directly;
(b) helper T cells (T_H), which amplify responses by secreting a variety of local chemical mediators (interleukins) that stimulate activated T cells to proliferate, help B cells to make antibodies, and activate macrophages;
(c) suppressor T cells (T_S), which inhibit the responses of helper T cells.

Helper and suppressor T cells are the principal regulators of immune responses, and the ratio between the two cell subsets is an important factor in ensuring immunocompetence in the living organism.

1. T-cell ontogeny

158. Most T cells bear an antigen receptor consisting of a heterodimer of transmembrane α and β polypeptide chains. Both chains are required for antigen recognition [D3]. These polypeptide molecules resemble antibodies in structure and interact with antigens that are presented to T cells as proteolytic digestion fragments associated with Class I and Class II major histocompatibility complex proteins on the surface of antigen-presenting cells. The specific binding of antigen histocompatibility complexes on the surface of a target cell to antigen receptors on the surface of the T cell is often not strong enough to mediate a functional interaction between the two cells. Various cell-cell adhesion glycoproteins on T cells help to stabilize such interactions by increasing the overall strength of cell-cell binding. The characteristics of these accessory glycoproteins are summarized in Table 21. Among the best characterized are the CD4 and CD8 glycoproteins, which are expressed on the surface of helper and cytotoxic T cells, respectively.

159. The T-cell receptor is also associated with a group of glycoproteins, collectively referred to as the CD3 complex, which is involved in receptor assembly [C10]. The T-cell receptor CD3-α/β heterodimer is expressed on the surface of the vast majority of mature $CD4^+$ or $CD8^+$ cells in peripheral human blood lymphocytes and lymphoid organs [A5, H6, M14]. (The presence or absence of accessory proteins on T cells is designated by the superscripts + and -, e.g. $CD4^+CD8^-$). A second group of glycoproteins consisting of γ and δ chains is also CD3-associated and is found mainly in the double-negative ($CD4^-CD8^-$) T-cell population [B17]. According to current understanding, $CD4^-CD8^-$ T cells are thought to be the precursors of thymocytes, which differentiate into $CD4^+CD8^+$ T cells, which in turn differentiate into $CD4^+CD8^-$ or $CD4^-CD8^+$ T cells [S36]. These last two cell types become the antigen-positive mature T cells.

160. Spontaneous loss and alteration in antigen receptor expression in mature human $CD4^+$ T cells obtained from healthy donors have been described [K11]. The T-cell receptor CD3 complex plays a central role in antigen recognition and activation of mature cells, so abnormalities in the expression of this complex should be related to the unresponsiveness of T cells to antigen stimulus. Using flow cytometry, variant T cells with loss or alteration of T-cell receptor CD3 expression among $CD4^+$ cells were detected and enumerated. Variance was demonstrated by defects in protein expression and partial protein deletion. The variant frequency in peripheral blood increased with age in normal donors and was highly elevated in patients with ataxia telangiectasia. Thus, alterations in antigen-receptor expression may be induced by the spontaneous somatic mutation of T-cell receptor genes, and these alterations could be important factors related to age-dependent, disease-associated T-cell dysfunction and radiation-induced T-cell damage.

161. Over the past decade, knowledge of T-cell renewal, differentiation and maturation in the mouse has been remarkably advanced by the development and use of specific monoclonal antibodies that identify T cells at the various stages of differentiation. It should be noted that a different terminology for identifying glycoproteins is sometimes used in the mouse. Thus, $CD4^+$ equates with $L3T4^+$, and $CD8^+$ equates with $lyt2^+$. It is now generally agreed that a small progenitor cell compartment exists in the thymus that expresses neither L3T4 nor lyt2 surface antigens (i.e. it is double-negative). The progeny of these cycling cells become the major cell population in the thymus and express both L3T4 and lyt2 antigens (double-positive). At present, there is insufficient evidence to determine whether the mature thymocytes, which express either L3T4 or lyt2, differentiate from a subset of double-positive cells or whether they derive directly from the double-negative cells. In either case, the majority of double-positive cells are not selected because of inappropriate major histocompatibility complex reactivity. The single-positive functional effector cells are exported to peripheral lymphoid tissue, where, in contrast to their behaviour

in the thymus, they become the predominant cell populations. A model for the various stages in T-cell ontogeny, as defined by L3T4 and lyt2 antigen expression, is schematically presented in Figure VIII.

2. Apoptosis and radiation-induced interphase death

162. The mechanism responsible for producing T cells is unusual in that the vast majority of differentiating cells are destroyed prior to their complete differentiation and release to the peripheral lymphoid tissues. This phenomenon, known as apoptosis, occurs within the thymus and results in the death of most of the developing thymocytes. The process is thought to ensure that differentiating cells with potential for reacting against the host are eliminated, thereby preventing an auto-immune response, and it essentially involves double-positive T cells. One feature of apoptosis is then an alteration in the differentiation rate of developing T cells in the thymus [L15].

163. Some types of lymphocytes die soon after exposure to radiation and before entering the mitotic phase of the cell cycle. The process is referred to as interphase cell death. To determine any similarity between apoptotic death and radiation-induced interphase death, thymocytes were collected from male C57Bl/6 mice between five and six weeks old, separated into $CD4^-CD8^-$, $CD4^+CD8^+$, $CD4^+CD8^-$ and $CD4^-CD8^+$ T-cell subsets and exposed to gamma-radiation [M19]. They were assayed 8 hours later for evidence of cell death by measuring the amount of DNA fragmentation. The results given in Table 22 show that double-positive $CD4^+CD8^+$ T cells are extremely sensitive to radiation. With a 50% fragmentation index (FD_{50}) at 0.22 Gy, it would appear that a small but significant fraction of the double-positive T cells die as a result of apoptosis following exposure to low doses of radiation. Since the double-positive T cells constitute the majority of the parenchymal cells in the thymus in an intermediate stage of differentiation, elimination of a fraction of them could conceivably alter the dynamic balance between cells in different stages of differentiation.

164. On this basis, interphase cell death caused by radiation may be considered similar to, if not identical with, apoptosis [D4, W20, W21]. The elimination of cells damaged by radiation ensures that the clonal expansion of renegade T cells is prevented.

165. A biochemical mechanism responsible for radiation-induced interphase cell death in lymphocytes has recently been proposed [Z2]. Within tens of minutes after exposing rat thymocytes to radiation, a sharp increase in mRNA polymerase II was observed. This was followed within 2 hours by an increase in intracellular calcium ion concentration and the appearance of newly synthesized polypeptides. In particular, there was a sharp increase in the phosphorylation of three proteins with molecular weights of 20, 35 and 48 kDa [Z3]. Different stress-related proteins of molecular weights of 48, 70 and 90 kDa were also detected after heat shock or after the addition of glucocorticoids to the culture medium [M16].

3. Signalling processes in thymocytes

166. Some consensus has emerged regarding the nature of the signalling process that couples antigen recognition to changes in lymphocyte behaviour. It is postulated that the activation of T cells requires two signals, one through the CD3 complex, the other from antigen-presenting cells. Antigenic, allogenic, mitogenic or monoclonal antibody stimulation of the T-cell antigen receptor (TCR/CD3 complex) leads to a series of biochemical processes, including the activation of phospholipase C, with subsequent hydrolysis of phosphatidyl inositol-4,5-diphosphate to generate two second messengers, inositol-1,4,5-triphosphate (IP3) and diacylglycerol (DG) [H7, I12, J5, T14]. These two messengers cause an increase in intracellular calcium ion concentration (Ca^{++}) and the activation of protein kinase, respectively. The two events are considered essential in the expression of the "early immediate" gene, *c-fos,* to be followed later by the expression of *c-myc*, gamma interferon, interleukins 1 and 2 (IL-1 and IL-2) and transferrin receptor, which are crucial in the activation and proliferation of T cells [K26]. Activated T cells secrete interleukin-2, which results in division in those cells expressing interleukin-2 receptors. The T cells produce a series of lymphokines that clonally expand activated B cells and cause their maturation into antibody-synthesizing cells [S37, W22]. Signals derived from the T-cell receptor could also regulate the maturation and selection of immature T-cell progenitors in the thymus [V5].

167. The kinases responsible for protein synthesis have not yet been fully identified, but it is known that T cells contain transcripts encoded by at least three members of the *src* family of protein-tyrosine kinase genes: namely *lck, fyn* and *yes* [C11, S38]. The $p56^{lck}$ protein was first identified by virtue of its over-expression in a murine lymphoma cell line. It is the product of the *lck* proto-oncogene, which is closely related to the *c-src* gene, and hence is a potential signal transduction element [P8]. Two pieces of evidence suggest the type of signalling event that may be mediated by $p56^{lck}$. First, expression of *lck* mRNA

and of $p56^{lck}$ is altered by stimuli that induce lymphokine release from T cells [M15]. Secondly, $p56^{lck}$ is associated with CD4 and CD8 glycoproteins [V6]. When $p56^{lck}$ is brought into proximity with the T-cell antigen recognition unit by CD4 or CD8 glycoproteins, its activity may increase to reflect receptor occupancy.

168. A second membrane-associated protein tyrosine kinase, $p59^{fyn}$, the product of the *fyn* proto-oncogene, is also abundant in T cells [C11]. Using transgenic mice in which a lymphocyte-specific protein-tyrosine kinase isoform, $p59^{fyn(T)}$, was 20-fold overexpressed in developing T-lineage cells, it was found that thymocytes from these animals were highly sensitive to stimulation [C12]. Furthermore, the phenotypes produced by overexpression of $p59^{fyn}$ were different from those produced by overexpression of $p56^{lyk}$. Although $p59^{fyn}$ may regulate early steps in the T-cell receptor signalling, it cannot by itself confer a mature, responsive phenotype upon otherwise refractory, immature thymocytes. Thus, $CD4^{+}CD8^{+}$ thymocytes from *lyk-fyn* transgenic mice exhibit enhanced calcium ion accumulation following activation, but they do not release interleukin-2 or proceed to proliferate. Hence, components of the signalling cascade that are essential for activation are apparently unavailable or inactive in immature thymocytes in these transgenic mice.

B. RESPONSE IN THE ORGANISM

169. The cell types involved in the immune response exhibit a broad spectrum of radiosensitivity with consequent effects in the organism [A6]. Some populations of lymphocytes are exceedingly radiosensitive, while plasma cells and macrophages are, by comparison, relatively radioresistant. The basis of these differences in radiosensitivity is not well understood, but the effects of radiation on the immune system have been extensively studied. Several reviews have been published that provide comprehensive background information [A6, A7, A8, S39, T6].

1. Effects in animals

170. Irradiation at high doses is known to inhibit the immune response. A dose-response relationship was observed almost three decades ago by Kennedy et al. [K12]. They exposed mice to x rays within the dose range 0.5-7 Gy and 10 days later injected the animals with antigenic sheep red blood cells. Four days after injection of the antigen, they measured the number of splenic plaque-forming cells. The effect of exposure to between 1 and 3 Gy from x rays was to reduce the number of splenic plaque-forming cells to about 30% and 1%, respectively, compared with the numbers of plaque-forming cells observed in sham-irradiated animals. The results are illustrated in Figure IX.

171. More recently, the immediate and long-term effects of radiation on the immune response of specific-pathogen-free mice were described after exposing mice to a dose of 3 or 6 Gy from x rays and measuring the surviving fraction of T-cell subsets [S39]. At high doses (>1 Gy), T_H and T_S cells were equally radiosensitive but there was a marked difference in the radiosensitivity of T cells between C3H, BALB/C, C57Bl/6 and B10BR strains of mice. No long-term effect of radiation at these high doses on the immune system was found in terms of accelerated ageing of the immune function in animals exposed as young adults.

172. Other experiments have shown an enhancement of the immune response following whole-body exposure to low doses. As an example, rabbits were exposed to small doses of x rays (about 0.25 Gy) before or after immunization with antigenic sheep red blood cells [T7, T8]. These small doses were able to stimulate the proliferation of plaque-forming cells. This finding has been confirmed in experiments in which human and murine T cells were exposed to low doses [A8, A9, D8, L9, L16, L17, M17, S40, S51, T9].

173. In contrast, four different strains of pathogen-free mice (B10/Sn, B10/SgSn, C3H/HeMsNrs and C57Bl/6JS1c) were exposed to doses of 0.025-0.25 Gy from x rays [K13]. Nine hours later, they were injected with sheep red blood cells; the number of direct plaque-forming cells per spleen was assessed after 4.5 days. There was no evidence of enhancement of the plaque-forming cells in any of the strains within the dose range used. In an extension of this experiment, C57Bl/6J mice were immunized with sheep red blood cells and exposed two days later to higher doses (1.5-3.0 Gy) from x rays. Numbers of indirect and direct plaque-forming cells per spleen were assessed at intervals thereafter. Mice exposed to 1.5 Gy were found to have a significant increase in the number of indirect plaque-forming cells 11 days after the injection of the sheep red blood cells. This was highly correlated with an increase in the $CD4^{+}/CD8^{+}$ cell ratio during the first three days after radiation exposure, followed by a rapid decline, as shown in Figure X. These results suggest that the ratio of T_H to T_S subsets changes in favour of helper cells shortly after exposure, possibly contributing to the observed enhancement of the indirect plaque-forming cell response.

174. Anderson and Lefkovits [A9] proposed that the enhancing effect within the dose range 1-7 Gy could be due to the elimination of the more highly radiosensitive T_S cells. These investigators questioned,

however, whether the same mechanism operated when doses below this range were used. Under these circumstances, they postulated that T_S cells could proliferate. The enhancing effect on murine splenic T cells, which is illustrated in Figure XI, is however, effective over only a narrow range of doses [M18]. The maximum effect occurs at about 0.25 Gy.

175. To determine the fate of proliferating T cells, animals from a normal C57Bl/6J+/+ strain of mouse were chronically exposed to gamma-radiation in the range 0.005-0.04 Gy d^{-1} during 20 days in a four-week period [J7]. Animals from an immunologically depressed C57Bl/6J lpr/lpr strain were exposed under similar conditions. A dose-related proliferation in thymocytes and splenocytes was noted in both strains up to cumulative doses of about 0.8 Gy. This is illustrated in Figure XII. The mitogen-responsive $L3T4^+lyt2^-$ and $L3T4^-lyt2^+$ T cells increased with dose, while the mitogen-unresponsive $L3T4^-lyt2^-$ and $L3T4^+lyt2^+$ T cells decreased in both strains relative to sham-irradiated animals.

176. However, the magnitudes of the changes in thymic and splenic T cells subsets were different, according to whether or not the animals were fed on a calorie-restricted diet. The immunologically depressed mice were more sensitive to these dietary changes. It was proposed that these experiments support the hypothesis that the stress of continuous low-dose irradiation is consistent with an adaptive mechanism for cell renewal and maintenance of mitogen-responsive cells.

177. Two recent reviews [L16, M19] provide the basis of a hypothesis with which to explain a radiation-induced T-cell response to low doses of radiation. At the cellular level, the question as to whether or not some subsets are more radiosensitive than others remains unresolved. The evidence points to the need for an intact thymus, in that thymectomy prevented the adaptive response [J6]. Mitogen-responsive T cells appear to be a prime target, and their stimulation is associated with enhanced expression of the HSP70 gene and its related proteins. There may be selective destruction of mitogen-unresponsive T cells. The interpretation is complicated by the evidence that the metabolic status of irradiated animals can influence the response and that there is a marked difference in radiosensitivity between the strains of animals used in the experiments.

178. Liu et al. [L8, L9, L16] have investigated several reactions of splenocytes in two strains (C57Bl/6 and Kunming) of mice irradiated with x rays. The dose-effect relationships for antigenic (plaque-forming cell) ability, mixed lymphocyte reaction, mutagenic (concanavalin A)-stimulation, antibody-dependent cell-mediated cytotoxicity, natural killer cell activity, interleukin-2 and interferon secretions, and the lipopolysaccharide reaction are given in Table 23. Interestingly, the maximum response in most of these reactions occurs at 0.075 Gy.

179. Some specific features of the reactions occurring at maximum response are summarized in Table 24. An analysis of cell-cycle progression by measuring the uptake of tritiated thymidine showed an increase in the number of thymocytes entering the S phase, with a corresponding decrease in the number of cells in the G_0 phase 3-7 days after exposing mice to 0.075 Gy. The T_H/T_S ratio of thymic lymphocytes was the same for the first three days, with a lowering of the ratio after seven days, which the authors claim was due to a decrease in the number of $CD4^+CD8^-$ cells (compared with $CD4^-CD8^+$ cells) in the thymus.

180. A parallel examination of the phenotypic changes of the thymocyte subsets, with flow cytometry using fluorescence-labelled monoclonal antibodies against surface antigens, showed a significant increase in the percentage of the double-negative ($CD4^-CD8^-$) cells in the thymus after 0.075 Gy, implying an enhanced renewal of thymocytes. These results are also included in Table 24. At the same time, the secretion of colony-stimulating factors by thymocytes was stimulated. It is known that the colony-stimulating factors secreted by the thymocytes act on macrophages to facilitate the production of interleukin-1, which, in turn, serves as a signal for the maturation of immature thymic lymphocytes. All these changes occurring in the thymus following low doses should increase the supply of mature lymphocytes.

181. Liu et al. [L26] proposed a mechanism to explain their findings after the whole-body irradiation of mice at low doses. This is summarized in Figure XIII, which outlines the changes in signal transduction of T lymphocytes. It is a process similar to that postulated for apoptosis and interphase death. How it might be influenced by neuroendocrine factors is being investigated. What is known is that exposure to 0.075 Gy causes a decrease in serum corticosterone in the course of a few weeks accompanied by an increase in 5-hydoxytryptamine from the hypothalamus and a decrease in adrenocorticotropic hormone [L27]. It is not known how these changes in endocrine function following low-dose irradiation affect the signal transduction process in T cells, but it is an area of research worth pursuing.

2. Effects in humans

182. Several parameters of cellular immune function have been assessed for 168 persons who survived the

atomic bombings of Hiroshima and Nagasaki but who now reside in the United States [B18, B19, B20]. Persons exposed to doses between 0.01 and 1 Gy were compared with persons exposed to less than 0.01 Gy. Lymphocytes were isolated from the peripheral blood of these individuals and assessed for the following parameters of cellular immunity: (a) mitogenic response to phytohaemagglutinin; (b) mitogenic response to allogenic lymphocytes (mixed lymphocyte response); (c) natural-cell-mediated cytotoxicity; (d) interferon production. In every case, the response of the group exposed to 0.01-1 Gy was greater than that of the group exposed to less than 0.01 Gy, although only the difference for natural cell-mediated cytotoxicity was statistically significant (Figure XIV). It is not possible to say whether the increase in natural-cell-mediated cytotoxicity observed in these survivors of the atomic bombings exposed to very low doses of radiation is a genuine adaptive response that was modulated by post-radiation environmental conditions or a chance finding.

183. In more recent studies carried out among 1,328 survivors of the atomic bombings and now living in Japan, there was an age-related decrease in the total number of T cells ($CD5^+$, $CD4^+$, $CD8^+$) [K14]. Although the differences were not statistically significant, the overall impression is that ageing of the T-cell-related immune system is accelerated in older persons receiving doses in excess of 1 Gy, compared with the control group, who received doses of less than 0.01 Gy.

184. The responsiveness of peripheral blood lymphocytes to allogenic antigens in mixed lymphocyte culture was measured in 139 survivors of the atomic bombings. In contrast to the findings in paragraph 182, the study revealed a significant decrease in mixed lymphocyte culture response with increasing dose of previous radiation exposure. This decline was marked in the survivors who were older than 15 years at the time of the bombings. It suggests a possible relationship between the recovery of T-cell-related function and the thymic function that processes mature T cells for the immune system. Thus it may be that in the persons who were older at the time of the bombings, the thymus function had started to involute, allowing less recovery of T-cell function than in younger survivors, who had adequate processing T-cell activity [A19].

185. As indicated in paragraph 160, the spontaneous loss and alteration of antigen receptor expression in mature CD4 cells has been found in blood cells taken from 127 healthy donors who had not been exposed to other than natural background radiation [K11]. The mean frequency of mutant T cells in males with altered T-cell receptor expression was 2.5 10^{-4}, equal to about 1.3 10^{-4} per single T-cell receptor locus among the CD4 cells, increasing with age.

186. The mean frequency of mutant T lymphocytes was measured in 203 survivors of the atomic bombings, 78 of whom received doses greater than 1.5 Gy and 125 of whom received doses less than 0.005 Gy [K16]. The results were 1.39 ± 0.63 10^{-4} mutants per cell in exposed males compared with 1.20 ± 0.35 10^{-4} in unexposed males and 0.99 ± 0.39 10^{-4} in exposed females compared with 0.89 ± 0.38 10^{-4} in unexposed females. There was no statistical difference in mutant frequency between the two groups, although the frequency was higher in males than females. The smoking habits of the males may have contributed to the higher values.

187. Antibody titres to Epstein-Barr virus antigens were determined in the sera of 372 atomic bomb survivors to evaluate the effect of the previous radiation exposure on immune competence against the latent infection of the virus. The proportion of persons with high titres (≥1:40) of IgG antibodies to the early antigen was significantly elevated in the exposed survivors. Furthermore, the distribution of IgM titres against the viral capsid antigen was significantly affected by radiation dose, with an increased occurrence of titres of 1:5 and 1:10 in the exposed survivors, although the dose effect was only marginally suggestive when persons with rheumatoid factor were eliminated from the analysis. These results suggest that reactivation of Epstein-Barr virus in the latent stage occurs more frequently in the survivors, even though this might not be affected by the radiation dose. Otherwise, there was neither an increased trend in the prevalence of high titres (>1:640) of IgG antibodies to the viral capsid antigen among the survivors nor a correlation between the radiation exposure and distributions of titres of IgA antibodies to the viral capsid antigen or antibodies to the nuclear antigen associated with anti-Epstein-Barr virus [A11].

188. The effects on the immune system of long-term low-level radiation exposure were measured in two areas in Gaungdong Province, China [Y8]. The annual effective dose in the area of low background radiation in Enping County was about 2.0 mSv. In the area of high background radiation in Yangjiang County, the annual effective dose was about 5.4 mSv. The annual equivalent doses from external gamma-radiation to the red bone marrow were estimated to be about 0.77 mSv and 2.1 mSv, respectively. Twenty-five healthy male subjects from the high-background-radiation area and 27 subjects from the control (low-background-radiation) area were divided into three age groups, and the frequency of interleukin-2-secreting cells was measured as an index of immune competence in peripheral blood lymphocytes.

189. The results shown in Table 25 indicate that the frequency of interleukin-2-secreting cells was significantly greater in subjects from the high-background-radiation area than in those from the control (low-background-radiation) area. The increased production of interleukin-2 in lymphocytes after low-dose irradiation has also been demonstrated [Z7], consistent with the view that long-term exposure to low doses of radiation may affect the immune system. However, other factors, particularly differences in smoking habits, could also have contributed to the differences.

190. A study of the T-cell subsets in occupationally exposed persons revealed no significant influence of radiation on their profile [T10]. Data were pooled into two groups, one with individual exposures less than 0.5 mSv and another with exposures in excess of 0.5 mSv in the previous three months. The average doses for the two groups were 0.36 ± 0.09 mSv and 1.06 ± 0.77 mSv, respectively. Natural background radiation was estimated to be 0.06 mSv during the period. The results shown in Figure XV indicate no statistically significant difference in the percentage of $CD2^+$, $CD4^+$, $CD8^+$ and HNK-1 (natural killer cells). However, as indicated by the data given in Table 26, significant differences were shown as an effect of cigarette smoking.

3. Effects on tumour growth

191. The precise involvement of the immune system in the natural course of cancer remains uncertain. It is known that white cells cultured *in vitro* in the presence of the cytokine interleukin-2 acquire the capacity to kill tumour cells [S50]. The phenomenon was termed lymphokine-activated killer activity. The use of *in vitro* techniques to study cytokine kinetics has given some insight into their role in tumour biology. Some cytokines (e.g. gamma-interferon) can render tumour cells resistant to cell-mediated killing, while others (e.g. transforming growth factor β) can enhance the growth of some tumour cells. Thus, cytokines interact to generate immune responses and modulate the outcome of effector mechanisms on target cells.

192. Thus, the response of a cell to a cytokine depends on the context in which the cytokine signal is received. For example, if white cells are mixed with tumour cells in culture, the ability of the white cells to kill the tumour cells in the presence of cytokines can be measured. This is shown in Figure XVI. Interleukin-2 can profoundly increase the cytotoxicity of the white cells. Conversely, if either interleukin-4 or transforming growth factor β is added at the same time as the interleukin-2, the cytotoxicity is significantly reduced. However, if the white cells are first stimulated with interleukin-2 and then interleukin-4 or transforming growth factor β is added, the reduction in cytotoxicity does not occur.

193. Low doses of radiation (0.25-1 Gy) have been used to augment the effect of immunization of animals to reduce tumour growth [A4]. The immunization procedure involved injecting animals with an antigen of non-active tumour cells before exposing them to live tumour cells. The antigen was prepared by treating the ascitic form of a methylcholanthrene-induced fibrosarcoma (SaI cells) with mitomycin and paraformaldehyde. These mitomycin-treated cells were then injected subcutaneously into the A/J strain of mouse and, to assess immunity, the mice were injected subcutaneously with an amount of living SaI cells known to induce subcutaneous tumours. The effect of irradiation upon this immunization process can be measured. Figure XVII shows the effect of whole-body exposure to 0.15 Gy from x rays on the response of A/J mice to mitomycin-treated SaI cells. Immediately after irradiation, the mice were injected with varying numbers of treated SaI cells, as indicated in the Figure. Twenty-one days later, they received a subcutaneous injection of 10^4 viable SaI cells into the left flank, and the size of the growing tumour was measured over a period of 26 days. The non-irradiated control group (solid line) did not receive the mitomycin-treated cells. Injection with 10^3 to 10^5 mitomycin-treated tumour cells resulted in variable degrees of immunity in both sham-irradiated and irradiated groups, expressed in terms of smaller tumour size than in the control group. However, the degree of immunity was almost always greater in the irradiated mice. Low-dose augmentation was less pronounced in recipients of thymus-derived T cells. From these results it was suggested that exposure to low doses of radiation reduces the number of T_S cells, which normally suppress tumour rejection.

194. In a follow-up to these studies, the effect of radiation was shown in an *in vitro* system in which the effect of irradiating donor spleen cells could be measured [A18]. One procedure (Winn assay) involved injecting viable SaI cells into A/J mice and killing them two days later. The spleens were sham-irradiated or irradiated with 0.15 Gy, and 10^4 spleen cells were mixed with 10^4 SaI cells. This mixture was injected subcutaneously into A/J mice and tumour size measured over 18 days. Figure XVIII shows the inhibitory affect of the radiation.

195. The effects of low-dose radiation on another type of experimental tumour, the Lewis cell carcinoma, have been measured [L29]. C57Bl/6 mice received whole-body x-irradiation giving doses in the range 0.05-0.15 Gy at a dose rate of 0.0125 Gy min^{-1}. All mice were injected intravenously with 7 10^5 cells of

Lewis lung carcinoma 24 hours after the irradiation and were killed 14 days later. The lungs were removed and the tumour nodules on the lung surface were counted. The mean number of lung tumour nodules was 16 in the 0.05 Gy dose group, 27 in the 0.075 Gy group, 20 in the 0.1 Gy group and 27 in 0.15 Gy group, all significantly lower than the 85 nodules counted in the sham-irradiated group ($p < 0.01$-0.001). This supports the view that there may be an inhibitory effect of low-dose radiation on the pulmonary dissemination of Lewis lung carcinoma in mice.

196. Miyamoto and Sakamoto [M28] demonstrated the anti-tumour effects of low doses of radiation in mice of the WHT/Ht strain. Mice were exposed to 250 kV x rays in the dose range 0.05-1 Gy at a dose rate of 1.23 Gy min^{-1}. They were then injected with squamous carcinoma cells, and the effect of the radiation on subsequent tumour growth was measured. For example, whole-body irradiation of 0.1 Gy to mice bearing primary leg tumours followed 12-24 hours later by localized tumour irradiation delayed the growth of the tumours more than did localized irradiation alone. The results for various times and amounts of local irradiation are given in Table 27. The *in vitro* immune responses of splenocytes from tumour-bearing mice exposed to 0.1 Gy are compared to those from sham-irradiated tumour-bearing mice in Figure XIX. The selective depression of T_S cells may have been responsible for this low-dose effect.

197. Immune response and tumour growth have also been studied in humans. In a study of non-Hodgkin's lymphoma patients, the effect of whole- or half-body x-irradiation on the various T-cell subsets was measured [T16]. Patients received 0.1-0.15 Gy per fraction, delivered two or three times per week, until the cumulative dose was 1.5 Gy. The changes in the T-cell subsets are shown in Table 28. A statistically significant increase in helper T cells and helper-inducer T cells was found, with no change in the suppressor cells or natural killer cells. The direct anti-tumour effect was evaluated in 10 patients. Two patients showed complete remission, seven partial remission and one, no change. In four patients given half-body irradiation, the tumours showed complete or almost complete remission even though the primary tumours in the tonsils or neck lymph nodes were outside the irradiation field. Anti-tumour activity in the other patients was obscured by chemotherapy received after the irradiation. The changes in T-cell subsets and tumour responses could be consistent with an adaptive feedback control signal that up-regulates stem cells, with preferential proliferation of differentiating T_H cells. These studies need to be substantiated to eliminate the possibility of spontaneous remission.

C. SUMMARY

198. Studies to characterize the role of the immune response to low doses of radiation are continuing. The mechanisms responsible for T-cell selection during normal differentiation in the thymus and activation of the signalling process that couples antigen recognition to changes in lymphocyte behaviour are gradually being elucidated. A sequential role for phospholipase C, intracellular calcium ion and protein kinase C, followed by transcription of genes such as *c-fos*, and interleukin-2 production in the activation of T cells has been proposed. Similar sequences of events appear to be associated with apoptosis and in low-dose-radiation-induced interphase death, which causes enhanced DNA replication and proliferation of T cells. Whole-body irradiation of mice results in a dose-related enhancement of the T-cell response to antigenic, allogenic and mitogenic stimuli at doses below about 0.1 Gy.

199. As an alternative explanation to radiation-induced interphase death, experimental studies using mice have shown selective depression of radiosensitive suppressor T cells within the dose range 0.025-0.25 Gy, with an optimum response at 0.075 Gy. This needs to be investigated further as a possible mechanism of the adaptive response.

200. The evidence for changes in the T_S/T_H ratio of T cells in humans exposed to radiation remains equivocal. Reported changes in the ratio of these cells in the blood of atomic bomb survivors over 40 years after exposure are difficult to interpret, the effects of the long time interval and of changes in the environment and cigarette smoking being confounding and inexplicable factors. A Chinese study of the chronic exposure of a large population to background levels of external radiation of about 5.4 mGy a^{-1} indicated that levels of interleukin-2-secreting cells were significantly increased in persons from this area compared with persons living in an area of lower radiation background where the annual effective dose from external exposure was about 2.0 mSv; the authors acknowledged, however, that differences in geochemical and other environmental factors provide an equally plausible explanation. A study of workers chronically exposed to low doses of radiation showed no statistical difference in the percentage of helper or suppressor T cells or of natural killer cells in groups with average doses of 0.36 and 1.06 mSv, however significant differences were found in relation to cigarette smoking.

201. A role for the neuroendocrine system in influencing T-cell proliferation after low doses of radiation has recently been proposed. The blood vessels in most lymphoid tissues are innervated with sympathetic nerves, and the lymphoid tissues are under the control of hormonal factors via the hypothalamus. There is evidence to support the view that radiation can activate the neuroendocrine system and enhance the immune response.

202. A cytotoxic effect of irradiated T cells on tumour cells has been demonstrated. It is postulated that low doses of radiation act by modulating antigen-stimulated clonal growth. There is evidence of a down-regulating effect on tumours in animals, but the evidence for a similar effect on human tumours is sparse. There is a need to investigate these effects, which could be of clinical significance.

IV. EXPERIMENTAL STUDIES OF RESPONSE IN MAMMALS

203. The effects of radiation in animals have been studied in numerous experiments. Evidence of possible adaptive responses have sometimes emerged in lifetime studies of animals exposed to acute or chronic low-LET radiation at doses well below those associated with bone marrow failure. Life-span and incidence of neoplastic diseases have been assessed. Other experiments have been done with animals exposed to non-lethal conditioning doses prior to acute exposures to potentially lethal doses. It should be noted that in many of these studies the conditioning doses were well above the range of doses producing the adaptive response in cellular systems.

A. SHORT-TERM SURVIVAL FOLLOWING ACUTE, HIGH-DOSE EXPOSURE

204. Experiments were carried out in the 1950s and 1960s to investigate the hypothesis that stimulating haematopoietic stem cells to proliferate before exposing animals to a potentially lethal dose of radiation could improve their chances of recovering from acute bone marrow failure. A historical review of these experiments was given by Dacquisto and Major [D2].

205. In one of these early experiments, adult female Swiss white mice (Bethesda Naval Medical Research Institute strain) were acutely exposed to a dose of about 1.2 Gy from 250 kV x rays at weekly intervals over a period of three weeks [C18]. Thirty days after the third exposure, the mice were acutely exposed to a challenge dose of about 5.6 Gy, which was known to be the dose approximating the $LD_{50/30}$. Of the animals in this group, 26% died after 28 days, compared with 41% in the control group, which received only the challenge dose.

206. This observation was confirmed by another study in which adult female Swiss white mice (Walter Reed strain) were acutely exposed to a conditioning dose of about 0.4 Gy from x rays, either 10 or 15 days before determining $LD_{50/30}$ values [D2]. The $LD_{50/30}$ values for the two irradiated groups were about 4.48 ± 0.25 Gy and 4.94 ± 0.25 Gy, respectively, compared with 3.90 ± 0.21 Gy for animals not exposed to the conditioning dose. There was no evidence of splenic or thymic hypertrophy in animals that received the conditioning doses, in contrast to an earlier observation that localized splenic irradiation to a dose of about 0.16 Gy caused transient hypertrophy [P13].

207. Age at exposure was shown to be an important factor influencing survival. Female mice of the BDF_1/J (C57Bl/6J × DBA/2J) strain were exposed to 0.8, 2.4 or 3.2 Gy from 250 kV x rays at 90 days of age, and $LD_{50/30}$ values were determined at various ages thereafter [S55]. The results, given in Table 29, showed that there was no statistically significant difference in LD_{50} values between controls and those pretreated with 0.8 Gy in animals below the age of 550 days. With higher conditioning doses, however, there was residual damage in the animals resulting in lower resistance to a lethal dose (expressed as reduced LD_{50} values).

208. As a recent example of these short-term survival studies, two-month-old SPF mice of the C57Bl strain were acutely exposed to a range of doses from x rays between 0.025 and 0.1 Gy [Y9]. Two months later, they were acutely exposed to 7.75 Gy, and 30-day survival was measured. The results, given in Table 30, showed a marginally significant increase in the life-span of animals receiving the lowest conditioning dose and a definite improvement in survival at conditioning doses of 0.05 and 0.1 Gy compared with animals not exposed to a conditioning dose. The improvement in survival rate coincided with an increase in endogenous colony-forming units, consistent with increased proliferation of haematopoietic stem cells following the conditioning doses.

209. In a subsequent experiment, mice were irradiated with x rays to doses of 0.025-0.5 Gy at six weeks of age and exposed two months later to a dose of 7 Gy from x rays [Y7]. A dose of 0.025 Gy was insufficient to affect survival after exposure to the high dose, but pretreatment with 0.05-0.1 Gy produced a significantly increased survival rate. However, the response to pretreatment with 0.05 Gy after two months was strain-specific. It occurred in C57Bl but not in BALB/c mice. Surprisingly, an increase in survival was not observed in animals pretreated with 0.2 Gy if exposure to the challenge dose occurred up to 1.5 months after the conditioning dose; and pretreatment with 0.5 Gy resulted in an increase in survival if the challenge dose was given two weeks later but not after two months. These findings persuaded the authors to postulate different mechanisms involving time-related stimulation of haematopoiesis, although they were not specified.

B. LONG-TERM SURVIVAL FOLLOWING SUB-LETHAL EXPOSURE

210. Studies using rodents and beagle dogs have provided some insight into the long-term effects of acute exposure to low doses and of chronic exposure at low dose rates. They date from the early pioneering work of Russ and Scott [R2, R3].

1. Experiments with rodents

211. The effect of chronic irradiation of mice was described by Lorenz et al. [L14, L19] in an early experiment in which both survival and tumour incidence were measured. Male and female mice of the (C57Bl × A)F_1 strain, referred to as LAF_1, were exposed from the age of one month for the duration of life to gamma rays from ^{226}Ra delivered at the rate of about 1 mGy for 8 hours each day. Control mice were housed in a room adjacent to the exposure room. Almost a year after the start of the experiment, the control mice developed dermatitis and had to be replaced by mice not directly related to the irradiated mice but from crosses of the same inbred strains.

212. The mean survival times of the irradiated male mice was significantly higher than those of the control mice, but there was no statistical difference between the two groups of females, although there was a tendency for the females to live longer than the males (Table 31). However, the differences in life-span in the males cannot be accepted without reservation, since the replacement control group may have been subject to different environmental conditions and may have possessed genetic characteristics slightly different from those of the original group.

213. Although no single cause of death predominated, pyelonephritis appeared to be a major contributing factor. Its occurrence is frequently associated with dermatitis, particularly among males. There was an increase in the incidence of lymphosarcomas and other tumours of the reticular tissues in both male and female irradiated mice, and mammary carcinoma incidence was also higher in the irradiated females ($p < 0.1$). The irradiated females showed an increase in ovarian tumours ($p < 0.05$), and the irradiated males showed an increase in lung tumours ($p < 0.05$). These results were essentially confirmed in a study in which guinea pigs of the Tumble Brook Farms strain were exposed to ^{60}Co gamma rays from the age of 100 days for the duration of life at a dose rate of about 1 mGy d^{-1} [R11].

214. Sacher and Grahn [S52], in a study of the survival of mice exposed for the duration of life, also reported an increase in the life-span of animals exposed to ^{60}Co gamma rays at a dose rate of up to about 30 times that of the background radiation. Male and female mice of the LAF_1 strain were exposed to a wide range of doses at different dose rates. Non-irradiated animals serving as controls were placed either in a corridor adjacent to the irradiation facility or in an adjoining room. It was subsequently discovered that the control group in the corridor were in fact exposed to stray radiation from the irradiation facility at a rate of between 1.6 and 4.9 mGy a^{-1}. This compared to about 0.14 mGy a^{-1} in the control group in the adjoining room. The mean survival times are shown in Table 32. For comparison, the results for animals irradiated at the rate of 40 mGy a^{-1} are given. There was a statistically significant increase in the mean survival times of males but not of females exposed to between 1.6 and 4.9 mGy a^{-1} compared with those exposed to dose rates of 0.14 mGy a^{-1}.

215. One interpretation of these studies is that any increase in survival for male mice receiving the higher doses of radiation reflects a decreased number of deaths in early life. A possible explanation could be that minimal injury to the haematopoietic system causes a rebound in stem cell proliferation. This regenerative hyperplasia could then create a larger mass of tissues devoted to the defence against intercurrent and potentially lethal infection. This mechanism, however, does not prevent the occurrence of tumours that appear in late life, which are not necessarily the cause of death. Why this response is confined to males cannot be explained.

216. Upton et al. [U12, U13] studied the late effects of low-LET radiation in RF/Un mice. Life-span and the incidence of neoplastic diseases were assessed. The effects of mean accumulated doses up to about 3 Gy at varying dose rates between 0.05 Gy d^{-1} and 0.8 Gy min^{-1} are shown in Table 33. Several interesting features emerge from these studies. In males, there was little effect on life-span (mean age at death) at dose rates of 0.8 Gy min^{-1} until the accumulated dose approached about 1 Gy; there was little effect of dose rate in the range 0.05-0.77 Gy d^{-1} at an accumulated dose of about 1.5 Gy; and, overall, gamma rays were less effective at low dose rates than at high dose rates at accumulated doses of about 3 Gy. In females, there was no significant change in the mean age at death at dose rates of 0.067 Gy d^{-1} until the accumulated dose reached about 2 Gy; reducing the dose rate to 0.01 Gy d^{-1} increased the mean age at death at accumulated doses of about 3 Gy.

217. The mean age at death of animals with neoplasms was also influenced by dose rate. Thus in males, irradiated animals died earlier than controls at dose rates of 0.8 Gy min^{-1}, even at the lowest accumulated dose of 0.25 Gy, but there was no significant difference compared with controls at dose rates of 0.15-0.77 Gy d^{-1} at an accumulated dose of about 1.5 Gy, with a marginal difference at a dose rate of 0.05 Gy d^{-1}. In females, earlier death from neoplasms occurred at dose rates of 0.067 Gy d^{-1} after an accumulated dose of about 2 Gy, and there was no difference from controls at doses of

3 Gy, if the dose rate was reduced to 0.01 Gy d^{-1}. In summary, these results support the view that at low dose rates and accumulated doses of a few gray there is no apparent life shortening or early appearance of neoplastic diseases causing death of the irradiated animals compared with unirradiated controls. There is, as well, no support for an apparent increase in life-span or decrease in tumour incidence due to the radiation exposures at the lowest doses.

218. Ullrich and Storer [U14, U15] also described the influence of radiation on life-span and tumour induction in mice of the RFM and BALB/c strains. Specifically, a study was made of the effects produced by ^{137}Cs gamma-radiation delivered at 0.4 Gy min^{-1} or 0.083 Gy d^{-1}, within the dose range 0.1-4 Gy for RFM mice and 0.5-4 Gy for BALB/c mice. At the lower dose rate, the life-shortening effect in BALB/c mice could be described as a linear or linear-quadratic function of dose, although the lowest dose used was 0.5 Gy. The influence of dose at a dose rate at 0.083 Gy d^{-1} on the incidence of neoplastic disease in female BALB/c mice is shown in Table 34. For RFM mice, thymic lymphomas were the predominant reticular tissue neoplasm, the solid tumours being lung adenomas or endocrine-related tumours. The incidence of tumours was at all doses higher than in the unirradiated controls.

219. Sato et al. [S42, S53] exposed male and female mice of the C57Bl/6J strain to ^{137}Cs gamma rays at rates of 0.029-0.38 Gy d^{-1}. Exposures at the rate of 0.029 Gy d^{-1} started at four weeks of age. The mean life-span of female mice was 628 ± 10 days, with a mean accumulated dose of 16.4 Gy; for male mice, the mean life-span was 723 ± 10 days, with a mean accumulated dose of 18 Gy. The mean life-spans of unirradiated females and males were 693 ± 11 days and 725 ± 15 days, respectively. In the irradiated females, this life shortening was statistically significant; in the males, it was not. In a follow-up study, life shortening at the higher dose rates was associated with the occurrence of thymic lymphomas, which occurred more frequently in exposed younger mice [O5]. At the highest dose rate, and at accumulated doses of 39 Gy over a period of 105 days, life shortening was also associated with haemorrhage and infectious diseases owing to depletion of the stem cells of bone marrow.

220. Covelli et al. [C20] studied the effects of x rays on life-span and tumour induction as a function of age at exposure. Male and female mice of the $BC3F_1$ strain were acutely exposed to 250 kV x rays, either *in utero* at 17 days after coitus, at 3 months, or at 19 months of age. The doses varied from 0.3 to 2.1 Gy for *in utero* irradiation and from 0.5 to 7 Gy at 3 months or 19 months of age. Prenatal irradiation or irradiation at 19 months of age did not result in clearly measurable life shortening (Table 35). There was, however, a trend suggestive of life shortening in animals exposed at 3 months of age at doses in excess of about 2 Gy, albeit the confidence limits were extremely wide. There was a non-significant trend suggesting an increase in life-span in animals exposed at 19 months of age at a dose of 2 Gy. Life shortening was associated with an increased incidence of reticulum cell sarcoma (a non-thymic malignant lymphoma that infiltrates the spleen, liver and mesenteric lymph nodes). Primary tumours of the lung and liver, mostly benign adenomas, and subcutaneous fibrosarcomas were also found. However, no difference could be detected in the patterns of total tumour incidence for irradiated and non-irradiated (control) animals.

221. Sasaki and Kasuga [S54] compared the effects of ^{137}Cs gamma rays on female mice of the $B6C3F_1$ strain exposed under specific-pathogen-free conditions until natural death. Mice were irradiated at 17 days *in utero*, or at birth, or at 15 weeks of age with doses of 1.9, 3.8 or 5.7 Gy. There was a decrease in life-span at 1.9 Gy in all groups, but no statistically significant increase in total tumour incidence between irradiated and control groups was observed (Table 36). Mice in the fetal period were found to be susceptible to pituitary tumours and liver and lung tumours. There was an excess of malignant lymphomas at a dose of 5.7 Gy, and the total time until incidence was less than with controls. The exposure of young adults was associated with the induction of myeloid leukaemias and Harderian gland tumours.

222. Maisin et al. [M29] measured life-span and disease incidence in the C57Bl/Cnb strain of mouse exposed to acute and fractionated doses of ^{137}Cs gamma rays. Twelve-week-old mice were exposed to doses ranging from 0.25 to 6 Gy at a dose rate of 0.3 Gy min^{-1} in a single session or in 10 sessions delivered 24 hours apart, or in 8 sessions, delivered 3 hours apart. The results of an acute exposure are shown in Table 37. There was no indication of an increase in life-span at these doses; decreases in life-span were not statistically evident until the dose approached 1 Gy. A fractionated exposure was less effective in reducing survival time than an acute exposure, and life shortening was not observed in the 8 and 10 fraction protocols until the dose approached 2-3 Gy and 4 Gy, respectively.

223. The causes of death in unirradiated mice could be segregated into late degenerative changes in lung and kidney (glomerulosclerosis) (~60%), leukaemias (~21%) and carcinomas (mainly liver adenocarcinoma) and sarcomas (~16%). The incidence of thymomas was low (~1%), and no radiation-induced excess of lymphoma was observed until the doses approached 2-4 Gy, but the tumours did result in death at an earlier age than in unirradiated animals. The incidence of all malignancies was less than in the unirradiated mice after acute doses between 0.25 Gy and 1 Gy, with no significant change in the alpha parameter of the Cox linear proportional hazard models within this dose range. Fractionation increased the

incidence of all malignancies: thus their incidence was 45% and 52% at 1 Gy and 2 Gy, respectively, for the 8-fraction protocol and 40%, 32% and 36% at 0.25 Gy, 0.5 Gy and 1 Gy, respectively, for the 10-fraction protocol. The authors concluded that these differences between fractionated and acute exposures were small and not well enough established from a statistical viewpoint.

224. The effects of gamma rays and fission neutrons on the life-span of mice has been studied for many years at the Argonne National Laboratory in the United States. Earlier studies of low-LET radiation principally involved ^{60}Co gamma rays delivered either as acute exposures or continuously [C19, T15]. For the most part, the acute exposures were delivered over a 20-minute period at dose rates ranging from 0.0083 to 0.094 Gy min^{-1}. The exposure regime began when the animals were 110 days of age.

225. The results obtained on the mean survival after irradiation of male mice of the B6CF$_1$ strain exposed for 22 hours each day, five days each week have been reported [T15]. The exposures were continued for either 23 or 59 weeks, the accumulated doses ranging from 2.1 to 24.6 Gy. The effects of these exposure conditions are shown in Table 38. There was no indication of an increase in life-span at any dose in this range. Life shortening from deaths due to all causes did not become significant until the total dose approached 4 Gy for a 23-week exposure or 5 Gy for a 59-week exposure. These values were not significantly altered when the analysis was restricted to mice dying from tumours. Above ~4 Gy, the dose-response relationship was linear and inversely dependent on the protraction period. The life-shortening coefficients for the 23-week continuous and 59-week continuous protocols were 0.16 and 0.08 days lost per 0.01 Gy, compared with 0.39 days for an acute single exposure, thus showing a marked dose-rate effect.

2. Experiments with beagle dogs

226. The effects on life-span of external whole-body exposure of beagle dogs to x or gamma rays has been studied. In one study, dogs were exposed to 0.16 Gy or 0.83 Gy from ^{60}Co gamma rays at 8, 28 or 55 days *in utero* or 2 days after birth. In addition, some dogs were exposed to 0.83 Gy at 70 days or 365 days after birth. The status of these studies in 1982 is shown in Table 39. Through 10 years of age, no differences in survival were evident in any of the exposure groups [B23]. Irradiation during the fetal period was, however, associated with abnormalities of skeletal, dental and central nervous system development. Perinatal irradiation resulted in kidney dysplasia and, in animals receiving higher doses, in chronic renal disease. The thymus gland, particularly the thymic epithelium, was found to be highly radiosensitive during fetal development.

227. In another study, dogs were exposed bilaterally to 250 kV x rays, delivered in different numbers of fractions and different fractionation intervals, to total doses of 1-3 Gy. The dogs were exposed between 8 and 15 months of age. Some dogs were bred after exposure, distinguished as parous or nulliparous. The dogs listed in each group shown in Table 40 are those surviving at least 90 days after the irradiation. Two general summaries of the data have been published, one before all the animals were dead [A13], the other emphasizing the effects on life-span and tumour induction [R12].

228. There was no increase in survival time, expressed as median survival after exposure, of any of the irradiated dogs compared with the unirradiated controls. Although life shortening was only marginal in animals given acute or fractionated doses up to 1 Gy, it occurred in some groups of animals given 2 or 4 fractionated doses of between 0.75 and 1.5 Gy total dose and fractionated doses of 3 Gy. The main causes of death were similar in irradiated and unirradiated dogs. The development of non-neoplastic diseases (essentially fibrosis) at an earlier age in irradiated animals at the higher dose explained, in large part, the observed life shortening. However at 1 Gy, the incidence of non-neoplastic diseases, mammary tumours and non-mammary tumours was broadly similar to that observed in the unirradiated group.

229. In a third study, dogs were exposed to ^{60}Co gamma rays continuously until death or, in a companion experiment, until they had received a predetermined total dose [F1, F22, G13]. Dose rates ranged from 0.003 to 0.054 Gy d^{-1}, and the accumulated doses at termination of exposures ranged from 4.5 to 30 Gy. At dose rates greater than 0.019 Gy d^{-1}, the responses were dose-rate-dependent and limited primarily to damage of the haematopoietic system (Table 40). At dose rates between 0.003 and 0.008 Gy d^{-1}, myeloproliferative diseases ultimately expressed as myelogenous leukaemia were the limiting factor for survival [K25].

C. SUMMARY

230. Observations of adaptive responses in *in vitro* cellular studies cannot be readily extrapolated to postulate adaptive response in the irradiated animal. The different exposure patterns in animal studies (acute versus chronic lifetime exposure) could involve mechanisms different from those involved in the cellular adaptive response to a low conditioning dose followed by a high challenge dose. The complexity of multicellular organisms, including the crucial role of immunosurveillance and endocrine factors in maintaining the healthy state, must be taken into account. The careful management of the animal colonies to avoid infections and stress and the need to recognize strain-specificity of tumours should not be overlooked in future experiments.

231. Stimulating the haematopoietic system to proliferate by exposing the bone marrow *in vivo* to between about 0.025 and 0.1 Gy has been shown to improve the short-term survival (expressed as $LD_{50/30}$) of mice subsequently exposed to a potentially lethal dose of radiation. Whether a similar mechanism is responsible for improving the longer term survival of animals chronically exposed at low dose rates is unclear. It is plausible to suggest that improvement in long-term survival could be a result of chronic minimal injury to the bone marrow, causing a rebound in stem cell proliferation and protection against infections, but occurring along with this effect is an increased risk of malignancy due to the proliferation of potentially malignant cells.

232. Experiments in the 1950s indicated that the mean life-span of male, but not female, mice could be increased if the animals were exposed daily to low-LET radiation from a few milligray to several hundred milligray per year above the level of background radiation. A feature of these observations was that the mean increase in life-span reflected a decreased number of deaths in younger irradiated animals. Why the response was confined to males is not apparent.

233. Other experiments related life-span to the incidence of neoplastic diseases. Taken together, the results of these experiments could be interpreted as demonstrating that, compared with the pattern observed in unirradiated controls, there was no significant effect on mean life-span following exposure to accumulated doses up to a few gray, at dose rates between 0.005 and 0.3 Gy d^{-1}, or on the time of appearance of tumours or on tumour incidence among irradiated animals. Why some experiments resulted in an increase in life-span is not easily explained. While the observed increase in life-span could be due to random variability, it is possible that the effect is real. If so, it is important to understand the precise conditions under which life-span is increased.

234. More recent experiments using mice have confirmed that there is no statistically significant change in tumour incidence in irradiated mice compared to unirradiated controls below about 2 Gy, depending on the strain of mouse used. At higher doses there is a dose-related increase in several tumour types causing death and a corresponding decrease in mean life-span.

235. The results of experiments using beagle dogs are in general agreement with those obtained using mice. In addition, they confirmed the increased sensitivity of the fetus and young animals. Observed life shortening at accumulated doses greater than a few gray could be related to the development of non-neoplastic diseases at an earlier age than in unirradiated animals. Irradiated animals were susceptible to the development of myeloproliferative disorders, which were frequently expressed as myelogenous leukaemia at dose rates above about 0.003 Gy d^{-1} and after accumulated doses of a several gray. Error-prone DNA repair mechanisms were proposed to explain the onset of myeloproliferative disorders.

V. EPIDEMIOLOGICAL STUDIES OF RESPONSE IN HUMANS

236. The effectiveness of a cellular adaptive response in humans, expressed in terms of a reduced rate of spontaneously occurring cancers or of a reduction in the expected numbers of radiation-induced cancers, would be most convincingly demonstrated if it were the outcome of epidemiological studies involving exposure to low doses. The general results of these studies are discussed in Annex A, "Epidemiological studies of radiation carcinogenesis", most of which indicate steadily increasing cancer incidence with increasing dose. Only those cases in which exposures were only marginally above the natural background level are re-examined in this Chapter. For inherent reasons, the evidence of an increase in the spontaneous incidence of cancer is in these cases equivocal.

237. There are theoretical reasons based solely on the nature of DNA damage and repair to expect that cancer can occur at the lowest doses without a threshold in the response, although this effect would perhaps not be statistically demonstrable. The dose-response relationship has been judged to be linear without threshold for all cancers excluding leukaemia. For leukaemia, the relationship that best fits the data is linear-quadratic and linear at low doses. An increase in total cancer mortality has been statistically demonstrated at doses above ~0.2 Gy of low-LET radiation but not at lower doses, except in special circumstances such as cancer in childhood following *in utero* irradiation and thyroid cancer after acute exposure of the thyroid to low-LET radiation during childhood.

238. Among the limitations on the confidence of low-dose epidemiological studies, by far the most consistent problem has been the lack of statistical power. This has been due to a combination of factors, including insufficient numbers in the irradiated population, inadequate follow-up and doses that are collectively too small to have any chance of providing a dose-response relationship.

A. INHABITANTS OF HIGH-BACKGROUND AREAS

239. Examples of background radiation studies in France, Japan, Sweden, the United Kingdom and the United States are reviewed in Annex A, "Epidemiological studies of radiation carcinogenesis". The conclusion from these studies is that there is no demonstrable association between leukaemia and other cancers and background radiation.

240. Updated estimates of the cancer risk for a population of about 80,000 persons continuously exposed to natural background radiation at a dose rate about three times higher than the world average have recently been reported [W29, W30]. This population lives in two areas in Yiangjiang County of Guangdong Province, China. A comparison is being made with a population of similar size and age structure in a nearby control area of low background radiation in Enping County. Details of the doses are given in paragraph 188. These populations are stable in that there is little movement into and out of the areas. Their age structure is biased towards younger age groups compared with the rest of China. Ascertainment of cause of death is of a high standard, and the dose measurements, based on several different techniques, are in good agreement. The rates of site-specific cancer mortality and estimates of excess risk in the high-background-radiation area, relative to the control area, are given in Table 41 [W29]. The high incidences of primary liver cancer and nasopharyngeal cancer are peculiar to the whole of Guangdong Province and are probably related to local environmental factors, which include infection with hepatitis B virus and smoking, respectively. These two cancer types comprise the largest number of cancers in the two areas and are highest in the low background area. The totals of all cancers other than leukaemia are not strictly comparable with these cancers included.

241. While there are slight differences in the site-specific cancer mortality rates between the two populations, the overall differences in leukaemia and other cancers combined are not statistically significant, and the results so far do not provide clear evidence of the presence or absence of deleterious effects due to low doses of radiation in the environment. This may not be surprising, considering the difference in cumulative dose between the two areas, based on a 50-year exposure, of about 60 mSv. The problem, then, is the lack of statistical power to detect an effect on risk at such low doses.

242. Thyroid nodularity following continuous low-dose exposure in China was determined in about 1,000 women aged 50-65 years living in the high-background-radiation area and in a similar number of controls living in the low-background control area [W28]. Cumulative doses to the thyroid were estimated to be about 0.14 Gy and 0.05 Gy, respectively. For multiple nodular disease, the prevalences in the high-background-radiation area and in the control area were 9.5% and 9.3%, respectively. For single nodules, the prevalences were 7.5% in the high-background area and 6.6% in the control area (prevalence ratio = 1.13; 95% CI: 0.82-1.55). No differences were found in serum levels of thyroid hormones. Women in the high-background region, however, had significantly lower concentrations of urinary iodine. The prevalence of mild diffuse goitre was higher in the high-background-radiation region, perhaps related to a low dietary intake of iodine. The authors suggested that continuous exposure throughout life to a total dose of about 0.1 Gy in excess of the low background level did not influence the risk of thyroid nodular disease; however, this study did not provide statistically significant results.

B. OCCUPATIONALLY EXPOSED INDIVIDUALS

243. Studies of the effects of radiation following occupational exposures are useful in clarifying the possible relationship between relatively low doses of gamma-radiation and the risk of cancer. One relevant study was of nuclear shipyard workers in the United States [M13]. From a database of almost 700,000 shipyard workers, including about 108,000 nuclear workers, three study groups were selected, consisting of 28,542 nuclear workers with working lifetime doses ≥5 mSv (many of them received doses well in excess of 5 mSv), 10,462 nuclear workers with doses <5 mSv and 33,352 non-nuclear workers. The type of work carried out by the three groups was identical, except that the nuclear workers were exposed additionally to ^{60}Co gamma-radiation. The median age of entry into employment for the three groups was similar, that is, about 34 years. The study included exposures received from the beginning of nuclear ship overhauls in the 1960s until the end of 1981. All groups worked in areas where intakes of asbestos were possible.

244. Deaths in each of the groups were classified as due to all causes, leukaemia, lymphatic and haematopoietic cancers, mesothelioma and lung cancer. The results, summarized in Table 42, demonstrate a statistically significant decrease in the standardized mortality ratio for the two groups of nuclear workers for "death from all causes" compared with the non-nuclear workers. This was due in part to a higher incidence in the non-nuclear workers of deaths from diseases other than cancer, which included cardiovascular disease and respiratory, genito-urinary and digestive tract disorders. Both groups of nuclear workers had lower death rates from leukaemia and from lymphatic and haematopoietic cancers than the non-nuclear workers, but this was not statistically significant. All three groups of workers had lower death rates from lymphatic and haematopoietic cancers than the general population in the United States. This has been referred to as the healthy worker effect. Mesothelioma was the only

cancer that showed a significantly higher incidence for all groups. The slightly higher, but non-significant, incidence of lung cancer for all three groups compared to the general population of the United States could have been associated with asbestos exposure.

245. Two features of this study need to be emphasized. First, the collective lifetime doses from occupational exposure in the two nuclear worker groups were estimated to be 1,450 man Sv and 26 man Sv, respectively. The estimated collective doses from natural background radiation in the same periods were 1,067 man Sv and 409 man Sv, respectively. The collective dose to the non-nuclear workers from background radiation was estimated to be 1,275 man Sv. It is difficult to draw conclusions about the incidence of diseases that might be associated with low-level occupational radiation exposures when the doses are comparable to, or less than, the doses from natural background radiation. Also, the statistically significant decrease in standardized mortality ratio for deaths from all causes cannot be due to the healthy worker effect alone, since the non-nuclear workers and the nuclear workers were similarly selected for employment and were afforded the same health care thereafter.

246. A study of about 95,000 radiation workers in the United Kingdom has recently been reported. The cohort received a collective lifetime dose of 3,200 man Sv, with an average individual dose of 34 mSv [K15]. Up to 1988, 6,600 workers had died. The standardized mortality ratio for all causes of death after excluding the first 10 years following the start of radiation work was 0.85 ($p < 0.001$). The standardized mortality ratios for all malignant neoplasms in 23 organs or tissues and for all known, non-violent causes other than malignant neoplasms were 0.86 ($p < 0.001$) and 0.84 ($p < 0.001$), respectively. For most other tissue-specific cancers, the standardized mortality ratios were below unity, but they were not statistically significant. The only cancer for which an elevated standardized mortality ratio reached statistical significance was thyroid cancer (SMR = 3.03). It is not unexpected that one organ or tissue showed an increased standardized mortality ratio by chance. The excess relative risks for all cancers and for leukaemia, excluding chronic lymphocytic leukaemia, were 0.47 (90% CI: -0.12-1.20) and 4.3 (0.4-13.6), respectively. A confounding factor in this study was that cancers could have been associated with exposure to chemical carcinogens.

247. Since the healthy worker effect complicates the interpretation of standardized mortality ratios, greater importance was attached by the authors to internal analysis or to testing for the trend in risk with dose, in which steps are taken to compensate for the factors that lead to the healthy worker effect. Positive trends with dose were found for all malignant neoplasms taken together and for leukaemias, excluding chronic lymphatic leukaemia (Figure XX). The former trend did not reach statistical significance ($p = 0.10$, one-sided test), while the latter was statistically significant ($p = 0.03$). Because it was considered possible that cigarette smoking may have influenced the results, Figure XX also shows the relative risk for the malignant neoplasms, excluding both lung cancer and leukaemia. The scatter of the points showing the average relative risk and the wide confidence intervals on these points provide no reliable information on the risks at doses below about 0.2 Gy.

248. Updated internal analyses of mortality in 33,000 workers monitored for six months or longer at the Hanford site in the United States provide no evidence of a correlation between cumulative occupational external dose and mortality from leukaemia and from other cancers [G12]. The relative risks and their confidence intervals are summarized in Table 43. Of 24 tissue-specific cancer categories evaluated, only cancer of the pancreas and Hodgkin's disease showed positive correlations with dose that approached statistical significance (one-tailed p-values of 0.03 and 0.04, respectively). These correlations were interpreted by the authors as probably spurious. A significant correlation ($p < 0.05$) was obtained at doses above 50 mSv but not at 10 mSv. Earlier analysis of mortality from multiple myeloma had shown a significant excess risk [G14], but the extended analysis of these data now show that the excess relative risk for this form of cancer is no longer statistically significant [P1, R10].

249. A study has been under way since 1980 on the mortality of past and present employees of Atomic Energy of Canada Ltd. [G8]. The study population consists of 13,491 persons, 9,997 males and 3,494 females, for a total of 262,403 person-years at risk until 1985. The number of female deaths (121) was too small for detailed analysis, but the 1,178 deaths in the male population gave a limited basis for study. Mortality patterns in the cohort between 1950 and 1985 were examined by comparing the observed mortality with that expected in the general population for three groups of workers: those with no occupational exposure, those with up to about 50 mSv and those with more than 50 mSv. The number of deaths was fewer than would have been expected on the basis of general population statistics for males in the three groups. The findings were similar for the groups "all cancer deaths" and "all other causes of death". In the occupationally exposed males, elevated standardized mortality ratios were seen for non-Hodgkin's lymphoma and for buccal cavity, rectum, rectosigmoid and prostate cancers. But in the

unexposed male group, there were elevated standardized mortality ratios for lymphatic and myeloid leukaemias and for large intestine, prostate, brain and biliary system cancers. The number of cases identified of all these cancers was small and the confidence limits were wide, such that none of the elevated standardized mortality ratios were statistically significant.

C. SURVIVORS OF THE ATOMIC BOMBINGS

250. The epidemiological study of the survivors of the atomic bombings in Japan has up to now been the primary source of data from which to estimate the effects of radiation on humans. A dose-response analysis of these data for doses less than 0.5 Sv was recently presented [S30]. The end-points measured were cancer mortality, cancer incidence and non-cancer mortality. The relative risks with 95% confidence intervals and the dose-response relationships fitted by the authors are illustrated in Figure XXI.

251. For mortality from leukaemia, the relative risks varied among the five dose groups (0.010-0.019, 0.020-0.049, 0.050-0.099, 0.100-0.199, 0.20-0.49 Sv), but they did not differ statistically from the control group, which received less than 0.010 Sv ($p > 0.10$). The relative risks for the three dose groups less than 0.1 Sv were less than unity but still within the range of what the authors considered to be random variation about a value equal to unity. For leukaemia mortality, a linear-quadratic model fitted marginally better than a linear model ($p = 0.07$ compared with 0.06).

252. For all cancers other than leukaemia, the relative risks generally increased with dose, as shown by the dose-response curve, although the lowest four dose groups had relative risks not significantly different from unity. The highest dose interval (0.20-0.49 Sv) showed a statistically significant increase in mortality.

253. Mortality from lung cancer and incidence of thyroid cancer by dose group, based on data from the Hiroshima and Nagasaki tumour registries during 1958-1987, were also analysed. A pattern similar to that observed for mortality from all cancers except leukaemia was seen, that is, the relative risk varied among comparison groups with wide confidence limits but did not differ statistically from unity within the lowest dose intervals.

254. For mortality from all diseases other than cancer, a significantly elevated risk was observed at higher doses (estimated threshold dose: 1.5 Sv) for younger survivors of the atomic bombings (age at the time of bombings <40 years). However, the relative risks for the various low-dose groups (<0.5 Sv) did not differ and were close to unity with the exception of one significant point at the dose interval 0.20-0.49 Sv (RR = 0.83). It is interesting to note that the decrease in relative risk at this point corresponds to a significant increase in relative risk for mortality from all cancers except leukaemia.

D. PATIENTS EXAMINED OR TREATED WITH RADIATION

255. There are several examples in clinical practice where low doses have been used for diagnostic purposes. These are x-ray examinations to detect fetal abnormalities, x-ray fluoroscopy to check the efficacy of artificial pneumothorax in the treatment of pulmonary tuberculosis, (e.g. [B2]), x-ray examinations to assess the progress of skeletal development during treatment for scoliosis and ^{131}I diagnostic tests to detect thyroid abnormalities.

256. The available studies [G3, H19, M8, M22, M26, S24] of *in utero* exposure are discussed in detail in Annex A, "Epidemiological studies of radiation carcinogenesis". The main conclusion is that the low-dose exposure of the fetus in diagnostic examinations is associated with a positive risk of cancer induction, but quantification of the risk is subject to much uncertainty. There is nothing to support the assumption that adaptive processes could be operating after irradiation that could reduce the incidence of radiation-induced cancers.

257. The evidence for radiation-induced breast cancer is discussed in Annex A, "Epidemiological studies of radiation carcinogenesis". Although exposure to radiation at high doses and high dose rates is associated with excess breast cancer, the potential hazard from low-dose, fractionated exposures during early breast development has not been thoroughly evaluated. The failure to detect increased breast cancer in several large studies is surprising, and no satisfactory explanation is forthcoming. However, many of the women were over 35 years of age at exposure.

258. A retrospective study of 35,074 Swedish patients receiving ^{131}I for suspected thyroid disorders between 1951 and 1969 has been reported [H17, H18, H21]. It is not possible to be precise about the doses delivered to individual thyroids because of differences in ^{131}I uptake and variations in mass of the thyroids. However, the average amount of ^{131}I activity administered was 1.92 MBq, delivering an estimated average dose of approximately 0.5 Gy to the thyroid. The data are given in Table 44. Patients given ^{131}I for reasons other than a suspected tumour were not at increased risk (standardized incidence ratio = 0.62; n = 16). Overall, the data provide no indication that exposure to ^{131}I for diagnostic purposes

increased the incidence of thyroid cancer in the follow-up period after 10 years. It was concluded by the authors that the observed increase in the 5-9 year period was most likely to be due to a high level of medical surveillance, leading to an increased detection of indolent tumours.

E. SUMMARY

259. The human epidemiological studies following exposures at low doses and low dose rates to low-LET radiation do not at present provide evidence of an adaptive response expressed as a decrease in the prevalence of spontaneously occurring human cancers. This is not surprising in view of the low statistical power of these studies. The results have been interpreted variously as being consistent with the upper bound on the confidence limits of total cancer risk at low doses obtained by extrapolating from high-dose and high-dose-rate data; or as indicating no additional risk at low doses compared with the spontaneously occurring rate. Statistical limitations do not permit a decisive choice at the present time. Caution is necessary when using isolated examples in the epidemiological literature to justify either an increased or a decreased risk at doses of a few hundred milligray, bearing in mind the statistical limitations of the data.

CONCLUSIONS

260. Adaptive response is the collective term used to describe the results of experiments in which a small dose of radiation can condition cells so as to induce repair processes and/or to stimulate proliferation. One consequence of DNA repair might be to reduce the natural incidence of cancer in its various forms or the likelihood of excess cancers being caused by further radiation exposure. A great deal of effort is being directed into characterizing these processes, and in recent years results of research especially at the cellular level have become available.

261. There is convincing evidence that the number of radiation-induced chromosome aberrations and mutations that occur in proliferating mammalian cells after an acute dose of low-LET radiation in the range 1-3 Gy can be reduced by exposing the cells to an acute dose of between a few milligray and a few tens of milligray several hours before the high dose. These experiments involving a low conditioning dose and a high challenge dose were designed to demonstrate the adaptive response as a laboratory phenomenon. They were carried out under clearly defined conditions using mitogen-stimulated lymphocytes, proliferating bone marrow cells, spermatocytes and fibroblasts. The response has not been demonstrated so far in other cell systems or convincingly in cells under conditions of chronic exposure.

262. The evidence that is becoming available indicates that following radiation-induced damage to cells, a number of changes occur. Among these changes are the activation of several classes of genes, including those coding for the synthesis of enzymes involved in the control of cell cycling, proliferation and repair. It is not entirely clear how these changes may specifically improve repair capacity. There is some evidence to indicate that radiation-induced enzymes, which remain to be isolated and characterized, are related to stress-response proteins. There seems to be some similarity in the types of damage induced by radiation and other toxic agents. The adaptive response may therefore be a common feature of cellular response to damage.

263. A multi-step process has been proposed to explain the cellular adaptive response. After acute doses of several hundreds of milligray, cell cycling in proliferating cells may be delayed. The period of delay allows enzymes induced by the radiation to repair damage before the cells proceed through cell cycle and undergo mitosis [Y5, Y6]. The adaptive response at these higher doses may therefore depend on whether or not the cells in cycle are temporarily blocked. There is no direct evidence, however, to suggest that cell cycle delay occurs after acute doses ranging from a few milligray to a few tens of milligray, even though the adaptive response has been observed in this range of doses. The fate of cells exposed to a radiation dose in the resting phase remains to be established.

264. In evaluating the effectiveness of the adaptive response in cells exposed to a conditioning dose of up to a few tens of milligray or to concentrations of toxic agents below that concentration known to produce a toxic reaction, it is important to recognize the unstable nature of DNA in living cells during normal metabolism [H2, L11, L21, V3, V4]. It has been estimated that the DNA molecule within each nucleus undergoes several thousand detectable changes every hour as a result of metabolism [B10, L12, S27, W12, W13, W14, W15]. Despite this high rate of spontaneous molecular change, few stable mutations accumulate in the genome. Thus, cells have evolved efficient processes for the correction of metabolically induced changes.

265. This inherent ability to repair DNA needs to be taken into account in assessing the ability of cells to repair the damage caused by doses of radiation in the wide range of a few milligray per year to a few tens of milligray

delivered in minutes. Just how capable the existing mechanisms are of coping with the additional radiation-induced damage is not readily obvious. But it would be reasonable to assume from the evidence that damage caused by natural background radiation, in which the energy deposition events in a particular nucleus are separated by weeks or months [B21], should be readily reparable by the available metabolically driven mechanisms.

266. However, errors in repair do occur, even during metabolism, such as small base-sequence changes (point mutations), gene deletions or rearrangements, although the overall DNA integrity may be retained [F3, F4, F5]. It needs to be recognized, therefore, that the effectiveness of DNA repair in irradiated mammalian cells is not absolute, some fraction of the cells retaining stable mutations. Thus, the same low conditioning doses that result in an adaptive response are likely also to result in malignant cellular transformations by the mechanisms discussed in Annex E, "Mechanisms of radiation onco-genesis" in the UNSCEAR 1993 Report [U1]. It would seem important to judge the balance between the fidelity of repair, residual damage and malignant transformations and whether indeed these effects interact with each other. The Committee hopes that more data will become available in the near future to address this point.

267. Alternative cellular mechanisms have been proposed to explain the adaptive response. These include the detoxification of reactive radicals, thereby reducing the potential for damage, and the activation of membrane-bound receptors stimulating cell proliferation. Efforts should be made to characterize the possible role of these processes.

268. It remains doubtful whether the immune system plays a significant role in any of the adaptive processes at low doses. In the UNSCEAR 1993 Report [U1], the Committee concluded that the immune system may not play a major role in moderating human radiation onco-genesis, although immune function in certain organs may ensure that some early neoplastic cells are eliminated before they become established. The data in this Annex are not in conflict with this generalized conclusion. Some transient effects on the ratio of subsets of T cells and in accelerating programmed cell death in damaged lymphocytes have been identified. It is interesting in this respect to note that a dose of a few hundreds of milligray can influence tumour growth kinetics, expressed as a transient reduction in tumour size in experimental animals. The evidence for changes in the human immune system long after exposure is not convincing.

269. Animal experiments in the 1950s and 1960s showed that chronic exposure of rodents at doses of up to a few milligray per day from low-LET radiation could result in increased life-span compared to controls exposed to only background radiation. However, some anomalies in these experiments need to be explained. Why was the response confined to male mice, and why was it not observed consistently in pathogen-free mice?

270. More recent experiments with rodents and beagle dogs exposed at various ages to low dose rates of low-LET radiation have generally been unable to demonstrate any statistically significant difference in life-span of irradiated and control groups after accumulated doses of up to about a gray. However, tumour incidence did not increase until the dose was in excess of about a gray, depending on the mouse strain and the susceptibility of the animals to developing spontaneous tumours. In some studies there was a non-statistical trend towards a lower-than-expected incidence of tissue-specific tumours, but in other studies there was a non-statistical trend towards a higher-than-expected incidence at doses of ≥2 Gy. A reduction in life-span or an increase in tumour incidence after fractionated exposures was not of statistical significance until accumulated doses exceeded a few gray.

271. The low statistical power of most human epidemiological cancer surveys with exposures at low doses makes it difficult to reach a decisive conclusion on the existence or absence of an adaptive response. Studies of exposure to higher-than-average levels of natural background radiation have made little contribution so far to estimating the risk of cancer from low-dose-rate, low-LET radiation. Studies of occupational exposures have recently shown more promise of yielding positive results, especially after moderate doses, but in the low-dose region the confidence limits are so broad that the results are still equivocal. The Life Span Study of survivors of the atomic bombings shows no significant excess of total cancer mortality below about 0.2 Gy. All cancers other than leukaemia are in excess but not statistically significant down to a dose range of 0.01-0.05 Gy. Leukaemia shows a deficit at doses less than 0.1 Gy, which again is not statistically significant. At present no conclusion can be drawn about the dose response below 0.2 Gy because of statistical limitations.

272. In conclusion, there is substantial evidence of an adaptive response in selected cellular systems following acute exposure to conditioning doses of low-LET radiation. The precise molecular processes involved in the adaptive response are not well understood at present, but cellular repair is likely to play a role by mechanisms similar to those involved in the generalized stress response. The presence of an adaptive response is not readily evident from the results of experiments in mammalian organisms in terms of reduced tumour induction. Interpretation of these experiments is complicated by variations in the susceptibility of different animal strains to spontaneous tumour induction. The low statistical power of the epidemiology studies also prevents a clear statement on the presence of an adaptive response in humans

exposed to low doses. It is to be hoped that better understanding of mechanisms of radiation effects obtained in molecular studies might provide a basis upon which to judge the role of adaptive response in the organism. In the meantime, it would be premature to conclude that cellular adaptive responses could convey possible beneficial effects to the organism that would outweigh the detrimental effects of exposures to low doses of low-LET radiation. The Committee recommends that this research be continued in order to clarify the nature and importance of the effects of radiation-induced adaptive response.

Table 1
Adaptive response in human lymphocytes conditioned with tritiated thymidine
[O2]

Conditioning treatment	Challenge treatment	Chromosome aberrations in 100 cells		
		Breaks	Exchanges	Deletions [a]
None (control)		3	0	0
Tritiated thymidine (370 Bq ml^{-1}) Tritiated thymidine (3700 Bq ml^{-1})		3 6	0 0	2 5
	1.5 Gy (x rays)	12	1	36
Tritiated thymidine (370 Bq ml^{-1}) Tritiated thymidine (3700 Bq ml^{-1})	1.5 Gy (x rays) 1.5 Gy (x rays)	9 10	3 3	23 (38) 13 [b] (41)

[a] Expected number in parentheses.
[b] $p < 0.01$.

Table 2
Adaptive response in human lymphocytes conditioned with x rays
[W5]

Conditioning treatment	Challenge treatment	Chromosome aberrations in 100 cells (for donors 1, 2, 3)			
		Breaks	Exchanges	Deletions	Total [a]
None (control)		1, 1, 1	0, 0, 0	0, 0, 0	1, 1, 1
0.1 Gy (x rays)		7, 1, 1	0, 0, 1	0, 0, 0	7, 1, 3
	1.5 Gy (x rays)	240, 34, 41	27, 5, 8	3, 1, 2	300, 54, 61
0.01 Gy (x rays)	1.5 Gy (x rays)	181, 18, 23	18, 7, 7	3, 0, 0	223 [b], 32 [b], 37 [b]

[a] Exchanges and deletions counted as two breaks.
[b] $p < 0.05$.

Table 3
Adaptive response in human lymphocytes inhibited by cycloheximide
[Y1]

Conditioning treatment	Challenge treatment	Chromatid deletions in 200 cells	
		Male donor	Female donor
None (control)		2	0
0.01 Gy (x rays) 10 μg ml^{-1} of cycloheximide 0.01 Gy (x rays) and cycloheximide		4 0 2	1 0 0
	1.5 Gy (x rays)	74	94
0.1 Gy (x rays) 0.1 Gy (x rays) and cycloheximide	1.5 Gy (x rays) 1.5 Gy (x rays)	39 [a] 75	59 [b] 78

[a] $p < 0.001$.
[b] $p < 0.01$.

Table 4
Inhibition of adaptive response in human lymphocytes by 3-aminobenzamide
[S12]

Conditioning treatment	*Challenge treatment*	*Chromosome aberrations in 200 cells* [a]	
		Breaks without 3-aminobenzamide	*Breaks with 3-aminobenzamide*
None (control)		4	4
0.005 Gy (x rays)		4	6
0.01 Gy (x rays)		6	6
	1.5 Gy (x rays)	81	79
0.005 Gy (x rays)	1.5 Gy (x rays)	45 [b] (81)	69 (81)
0.01 Gy (x rays)	1.5 Gy (x rays)	48 [b] (83)	73 (81)

[a] Expected number in parentheses.
[b] $p < 0.005$.

Table 5
Adaptive response in human lymphocytes following successive conditioning treatments
[F2]

Conditioning treatment	*Challenge treatment*	*Total breaks in 400 cells* [a]
None (control)		3
0.01 Gy (x rays)		6
0.01 Gy (x rays) after 4 h		7
0.01 Gy + 0.01 Gy (x rays) after 4 h		9
	1 Gy (x rays)	117
0.01 Gy (x rays)	1 Gy (x rays)	66 [b] (120)
0.01 Gy (x rays) after 4 h	1 Gy (x rays)	71 [b] (121)
0.01 Gy + 0.01 Gy (x rays) after 4 h	1 Gy (x rays)	75 [b] (127)

[a] Expected number in parentheses.
[b] $p < 0.005$.

Table 6
Influence of extracellular pH on adaptive response in human lymphocytes
[B8]

Conditioning treatment	*Challenge treatment*	*Chromatid deletions for challenge dose at 50 hours*		
		pH at time of treatment 7.2-7.4	*pH at time of treatment 7.0-7.2*	*pH at time of treatment 6.2-6.9*
None (control)		10	10	13
0.02 Gy (x rays)		8	9	12
	0.3 Gy (x rays)	372	518	1219
0.02 Gy (x rays)	0.3 Gy (x rays)	400	593 [a]	1155
		Chromatid deletions for challenge dose at 78 hours		
		pH at time of treatment 7.2-8.2	*pH at time of treatment 6.8-6.9*	*pH at time of treatment 6.4-6.8*
None (control)		14	9	11
0.02 Gy (x rays)		7	11	15
	0.3 Gy (x rays)	732	642	1137
0.02 Gy (x rays)	0.3 Gy (x rays)	696	572 [b]	983 [c]

[a] Synergistic response.
[b] $p < 0.5$.
[c] $p < 0.001$.

Table 7
Influence of cell cycle phase on adaptive response in human lymphocytes
[K4]

Conditioning treatment	*Challenge treatment*	*Chromosome aberrations in 100 cells* [a]
0.03 Gy (x rays) in G_0 phase 0.03 Gy (x rays) in G_1 phase 0.03 Gy (x rays) in S phase		4 6 3
Tritiated water		6
	3 Gy (x rays) in G_1 phase 1.5 Gy (x rays) in G_2 phase 1.2 Gy (x rays) in G_2 phase	88 22 11
0.03 Gy (x rays) in G_0 phase 0.03 Gy (x rays) in G_1 phase 0.03 Gy (x rays) in G_0 phase 0.03 Gy (x rays) in G_1 phase 0.03 Gy (x rays) in S phase	3 Gy (x rays) in G_1 phase 3 Gy (x rays) in G_1 phase 1.5 Gy (x rays) in G_2 phase 1.5 Gy (x rays) in G_2 phase 1.5 Gy (x rays) in G_2 phase	107 (92) 96 (96) 14 [b] (26) 20 [b] (28) 12 [b] (24)
Tritiated water	1.5 Gy (x rays) in G_2 phase	20 [b] (28)

[a] Expected number in parentheses.
[b] $p < 0.05$.

Table 8
Adaptive response in rabbit lymphocytes conditioned *in vivo* and challenged *in vitro*
[L8]

Conditioning treatment in vivo	*Challenge treatment* in vitro	*Cells with aberrations (%)* [a]	
		Challenge dose in G_0 phase	*Challenge dose in G_2 phase*
None (control)	*1.5 Gy*	12.8 ± 3.2	17.3 ± 1.3
0.30 Gy (gamma rays)	1.5 Gy (x rays)	7.8 ± 2.2 [b] (14.1)	11.8 ± 1.9 [b] (18.6)
0.75 Gy (gamma rays)	1.5 Gy (x rays)	6.3 ± 1.9 [b] (14.3)	13.3 ± 2.0 [b] (18.8)
0.90 Gy (gamma rays)	1.5 Gy (x rays)	7.5 ± 1.7 [b] (15.0)	11.8 ± 3.5 [b] (19.5)
1.20 Gy (gamma rays)	1.5 Gy (x rays)	8.0 ± 2.2 [b] (16.5)	13.5 ± 4.9 [b] (21.0)
1.50 Gy (gamma rays)	1.5 Gy (x rays)	10.5 ± 4.1 [c] (16.5)	17.8 ± 5.0 (21.0)
1.80 Gy (gamma rays)	1.5 Gy (x rays)		23.5 ± 3.3 (22.5)

[a] Includes all types of chromosome aberrations; 400 cells scored; expected number of aberrations in parentheses.
[b] $p < 0.01$.
[c] $p < 0.05$.

Table 9
Relative survival and mutation frequency in human lymphocytes conditioned with tritiated thymidine
[S18]

Conditioning treatment	*Challenge treatment*	*Relative survival of clone-forming cells*	*Mutation frequency (10^{-6})*
None (control)		1.00	2.2 ± 0.5
Tritiated thymidine (37 Bq ml^{-1})		0.86 ± 0.07	2.3 ± 0.7
Tritiated thymidine (370 Bq ml^{-1})		0.99 ± 0.07	3.2 ± 0.9
Tritiated thymidine (3700 Bq ml^{-1})		1.00 ± 0.1	2.2 ± 0.5
Tritiated thymidine(37000 Bq ml^{-1})		1.00 ± 0.1	3.3 ± 0.5
	1.5 Gy (x rays)	0.80 ± 0.05	15.2 ± 2.0
Tritiated thymidine (3700 Bq ml^{-1})	1.5 Gy (x rays)	0.82 ± 0.04	3.9 ± 0.9 [a]
Tritiated thymidine(37000 Bq ml^{-1})	1.5 Gy (x rays)	0.84 ± 0.06	6.3 ± 1.8 [a]
	3.0 Gy (x rays)	0.52 ± 0.03	19.7 ± 2.8
Tritiated thymidine (37 Bq ml^{-1})	3.0 Gy (x rays)	0.64 ± 0.06	14.4 ± 6.2
Tritiated thymidine (370 Bq ml^{-1})	3.0 Gy (x rays)	0.56 ± 0.06	21.5 ± 7.8
Tritiated thymidine (3700 Bq ml^{-1})	3.0 Gy (x rays)	0.55 ± 0.08	7.2 ± 3.7 [a]
Tritiated thymidine(37000 Bq ml^{-1})	3.0 Gy (x rays)	0.57 ± 0.05	9.5 ± 1.7 [a]

[a] Significant difference.

Table 10
Mutation frequency in human lymphocytes conditioned with x rays
[K5]

Conditioning treatment	*Challenge treatment*	*Cloning efficiency (%)*	*Mutation frequency (10^{-6})*
None (control)		24.4 ± 7.9	2.6 ± 0.3
0.01 Gy (x rays)		24.0 ± 12.0	4.5 ± 2.3
	3.0 Gy (x rays)	21.0 ± 8.4	15.5 ± 8.0
0.01 Gy (x rays)	3.0 Gy (x rays)	18.4 ± 6.6	5.2 ± 2.8 [a]

[a] Significant difference, $p < 0.5$.

Table 11
Adaptive response in human lymphocytes conditioned with hydrogen peroxide
[W8]

Conditioning treatment	*Challenge treatment*	*Aberrations in 100 cells*
None (control)		0.5
Hydrogen peroxide (0.1 μM) Hydrogen peroxide (1 μM) Hydrogen peroxide (10 μM) Hydrogen peroxide (25 μM) Hydrogen peroxide (50 μM)		1.0 0 0.5 1.0 0.5
	1.5 Gy (x rays)	34.0
Hydrogen peroxide (0.1 μM) Hydrogen peroxide (1 μM) Hydrogen peroxide (10 μM) Hydrogen peroxide (25 μM) Hydrogen peroxide (50 μM)	1.5 Gy (x rays) 1.5 Gy (x rays) 1.5 Gy (x rays) 1.5 Gy (x rays) 1.5 Gy (x rays)	21.5 [a] 20.4 [b] 21.5 [b] 20.5 [a] 25.0 [c]

[a] $p < 0.01$.
[b] $p < 0.0001$.
[c] $p < 0.05$.

Table 12
Synergistic response in human lymphocytes conditioned with x rays or methylated chemicals
[W4]

Conditioning treatment	*Challenge treatment*	*Aberrations in 200 cells*
None (control)		3
0.01 Gy (x rays) 0.01 Gy (x rays)	 Methyl methane sulphonate (0.42 mM) Methyl methane sulphonate (0.42 mM)	3 114 169 [a]
Methyl methane sulphonate (0.018 mM) Methyl methane sulphonate (0.018 mM)	 Methyl methane sulphonate (0.42 mM) Methyl methane sulphonate (0.42 mM)	2 97 133 [a]
N-methyl-N′-nitro-N-nitrosoguanidine (600 ng ml^{-1}) N-methyl-N′-nitro-N-nitrosoguanidine (600 ng ml^{-1})	 Methyl methane sulphonate (0.42 mM) Methyl methane sulphonate (0.42 mM)	1 97 147 [b]

[a] Synergistic response, $p < 0.01$.
[b] Synergistic response, $p < 0.001$.

Table 13
Adaptive response in human lymphocytes exposed to drugs
[S19]

Conditioning treatment	*Challenge treatment*	*Chromatid exchanges (%)*	
		Donor A	*Donor B*
None (control)		3.58 ± 0.6	4.2 ± 0.2
0.05 Gy (x rays)		3.7 ± 0.3	4.5 ± 0.2
 0.05 Gy (x rays)	VP-16 (etoposide) VP-16 (etoposide)	11.1 ± 1.3 9.8 ± 1.2	13.0 ± 0.7 10.7 ± 0.8
 0.05 Gy (x rays)	BCNU (1,3-bis (2-chloroethyl)-1-nitrourea) BCNU (1,3-bis (2-chloroethyl)-1-nitrourea)	8.7 ± 0.4 8.0 ± 0.5	10.1 ± 0.4 9.0 ± 1.2
 0.05 Gy (x rays)	cis-Pt (cis-diaminodichloroplatinum(II) cis-Pt (cis-diaminodichloroplatinum(II)	12.8 ± 1.2 11.9 ± 1.1	14.3 ± 1.3 13.8 ± 1.4

Table 14
Adaptive response in mouse bone marrow cells and spermatocytes

Conditioning treatment in vivo	*Challenge treatment*	*Chromosome aberrations (%)*	
		Bone marrow cells	*Spermatocytes*
Male Kunming mouse [C8]			
Control 0.01 Gy (x rays)	0.75 Gy (x rays) 0.75 Gy (x rays)	38.5 19.5 [a]	12.6 8.4 [a]
Female C57Bl/6 mouse [Y2]			
None (control)		0.5	
	0.65 Gy (x rays)	27.1	
0.002 Gy (x rays) 0.005 Gy (x rays) 0.010 Gy (x rays) 0.050 Gy (x rays)	0.65 Gy (x rays) 0.65 Gy (x rays) 0.65 Gy (x rays) 0.65 Gy (x rays)	2.8 8.8 3.5 9.5	

[a] $p < 0.01$.

Table 15
Investigation of response in pre-implantation mouse embryos [W5]

Conditioning treatment	*Challenge treatment*	*Aberrations per 100 metaphases*				*Aberrant cells*
		Breaks	*Exchanges*	*Rearrangements*	*Total*	*(%)*
None (control)		9.4	0	0	9.4	6.9
0.05 Gy (x rays) *in vitro* 0.05 Gy (x rays) *in vitro*	 1.5 Gy (x rays) *in vitro* 1.5 Gy (x rays) *in vitro*	7.0 39.7 27.1	0 0 5	0 0 0.5	7.0 39.7 37.8	6.4 22.8 26.4
None (control)		7.3	0	0.6	7.9	6.4
0.05 Gy (x rays) *in vivo* 0.05 Gy (x rays) *in vivo*	 1.5 Gy (x rays) *in vitro* 1.5 Gy (x rays) *in vitro*	10.2 24.3 25.2	0 6.8 6.2	1.3 1.7 2.7	11.5 39.6 40.3	7.6 27.0 28.0
Repair *in situ*						
None (control) 0.05 Gy (x rays) *in vivo*	 2 Gy (x rays) *in vivo*	12.2 50.7	0 32.3	0 10.7	12.2 126.0	4.9 52.3
None (control) 0.05 Gy (x rays) *in vitro*	 2 Gy (x rays) *in vitro*	9.4 58.2	0 20.0	0 16.4	9.4 114.6	6.9 56.4

Table 16
Investigation of response in two-cell mouse embryos [M6]

Conditioning treatment [a]	*Challenge treatment* [a]	*Results at 144 hours after conception* [b c]		*Results at 192 hours after conception* [b c]		
		Cells per embryo	*Hatched blastocysts (%)*	*Trophoblast outgrowth (%)*	*Inner cell mass (%)*	*Two-cell layers (endoderm and ectoderm) (%)*
None (control)		79	77	92	88	48
0.05 Gy (x rays) 0.05 Gy (x rays)	 2 Gy (x rays) 2 Gy (x rays)	87 40 44 (44)	81 20 29 (21)	84 31 31 (28)	71 22 18 (18)	36 8 7 (6)

[a] Conditioning treatment given in the early G_2 phase, challenge treatment given in the late G_2 phase.
[b] Results for 144 hours after conception and 192 hours after conception were obtained in independent experimental series.
[c] Expected results in parentheses.

Table 17
Some DNA-damage-induced genes identified in mammalian cells
[F19]

Gene	Inducing agents [a]		Dose	Detection	Detection time after exposure (hours)	Species	Cell	Ref.
	For positive response	For negative response						
Coding transcription factor								
c-fos	x rays		0.75-0.9 Gy			Syrian hamster	SHE fibroblast cell line	[W16]
	x rays		20 Gy	mRNA	3	Human	HL-60	[S48]
	x rays		20 Gy	mRNA	0.5	Mouse	Myeloid cell	[I9]
		x rays	20 Gy	mRNA		Human	Epithelium cell lines	[H12]
	MMS		100 $\mu g\ ml^{-1}$	mRNA	2-8	Chinese hamster	CHO-K1	[H15]
	H_2O_2		0.4 mM	mRNA	3	Chinese hamster	CHO-K1	[H15]
	H_2O_2		0.25 mM	mRNA	0.75	Human	HeLa-3S	[D7]
c-jun	H_2O_2		0.25 mM	Activity		Human	HeLa-3S	[D7]
	x rays		20 Gy	mRNA	1-6	Human	HL-60	[S48]
	x rays		20 Gy	mRNA	3-6	Human	U937	[S48]
	x rays		20 Gy	mRNA	1-4	Human	AG1522	[S48]
	x rays		20 Gy	mRNA	>0.5	Human	Epithelium cell lines	[H12]
	Gamma rays		2 Gy	mRNA	1	Syrian hamster	SHE-fibroblast cell line	[W26]
junB	H_2O_2		0.25 mM	mRNA	0.75	Human	HeLa-3S	[D7]
junD	x rays		20 Gy	mRNA	3-6	Human	HL-60	[S48]
EGR1	H_2O_2		0.25 mM	mRNA	0.75	Human	HeLa-3S	[D7]
NF-κB gene	x rays		20 Gy	mRNA		Human	Epithelium cell lines	[H12]
	Gamma rays		20 Gy	mRNA	3-6	Human	KG-1	[B24]
	Gamma rays		20 Gy	Activity	0.5-6	Human	KG-1	[B24]
	H_2O_2		0.1 mM	Activity	0.5-1	Human	Burkatt T	[S47]
Coding nuclear protein								
p53 gene	Gamma rays		4 Gy	Protein	1	Human	Colon carcinoma	[K22]
	Gamma rays		4 Gy	G_1-arrest		Human	Colon carcinoma	[K22]
	Gamma rays		2 Gy	Protein	1	Human	Lymphocyte cell line	[K23]
Other oncogenes								
MoMuSV-LTR	x rays		0.7 Gy	Activity	3-24	Mouse	(Transfectant) NIH?3T3	[L24]
N,K and H-ras	x rays		1-10 Gy	Survival		Mouse	NIH/3T3	[G10]
N,K-ras		Gamma rays	1-5 Gy	Survival		Human	Retinoblasts	[G10]
v-erb, B, v-sis, v-abl, v-svc	Gamma rays			Survival		Mouse	32Dcl3, NIH/3T3	[G10]
Ha-ras	Gamma rays		1-40 Gy	Survival		Rat	Embryo cells	[G10]
Coding cytokine or receptor								
IL-1β gene	Neutrons		0.021 Gy	mRNA	3	Syrian hamster	SHE-fibroblast cell line	[W26]
	x rays		20 Gy	mRNA	0.5	Mouse	Myeloid cell	[I9]
	Gamma rays		8.5 Gy	mRNA	24-48	Mouse	Spleen	[I9]
TNF-α gene	x rays		20 Gy	mRNA	3	Human	HL-60 derivative	[H14]
	x rays		5 Gy	mRNA	3	Human	Sarcoma cell line	[H13]

Table 17, continued

Gene	Inducing agents [a]		Dose	Detection	Detection time after exposure (hours)	Species	Cell	Ref.
	For positive response	For negative response						
TNF-α gene	x rays		5 Gy	Activity	>96	Human	Sarcoma cell line	[H13]
	Gamma rays		20 Gy	mRNA	3	Human	HL-60 derivative	[H14]
	Gamma rays		20 Gy	mRNA	1-3	Human	Monocyte	[S49]
	Gamma rays		20 Gy	mRNA	1-6	Human	HL-60, U937	[S49]
Coding enzyme or functional protein								
Protein kinase C gene	x rays		0.075 Gy	mRNA	0.3-12	Syrian hamster	SHE	[W17]
	Gamma rays		0.12 Gy	mRNA	1-3	Syrian hamster	SHE	[W17]
	Neutrons		0.06-0.48 Gy	mRNA	1-3	Syrian hamster	SHE	[W17]
C-kinase function	Gamma rays		20 Gy	Activity	1-60	Human	HL-60 derivative	[H14]
β-polymerase gene	H_2O_2		0.4 mM	mRNA	4	Chinese hamster	CHO	[F25]
O^6methylguanine	Gamma rays		3 Gy	mRNA	24-48	Rat	H4	[L23]
Transferase gene	Gamma rays		3 Gy	Protein	48	Rat	H4	[L23]
	x rays		3 Gy	mRNA	24	Rat	H4	[F24]
			0.2 mM	mRNA	24	Rat	H4	[F24]
N^3methyladenine	Gamma rays		3 Gy	mRNA	27-72	Rat	H4	[L23]
Glycosylase gene	Gamma rays		3 Gy	Activity	48	Rat	H4	[L23]
Orn decarboxylase gene	Neutrons		0.21 Gy	mRNA	<1	Syrian hamster	SHE fibroblast cell line	[W26]
Heme oxygenase gene	x rays		0.1 Gy	mRNA	3-4	Human	Skin-fibroblast cell line	[A15]
	H_2O_2		0.1 mM	mRNA	2-6	Human	Skin-fibroblast cell line	[A15]
	H_2O_2		0.1 mM	mRNA	4-8	Human	Lymphoblast	[A15]
Metallothionine IIA gene	Gamma rays		2 Gy	Activity	24	Chinese hamster	CHO-x-ray^s cell line	[K24]
	MNU		0.4 mM	Activity	24	Chinese hamster	CHO-x-ray^s cell line	[K24]
		MMS		Activity	24	Chinese hamster	CHO-x-ray^s cell line	[K24]
Tissue plasminogen activator	x rays		2 Gy	Activity	2-15	Human	Fibroblast	[B13]
	x rays		1 Gy	mRNA	4	Human	Melanoma	[B13]
Virus-like 30S DNA	Gamma rays		3 Gy	Amplification [b]	1	Mouse [d]	Liver, gut	[P12]
	Neutrons		0.5 Gy	Decrease [c]	1	Mouse [d]	Liver, gut	[P12]
Virus-like 40S DNA	Gamma rays		1-5 Gy	Survival		Human	Retinoblasts	[G10]
Undefined								
gadd45	x rays		2 Gy	mRNA	2-4	Human	Lymphoblast cell line	[P6]
	MMS		0.1 mg ml^{-1}	mRNA	2-4	Human	Lymphoblast cell line	[P6]
	MMS		0.1 mg ml^{-1}	mRNA	2-10	Human	Distinct cell lines	[F16]
gadd153	MMS	x rays	0.1 mg ml^{-1}	mRNA	2-4	Human	Lymphoblast cell line	[P6]
	MMS		0.1 mg ml^{-1}	mRNA	2-10	Human	Distinct cell lines	[F16]
gadd34	MMS		0.1 mg ml^{-1}	mRNA	2-10	Human	Distinct cell lines	[F16]
gadd33	MMS		0.1 mg ml^{-1}	mRNA	2-10	Human	Distinct cell lines	[F16]
gadd7	MMS		0.1 mg ml^{-1}	mRNA	2-10	Human	Distinct cell lines	[F16]
RP-2, RP-8	Gamma rays		9 Gy	mRNA	1-4	Rat	Thymocyte	[O7]

[a] Alkylating agents: MMS = methylmethane sulfonate; MNU = methylnitrosourea.
[b] Amplification of gene copy number.
[c] Decrease of gene copy number.
[d] (BALB/c × C57Bl)F_1 mouse.

Table 18
Accumulation of α-interferon mRNA and protein kinase C following gamma-ray exposure of Syrian hamster embryo fibroblasts
[W16]

Dose [a] *(Gy)*	*Relative amount* [b]	
	1 hours [c]	*3 hours* [c]
α-interferon mRNA		
0.02	1.0	1.0
0.12	1.0	1.5
0.25	1.0	3.9
0.5		6.9
1.0	1.5	3.7
2.0	0.9	3.8
β-protein kinase C		
0.02	1.0	1.0
0.12	1.7	1.9
0.25	2.1	2.4
0.50		2.4
1.0	2.9	3.4
2.0	3.5	4.3

a Gamma rays administered at a dose rate of 0.14 Gy min^{-1}.
b Amount of α-interferon mRNA in untreated cells was set at 1.0. All other values are expressed relative to that. Standard deviation for all values was ≤0.1.
c Incubation time following completion of the radiation exposure.

Table 19
Relative expression of transcripts encoding nuclear proteins following gamma-ray exposure [a]
[W18]

Dose (Gy)	*Dose rate (Gy min^{-1})*	*Relative amount of RNA for transcripts* [b]				
		c-jun	*Rb*	*H4-histone*	*p53*	*c-myc* [c]
0	0	1.0 (0.04)	1.0 (0.01)	1.0 (0.04)	1.0 (0.11)	1.0 (0.03)
0.06	0.01	1.6 (0.04) [d]	0.9 (0.08)	1.4 (0.08)	1.1 (0.09)	HND
0.25		2.2 (0.07) [c]	1.3 (0.03)	1.4 (0.03)	1.3 (0.09)	HND
0.50		1.8 (0.12) [c]	1.7 (0.03) [c]	1.7 (0.08) [c]	1.3 (0.06)	HND
0.75		1.7 (0.17) [c]	1.3 (0.08)	2.0 (0.07) [c]	1.4 (0.02)	HND
0.25	0.14	1.5 (0.11) [c]	0.9 (0.09)	1.4 (0.02)	1.1 (0.02)	HND
0.50		0.8 (0.21)	0.7 (0.11)	1.6 (0.04) [c]	1.3 (0.07)	HND
0.75		1.1 (0.12)	1.3 (0.06)	1.1 (0.05)	1.3 (0.02)	HND
2.00		4.7 (0.03) [c]	1.7 (0.19) [c]	1.4 (0.05)	0.9 (0.11)	HND

a Cycling cells were irradiated with ^{60}Co gamma rays at the doses or dose rates indicated 1 h prior to RNA harvest.
b Standard deviations are in parentheses.
c HND = Hybridizations not detected.
d Significantly different from control, $p < 0.05$.

Table 20
Relative constitutive levels of mRNAs and proteins in splenocytes of mice conditioned with x rays
[M19]

Constituents	*Relative level of constituent*	
	Controls	*Irradiated cells* [a]
HSP70 (heat shock protein) mRNA	0.38	0.72
GAPD (glyceraldehyde-3-phosphate dehydrogenase) mRNA	1.98	2.01
HSP70 (heat shock protein)	0.24	0.54

a Total dose was 0.8 Gy from 0.04 Gy d^{-1} for 20 days.

Table 21
Accessory glycoproteins on the surface of T cells in the blood
[A12]

Protein [a]	*Alternative name*	*Approximate molecular weight (kDa)*	*Expressed on*	*Putative function*
CD2	T11 Thy 1 in mice	50	All T cells	Promotes adhesion between T cells and their target cells by binding to LFA-3 on target cells
CD3	T3 Thy 1 in mice	γ chain: 25; δ chain: 20 ε chain: 20; ξ chain: 16	All T cells	Helps transduce signal when antigen-MHC complex binds to T cell receptors
CD4	T4 in humans L3T4 in mice	50	Helper T cells	Promotes adhesion to antigen-presenting cells and B cells, probably by binding to class II MHC molecules
CD8	T8 in humans Lyt2, Lyt3 in mice	60 (homodimer) 70 (heterodimer)	Cytotoxic T cells	Promotes adhesion to virus-infected target cells, probably by binding to class I MHC molecules
LFA-1	-	α chain: 190; β chain: 95	Most white blood cells	Promotes cell-cell and cell-matrix adhesion

[a] CD stands for cluster of differentiation, as each of the CD proteins was originally defined as T cell "differentiation antigen" recognized by multiple monoclonal antibodies. Their identification depended on large-scale collaborative studies in which hundreds of such antibodies, generated in many laboratories, were compared and found to consist of relatively few groups (or "clusters"), each recognizing a single cell-surface protein.

Table 22
Gamma-ray-induced DNA fragmentation of thymic T cells
[M19]

Thymocytes	*FD_{50} [a] (Gy)*	*95% CI*
$CD4^-CD8^-$	1.98	1.86-2.10
$CD4^+CD8^+$	0.22	0.21-0.23
$CD4^+CD8^-$	2.06	1.81-2.31
$CD4^-CD8^+$	> 8.00	(Not determined)

[a] Gamma-radiation dose to induce 50% fragmentation.

Table 23
Dose-effect relationship for immune functions in mouse splenocytes after x-irradiation [a]
[L16]

Immunologic parameter	*Percentage of control at dose*				
	0.025 Gy	*0.05 Gy*	*0.075 Gy*	*0.10 Gy*	*0.25 Gy*
PFC reaction	109.6	143.4	173.8 [c]	61.3	84.1
MLC reaction	108.8	133.3 [c]	122.2	124.8	110.7
Reaction to Con A	191.1	254.8	529.7 [c]	104.9	44.6
NK activity [b]	111.8	109.0	118.8 [c]	115.4	120.1
ADCC activity	109.0	127.6	131.7 [c]	150.3 [c]	111.7
LPS reaction	129.4	171.9	169.5	65.9	54.5
IFN-r secretion	103.8	110.8	130.3	130.9 [d]	87.3
IL2 secretion	-	-	166.0 [d]	-	-

[a] The immunologic parameters were examined on splenocytes of mice after whole-body irradiation expressed as a percentage of controls.
[b] The peak stimulation was found at 0.5 Gy, where the relative value was 145.0% of control.
[c] $p < 0.5$ compared to sham-irradiated control.
[d] $p < 0.01$ compared to sham-irradiated control.

Table 24
Response of mouse splenic and thymic lymphocytes to whole-body x-irradiation of 0.075 Gy
[L16, L18]

Time after irradiation (days)	*Cell cycle progression*		
	Percentage in G_0 and G_1 phases	*Percentage in S phase*	*Percentage in G_2 and M phases*
Sham-control	86.0 (0.6) [a]	10.8 (0.3)	3.3 (0.3)
3	82.1 (0.7) [b]	14.7 (0.7) [c]	3.1 (0.3)
4	81.9 (0.7) [b]	14.8 (0.8) [c]	3.4 (0.2)
7	82.2 (0.5) [b]	14.4 (0.6) [c]	3.4 (0.3)

Time after irradiation (days)	*Ratio of $T_H(CD4^+)$ to $T_S(CD8^+)$ cells*	
	Thymus (number of samples: 12)	*Spleen (number of samples: 6)*
Shamcontrol	2.99 (0.14) [a]	1.14 (0.04)
3	2.85 (0.14)	-
4	3.41 (0.20)	1.17 (0.09)
7	2.44 (0.13) [c]	1.17 (0.05)

Time after irradiation (days)	*Percentage change in thymocyte subsets*			
	$CD4^-CD8^-$	*$CD4^+CD8^+$*	*$CD4^+CD8^-$*	*$CD4^-CD8^+$*
Sham-control	2.09 (0.06) [a]	80.89 (1.43)	12.65 (0.98)	4.37 (0.45)
3	3.56 (0.29) [b]	80.25 (1.20)	11.84 (1.22)	4.37 (0.61)
4	3.50 (0.15) [b]	80.68 (1.33)	12.14 (0.94)	3.68 (0.33)
7	3.16 (0.16) [b]	83.64 (1.06)	9.36 (0.73) [d]	3.84 (0.25)

[a] Mean value; standard error in parentheses.
[b] $p < 0.001$.
[c] $p < 0.01$.
[d] $p < 0.05$.

Table 25
Comparison between frequency of interleukin-2-secreting cells from human lymphocytes in inhabitants of a high-background-radiation area and a control (low-background-radiation) area in China
[Y8]

Age group	*Inhabitants of high-background-radiation area (Yangjiang County, Gaungdong Province, China)*		*Inhabitants of low-background-radiation area (Enping County, China) (Control population)*		*P value*
	Number analysed	*Frequency (%)*	*Number analysed*	*Frequency (%)*	
< 20 years	9	20.11 ± 0.66	9	17.02 ± 0.69	>0.05
20-50 years	9	19.89 ± 0.58	9	17.17 ± 0.79	<0.05
> 50 years	7	20.71 ± 0.32	9	15.78 ± 0.52	<0.01
Total	25	20.20 ± 0.32	27	16.63 ± 0.53	<0.01

Table 26
Analysis of the effect of cigarette smoking on the percentage of lymphocytic subsets in occupationally exposed persons
[T10]

	Number of persons [a]	*CD2*$^+$ *(%)*	*CD4*$^+$ *(%)*	*CD8*$^+$ *(%)*	*HNK-1*$^+$ *(%)*	*Ratio CD4/CD8*
Non-smokers	120	77.6 ± 5.7	39.9 ± 8.1	27.2 ± 9.5	17.7 ± 5.1	1.7
Smokers	39	77.3 ± 6.7	39.1 ± 8.6	30.3 ± 8.9	18.4 ± 5.3	1.4
p-value (Wilcoxon two-sample test)		0.63	0.45	0.03	0.59	0.05

[a] Data were pooled for all persons tested in each category.

Table 27
Influence of a whole-body pre-irradiation of 0.1 Gy on the growth of primary murine squamous carcinoma subjected to localized irradiation
[M28]

Dose of local irradiation (Gy)	*Growth delay compared to controls (days)*	
	4-5 mm tumour	*9-10 mm tumour* [a]
6	0	4
10	1.8	5
15		6-7
35	3.0	

[a] Tumour size at time of irradiation.

Table 28
Change in the percentage of functional subsets of peripheral blood lymphocytes in non-Hodgkin's lymphoma patients before and after low-dose whole-body or half-body irradiation
[T16]

Subsets of lymphocytes	*Before irradiation (% ± SE)*	*After irradiation (% ± SE)*
Suppressor-inducer T (CD$^+$2H4$^+$)	8.9 ± 4.3	7.5 ± 5.7
Helper T (CD4$^+$2H4$^-$)	27.1 ± 7.7	33.1 ± 7.7 [a]
Helper-inducer T (CD4$^+$4B4$^+$)	23.7 ± 6.8	28.7 ± 7.2 [b]
Suppressor T (CD8$^+$CD11$^+$)	9.7 ± 6.6	9.3 ± 4.9
Cytotoxic T (CD8$^+$CD11$^-$)	21.0 ± 8.5	21.4 ± 6.9
Active helper/inducer T (CD4$^+$HLA-DR$^+$)	4.1 ± 1.6	6.8 ± 2.5 [b]
Active suppressor/cytotoxic T (CD8$^+$HLA-DR$^+$)	8.3 ± 5.6	10.8 ± 7.4
NK activity (++) (CD16$^+$Leu7$^-$)	5.9 ± 2.4	5.4 ± 2.5
NK activity (+) (CD16$^+$Leu7$^+$)	14.3 ± 9.4	14.0 ± 7.6
NK activity (+) (CD16$^-$Leu7$^+$)	21.0 ± 9.2	20.0 ± 8.0
Normalized ratio T_H:T_S	1.0	1.42 ± 0.86 [b]

[a] $p < 0.05$.

Table 29
$LD_{50/30}$ for mice pretreated with non-lethal doses of 250 kV x rays
[S55]

Age at testing [a] (d)	*Treatment group*							
	No treatment (controls)		*0.8 Gy (100 R) [b]*		*2.4 Gy (300 R) [b]*		*3.2 Gy (500 R) [b]*	
	Number of mice	*$LD_{50/30}$ ± SE (Gy)*	*Number of mice*	*$LD_{50/30}$ ± SE (Gy)*	*Number of mice*	*$LD_{50/30}$ ± SE (Gy)*	*Number of mice*	*$LD_{50/30}$ ± SE (Gy)*
120	90	6.10 ± 0.12	90	6.31 ± 0.09	90	5.90 ± 0.13	89	5.67 ± 0.10
150	40	6.42 ± 0.11	-	-	40	6.65 ± 0.16	40	6.60 ± 0.13
240	40	6.80 ± 0.14	40	6.74 ± 0.17	40	6.60 ± 0.06	40	6.40 ± 0.10
430	40	6.72 ± 0.13	-	-	-	-	-	-
550	40	6.58 ± 0.27	40	6.53 ± 0.19	40	6.08 ± 0.22	40	5.18 ± 0.34
670	-	-	-	-	48	5.69 ± 0.19	48	5.18 ± 0.30
730	40	6.55 ± 0.15	-	-	-	-	-	-
760	-	-	48	4.96 ± 0.34	-	-	45	4.33 ± 0.24
820	50	5.96 ± 0.15	-	-	49	4.86 ± 0.26	-	-
930	-	-	50	4.16 ± 0.21	-	-	-	-
960	57	4.53 ± 0.23						

[a] The mice were pre-treated at age 90 days.
[b] Converted to SI units: 100 R = 0.8 Gy.

Table 30
Short-term survival and acquired resistance in mice conditioned with low-dose x-irradiation
[Y9]

Group	*Number of mice*	*Conditioning treatment [a] (Gy)*	*Challenge treatment [b] (Gy)*	*30-day survival*		*Colony-forming units*	
				Percentage	*Statistical significance*	*Average per mouse ± SE*	*Ratio to control*
Control	70	0	7.75 (x rays)	14.3	-		
1	60	0.025 (x rays)	7.75 (x rays)	16.7	$p > 0.1$		
2	60	0.05 (x rays)	7.75 (x rays)	42.4	$p < 0.001$		
3	60	0.10 (x rays)	7.75 (x rays)	40.0	$p < 0.01$		
Control	40	0	7.0 (x rays)			1.48 ± 0.39	
1	40	0.025 (x rays)	7.0 (x rays)			1.98 ± 0.50	1.3
2	39	0.05 (x rays)	7.0 (x rays)			2.46 ± 0.53	1.7
3	39	0.10 (x rays)	7.0 (x rays)			2.51 ± 0.50	1.7

[a] Given at two of age.
[b] Given at four months of age.

Table 31
Mean survival time of LAF_1 mice exposed to gamma rays from a ^{226}Ra source at a dose rate of about 1 mGy each day for the duration of life
[L19]

Controls			*Irradiated*			*Difference between irradiated and control groups (days)*	*p Value comparing irradiated with controls*
Number of mice	*Sex*	*Mean survival time ± SE (days)*	*Number of mice*	*Sex*	*Mean survival time ± SE (days)*		
110	Male	683.5 ± 14.3	111	Male	783.1 ± 14.0	99.6	0.01
116	Female	802.9 ± 16.1	120	Female	820.3 ± 17.6	17.4	Not significant

Table 32
Survival of LAF_1 mice after lifetime irradiation
[S52]

Dose rate (mGy a^-1)	*Number of mice in sample*	*Male*		*Female*	
		Mean survival time ± SE (days)	*Standard deviation (days)*	*Mean survival time ± SE (days)*	*Standard deviation (days)*
0.14	90	464 ± 23	222	662 ± 17	164
1.6-4.9	90	528 ± 22	213	630 ± 20	193
40	120	501 ± 16	178	548 ± 11	125

Table 33
Life-span and neoplasms in RF mice exposed to x and gamma rays
[U13]

Mean accumulated dose (Gy d^-1)	*Average dose rate (Gy)*	*Duration of exposure (days)*	*Number of mice exposed*	*Surviving irradiation (%)*	*Number of mice necropsied*	*Mean age at death (days)*	*Mice with neoplasms*	
							Incidence at necropsy ± SE (%)	*Mean age at death ± SE (days)*
Males								
0	-	-	623	-	612	578	52 ± 2	606 ± 8
0.25	1150 [a]	-	201	100	198	565	64 ± 3	568 ± 13
0.50	1150 [a]	-	189	100	183	550	65 ± 4	562 ± 15
0.75	1150 [a]	-	188	100	186	548	71 ± 3	571 ± 15
1.00	1150 [a]	-	197	100	194	494	63 ± 4	514 ± 14
1.50	1150 [a]	-	427	100	417	484	75 ± 2	485 ± 8
1.48	0.052	30	98	97	96	606	67 ± 5	651 ± 20
1.53	0.15	10	178	100	175	581	68 ± 4	594 ± 13
1.55	0.77	2	80	100	79	590	75 ± 5	611 ± 17
3.00	1150 [a]	-	241	100	238	408	81 ± 3	415 ± 11
3.29	0.05	63	118	99	118	619	76 ± 4	638 ± 16
3.03	0.14	22	119	100	116	573	72 ± 4	580 ± 16
3.08	0.22	14	79	100	76	538	72 ± 5	552 ± 19
3.05	0.31	10	117	100	109	610	72 ± 4	629 ± 20
3.15	0.79	4	79	100	79	536	63 ± 5	568 ± 23
Females								
0	-	-	554	-	537	586	66 ± 2	608 ± 8
0.25	0.067	-	95	100	94	603	70 ± 5	597 ± 13
0.50	0.067	-	95	100	95	599	77 ± 4	612 ± 15
1.00	0.067	-	95	100	95	551	86 ± 4	549 ± 15
1.04	0.052	20	99	100	95	578	75 ± 4	592 ± 14
1.01	0.15	7	119	100	117	599	74 ± 4	605 ± 8
2.00	0.067	-	94	100	92	520	87 ± 4	525 ± 20
2.40	0.005	516	99	48	98	576	79 ± 4	587 ± 17
3.00	0.067	-	92	100	92	470	89 ± 3	473 ± 18
3.06	0.010	300	125	93	124	611	82 ± 4	625 ± 14
3.06	0.011	300	89	92	87	595	79 ± 4	609 ± 14
3.13	0.051	61	125	98	122	549	79 ± 4	549 ± 16
3.10	0.15	21	110	100	107	564	86 ± 3	571 ± 14
3.13	0.21	15	99	100	97	504	84 ± 4	534 ± 18
3.05	0.31	10	99	100	97	587	80 ± 4	597 ± 16
3.06	1.02	3	99	100	97	575	80 ± 4	582 ± 17

[a] X rays; all other groups exposed to gamma rays.

Table 34
Influence of dose on induction of neoplastic diseases in female BALB/c mice exposed to gamma rays
[U14]

Type of tumour	*Dose [a] (Gy)*	*Age-adjusted incidence (% ± SE)*
Ovarian tumour	0	6.4 ± 1.4
	0.5	9.9 ± 1.8
	1.0	21.9 ± 2.6
	2.0	42.5 ± 2.8
Mammary adenocarcinoma	0	7.6 ± 0.9
	0.5	9.0 ± 0.9
	1.0	13.2 ± 1.2
	2.0	13.9 ± 1.3
Lung adenocarcinoma	0	12.8 ± 2.2
	0.5	14.5 ± 1.8
	1.0	16.5 ± 2.1
	2.0	21.4 ± 2.6

[a] Dose rate of 0.083 Gy d^{-1}.

Table 35
Survival of male $BC3F_1$ mice after exposure to x rays
[C20]

Dose (Gy)	*Number of mice (Gy)*	*Mean long-term survival ± SD (d)*
	Mice irradiated *in utero*	
0.0	34 [a]	852 ± 247
0.3	48 [a]	798 ± 180
0.9	61 [a]	824 ± 212
1.5	46 [a]	897 ± 173
2.1	45 [a]	832 ± 167
	Mice irradiated at 3 months of age	
0	203	827 ± 188
0.5	44	828 ± 182
1.0	48	797 ± 238
2.0	50	767 ± 225
3.0	50	731 ± 176
4.0	48	792 ± 166
5.0	72	701 ± 210
6.0	95	718 ± 175
7.0	249	682 ± 222
	Mice irradiated at 19 months of age	
0	46	849 ± 170
0.5	48	865 ± 170
1.0	48	865 ± 141
2.0	48	893 ± 160
3.0	50	859 ± 125
4.0	50	834 ± 134
5.0	72	881 ± 167
6.0	95	862 ± 183
7.0	142	759 ± 184

[a] Number of mice at weaning.

Table 36
Life-span and development of neoplasms in irradiated mice
[S54]

Stage in life of mouse	*Dose (Gy)*	*Number of mice*	*Mean life-span ± SE (days)*	*Mice bearing neoplasms (%)*	*Mean number of neoplasms per mouse ± SE*
Control	0	198	869 ± 8	90.4	1.09 ± 0.04
Fetus	1.9	93	831 ± 14 [a]	91.4	1.30 ± 0.08 [a]
	3.8	81	741 ± 18 [a]	86.0	1.18 ± 0.09
	5.7	65	584 ± 21 [a]	70.8 [a]	0.95 ± 0.08
Neonate	1.9	85	784 ± 17 [a]	90.6	1.38 ± 0.09 [a]
	3.8	81	642 ± 16 [a]	96.3	1.49 ± 0.08 [a]
	5.7	91	453 ± 16 [a]	91.2	1.21 ± 0.07
Adult	1.9	81	789 ± 17 [a]	95.1	1.44 ± 0.08 [a]
	3.8	80	723 ± 19 [a]	91.3	1.56 ± 0.10 [a]
	5.7	83	586 ± 19 [a]	91.6	1.28 ± 0.09

[a] $p < 0.05$.

Table 37
Life-span and disease incidence after a single exposure of three-months-old C57Bl mice to ^{137}Cs gamma-radiation
[M29]

Dose (Gy)	*Survival*		*Thymoma*		*All leukaemias*		*Carcinoma and sarcoma*		*All malignancies*	
	Number of mice	*Life-span ± SE [a] (days)*	*Incidence (%)*	*Life-span ± SE [a] (days)*	*Incidence (%)*	*Life-span ± SE [a] (days)*	Incidence (%)	Life-span ± SE [a]	Incidence (%)	Life-span ± SE [a]
Control	473	606 ± 29	1.27	420 ± 204	20.93	624 ± 65	16.3	676 ± 78	33.19	648 ± 53
0.25	242	578 ± 38	1.65	500 ± 304	18.18	581 ± 92	14.04	636 ± 113	28.51	598 ± 74
0.5	239	558 ± 38	1.26	405 ± 303	15.48	582 ± 100	8.79	626 ± 143	23.01	599 ± 84
1	246	540 ± 36	0	-	15.04	546 ± 93	8.94	593 ± 134	21.95	566 ± 80
2	217	532 ± 38	2.3	174 ± 95	13.82	513 ± 102	14.29	584 ± 109	26.27	543 ± 76
4	143	478 ± 43	13.29	190 ± 46	26.57	375 ± 72	16.08	594 ± 129	39.16	444 ± 67
6	188	400 ± 32	30.32	253 ± 36	44.68	305 ± 37	14.36	538 ± 108	55.32	352 ± 39

[a] Total, corrected for competing risks.

Table 38
Survival of male B6CF$_1$ mice after acute, 23-week and 59-week exposures to gamma-radiation [a]
[C19, T15]

Total dose (Gy)	*Average dose rate (Gy week^{-1})*	*Mean survival ± SE (days)*	
		Death from all causes	*Death from cancers*
Single acute exposure			
0		863 ± 10	
0.9		834 ± 12	
1.4		827 ± 16	
2.1		807 ± 15	
2.7		727 ± 13	
Continuous exposure for 23 weeks			
0	0	857 ± 15	893 ± 15
2.1	0.09	830 ± 13	871 ± 15
4.2	0.18	806 ± 22	842 ± 21
9.6	0.42	675 ± 23	732 ± 24
19.2	0.83	579 ± 32	622 ± 35
Continuous exposure for 59 weeks			
0	0	803 ± 16	844 ± 14
5.3	0.09	768 ± 15	829 ± 15
10.7	0.18	719 ± 16	750 ± 16
24.6	0.42	616 ± 21	656 ± 20

[a] Mice were 107-114 days old at the beginning of irradiation.

Table 39
Survival of beagle dogs exposed to whole-body gamma-irradiation
[B23]

Age at exposure (d)	*Dose group (Gy)*	*Number in group*	*Percentage dead in 1982*	*Age of survivors (years)*	
				Mean	*Range*
Controls (all ages)	0	360	66	12.6	9.9-15.0
8 *(in utero)*	0.16	120	47	11.5	10.6-13.6
	0.83	120	51	11.4	10.4-13.4
28 *(in utero)*	0.16	120	67	11.8	10.5-13.9
	0.83	120	50	11.7	10.7-13.6
55 *(in utero)*	0.16	120	63	12.2	11.3-14.1
	0.83	120	68	12.1	11.3-14.2
2 *(post-partum)*	0.16	120	67	12.2	10.2-13.8
	0.83	120	60	11.9	11.2-14.2
70 *(post-partum)*	0.83	120	65	12.2	11.3-14.2
365 *(post-partum)*	0.83	240	73	12.5	11.3-15.0
Total		1680	63		

Table 40
Survival and disease incidence in beagle dogs exposed to x- and gamma-irradiation [T4]

Acute and fractionated whole-body irradiation

Number of dogs	*Exposure*			*Median post-exposure survival (years)*	*Incidence (%)*		
	Dose (Gy)	*Fractions*	*Interval between fractions (days)*		*Non-neoplastic diseases*	*Mammary tumours*	*Non-mammary tumours*
57	None (controls)			11.6	44	21	35
22	0.25	4	28	11.0			
25	0.25	4	14	10.0			
20	0.25	4	7	11.3			
21	0.50	2	28	10.6	46	23	31
21	0.50	2	14	9.8			
20	0.50	2	7	10.7			
23	1.00	1		10.8			
22	0.75	4	28	10.3			
23	0.75	4	14	9.3			
26	0.75	4	7	9.0			
25	1.50	2	28	9.2	54	17	29
21	1.50	2	14	8.5			
23	1.50	2	7	8.7			
11	3.00	1		10.4			

Continuous gamma-irradiation

Number of dogs	*Dose rate ($Gy\ d^{-1}$)*	*Number of dogs dead*	*Mean survival (years)*		*Cause of death (% incidence in parentheses)*		
			Of dead dogs	*Of living dogs*	*Septicaemia*	*Anaemia*	*Myeloproliferative disorder*
46	None	12	8.4	10.0	0 (0)	0 (0)	0 (0)
92	0.003	33	8.2	9.7	0 (0)	0 (0)	0 (0)
46	0.008	28	8.0	10.0	0 (0)	0 (0)	1 (4)
46	0.019	42	7.3	8.9	0 (0)	1 (2)	7 (17)
24	0.038	24	5.2		0 (0)	2 (8)	11 (46)
16	0.075	16	1.9		0 (0)	7 (44)	7 (44)
13	0.013	13	0.8		7 (54)	4 (31)	2 (12)
16	0.026	16	0.15		16 (100)	0 (0)	0 (0)
8	0.038	8	0.10		8 (100)	0 (0)	0 (0)
4	0.054	4	0.07		4 (100)	0 (0)	0 (0)

Table 41
Cancer mortality rates and estimates of excess relative risk of cancer for a high-background-radiation area and a control (low-background-radiation) area in China
[W29]

Site of cancer	*Age range (a)*	*Number of cancer cases in 80,000 persons* [a]		*Age adjusted mortality rates (10^{-5} PY^{-1})* [b]		*Excess relative risk (90% CI)*
		High-background areas (5.4 mSv a^{-1}) [c]	*Low-background areas (2 mSv a^{-1})* [d]	*High-background areas* [c]	*Low-background areas* [d]	
Leukaemia	0-69	31	32	3.0	3.4	-0.08 (-0.40,0.39)
Non-leukaemia	10-79	412	484			-0.06 (-0.16,0.05)
Non-leukaemia	0-39	40	29			0.44 (-0.03,1.17)
Nasopharynx	10-79	94	108	9.8	10.5	-0.07 (-0.26,0.18)
Oesophagus	50-89	12	16	1.4	1.5	-0.16 (-0.56,0.56)
Stomach	20-79	52	47	5.6	4.5	0.21 ((-0.13,0.68)
Liver	20-79	113	138	12.5	13.9	-0.13 (-0.30,0.07)
Intestine	30-79	16	23	1.7	2.3	-0.24 (-0.56,0.29)
Lung	20-79	25	35	2.7	3.3	-0.21 (-0.49,0.21)
Breast	30-79	7	13	0.75	0.59	-0.37 (-0.72,0.34)
Cervix uteri	40-79	13	5	1.4	1.1	2.12 (0.37,6.96)
Bone	-	4	3	0.52	0.45	-
Others	0-89	95	99			0.03 (-0.19,0.31)

[a] Latent period assumed is 10 years for solid tumours and 2 years for leukaemia.
[b] Adjusted with the combined population of both high- and low-background radiation areas.
[c] 1,008,769 person-years were observed in high-background radiation area.
[d] 995,070 person-years were observed in the control (low-background) area.

Table 42
Standardized mortality ratios for selected causes of death among shipyard workers in the United States
[M13]

Cause of death	*Standard mortality ratio (95% CI)*		
	>5 mSv	*<5 mSv*	*Non-nuclear radiation workers*
All causes	0.76 (0.73,0.79)	0.81 (0.76,0.86)	1.00 (0.97,1.03)
Leukaemia	0.91 (0.56,1.39)	0.42 (0.11,1.07)	0.97 (0.65,1.39)
Lymphatic and haemopoietic cancers	0.82 (0.61,1.08)	0.53 (0.28,0.91)	1.10 (0.88,1.37)
Mesothelioma	5.11 (3.03,8.08)	5.75 (2.48,11.33)	2.41 (1.16,4.43)
Lung cancer	1.07 (0.94,1.21)	1.11 (0.90,1.35)	1.15 (1.02,1.29)

Table 43
Relative risk of mortality for workers who were employed for at least six months at the Hanford nuclear materials production plant in the United States
[G12]

External dose (mSv)	*Relative risk (90% CI)* [a]			
	All cancers	*All diseases except cancer* [b]	*Leukaemia* [c]	*Multiple myeloma*
0-9	1.00	1.00	1.0	1.0
10-49	1.04 (0.9,1.2)	0.89 (0.8,1.0)	0.8 (0.4-1.6)	0.4 (0.1,1.3)
50-99	1.01 (0.8,1.3)	0.85 (0.7,1.0)	0.3 (0.03-1.6)	4.2 (0.7,19)
100-199	1.17 (0.9,1.5)	0.83 (0.7,1.0)	1.5 (0.4-4.8)	5.9 (0.5,41)
>200	0.93 (0.7,1.3)	0.96 (0.8,1.2)	0.3 (0.02-1.3)	21 (2.1,270)

[a] The relative risk is the ratio of the risk for the indicated dose category relative to that for the lowest dose category (0-9 mSv).
[b] Mainly deaths from cardiovascular diseases, genito-urinary diseases and diseases of the respiratory tract.
[c] Excluding chronic lymphatic leukaemia.

Table 44
Incidence of thyroid cancer after administration of ^{131}I in medical diagnostic examinations
[H18]

Activity administered	*Follow-up*											
	5-9 years			*10-14 years*			*15-19 years*			*≥20 years*		
(MBq)	*Observed*	*SIR*	*95% CI*	*Observed*	*SIR*	*95% CI*	*Observed*	*SIR*	*95% CI*	*Observed*	*SIR*	*95% CI*
<0.9	8	1.90	0.82-3.76	2	0.39	0.05-1.42	3	0.85	0.18-2.50	1	0.56	0.01-3.10
0.9-2.2	5	1.22	0.40-2.85	7	1.47	0.59-3.03	3	0.72	0.14-2.10	4	1.16	0.32-2.96
>2.2	10	4.88	2.34-8.97	2	0.85	0.10-3.07	3	1.44	0.30-4.22	2	1.08	0.13-3.88
All	23	2.22	1.14-3.34	11	0.90	0.45-1.62	9	0.92	0.42-1.75	7	0.99	0.45-2.03

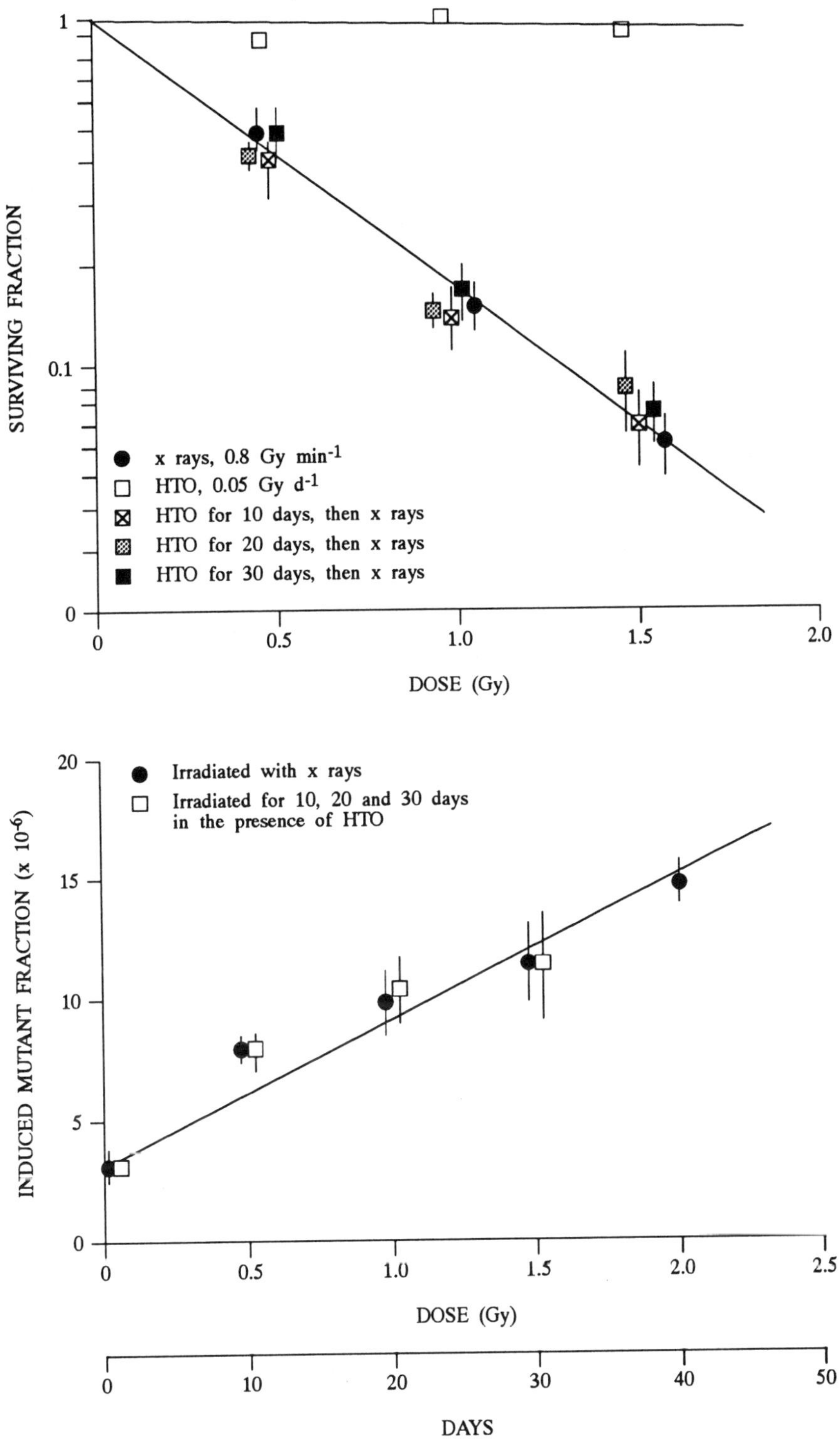

Figure I.
No evidence of adaptive response in human lymphoblastoid cells conditioned with tritiated water (HTO).
[T1]

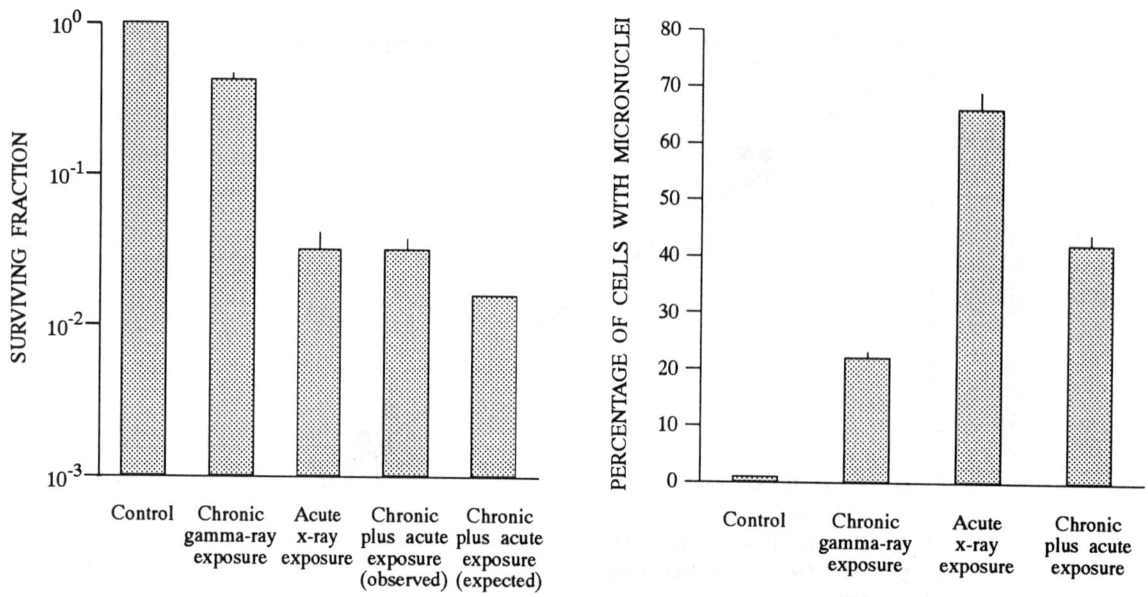

Figure II.
Adaptive response of skin fibroblast (AG1522) cells.
The cells were exposed to 4.25 Gy of chronic gamma-radiation and/or 4.25 Gy of acute x-radiation.
[A2]

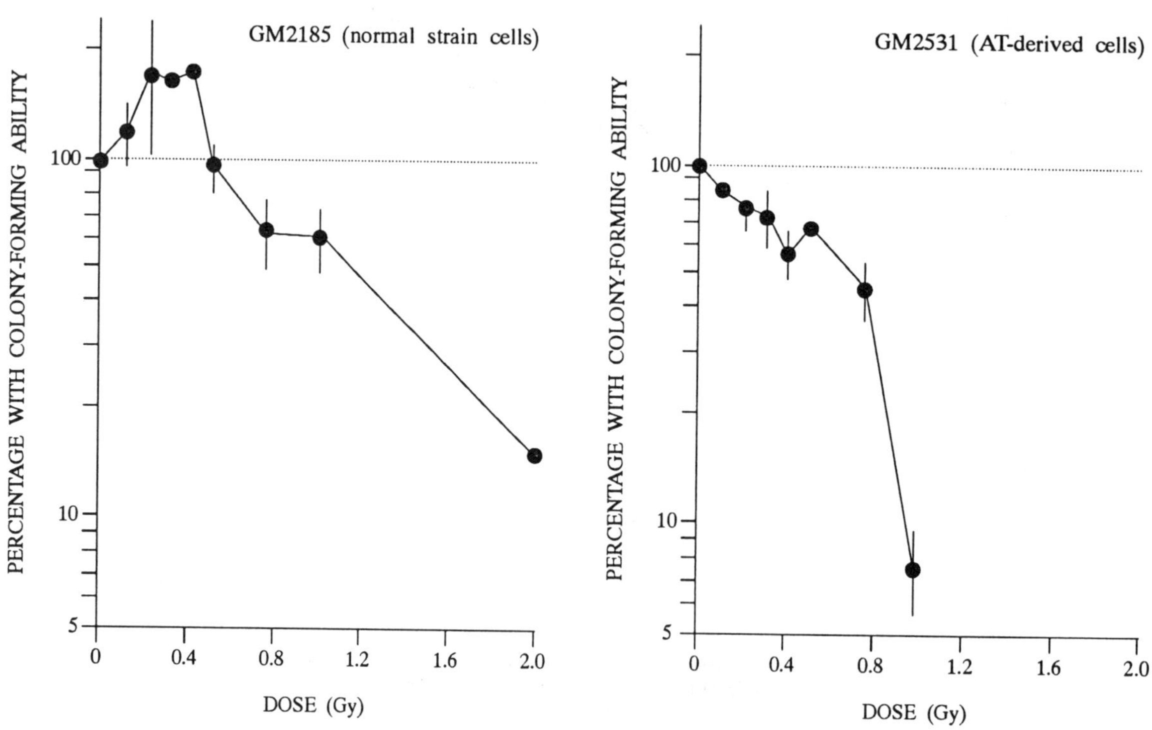

Figure III.
Colony-forming ability of human fibroblasts: normal GM2185 strain cells compared with ataxia-telangiectasia-derived [GM2531 (AT)] strain cells.
[S26]

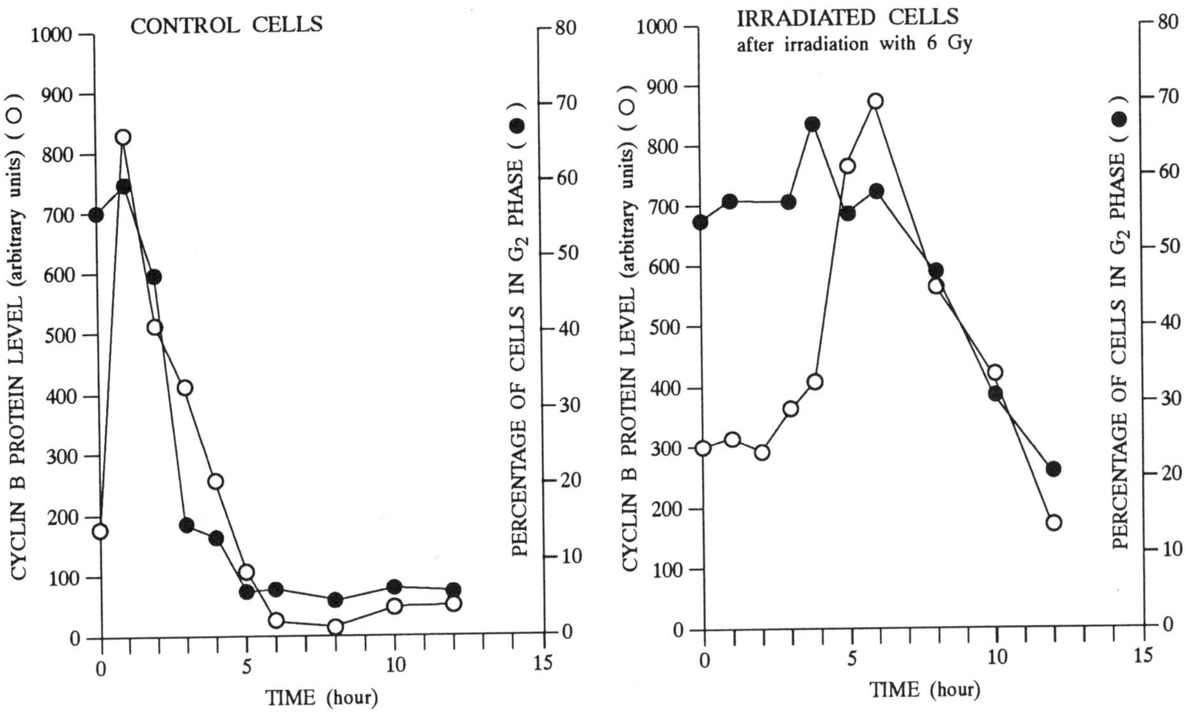

Figure IV.
Cyclin B protein levels in HeLa cells after irradiation in G_2 phase.
[M7]

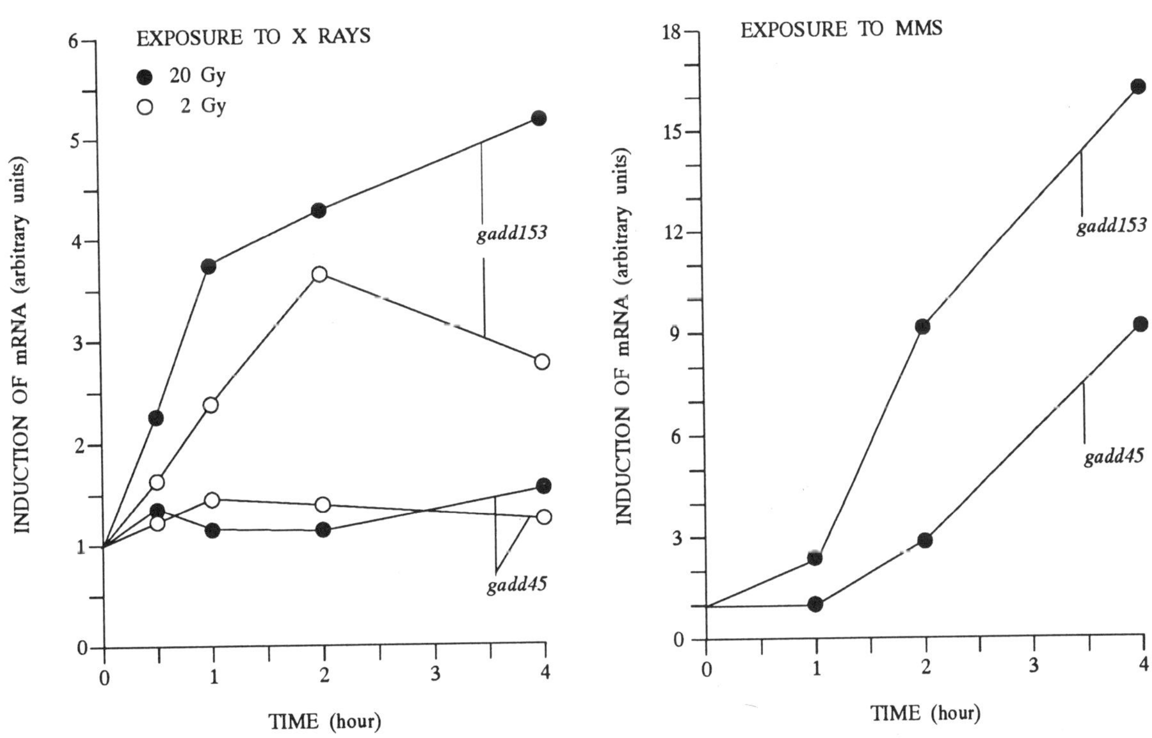

Figure V.
Relative levels of induced *gadd45* and *gadd153* mRNA in human lymphoblast (GM0536) cells following exposure to x rays or to methyl methane sulphonate (MMS, 100 μg ml-1)
[S26]

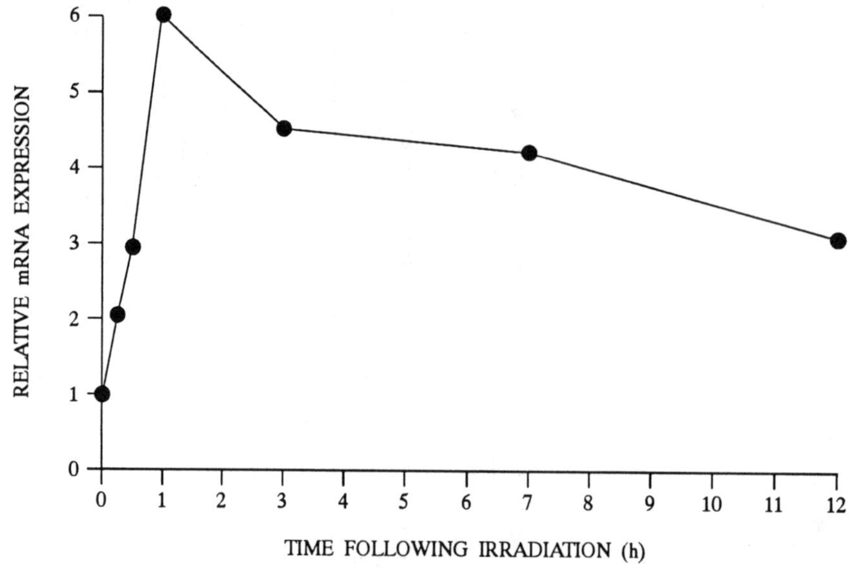

Figure VI.
Increased expression of β-protein kinase C in Syrian hamster embryo cells at various times following acute exposure to 0.75 Gy from x rays.
[W17]

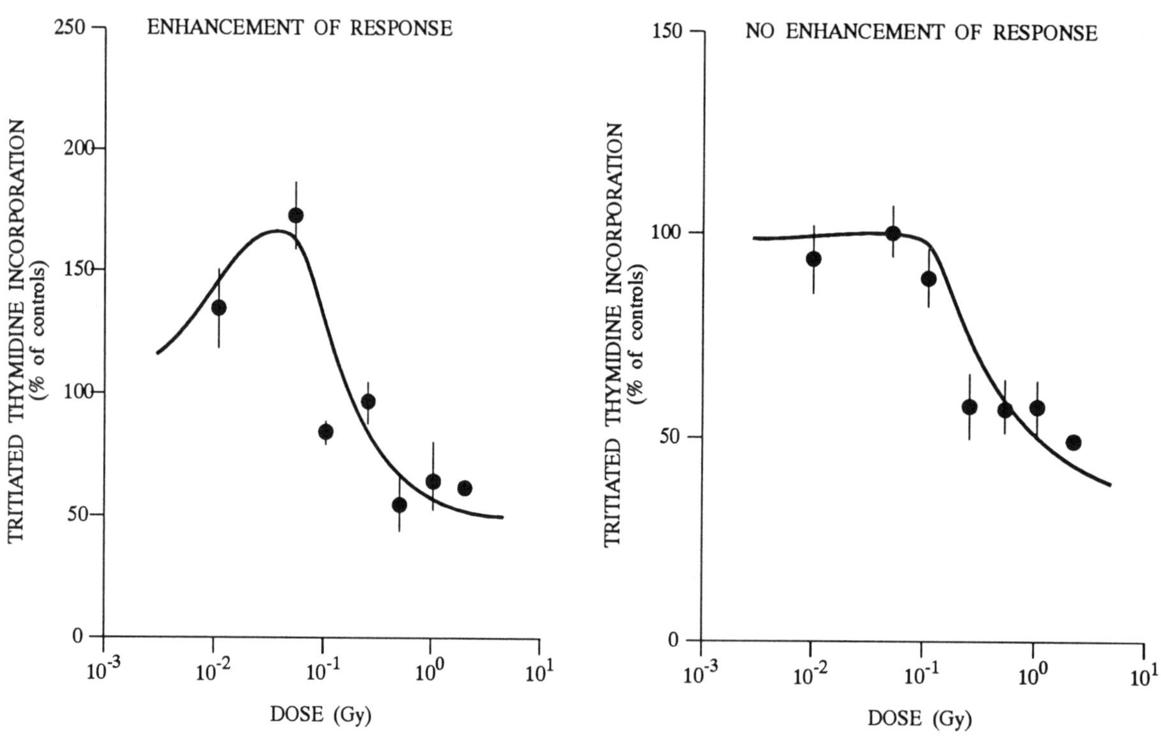

Figure VII.
Concanavalin-A-induced proliferation with small doses of whole-body x-irradiation leading to enhancement of response in rat splenocytes and no enhancement of response in rat thymocytes.
[I3]

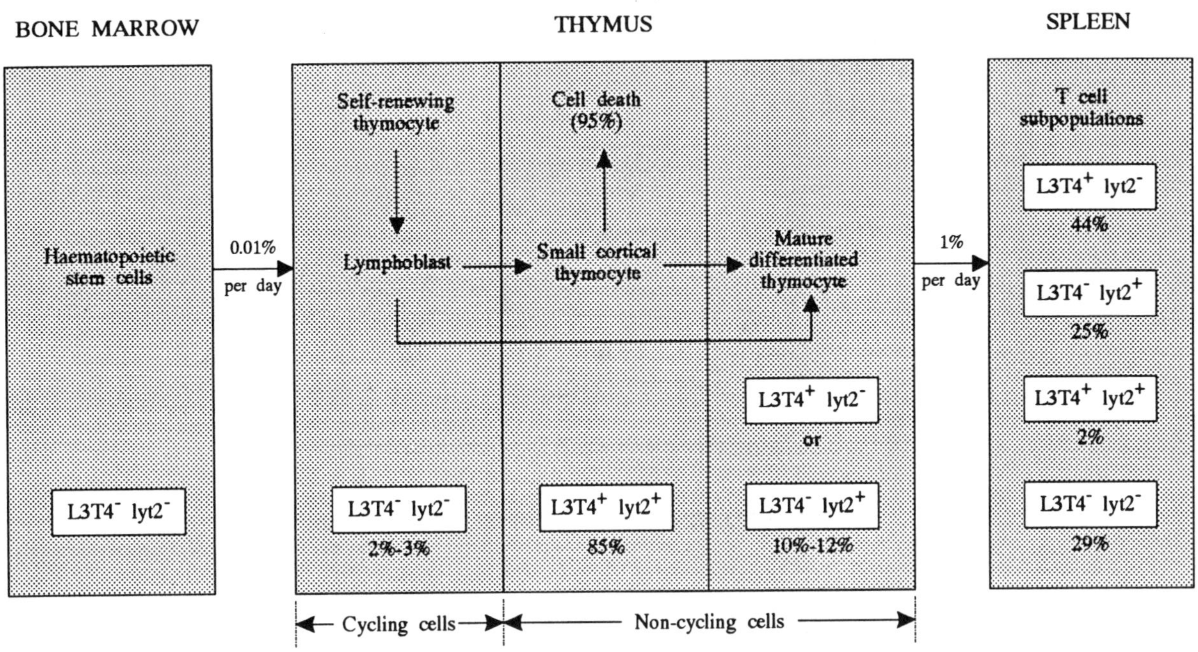

Figure VIII.
A model of T-cell ontogeny from bone marrow and thymic T-cell precursors to fully differentiated peripheral T cells in terms of L3T4 and lyt2 surface antigen expression.
[J7]

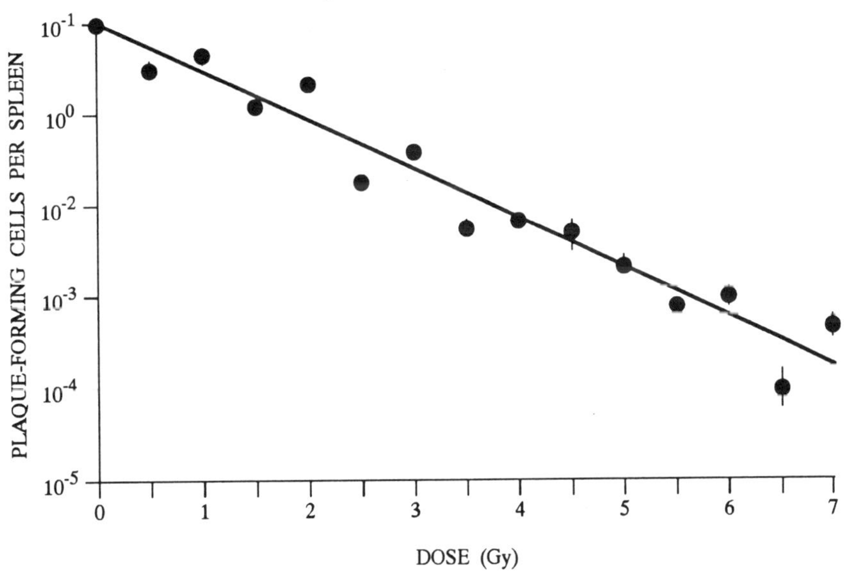

Figure IX.
Relationship between whole-body dose from x-irradiation and plaque-forming capacity in mice. Antigenic sheep red blood cells (4 10^8) were injected 10 days after the irradiation, and assays for plaque-forming spleen cells were performed 4 days later.
[K12]

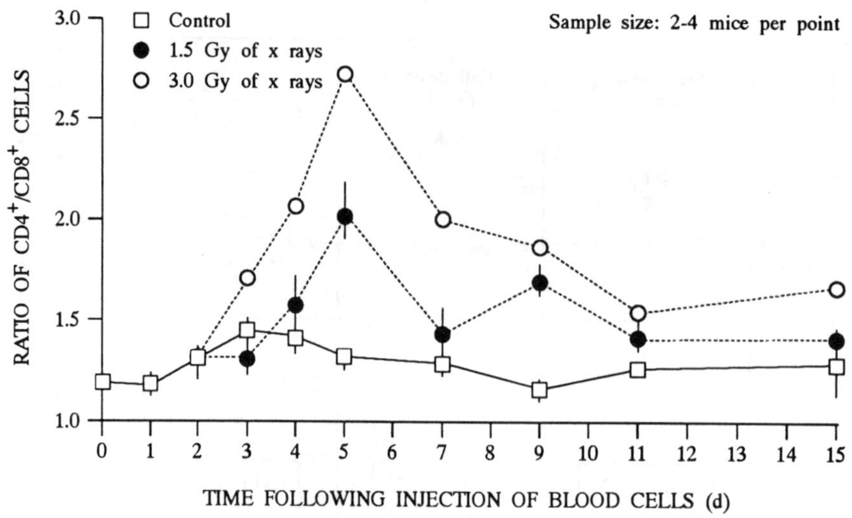

Figure X.
Effects of radiation given two days after injection of antigenic sheep red blood cells on the CD4+/CD8+ cell ratios in the spleens of C57Bl/6J mice.
[K13]

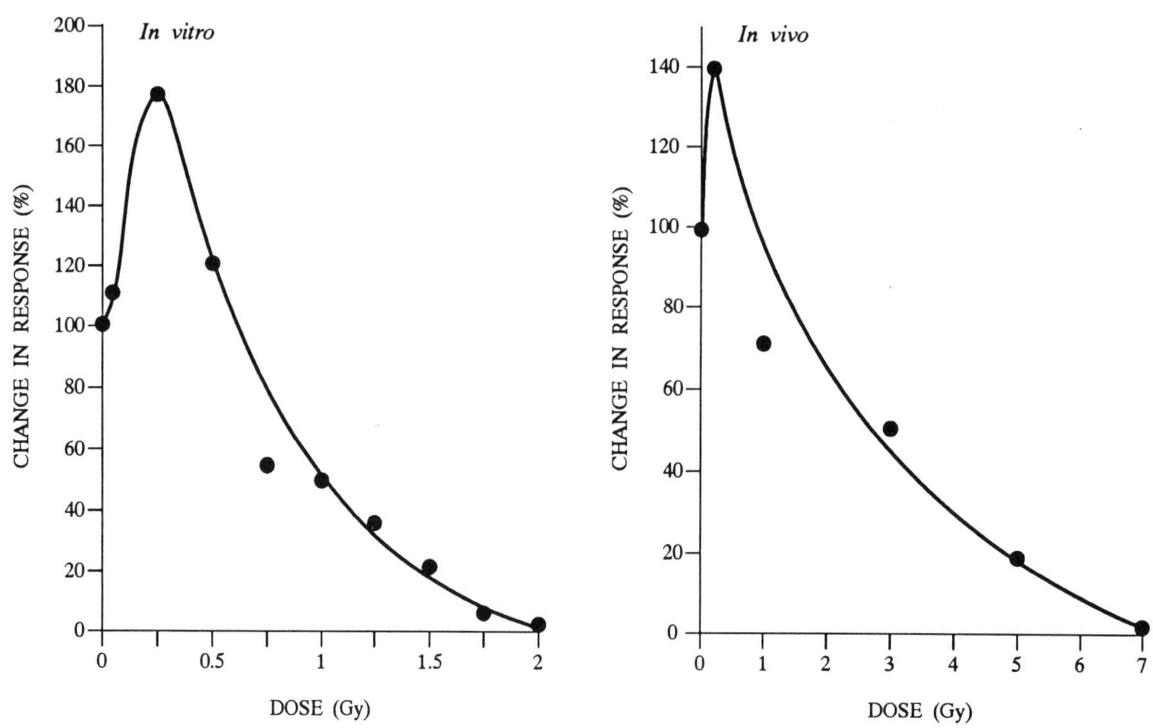

Figure XI.
Effect of radiation on mouse splenic cells primed with antigenic sheep red blood cells.
Results are expressed as the mean per cent change in peak antigen response relative to the highest responder group.
[M18]

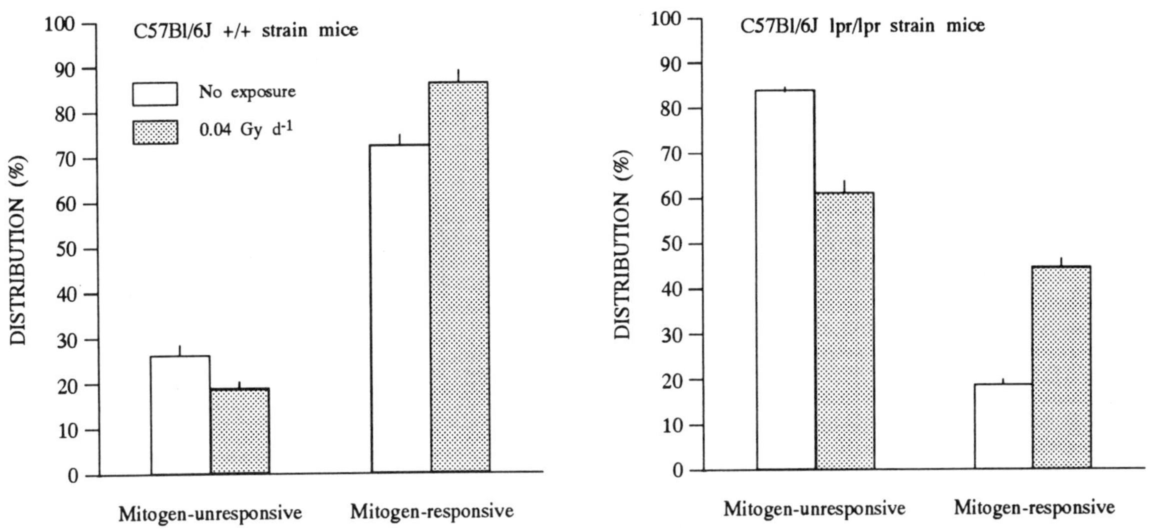

Figure XII.
Effect of low dose on the percentage distribution of mitogen-responsive and mitogen-unresponsive splenic cells in normal (C57Bl/6J +/+ strain) and immunologically deficient (C57Bl/6J lpr/lpr strain) mice.
[J7]

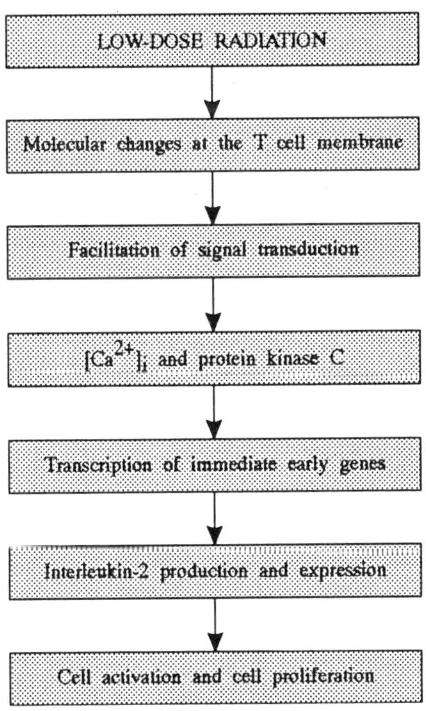

Figure XIII.
Proposed scheme of signal transduction in lymphocytes after low-dose radiation.
[L26]

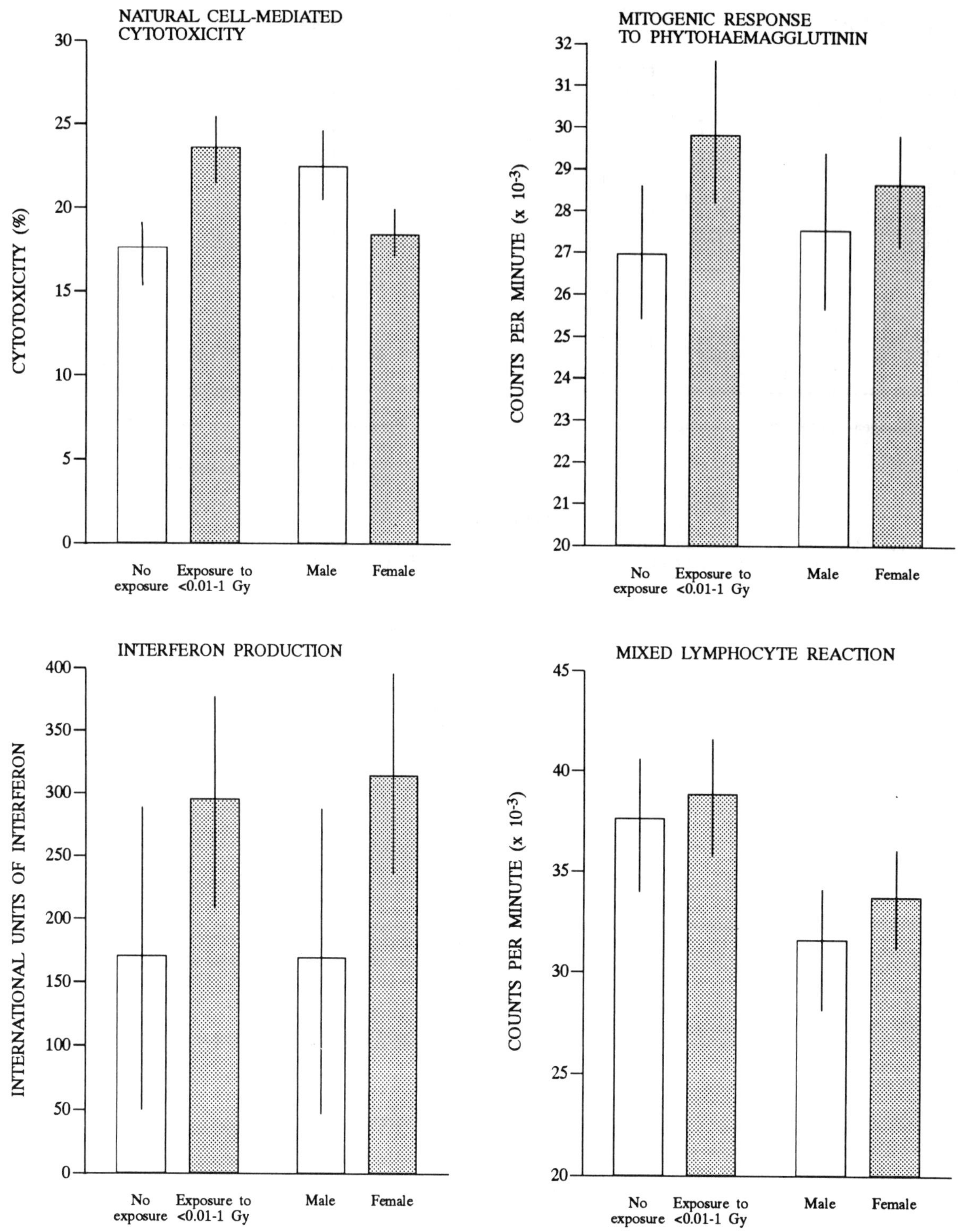

Figure XIV.
Changes in immunological function in atomic bomb survivors residing in the United States. [B19]

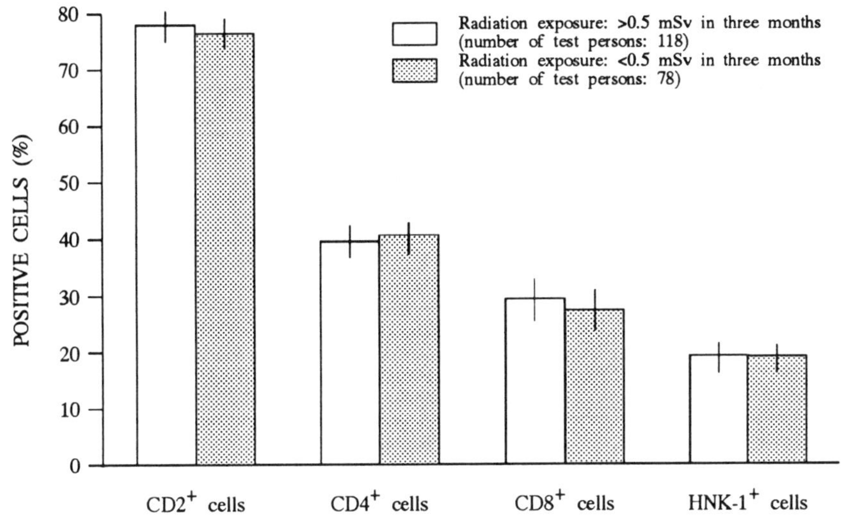

Figure XV.
Percentage of $CD2^+$, $CD4^+$, $CD8^+$ and $HNK\text{-}1^+$ cells within the peripheral lymphocytes of persons occupationally exposed to low-dose radiation.
[T10]

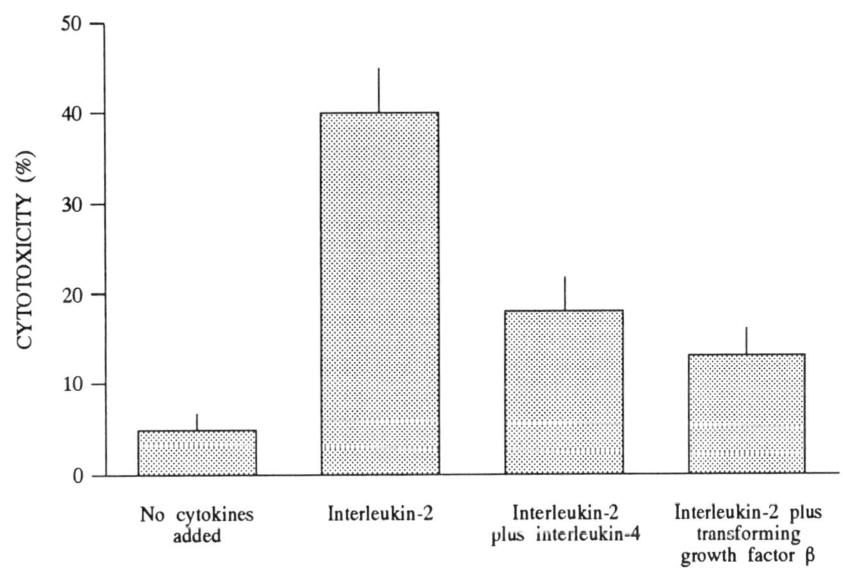

Figure XVI.
Cytotoxicity of white blood cells against tumour cells in the presence of cytokines.
[B25]

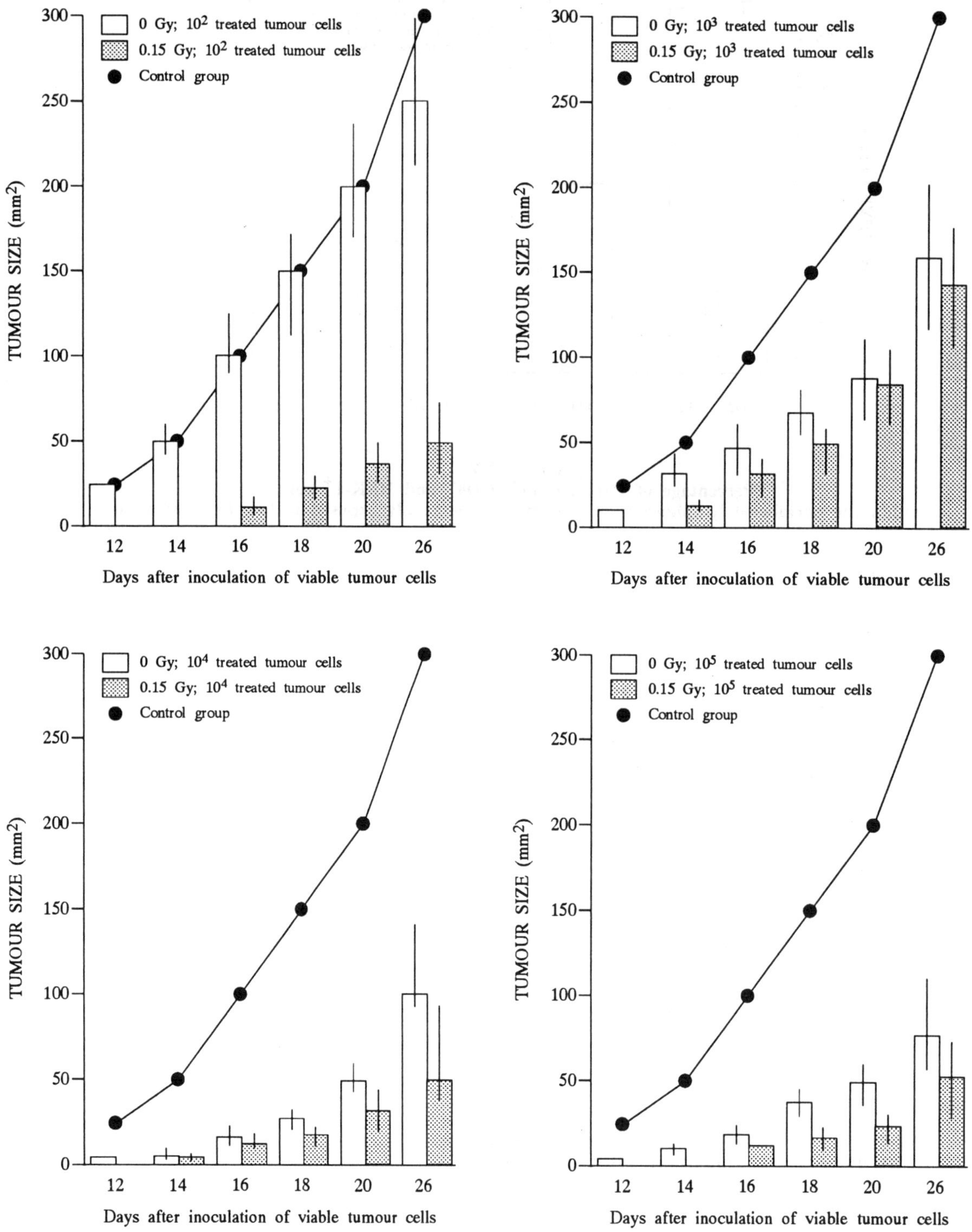

Figure XVII.
Effect of 0.15 Gy upon response of A/J mice to varying numbers of mitomycin-treated SaI cells. Groups of 60 mice were exposed to whole-body irradiation or sham-irradiated and inoculated subcutaneously with the indicated numbers of mitomycin-treated tumour cells. Twenty-one days later, all animals received 10^4 untreated SaI cells and were followed for tumour size. A control group did not receive mitomycin-treated cells.
[A4]

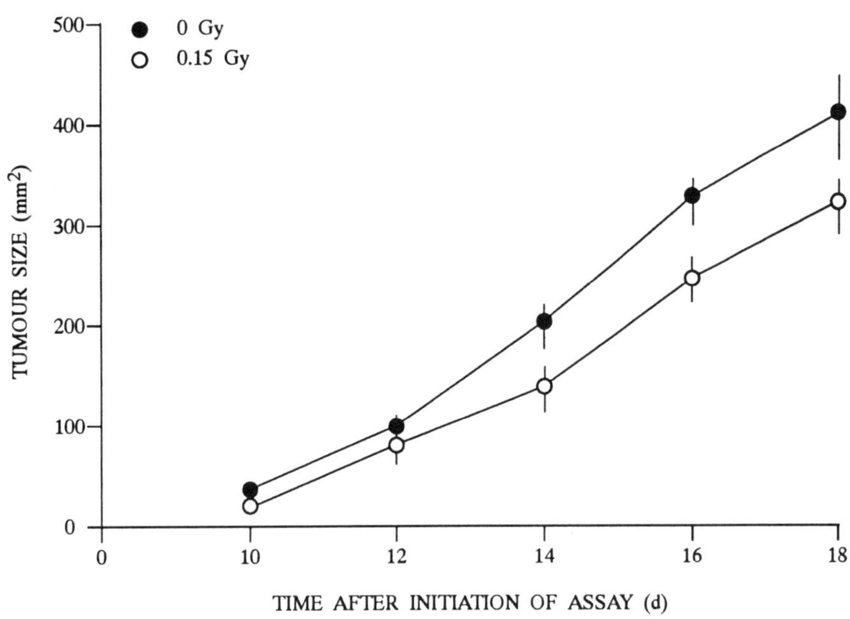

Figure XVIII.
Effect of *in vitro* irradiation of donor spleen cells on tumour size.
Mice were inoculated subcutaneously with 10^4 SaI cells; after two days they were killed and their spleens irradiated or sham-irradiated and then used in the Winn assay technique with an equal number of SaI cells.
[A18]

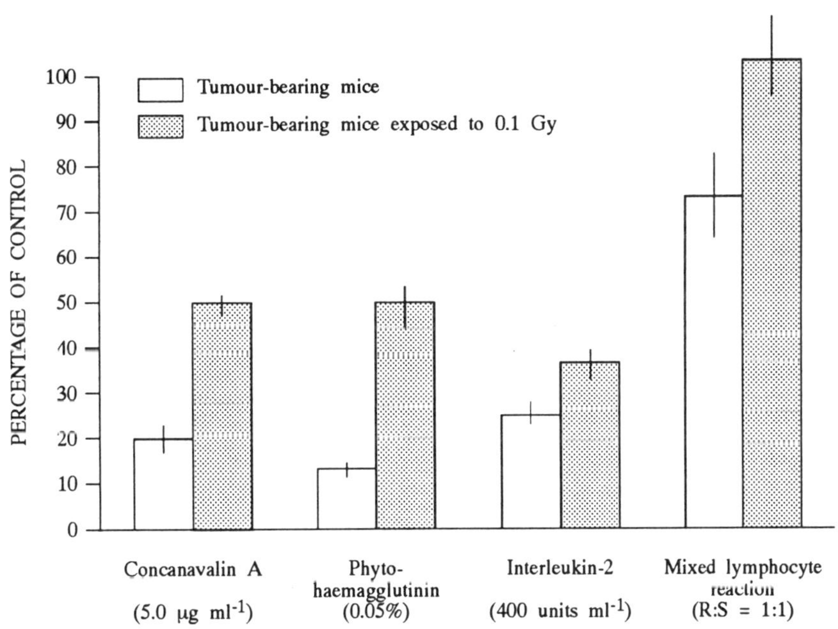

Figure XIX.
***In vitro* immune response of splenocytes from tumour-bearing mice. The control value of 100% represents the amount of [^{3}H]Tdr incorporated into the splenocytes of non-tumour-bearing mice.**
[M28]

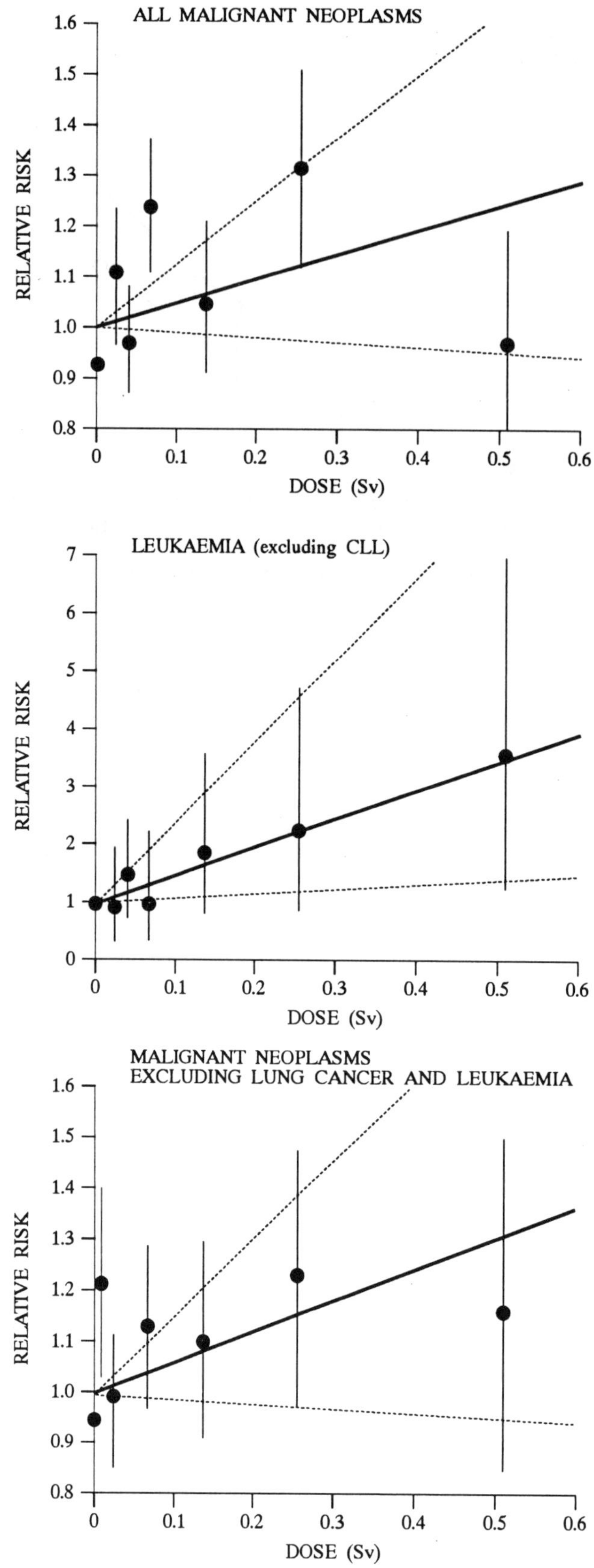

Figure XX.
Relative risk compared with dose for radiation workers in the United Kingdom.
The solid line shows the mean, the dashed lines 90% CIs.
[K15]

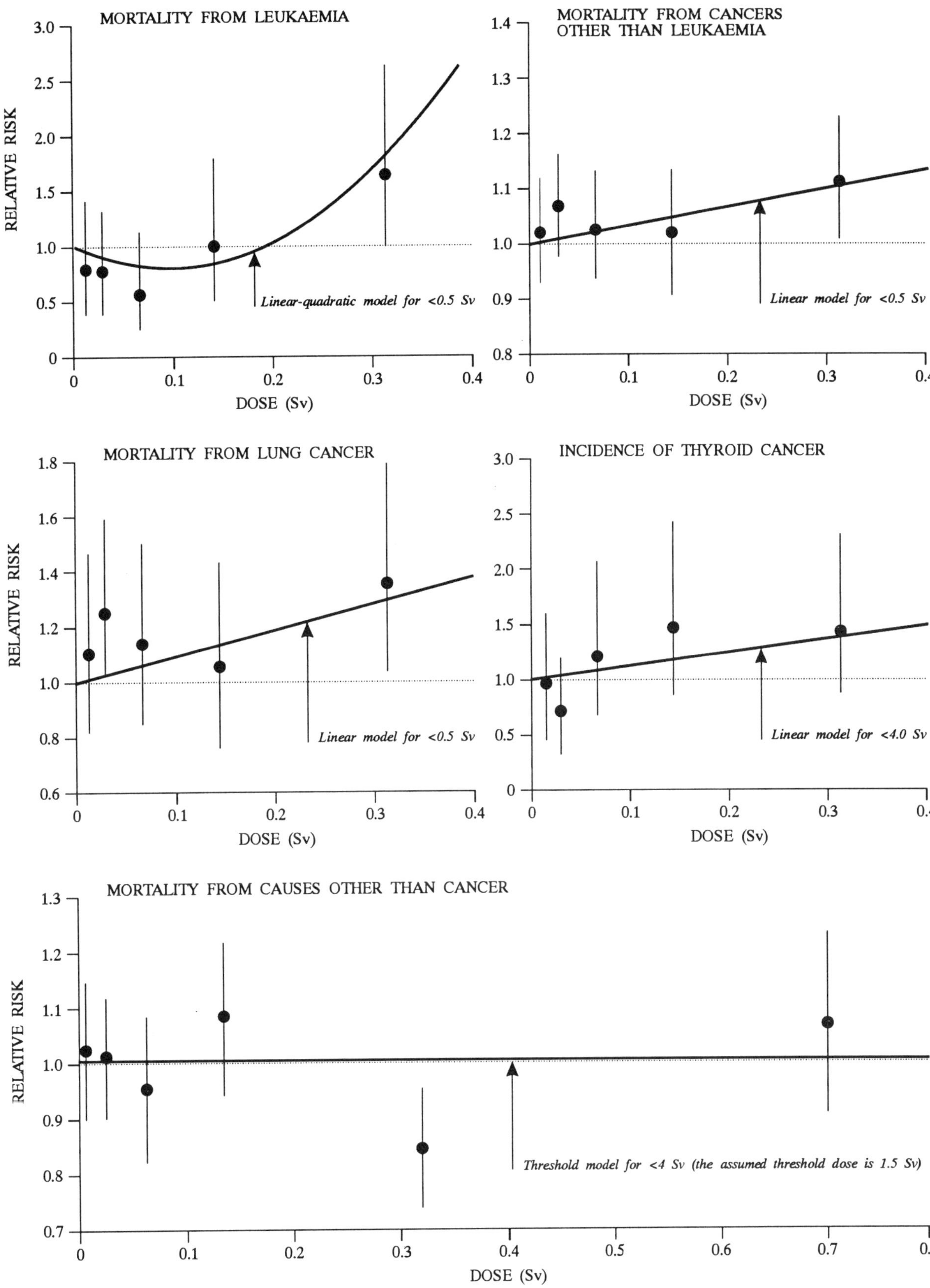

Figure XXI.
Relative risk and fitted dose-response curves for mortality and incidence from various causes.
[S30]

References

A1 Adams, G.E., W.S. Chan and I.J. Stratford. Radiation and other stress responses and their relevance to cancer: some general features. p. 389-397 in: Low Dose Irradiation and Biological Defense Mechanisms. (T. Sugahara, L.A. Sagan and T. Aoyama, eds.) Elsevier Science Publishers, The Netherlands, 1992.

A2 Azzam, E.I., S.M. de Toledo, G.P. Raaphorst et al. Radiation-induced radioresistance in a normal human skin fibroblast cell line. p. 291-294 in: Low Dose Irradiation and Biological Defense Mechanisms. (T. Sugahara, L.A. Sagan and T. Aoyama, eds.) Elsevier Science Publishers, The Netherlands, 1992.

A3 Adamo, S., C. Caporale, S. Aguanno et al. Proliferating and quiescent cells exhibit different subcellular distribution of protein kinase C activity. FEBS Lett. 195: 352-356 (1986).

A4 Anderson, R.E., S. Tokuda, W.L. Williams et al. Low dose irradiation permits immunization of A/J mice with subimmunogenic numbers of SaI cells. Br. J. Cancer 54: 505-509 (1986).

A5 Allison, J.P. and L.L. Lanier. The structure, function and serology of the T cell antigen receptor complex. Annu. Rev. Immunol. 5: 503-540 (1987).

A6 Anderson, R.E. and N.L. Warner. Ionizing radiation and the immune response. Adv. Immunol. 24: 215-335 (1976).

A7 Anderson, R.E. Effects of low-dose radiation on the immune response. p. 95-112 in: Biological Effects of Low Level Exposures to Chemicals and Radiation. (E.J. Calabrese, ed.) Lewis Publishers, Boca Raton, 1992.

A8 Anderson, R.E., J.C. Standifer and S. Tokuda. The structural and functional assessment of cytotoxic injury of the immune system with particular reference to the effect of radiation and cyclophosphamide. Br. J. Cancer 53: 140-160 (1986).

A9 Anderson, R.E. and I. Lefkovits. In vitro evaluation of radiation-induced augmentation of the immune response. Am. J. Pathol. 97: 456-472 (1979).

A10 Aghamohammadi, S.Z. and J.R.K. Savage. A BrdU pulse double-labelling method for studying adaptive response. Mutat. Res. 251: 133-143 (1991).

A11 Akiyama, M., Y. Kusunoki, S. Kyoizumi et al. Study of the titers of anti-epstein-barr virus antibodies in the sera of atomic bomb survivors. Radiat. Res. 133: 297-302 (1993).

A12 Alberts, B., D. Bray, J. Lewis et al. p. 1051 in: Molecular Biology of the Cell. Second edition. Garland Publishing Co., New York, 1989.

A13 Andersen, A.C. and L.S. Rosenblatt. The effect of whole-body x-irradiation on the median lifespan of female dogs (beagles). Radiat. Res. 39: 177-200 (1969).

A14 Anderson, R.L., E. Shiu, G.A. Fisher et al. DNA damage does not appear to be a trigger for thermotolerance in mammalian cells. Int. J. Radiat. Biol. 54: 285-298 (1988).

A15 Applegate, L.A., P. Luscher and R.M. Tyrrell. Induction of heme oxygenase: a general response to oxidant stress in cultured mammalian cells. Cancer Res. 51: 974-978 (1991).

A16 Althaus, F.R., H.E. Kleizkowska, P.L. Panzeter et al. Molecular mechanisms of low dose adaptation: the role of the poly ADP-ribosylation system. p. 68 in: International Symposium on Biological Effects of Low Level Exposures to Radiation and Related Agents (Isbelles '93), Abstracts, Changchun, China, 1993. [To be published in the Chinese Medical Journal (1994)].

A17 Azzam, E.I., G.P. Raaphorst and R.E.J. Mitchel. Radiation-induced adaptive response for protection against micronucleus formation and neoplastic transformation in C3H 10T½ mouse embryo cells. Radiat. Res. 138 (1): S28-S31 (1994).

A18 Anderson, R.E., W.L. Williams and S. Tokuda. Effect of low-dose irradiation upon T cell subsets involved in the response of primed A/J mice to SaI cells. Int. J. Radiat. Biol. 53 (1): 103-118 (1988).

A19 Akiyama, M., O.L. Zhou, Y. Kusunoki et al. Age and dose related alteration of *in vitro* mixed lymphocyte culture response of blood lymphocytes from A-bomb survivors. Radiat. Res. 117: 26-34 (1989).

B1 Brennand, J. and G.P. Margison. Reduction of the toxicity and mutagenicity of alkylating agents in mammalian cells harboring the Escherichia coli alkyltransferase gene. Proc. Natl. Acad. Sci. (USA) 83: 6292-6296 (1986).

B2 Boice, J.D., D. Preston, F.G. Davis et al. Frequent chest x-ray fluoroscopy and breast cancer incidence among tuberculosis patients in Massachusetts. Radiat. Res. 125: 214-222 (1991).

B3 Brunet, S. and P.U. Giacomoni. Heat shock mRNA in mouse epidermis after UV irradiation. Mutat. Res. 219: 217-224 (1989).

B4 Burdon, R.H., V. Gill and C. Rice-Evans. Cell proliferation and oxidative stress. Free Radic. Res. Commun. 7: 149-159 (1989).

B5 Benjamin, I.J., B. Kröger and R.S. Williams. Activation of the heat shock transcription factor by hypoxia in mammalian cells. Proc. Natl. Acad. Sci. (USA) 87: 6263-6267 (1990).

B6 Bosi, A. and G. Olivieri. Variability of the adaptive response to ionizing radiations in humans. Mutat. Res. 211: 13-17 (1989).

B7 Bauchinger, M., E. Schmid, H. Braselmann et al. Absence of adaptive response to low-level irradiation from tritiated thymidine and X-rays in lymphocytes of two individuals examined in serial experiments. Mutat. Res. 227: 103-107 (1988).

B8 Bosi, A., A. Micheli, S. Pietrosanti et al. Effect of pH shifts on radiosensitivity of human lymphocytes irradiated in the G_2 stage. Mutat. Res. 250: 325-329 (1991).

B9 Boerrigter, M.E. DNA strand break metabolism in human lymphocytes: a reevaluation. Exp. Cell Res. 196: 1-5 (1991).

B10 Billen, D. Spontaneous DNA damage and its significance for the "negligible dose" controversy in radiation protection. Radiat. Res. 124: 242-245 (1990).

B11 Bouzyk, E., O. Rosiek and M. Wojewódzka. Small doses of x-rays make human peripheral blood lymphocytes less sensitive to induction of chromatid aberrations by a subsequent large dose of this radiation. Nukleonika 34: 10-12 (1989).

B12 Boothman, D.A., I. Bouvard and E.N. Hughes. Identification and characterization of x-ray-induced proteins in human cells. Cancer Res. 49: 2871-2878 (1989).

B13 Boothman, D.A., M. Wang and S.W. Lee. Induction of tissue-type plasminogen activator by ionizing radiation in human malignant melanoma cells. Cancer Res. 51: 5587-5595 (1991).

B14 Boothman, D.A., S. Lee, D.K. Trask et al. X ray induced proteins and genes in human cells. p. 136 and 309-317 in: UCLA Symposia On Molecular and Cellular Biology: Ionizing Radiation Damage to DNA. New series. (S. Wallace and R. Painter, eds.) Wiley-Liss, Inc., 1990.

B15 Boothman, D.A., D. Trask and A.B. Pardee. Inhibition of potentially lethal DNA damage repair in human tumor cells by beta-lapachone, an activator of topoisomerase 1. Cancer Res. 49: 605-612 (1989).

B16 Bukh, A., H. Martinez-Valdez, S.J. Freedman et al. The expression of c-fos, c-jun and c-myc genes is regulated by heat shock in human lymphoid cells. J. Immunol. 144: 4835-4840 (1990).

B17 Brenner, M.B., J. McLean, D.P. Dialynas et al. Identification of a putative second T-cell receptor. Nature 332: 145-149 (1986).

B18 Bloom, E.T., E.L. Korn, M. Takasugi et al. Immune function in aging atomic bomb survivors residing in the United States. Radiat. Res. 96: 399-410 (1983).

B19 Bloom, E.T., M. Akiyama, Y. Kusunoki et al. Delayed effects of low-dose radiation on cellular immunity in atomic bomb survivors residing in the United States. Health Phys. 52: 585-591 (1987).

B20 Bloom, E.T., M. Akiyama, E.L. Korn et al. Immunological responses of aging Japanese A-bomb survivors. Radiat. Res. 116: 343-355 (1988).

B21 Burkart, W., P. Heusser and Vijayalaxmi. Microdosimetric constraints on specific adaptation mechanisms to reduce DNA damage caused by ionising radiation. Radiat. Prot. Dosim. 31: 269-274 (1990).

B22 Bai, Y. and D. Chen. Accumulative effect of two low doses of irradiation in inducing an adaptive response in human lymphocytes. Mutat. Res. 302: 191-196 (1993).

B23 Benjamin, S.A., A.C. Lee, G.M. Angleton et al. Life-span radiation effects studies in prenatally and postnatally exposed beagle dogs at Colorado State University. p. 1-13 in: Life-Span Radiation Effects Studies in Animals: What Can They Tell Us? (R.C. Thompson and J.A. Mahaffey, eds.) CONF-830951 (1986).

B24 Brach, M.A., R. Hass, M.L. Sherman et al. Ionizing radiation induces expression and Pd binding activity of the nuclear factor kappa B. J. Clin. Invest. 88: 691-695 (1991).

B25 Brooks, B. Immunology and cancer. Biologist 40 (4): 162-166 (1993).

B26 Boothman, D.A. Role of x-ray-inducible genes and proteins in adaptive survival responses. p. 14 in: International Symposium on Biological Effects of Low Level Exposures to Radiation and Related Agents (Isbelles '93), Abstracts, Changchun, China, 1993. [To be published in the Chinese Medical Journal (1994)].

C1 Conter, A., D. Dupouy and H. Planel. Demonstration of a biological effect of natural ionizing radiation. Int. J. Radiat. Biol. 43: 421-432 (1983).

C2 Conter, A., D. Dupouy, M. Murat et al. Radiation stimulation during the early stationary growth phase in Syncchoccus lividus and its correlation with photo-oxidative stress occurring before the stationary phase. Environ. Exp. Bot. 27: 21-28 (1987).

C3 Conter, A., D. Dupouy and H. Planel. Effects of dose rate on response of Synechococcus lividus to very low doses of chronic gamma-radiation: influence of enzymatic equipment of starting cells. Radiat. Res. 105: 379-386 (1986).

C4 Conter, A. Pre-irradiation of medium induces a subsequent stimulation or inhibition of growth according to the physiological state in synechococcus lividus in culture. Radiat. Res. 109: 342-346 (1987).

C5 Conter, A., D. Dupouy, C. Vincent et al. Radiation stimulation during the early stationary growth phase in synechococcus lividus and its correlation with photooxidative stress occurring before the stationary phase. Health Phys. 53: 281-286 (1987).

C6 Croute, F., J.P. Soleilhavoup, S. Vidal et al. Paramecium tetradurelia growth stimulation under low-level chronic irradiation: investigations on a possible mechanism. Radiat. Res. 92: 560-567 (1982).

C7 Calabrese, E.J., M.E. McCarthy and E. Kenyon. The occurrence of chemically induced hormesis. Health Phys. 52: 531-542 (1987).

C8 Cai, L. and S.Z. Liu. Induction of cytogenetic adaptive response of somatic and germ cells in vivo and in vitro by low-dose X-irradiation. Int. J. Radiat. Biol. 58 (1): 187-194 (1990).

C9 Clerget, M. and B.S. Polla. Erythrophagocytosis induces heat shock protein synthesis by human monocytes-macrophages. Proc. Natl. Acad. Sci. (USA) 87: 1081-1085 (1990).

C10 Clevers, H., B. Alarcon, T. Wileman et al. The T cell receptor/CD3 complex: a dynamic protein ensemble. Annu. Rev. Immunol. 6: 629-662 (1988).

C11 Cooke, M.P. and R.M. Perimutter. Expression of a novel form of the fyn proto-oncogene in hematopoietic cells. New Biol. 1: 66-74 (1989).

C12 Cooke, M.P., K.M. Abraham, K.A. Forbush et al. Regulation of T cell receptor signalling by a src family protein-tyrosine kinase ($p59^{fyn}$). Cell 65: 281-291 (1991).

C13 Cortés, F., I. Dominguez, J. Pinero et al. Adaptive response in human lymphocytes conditioned with hydrogen peroxide before irradiation with x rays. Mutagenesis 5: 555-557 (1990).

C14 Cortés, F., I. Dominguez, S. Mateos et al. Evidence for an adaptive response to radiation damage in plant cells conditioned with x-rays or incorporated tritium. Int. J. Radiat. Biol. 57: 537-541 (1990).

C15 Cerutti, P.A. and B.F. Trump. Inflammation and oxidative stress in carcinogenesis. Cancer Cells 3: 1-7 (1991).

C16 Caldecott, K.W., J.D. Tucker and L.H. Thompson. Overcorrection of the CHO EM9 DNA repair mutant by overexpression of a human XRCC1 minigene. Mol. Cell. Biol. 14 (1): 68-76 (1994).

C17 Cai, L., J. Jiang, B. Wang et al. Induction of an adaptive response to dominant lethality and to chromosome damage of mouse germ cells by low dose radiation. Mutat. Res. (1994, in press).

C18 Cronkite, E.P., C.R. Sipe, D.C. Eltholtz et al. Increased tolerance of mice to lethal x radiation as a result of previous sublethal exposures. Proc. Soc. Exp. Biol. Med. 73: 184-186 (1950).

C19 Carnes, B.A., D. Grahn and J.F. Thomson. Dose-response modeling of life shortening in a retrospective analysis of the combined data from the JANUS program at Argonne National Laboratory. Radiat. Res. 119: 39-56 (1989).

C20 Covelli, V., V. Di Majo, B. Bassani et al. Influence of age on life shortening and tumor induction after x-ray and neutron irradiation. Radiat. Res. 100: 348-364 (1984).

C21 Chernikova, S.V., V.Ya. Gotlib and I.I. Pelevina. Influence of low dose radiation on the radiosensitivity. Radiatsionnaya Biologia Radioekologia 33: 537-541 (1993).

D1 Demple, B. and J. Halbrook. Inducible repair of oxidative DNA damage in Escherichia coli. Nature 304: 466-468 (1983).

D2 Dacquisto, M.P. and M.C. Major. Acquired radio-resistance. A review of the literature and report of a confirmatory experiment. Radiat. Res. 10: 118-129 (1959).

D3 Davis, M.M. and P.J. Bjorkman. T-cell antigen receptor genes and T-cell recognition. Nature 334: 395-402 (1988).

D4 Duvall, E., A.H. Wyllie and R.G. Morris. Macrophage recognition of cells undergoing programmed cell death (apoptosis). Immunology 56 (2): 351-358 (1985).

D5 Demple, B. Regulation of bacterial oxidative stress genes. Annu. Rev. Genet. 25: 315-337 (1991).

D6 Dominguez, I., E. Panneerselvam, P. Escalza et al. Adaptive response to radiation damage in human lymphocytes conditioned with hydrogen peroxide as measured by the cytokinesis-block micronucleus technique. Mutat. Res. 301: 135-141 (1993).

D7 Devary, Y., R.A. Gottlieb, L.F. Lau et al. Rapid and preferential activation of the c-jun gene during the mammalian UV response. Mol. Cell. Biol. 11: 2804-2811 (1991).

D8 Dixon, F.J. and P.J. McConahey. Enhancement of antibody formation by whole body x-irradiation. J. Exp. Med. 117: 833-848 (1963).

F1 Fritz, T.E., T.M. Seed, D.V. Tolle et al. Late effects of protracted whole-body irradiation of beagles by cobalt-60 gamma rays. p. 116-141 in: Life-Span Radiation Effects Studies in Animals: What Can They Tell Us? (R.C. Thompson and J.A. Mahaffey, eds.) CONF-830951 (1986).

F2 Fan, S., Vijayalaxmi, G. Mindek et al. Adaptive response to two low doses of X-rays in human blood lymphocytes. Mutat. Res. 243: 53-56 (1990).

F3 Frankenberg-Schwager, M., D. Frankenberg and R. Harbich. Exponential or shouldered survival curves result from repair of DNA double-strand breaks depending on post-irradiation conditions. Radiat. Res. 114: 54-63 (1988).

F4 Frankenberg-Schwager, M. Review of repair kinetics for DNA damage induced in eukaryotic cells in vitro by ionizing radiation. Radiother. Oncol. 14: 307-320 (1989).

F5 Frankenberg-Schwager, M. and D. Frankenberg. DNA double-strand breaks: their repair and relationship to cell killing in yeast. Int. J. Radiat. Biol. 58: 569-575 (1990).

F6 Fornace, A.J., B. Zmudzka, M.C. Hollander et al. Induction of mammalian β-polymerase mRNA by DNA damaging agents in Chinese hamster ovary cells. Mol. Cell. Biol. 9: 851-853 (1989).

F7 Ferris, D.K., A. Harel-Bellan, R.I. Morimoto et al. Mitogen and lymphokine stimulation of heat shock proteins in T lymphocytes. Proc. Natl. Acad. Sci. (USA) 85: 3850-3854 (1988).

F8 Feinendegen, L.E., V.P. Bond, J. Vooz et al. Biochemical and cellular mechanisms of low-dose radiation effects. Int. J. Radiat. Biol. 53: 23-27 (1988).

F9 Feinendegen, L.E. Radiation risk of tissue late effects, a net consequence of probabilities of various cellular responses. Eur. J. Nucl. Med. 18: 740-751 (1991).

F10 Feinendegen, L.E., H. Mühlensiepen, H. Lindberg et al. Acute and temporary inhibition of thymidine kinase in mouse bone marrow after low-dose exposure. Int. J. Radiat. Biol. 45: 205-215 (1984).

F11 Feinendegen, L.E. and H. Mühlensiepen. In vivo enzyme control through a strong stationary magnetic field - the case of thymidine kinase in mouse bone marrow cells. Int. J. Radiat. Biol. 52: 469-479 (1987).

F12 Feinendegen, L.E. The cell dose concept; potential applications in radiation protection. Phys. Med. Biol. 35: 597-612 (1990).

F13 Feinendegen, L.E., J. Booz, V.P. Bond et al. Microdosimetric approach to the analysis of cell responses at low dose and dose rate. Radiat. Prot. Dosim. 13: 299-306 (1985).

F14 Feinendegen, L.E., H. Mühlensiepen, V.P. Bond et al. Intracellular stimulation of intracellular control mechanisms by low-dose, low-LET irradiation. Health Phys. 52: 663-669 (1987).

F15 Feinendegen, L.E. and H. Muhlensiepen. Effect of static magnetic field on cellular metabolism in the living mouse. Endeavour 12: 119-123 (1988).

F16 Fornace, A.J.J., D.W. Nebert, M.C. Hollander et al. Mammalian genes coordinately regulated by growth arrest signals and DNA-damaging agents. Mol. Cell. Biol. 9: 4196-4203 (1989).

F17 Fornace, A.J. Jr., I. Alamo Jr. and M.C. Hollander. DNA damage-inducible transcripts in mammalian cells. Proc. Natl. Acad. Sci. (USA) 85: 8800-8804 (1988).

F18 Farr, S.B. and T. Kogoma. Oxidative stress responses in Escherichia coli and Salmonella typhimurium. Micro. Rev. 55 (4): 561-585 (1991).

F19 Fornace, A.J. Mammalian genes induced by radiation: activation of genes associated with growth control. Annu. Rev. Genet. 26: 507-526 (1992).

F20 Fang, F. and J.W. Newport. Evidence that the G1-S and G2-M transitions are controlled by different cdc2 proteins in higher eukaroytes. Cell 66: 731-742 (1991).

F21 Fitzgerald, T.J., S. Henault, M. Sakakeeny et al. Expression of transfected recombinant oncogenes increases radiation resistance of clonal hematopoietic and fibroblast cell lines selectively at clinical low dose rate. Radiat. Res. 122: 44-52 (1990).

F22 Fritz, T.E., D.V. Tolle and T.M. Seed. The preleukemic syndrome in radiation-induced myelogenous leukemia and related myeloproliferative disorders. Chapter 7. p. 87-100 in: The Preleukemic Syndrome (Hemopoietic Dysplasia). (G.C. Bagby, ed.) CRC Press, Cleveland, 1985.

F23 Farooqi, Z. and P.C. Kesavan. Low-dose radiation-induced adaptive response in bone marrow cells of mice. Mutat. Res. 302: 83-89 (1993).

F24 Fritz, G. K. Tano, S. Mitra et al. Inducibility of the DNA repair gene encoding 06-methylguanine-DNA methyltransferase in mammalian cells by DNA-damaging treatments. Mol. Cell. Biol. 11: 4660-4668 (1991).

F25 Fornace, A.J.J., B. Zmudzka, M.C. Hollander et al. Induction of beta-polymerase mRNA by DNA-damaging agents in Chinese hamster ovary cells. Mol. Cell. Biol. 9: 851-853 (1989).

G1 Greenberg, J.T., P. Monach, J.H. Chou et al. Positive control of a global antioxidant defense regulon activated by superoxide-generating agents in Escherichia coli. Proc. Natl. Acad. Sci. (USA) 87: 6181-6185 (1990).

G2 Gaziev, A.I., L.A. Fomenko, I.K. Kozhanowskaja et al. Radio-adaptive response in mice chronically exposed to low dose rate of gamma-rays. p. 259-262 in: Low Dose Irradiation and Biological Defense Mechanisms. (T. Sugahara, L.A. Sagan and T. Aoyama, eds.) Elsevier Science Publishers, The Netherlands, 1992.

G3 Gibson, R., S. Graham, A.M. Lilienfeld et al. Irradiation in the epidemiology of leukaemia among adults. J. Natl. Cancer Inst. 48: 301-311 (1972).

G4 Gould, M.N. and K.H. Clifton. Evidence for a unique in situ component of the repair of radiation damage. Radiat. Res. 77: 149-155 (1979).

G5 Gentner, N.E. Commentary 3 to Schull and Weiss - Increased definition of abnormal sensitivity using low dose rate testing. Adv. Radiat. Biol. 16: 293-302 (1992).

G6 Greer, W.L. and J.G. Kaplan. Early nuclear events in lymphocyte proliferation. The role of DNA strand break repair and ADP ribosylation. Exp. Cell Res. 166: 399-415 (1986).

G7 Gee, D.E., J. Griffin, L. Sastre et al. Differentiation of myeloid cells is accompanied by increased levels of $p60^{c-src}$ protein and kinase activity. Proc. Natl. Acad. Sci. (USA) 83: 5131-5135 (1986).

G8 Gribbin, M.A., G.R. Howe and J.L. Weeks. A study of the mortality of AECL employees. V. The second analysis: mortality during the period 1950-1985. AECL-10615 (1992).

G9 Girard, F., U. Stransfeld, A. Fernandez et al. Cyclin A is required for the onset of DNA replication in mammalian fibroblasts. Cell 67: 1169-1179 (1991).

G10 Grant, M.L., R.K. Bruton, P.J. Byrd et al. Sensitivity to ionising radiation of transformed human cells containing mutant ras genes. Oncogene 5: 1159-1164 (1990).

G11 Gupta, S.S. and S.B. Bhattacharjee. Induction of repair functions by hydrogen peroxide in Chinese hamster cells. Int. J. Radiat. Biol. 53 (6): 935-942 (1988).

G12 Gilbert, E.S., E. Omonhundro, J.A. Buchanan et al. Mortality of workers at the Hanford site: 1945-1986. Health Phys. 64 (6): 577-590 (1993).

G13 Grahn, D. and T.E. Fritz. Chronic radiation injury with mice and dogs exposed to external whole-body irradiation at the Argonne National Laboratory. p. 14-31 in: Life-Span Radiation Effects Studies in Animals: What Can They Tell Us? (R.C. Thompson and J.A. Mahaffey, eds.) CONF-830951 (1986).

G14 Gilbert, E.S., G.R. Petersen and J.A. Buchanan. Mortality of workers at the Hanford site, 1945-1981. Health Phys. 56: 11-25 (1989).

H1 Hasan, N.M., R.B. Cundall and G.E. Adams. Effects of hypoxia and reoxygenation on the conversion of xanthine delydrogenase to oxidase in Chinese hamster. Free Radic. Biol. Med. 11: 179-185 (1991).

H2 Hart, R., M. Chou, R. Feuers et al. DNA repair: As influenced by age, nutrition and exposure to toxic substances. p. 53-58 in: Biological Effects of Low Level Exposures to Chemicals and Radiation. (E.J. Calabrese, ed.) Lewis Publishers, Boca Raton, 1992.

H3 Hughes, E.N. and D.A. Boothman. Effect of caffeine on the expression of a major x ray induced protein in human tumor cells. Radiat. Res. 125: 313-317 (1991).

H4 Haire, R.N., M.S. Peterson and J.J. O'Leary. Mitogen activation induces the enhanced synthesis of two heat-shock proteins in human lymphocytes. J. Cell Biol. 106: 883-891 (1988).

H5 Hohn-Elkarim, K., H. Mühlensiepen, K.I. Altman et al. Modification of effects on radiation on thymidine kinase. Int. J. Radiat. Biol. 58: 97-100 (1990).

H6 Haskins, K., R. Kubo, J. White et al. The major histocompatibility complex-restricted antigen receptor on T cells. I. Isolation with a monoclonal antibody. J. Exp. Med. 157: 1149 (1983).

H7 Hsi, E.D., J.N. Siegel, Y. Minami et al. T cell activation induces rapid tyrosine phosphorylation of a limited number of cellular substrates. J. Biol. Chem. 264: 10836-10842 (1989).

H8 Hain, J., R. Jaussi and W. Burkart. Lack of adaptive response to low doses of ionizing radiation in human lymphocytes from five different donors. Mutat. Res. 283: 137-144 (1992).

H9 Heindorff, K., R. Rieger, A. Schubert et al. Clastogenic adaptation of plant cells - reduction of the yield of clastogen-induced chromatid aberrations by various pretreatment procedures. Mutat. Res. 181: 157-171 (1987).

H10 Heindorff, K., R. Rieger, A. Michaelis et al. Clastogenic adaptation triggered by S-phase-independent clastogens in Vicia faba. Mutat. Res. 190: 131-135 (1987).

H11 Herrlich, P., H. Ponta and H.J. Rahmsdorf. DNA damage-induced gene expression: signal transduction and relation to growth factor signaling. Rev. Physiol. Biochem. Pharmacol. 119: 187-223 (1992).

H12 Hallahan, D.E., V.P. Sukhatme, M.L. Sherman et al. Protein kinase C mediates x-ray inducibility of nuclear signal transducers EGR1 and JUN. Proc. Natl. Acad. Sci. (USA) 88: 2156-2160 (1991).

H13 Hallahan, D.E., D.R. Spriggs, M.A. Beckett et al. Increased tumor necrosis factor alpha mRNA after cellular exposure to ionizing radiation. Proc. Natl. Acad. Sci. (USA) 86: 10104-10107 (1989).

H14 Hallahan, D.E., S. Virudachalam, M.L. Sherman et al. TNF gene expression is mediated by PKC following activation by ionizing radiation. Cancer Res. 51: 4565-4569 (1991).

H15 Hollander, M.C. and A.J.J. Fornace. Induction of fos RNA by DNA-damaging agents. Cancer Res. 49: 1687-1697 (1989).

H16 Holmberg, K., S. Falt, A. Johanssen et al. Clonal chromosome aberrations and genomic instability in x-irradiated human T lymphocyte cultures. Mutat. Res. 286: 321-330 (1993).

H17 Holm, L.E. Cancer risks after diagnostic doses of ^{131}I with special reference to thyroid cancer. Cancer Detect. Prev. 15 (1): 27-30 (1991).

H18 Holm, L.E., K.E. Wiklund, G.E. Lundell et al. Thyroid cancer after diagnostic doses of iodine-131: a retrospective cohort study. J. Natl. Cancer Inst. 80: 1132-1138 (1988).

H19 Harvey, E.B., J.D. Boice Jr., M. Honeyman et al. Prenatal x-ray exposure and childhood cancer in twins. N. Engl. J. Med. 312: 541-545 (1985).

H20 Hays, S.R., X. Li and B.F. Kimler. Is there an adaptive response to radiation in the developing brain of the fetal rat. Radiat. Res. 136: 293-296 (1993).

H21 Holm, L.E., K.E. Wiklund, G.E. Lundell et al. Cancer risk in population examined with diagnostic doses of ^{131}I. J. Natl. Cancer Inst. 81: 302-306 (1989).

I1 Imlay, J.A. and S. Linn. DNA damage and oxygen radical toxicity. Science 240: 1302-1309 (1988).

I2 Imlay, J.A., S.M. Chin and S. Linn. Toxic DNA damage by hydrogen peroxide through the Fenton reation in vivo and in vitro. Science 240: 640-642 (1988).

I3 Ishii, K., N. Muto and I. Yamamoto. Augmentation in mitogen-induced proliferation of rat splenocytes by low dose whole body X-irradiation. Nippon Acta Radiologica 50 (10): 1262-1267 (1990).

I4 Ishii, K., J. Misonoh, Y. Hosoi et al. Effects of low-dose irradiation on proliferative response and IL-1 production of mouse splenocytes. p. 191-194 in: Low Dose Irradiation and Biological Defense Mechanisms. (T. Sugahara, L.A. Sagan and T. Aoyama, eds.) Elsevier Science Publishers, The Netherlands, 1992.

I5 Ikushima, T. Chromosomal response to ionizing radiation reminiscent of an adaptive response in cultured Chinese hamster cells. Mutat. Res. 180: 215-221 (1987).

I6 International Commission on Radiological Protection. 1990 Recommendations of the International Commission on Radiological Protection. ICRP Publication 60. Annals of the ICRP 21(1-3). Pergamon Press, Oxford, 1991.

I7 Ikushima, T. Radio-adaptive response: a novel chromosomal response in Chinese hamster cells in vitro. 151-161 in: Chromosome Aberrations. (G. Obe and A.T. Natarajan, eds.) Springer Verlag, Berlin, 1990.

I8 International Symposium on Biological Effects of Low Level Exposures to Radiation and Related Agents (Isbelles '93), Abstracts, Changchun, China, 1993. [To be published in the Chinese Medical Journal (1994)].

I9 Ishihara, H., K. Tsuneoka, A.B. Dimchev et al. Induction of the expression of the interleukin-1 beta gene in mouse spleen by ionizing radiation. Radiat. Res. 133: 321–326 (1993).

I10 Ishii, K., J. Misonoh, Y. Hosoi et al. Acquired radioresistance on human embryo cells by low dose x ray irradiation. p. 49 in: International Symposium on Biological Effects of Low Level Exposures to Radiation and Related Agents (Isbelles '93), Abstracts, Changchun, China, 1993. [To be published in the Chinese Medical Journal (1994)].

I11 Ikushima, T. Radio-adaptive response. Involvement of induction of specific gene expression by low doses of ionising radiation. p. 255-258 in: Low Dose Irradiation and Biological Defense Mechanisms. (T. Sugahara, L.A. Sagan and T. Aoyama, eds.) Elsevier Science Publishers, The Netherlands, 1992.

I12 Isakov, N., W. Scholtz and A. Altman. Signal transduction and intracellular events in T-lymphocyte activation. Immunol. Today 7: 271-277 (1986).

J1 Jacobsen-Kram, D. and J.R. Williams. Failure to observe adaptive response to ionizing radiation in mouse bone marrow cells in vivo. Environ. Mol. Mutagenesis 2 (Suppl. 11): 49-50 (1988).

J2 Jiang, T., D. Chen and C. Wang. Adaptive response in mouse bone marrow cells to chronic gamma- and beta-rays. p. B-2 in: International Conference on Low Dose Irradiation and Biological Defense Mechanisms, Abstracts, Kyoto, Japan, 1992.

J3 Johnson, M.H. The molecular and cellular basis of preimplantation mouse development. Biol. Rev. 56: 463-498 (1981).

J4 Johnstone, A.P. and G.T. Williams. Role of DNA breaks and ADP-ribosyl transferase activity in eukaryotic differentiation demonstrated in human lymphocytes. Nature 300: 368-370 (1982).

J5 June, C.H., M.C. Fletcher, J.A. Ledbetter et al. Increases in tyrosine phosphorylation are detectable before phospholipase C activation after T cell receptor stimulation. J. Immunol. 144: 1591-1599 (1990).

J6 James, S.J., S.M. Enger, W.J. Peterson et al. Immune potentiation after fractionated exposure to very low doses of ionizing radiation and/or caloric restriction in autoimmune-prone and normal C57Bl/6 mice. Clin. Immunol. Immunopathol. 55: 427-437 (1990).

J7 James, S.J. and T. Makinodan. T cell potentiation in normal and autoimmune-prone mice after extended exposure to low doses of ionizing radiation and low caloric restriction. Int. J. Radiat. Biol. 53 (1): 137-152 (1988).

K1 Karran, P., T. Lindahl and B. Griffin. Adaptive response to alkylating agents involves alteration in situ of O^6-methylguanine residues in DNA. Nature 280: 76-77 (1979).

K2 Karran, P., T. Hjelmgren and T. Lindahl. Induction of a DNA glycosylase for N-methylated purines is part of the adaptive response to alkylating agents. Nature 296: 770-773 (1982).

K3 Katoaka, H., J. Hall and P. Karran. Complementation of sensitivity to alkylating-agents in E.Coli and Chinese hamster ovary cells by expression of cloned bacterial-DNA repair gene. EMBO J. 5: 3195-3200 (1986).

K4 Khandogina, E.K., G.R. Mutovin, S.V. Zvereva et al. Adaptive response in irradiated human lymphocytes: radiobiological and genetical aspects. Mutat. Res. 251: 181-186 (1991).

K5 Kelsey, K.T., A. Memisoglu, D. Frenkel et al. Human lymphocytes exposed to low doses of X-rays are less susceptible to radiation-induced mutagenesis. Mutat. Res. 263: 197-201 (1991).

K6 Kano, Y. and J.B. Little. Mechanisms of human cell neoplastic transformation: relationship of specific abnormal clone formation to prolonged lifespan in X-irradiated human diploid fibroblasts. Int. J. Cancer 36: 407-413 (1985).

K7 Kano, Y. and J.B. Little. Mechanisms of human cell neoplastic transformation: X-ray-induced abnormal clone formation in long-term cultures of human diploid fibroblasts. Cancer Res. 45: 2550-2555 (1985).

K8 Kleczkowska, H.E. and F.R. Althaus. Adaptive response of human keratinocytes to low doses of MNNG. p. 275-278 in: Low Dose Irradiation and Biological Defense Mechanisms. (T. Sugahara, L.A. Sagan and T. Aoyama, eds.) Elsevier Science Publishers, The Netherlands, 1992.

K9 Keyse, S.M. and R.M. Tyrrell. Induction of the heme oxygenase gene in human skin fibroblasts by hydrogen peroxide and UVA (365 mn) radiation: evidence for the involvement of the hydroxyl radical. Carcinogenesis 11: 787-791 (1989).

K10 Koppenol, W.H. and P.L. Bounds. Hormesis. Science 245: 311 (1989).

K11 Kyoizumi, S., M. Akiyama, Y. Hirai et al. Spontaneous loss and alteration of antigen receptor expression in mature $CD4^+$ T cells. J. Exp. Med. 171: 1981-1999 (1990).

K12 Kennedy, J.C., J.E. Till, L. Simonovitch et al. Radiosensitivity of the immune response to sheep red cells in the mouse, as measured by the hemolytic plaque methods. J. Immunol. 94: 715-722 (1965).

K13 Kamisaku, H., T. Sado, P. Pongpiachan et al. Low dose irradiation does not stimulate in vivo prima response in specific-pathogen-free mice. Yokohama Med. Bull. 42: 3-4 (1992).

K14 Kusunoki, Y., M. Akiyama, S. Kyoizumi et al. Age-related alteration in the composition of immuno-competent blood cells in atomic bomb survivors. Int. J. Radiat. Biol. 53: 189-198 (1988).

K15 Kendall, G.M., C.R. Muirhead, B.H. MacGibbon et al. First analysis of the national registry for radiation workers: occupational exposure to ionising radiation and mortality. NRPB-R251 (1992).

K16 Kyoizumi, S., S. Umeki, M. Akiyama et al. Frequency of mutant T lymphocytes defective in the expression of the T-cell antigen receptor gene among radiation-exposed people. Mutat. Res. 265: 173-180 (1992).

K17 Kuzin, A.M., L.V. Slozhenikina and T.E. Ushakova. Effect of gamma-irradiation of isolated biomembranes of rat embryo cells on the activity of adenylate cyclase. Dokl. Akad. Nauk SSSR 233: 978-980 (1977). (in Russian)

K18 Kuzin, A.M., V.P. Ruda and E.G. Mozgovoi. The role of receptors in radiation hormesis. Radiat. Environ. Biophys. 30: 259-266 (1991).

K19 Kusen, S.I. and R.S. Stoika. Molecular mechanisms in the influence of polypeptide growth factors. Nauka Moscow: 240 (1985).

K20 Koff, A., F. Cross, A. Fisher et al. Human cyclin E, a new cyclin that interacts with two members of the CDC2 gene family. Cell 66: 1217-1228 (1991).

K21 Kastan, M.B., O. Onyekwere, D. Sidransky et al. Participation of p53 protein in the cellular response to DNA damage. Cancer Res. 51: 6304-6311 (1991).

K22 Kuerbitz, S.J., B.S. Plunkett, W.V. Walsh et al. Wild-type p53 is a cell cycle checkpoint determinant following irradiation. Proc. Natl. Acad. Sci. (USA) 89: 7491-7495 (1992).

K23 Kastan, M.B., Q. Zhan, W.S. El-Deiry et al. A mammalian cell cycle checkpoint pathway utilizing p53 and GADD45 is defective in ataxia-telangiectasia. Cell 71: 587-597 (1992).

K24 Kaina, B., H. Lohrer, M. Karin et al. Overexpressed human metallothione in IIA gene protects Chinese hamster ovary cells from killing by alkylating agents. Proc. Natl. Acad. Sci. 87: 2710-2714 (1990).

K25 Klein, A.K., J.A. Dyck, J.A. Shimizu et al. Effect of continuous, whole-body gamma irradiation upon canine lymphohematopoietic (CFU-GM, CFU-L) progenitors and a possible hematopoietic regulatory population. Radiat. Res. 101: 332-350 (1985).

K26 Karanta, A., E. Konitny, M. Jondal et al. Mitogen stimulation increases c-fos and c-jun protein levels, AP-1 binding and AP-1 transcriptions activity. Cell Signal 4: 275-286 (1992).

L1 Luckey, T.D. Hormesis with Ionizing Radiation. CRC Press, Boca Raton, Florida, 1980.

L2 Luckey, T.D. Radiation Hormesis. CRC Press, Boca Raton, Florida, 1991.

L3 Lavin, M.F. and A.L. Schroeder. Damage-resistant DNA synthesis in eukaryotes. Mutat. Res. 193: 193-206 (1988).

L4 Luckey, T.D. Ionizing radiation promotes protozoan reproduction. Radiat. Res. 108: 215-221 (1986).

L5 Lindquist, S. The heat shock response. Annu. Rev. Biochem. 55: 1151-1191 (1986).

L6 Lepock, J.R., K.H. Cheng, H. Al-Qysi et al. Thermotropic lipid and protein transitions in Chinese hamster lung cell membranes: relationship to hyperthermic cell killing. Can. J. Biochem. Cell Biol. 61: 421-427 (1983).

L7 Luk, C.K., L. Veinot-Drebot, E. Tjan et al. Effect of transient hypoxia on sensitivity to doxorubicin in human and murine cell lines. J. Natl. Cancer Inst. 82: 684-692 (1990).

L8 Liu, S.Z., L. Cai and S.Q. Sun. Induction of cytogenetic adaptive response by exposure of rabbits to very low dose-rate gamma-radiation. Int. J. Radiat. Biol. 62 (2): 198-190 (1992).

L9 Liu, S.Z., W.H. Liu and J.B. Sun. Radiation hormesis: its expression in the immune system. Health Phys. 52: 579-583 (1987).

L10 Liu, S.Z., L. Cai and J.B. Sun. Effect of low-dose radiation in repair of DNA and chromosome damage. Acta Biol. Hung. 41 (1-3): 149-157 (1990).

L11 Lindahl, T. and B. Nyberg. Rate of depurination of native deoxyribonucleic acid. Biochemistry 11: 3610-3621 (1972).

L12 Leadon, S.A. Production and repair of DNA damage in mammalian cells. Health Phys. 59: 15-22 (1990).

L13 Lock, R.B. and W.E. Ross. Inhibition of p34cdc2 kinase activity by etoposide or irradiation as a mechanism of G_2 arrest in Chinese hamster ovary cells. Cancer Res. 50: 3761-3766 (1990).

L14 Lorenz, E., L.O. Jacobson, W.E. Henston et al. Effects of long-continued total-body gamma irradiation on mice, guinea pigs and rabbits. III. Effects on lifespan, weight, blood picture and carcinogenesis and the role of the intensity of radiation. p. 24-148 in: Biological Effects of External X and Gamma Radiation. National Nuclear Energy Series IV-22B. (R.E. Zirkle, ed.) McGraw-Hill, New York, 1954.

L15 Lubinski, J.M., A. Serrano-Williams, J. Chung et al. Varying susceptibility of thymic T cell subpopulations to radiation-induced DNA fragmentation. p. 365 in: Low Dose Irradiation and Biological Defense Mechanisms. (T. Sugahara, L.A. Sagan and T. Aoyama, eds.) Elsevier Science Publishers, The Netherlands, 1992.

L16 Liu, S.Z. Multilevel mechanisms of stimulatory effect of low dose radiation on immunity. p. 225-232 in: Low Dose Irradiation and Biological Defense Mechanisms. (T. Sugahara, L.A. Sagan and T. Aoyama, eds.) Elsevier Science Publishers, The Netherlands, 1992.

L17 Liu, W.H. and S.Z. Liu. Effect of continuous low level gamma-irradiation on plaque forming cells in mouse spleen and its mechanism. Chin. J. Radiol. Med. Prot. 7: 336-340 (1987).

L18 Liu, S.Z., Y.C. Zhang and J.J. Qi. Low dose ionising radiation does not cause ratio changes in regulatory T subsets in mouse spleen. J. Norman Bethune Univ. Med. Sci. 18: 209-213 (1992).

L19 Lorenz, E., J.W. Hollcroft, E. Miller et al. Long-term effects of acute and chronic irradiation in mice. I. Survival and tumor incidence following chronic irradiation of 0.11 r per day. J. Natl. Cancer Inst. 15 (4): 1049-1057 (1955).

L20 Laval, F. Pretreatment with oxygen species increases the resistance of mammalian cells to hydrogen peroxide and gamma-rays. Mutat. Res. 201: 73-79 (1988).

L21 Lindahl, T. Instability and decay of the primary structure of DNA. Nature 362: 709-715 (1993).

L22 Levine, A.J., J. Momand and C.A. Finlay. The p53 tumour suppressor gene. Nature 351: 543-456 (1991).

L23 Laval, F. Increase of 06-methylguanine-DNA methyltransferase and N3-methyladenine glycosidase RNA transcripts in rat hepatoma cells treated with DNA damaging agents. Biochem. Biophys. Res. Commun. 176: 1086-1092 (1991).

L24 Lin, C.S., D.A. Goldthwait and D. Samols. Induction of transcription from the long terminal repeat of Moleney murine sarcoma provirus by UV-irradiation, x-irradiation, and phorbor ester. Proc. Natl. Acad. Sci. (USA) 87: 36-40 (1990).

L25 Lu, C. and S.Z. Liu. Study on the mechanisms of cytogenetic adaptive response induced by low dose radiation. Chin. Med. J. 105 (4): 277-283 (1992).

L26 Liu, X.Z., S. Xu, Y.C. Zhang et al. Signal transduction in lymphocytes after low dose radiation. p. 15 in: International Symposium on Biological Effects of Low Level Exposures to Radiation and Related Agents (Isbelles '93), Abstracts, Changchun, China, 1993. [To be published in the Chinese Medical Journal (1994)].

L27 Liu, S.Z., Y. Zhao, Z.B. Han et al. Role of changes in functional status of hypothalamic-pituitary-adrenocortical axis in immunoenhancement after low dose radiation. p. 83 in: International Symposium on Biological Effects of Low Level Exposures to Radiation and Related Agents (Isbelles '93), Abstracts, Changchun, China, 1993. [To be published in the Chinese Medical Journal (1994)].

L28 Liu, S.Z. and X.K. Li. Induction of proteins in lymphocytes by in vitro and in vivo irradiation with low dose x- and gamma rays. p. 28 in: International Symposium on Biological Effects of Low Level Exposures to Radiation and Related Agents (Isbelles '93), Abstracts, Changchun, China, 1993. [To be published in the Chinese Medical Journal (1994)].

L29 Li, X.Y., Y.J. Li, H.Q. Fu et al. Effects of low dose radiation on pulmonary dissemination of Lewis cell carcinoma in mice. p. 103 in: International Symposium on Biological Effects of Low Level Exposures to Radiation and Related Agents (Isbelles '93), Abstracts, Changchun, China, 1993. [To be published in the Chinese Medical Journal (1994)].

M1 Miller, R.V. and T.A. Kokjohn. General microbiology of recA; environmental and evolutionary significance. Annu. Rev. Microbiol. 44: 365-394 (1990).

M2 Morohoshi, F. and N. Munakata. Two classes of Bacillus subtilis mutants deficient in the adaptive response to simple alkylating agents. Mol. Gen. Genet. 202: 200-206 (1986).

M3 McClintock, B. The significance of responses of the genome to challenge. Science 226: 792-801 (1984).

M4 Murphy, B.J., K.R. Laderoute, S.M. Short et al. The identification of heme oxygenase as a major hypoxic stress protein in Chinese hamster ovary cells. Br. J. Cancer 64: 69-73 (1991).

M5 Moquet, J.E., D.C. Lloyd, J.S. Prosser et al. Sister-chromatid exchanges induced by mitomycin C after exposure of human lymphocytes in G_0 to a low dose of X-radiation. Mutat. Res. 176: 143-146 (1989).

M6 Müller, W.U., C. Streffer and F. Niedereichholz. Adaptive response in mouse embryos?. Int. J. Radiat. Biol. 62 (2): 169-175 (1992).

M7 Muschel, R.J., H.B. Zhang, G. Iliakis et al. Effects of ionizing radiation on cyclin expression in HeLa cells. Radiat. Res. 132: 153-157 (1992).

M8 MacMahon, B. Prenatal x-ray exposure and childhood cancer. J. Natl. Cancer Inst. 28: 1173-1191 (1962).

M9 Meyers, M., R.A. Schea, A.E. Pettrowski et al. Role of x-ray-inducible genes and proteins in adaptive survival responses. p. 263-266 in: Low Dose Irradiation and Biological Defense Mechanisms. (T. Sugahara, L.A. Sagan and T. Aoyama, eds.) Elsevier Science Publishers, The Netherlands, 1992.

M10 Matsui, M.S. and V.A. DeLeo. Induction of protein kinase C activity by ultraviolet radiation. Carcinogenesis 11: 229-234 (1990).

M11 Melkonyan, H., T. Ushakova and S. Umansky. Chronic irradiation increases the expression of hsp70 genes in mouse tissues. p. 243-246 in: Low Dose Irradiation and Biological Defense Mechanisms. (T. Sugahara, L.A. Sagan and T. Aoyama, eds.) Elsevier Science Publishers, The Netherlands, 1992.

M12 Mühlensiepen, H., K.I. Altman and L.E. Feinendegen. The effect of low dose gamma irradiation and magnetic field exposure on the caution transport in bone marrow cells of mice (Abstract). 23rd Annual Meeting, European Society for Radiation Biology, Dublin, 1990.

M13 Matanoski, G.M. Health effects of low-level radiation in shipyard workers final report, June 1991. DOE DE-AC02-79 EV10095 (1991).

M14 Meuer, S.C., K.A. Fitzgerald, R.E. Hussey et al. Clonotpic structures involved in antigen-specific human T cell function: relationship to the T3 molecular complex. J. Exp. Med. 157: 705 (1983).

M15 Marth, J.D., J.A. Cooper, C.S. King et al. Neoplastic transformation induced by an activated lymphocyte-specific protein tyrosine kinase ($p56^{lck}$). Mol. Cell. Biol. 8: 540-550 (1988).

M16 Margulis, B.A., B.D. Zhivotovski, T.V. Pospelova et al. Patterns of protein synthesis in various cells after extreme heat shock. Exp. Cell Res. 193: 219-222 (1991).

M17 Makinodan, T., J.M. Lubinski and T.C. Fong. Cellular, biochemical and molecular basis of T cell senescence. Arch. Pathol. Lab. Med. 111: 910-914 (1987).

M18 Makinodan, T. and S.J. James. T cell potentiation by low dose ionizing radiation: possible mechanisms. Health Phys. 59 (1): 29-34 (1990).

M19 Makinodan, T. Cellular and subcellular alterations in immune cells induced by chronic, intermittent exposure in vivo to very low doses of ionizing radiation (LDR), and its ameliorating effects on progression of autoimmune disease and mammary tumor growth. p. 233-237 in: Low Dose Irradiation and Biological Defense Mechanisms. (T. Sugahara, L.A. Sagan and T. Aoyama, eds.) Elsevier Science Publishers, The Netherlands, 1992.

M20 Misonoh, J., M. Yoshida, Y. Okumura et al. Effects of low-dose irradiation of x-rays on 1UdR incorporation into mouse tissues. p. 323-326 in: Low Dose Irradiation and Biological Defense Mechanisms. (T. Sugahara, L.A. Sagan and T. Aoyama, eds.) Elsevier Science Publishers, The Netherlands, 1992.

M21 Mindek, G. No adaptive response in G_2 phase x-irradiated human lymphocytes and adenocarcinoma cells. Int. J. Radiat. Biol. (1994, in press).

M22 Monson, R.R. and B. MacMahon. Prenatal x-ray exposure and cancer in children. p. 97-106 in: Radiation Carcinogenesis: Epidemiology and Biological Significance. (J.D. Boice Jr. and J.F. Fraumeni Jr., eds.) Raven Press, New York, 1984.

M23 Murray, A.W. Creative blocks: cycle checkpoints and feedback controls. Nature 359: 599-604 (1992).

M24 Matsushime, H., M.E. Ewen, D.K. Strom et al. Identification and properties of an atypical catalytic subunit ($p34^{PSK-J3}$/cdc4) for mammalian D type G1 cylcins. Cell 71: 323-334 (1992).

M25 McKenna, W.G., M.C. Weiss, B. Endlich et al. Synergistic effect of the v-myc oncogene with H-ras on radioresistance. Cancer Res. 50: 97-102 (1990).

M26 Mole, R.H. Childhood cancer after prenatal exposure to diagnostic x ray examinations in Britain. Br. J. Cancer 622: 152-168 (1990).

M27 Meng, Q.Y. and Y.K. Jin. Cross induction of adaptive response by low level radiation and chemical mutagens in mouse germ cells. p. 43 in: International Symposium on Biological Effects of Low Level Exposures to Radiation and Related Agents (Isbelles '93), Abstracts, Changchun, China, 1993. [To be published in the Chinese Medical Journal (1994)].

M28 Miyamoto, M. and K. Sakamoto. Anti-tumour effect of total body irradiation of low doses in WHT/Ht mice. Jpn. J. Cancer Clin. 33 (10): 1211-1220 (1987).

M29 Maisin, J.R., A. Wambersie, G.B. Gerber et al. Life-shortening and disease incidence in C57B1 mice after single and fractionated γ and high-energy neutron exposure. Radiat. Res. 113: 300-317 (1988).

M30 Makedonov, G.P., N.I. Alekhina, L.V. Tskhovrebova et al. Comparison of expression of anticlastogenic action of interferon and radioadaptive response. Radiatsionnaya Biologia Radioekologia 33: 259-264 (1993).

N1 Nurse, P. Universal control mechanism regulating onset of M-phase. Nature 344: 503-508 (1990).

N2 Norburg, C. and P. Nurse. Animal cell cycles and their control. Annu. Rev. Biochem. 61: 441-470 (1992).

N3 Naegeli, H. and F.R. Althaus. Regulation of poly(ADP-ribose) polymerase. Histone-specific adaptations of reaction products. J. Biol. Chem. 266 (16): 10596-10601 (1991).

N4 Nogami, M., J.T. Huang, S.J. James et al. Mice chronically exposed to low dose ionizing radiation possess splenocytes with elevated levels of HSP70 mRNA, HSC70 and HSP72 and with an increased capacity to proliferate. Int. J. Radiat. Biol. 63 (6): 775-783 (1993).

N5 Nasmyth, K. and T. Hunt. Dams and sluices. Nature 366: 634-635 (1993).

N6 Neta, R., S. Douches and J.J. Oppenheim. Interleukin-1 is a radioprotector. J. Immunol. 136 (7): 2483-2485 (1986).

O1 Olsson, M. and T. Lindahl. Repair of alkylated DNA in Escherichia coli. Methyl group transfer from O6-methylguanine to a protein cysteine residue. J. Biol. Chem. 255: 10569-10571 (1980).

O2 Olivieri, G., J. Bodycote and S. Wolff. Adaptive response of human lymphocytes to low concentrations of radioactive thymidine. Science 223: 594-597 (1984).

O3 Olivieri, G., A. Bosi, R. Grillo et al. Interaction of low dose irradiation with subsequent mutagenic treatment. Studies with human lymphocytes. p. 279-282 in: Low Dose Irradiation and Biological Defense Mechanisms. (T. Sugahara, L.A. Sagan and T. Aoyama, eds.) Elsevier Science Publishers, The Netherlands, 1992.

O4 Ohnishi, T., M. Ihara and H. Utsumi. Induced proteins in human melanomas by gamma-ray. p. 247-249 in: Low Dose Irradiation and Biological Defense Mechanisms. (T. Sugahara, L.A. Sagan and T. Aoyama, eds.) Elsevier Science Publishers, The Netherlands, 1992.

O5 Otsu, H., S. Kobayashi, T. Furuse et al. Age and sex dependence in tumorigenesis in mice by continuous low-dose-rate gamma-ray whole-body irradiation. p. 211-216 in: International Conference on Radiation Effects and Protection, Mito, 1992. Japan Atomic Energy Research Institute, Tokyo, 1992.

O6 Ong, A., W.X. Li and C.C. Ling. Low-dose-rate irradiation of rat embryo cells containing the ha-ras oncogene. Radiat. Res. 134: 251-255 (1993).

O7 Owens, G.P., W.E. Hahn and J.J. Cohen. Identification of mRNA associated with programmed cell death in immature thymocytes. Mol. Cell. Biol. 11: 4177-4188 (1991).

P1 Preston, D.L., S. Kusumi, M. Tomonaga et al. Cancer incidence in atomic bomb survivors. Part III: leukaemia, lymphoma and multiple myeloma, 1950-87. RERF TR/24-92 (1992) and Radiat. Res. 137: S68-S97 (1994).

P2 Planel, H., J.P. Soleilhavoup, R. Tixador et al. Influence on cell proliferation of background radiation or exposure to very low, chronic gamma radiation. Health Phys. 52: 571-578 (1987).

P3 Planel, H., C. Caratero, F. Croute et al. Investigations on the biological effects of very low doses of ionizing radiations. p. 13-20 in: Low Dose Irradiation and Biological Defense Mechanisms. (T. Sugahara, L.A. Sagan and T. Aoyama, eds.) Elsevier Science Publishers, The Netherlands, 1992.

P4 Pereira Luis, J.H. and V.L. Povoa. Individual variability of adaptive response of human lymphocytes primed with low dose gamma rays. p. 315-317 in: Low Dose Irradiation and Biological Defense Mechanisms. (T. Sugahara, L.A. Sagan and T. Aoyama, eds.) Elsevier Science Publishers, The Netherlands, 1992.

P5 Pantazis, P., E. Sariban, D. Kufe et al. Induction of c-sis gene expression and synthesis of platelet-derived growth factor in human myeloid leukaemia cells during monocytic differentiation. Proc. Natl. Acad. Sci. (USA) 83: 6455-6459 (1986).

P6 Papathanasiou, M.A., N.C.K. Kerr, J.H. Robbins et al. Induction by ionizing radiation of the gadd45 gene in cultured human cells: lack of mediation by protein kinase C. Mol. Cell. Biol. 11: 1009-1016 (1991).

P7 Polla, B.S., A.M. Healy, W.C. Wojno et al. Hormone 1 alpha, 25-dihydroxyvitamin D3 modulates heat shock response in monocytes. Am. J. Physiol. 252: C640-C649 (1987).

P8 Perlmutter, R.M., J.D. Marth, S.F. Ziegler et al. Specialized protein tyrosine kinase proto-oncogenes in hematopoietic cells. Biochem. Biophys. Acta 948: 245-262 (1988).

P9 Peterson, K.R., N. Ossanna, A.T. Thliveris et al. Derepression of specific genes promotes DNA repair and mutagenesis in Escherichia coli. J. Bacteriol. 170: 1-4 (1988).

P10 Pilinskaya, M.A., S.S. Dibskiy and L.P. Pedan. Modification by dimatyph of cytogenetic damages in blood lymphocytes of children exposed to constant low-rate irradiation. Cytology and Genetics 27 (4): 87-90 (1993).

P11 Pines, J. and T. Hunter. Human cyclin A is adeno-virus EIA-associated protein p60 and behaves differently from cyclin B. Nature 346: 760-763 (1990).

P12 Panozzo, J., D. Bertoncini, D. Miller et al. Modulation of expression of virus-like elements following exposure of mice to high- and low-LET radiations. Carcinogenesis 12: 801-804 (1991).

P13 Pape, R. New observations and experimental findings on the effect of very small roentgen doses. Radiol. Austriaca 4: 35-51 (1951).

R1 Radman, M. SOS repair hypotheses. Phenomenology of an inducible DNA repair which is accompanied by mutagenesis. p. 355-367 in: Molecular Mechanisms for DNA Repair, Part A. (P. Hanawalt and R.B. Setlow, eds.) Plenum Publishing Co., New York (1975).

R2 Russ, S. and G.M. Scott. Biological effects of gamma radiation (series II). Br. J. Radiol. 12: 440-441 (1939).

R3 Russ, S. and G.M. Scott. Some biological effects of continuous gamma radiation, with a note on protection. Br. J. Radiol. 10: 619-628 (1937).

R4 Ruda, V.P. and A.M. Kuzin. Effect of chronic low-level gamma-irradiation on the activity of plasma membrane adenylate cyclase from rat lung tissue. Radiobiologiya 31: 743-746 (1991).

R5 Ruda, V.P., A.A. Narimanov and A.M. Kuzin. The motivation of sensitivity of membrane receptors to natural effectors by low dose gamma-radiation. p. 327-329 in: Low Dose Irradiation and Biological Defense Mechanisms. (T. Sugahara, L.A. Sagan and T. Aoyama, eds.) Elsevier Science Publishers, The Netherlands, 1992.

R6 Rieger, R.A., A. Michaelis and H. Nicoloff. Clastogenic adaptation of the Vicia faba root tip meristem as affected by various treatment parameters. Mutat. Res. 140: 99-102 (1984).

R7 Rieger, R., A. Michaelis and S. Takehisa. Low temperature between conditioning and challenge treatment prevents the "adaptive response" of Vicia faba root tip meristem cells. Mutat. Res. 282: 69-72 (1992).

R8 Rigaud, O., D. Papadopoulo and E. Monstacchi. Decreased deletion mutation in radioadapted human lymphoblasts. Radiat. Res. 133: 94-101 (1993).

R9 Ron, D., A.R. Brasier and J.F. Habener. Transcriptional regulation of hepatic angiotensinogen gene expression by the acute-phase response. Mol. Cell. Endocrinol. 74: C97-C104 (1990).

R10 Ron, E., D.L. Preston, K. Mabuchi et al. Cancer incidence in atomic bomb survivors. Part IV: Comparison of cancer incidence and mortality. RERF TR/11-93 (1993) and Radiat. Res. 137: S98-S112 (1994).

R11 Rust, J.H., R.J. Robertson, E.F. Staffeldt et al. Effects of lifetime periodic gamma-ray exposure on the survival and pathology of guinea pigs. p. 217-244 in: Radiation & Ageing, Proceedings of a Colloquium, Semmering, Austria. (P.J. Lindop and G.A. Sacher, eds.) Taylor & Francis, London, 1966.

R12 Rosenblatt, L.S., S.A. Book and M. Goldman. Effects of x-irradiation of young female beagles on life span and tumor incidence. p. 628-645 in: Life-Span Radiation Effects Studies in Animals: What Can They Tell Us? (R.C. Thompson and J.A. Mahaffey, eds.) CONF-830951 (1986).

S1 Sagan, L.A. Special issue on radiation hormesis. Health Phys. 52 (5): 517-678 (1987).

S2 Sugahara, T., L.A. Sagan and T. Aoyama (eds.). Low Dose Irradiation and Biological Defense Mechanisms. Elsevier, The Netherlands, 1992.

S3 Smith-Sonneborn, J. The role of the "stress protein response" in hormesis. p. 41-52 in: Biological Effects of Low Level Exposures to Chemicals and Radiation. (E.J. Calabrese, ed.) Lewis Publishers, Boca Raton, 1992.

S4 Sekiguchi, M. and Y. Nakabeppu. Adaptive response: induced synthesis of DNA repair enzymes by alkylating agents. Trends Genet. 3 (2): 51-54 (1987).

S5 Samson, L. and J. Cairns. A new pathway for DNA repair in Escherichia coli. Nature 267: 281-283 (1977).

S6 Samson, L., B. Derfler and E.A. Waldstein. Suppression of human DNA alkylation-repair defects by Escherichia coli DNA-repair genes. Proc. Natl. Acad. Sci. (USA) 83: 5607-5610 (1986).

S7 Storz, G., L.A. Tartaglia and B.N. Ames. Transcriptional regulator of oxidative stress-inducible genes: direct activation by oxidation. Science 248: 189-194 (1990).

S8 Shibahara, S., R.M. Müller and H. Taguchi. Transcriptional control of rat heme oxygenase by heat shock. J. Biol. Chem. 262: 12889-12892 (1987).

S9 Storch, T.G. and G.D. Talley. Oxygen concentration regulates the proliferative response of human fibroblasts to serum and growth factors. Exp. Cell Res. 175: 317-325 (1988).

S10 Sakata, K., T.T. Kwok, B.J. Murphy et al. Hypoxia-induced drug resistance comparison to P-glycoprotein-associated drug resistance. Br. J. Cancer 64: 809-814 (1991).

S11 Shen, J.W., J.R. Subjeck, R.B. Lock et al. Depletion of topoisomerase II in isolated nuclei during a glucose-related stress response. Mol. Cell. Biol. 9: 3284-3291 (1989).

S12 Shadley, J.D. and S. Wolff. Very low doses of X-rays can cause human lymphocytes to become less susceptible to ionizing radiation. Mutagenesis 2: 95-96 (1987).

S13 Shadley, J.D., V. Afzal and S. Wolff. Characterization of the adaptive response to ionizing radiation induced by low doses of x rays to human lymphocytes. Radiat. Res. 111: 511-517 (1987).

S14 Shadley, J.D. and J.K. Wiencke. Induction of the adaptive response by X-rays is dependent on radiation intensity. Int. J. Radiat. Biol. 56: 107-118 (1989).

S15 Sankaranarayanan, K., A.v. Duyn, M.J. Loos et al. Adaptive response of human lymphocytes to low-level radiation from radioisotopes or X-rays. Mutat. Res. 211: 7-12 (1989).

S16 Schmid, E., M. Bauchinger and U. Nahrstedt. Adaptive response from X-irradiation of human lymphocytes. Mutagenesis 4 (2): 87-89 (1989).

S17 Shadley, J.D. and G. Dai. Cytogenetic and survival adaptive responses in G_1 phase human lymphocytes. Mutat. Res. 265: 273-281 (1992).

S18 Sanderson, B.J.S. and A.A. Morley. Exposure of human lymphocytes to ionizing radiation reduces mutagenesis by subsequent ionizing radiation. Mutat. Res. 164: 347-351 (1986).

S19 Shadley, J.D. and G. Dai. Low dose x-ray-induced cytogenetic adaptive response in human lymphocytes. p. 283-286 in: Low Dose Irradiation and Biological Defense Mechanisms. (T. Sugahara, L.A. Sagan and T. Aoyama, eds.) Elsevier Science Publishers, The Netherlands, 1992.

S20 Semenetc, T.N., O.V. Semina and A.S. Saenko. Adaptive response of colony forming units (CFU-S) after preirradiation: the condition of manifestation in exotest. Radiatsionnaya Biologia Radioekologia 33: 525-528 (1993).

S21 Sagan, L.A., L.E. Feinendegen and G.B. Gerber (eds.). Low dose radiation and the immune system. Int. J. Radiat. Biol. 53 (1): 1-200 (1988).

S22 Sengupta, S. and S.B. Bhattacharjee. Effects of low-dose hydrogen peroxide treatment of V79 cells on killing and mutation by different agents. Mutat. Res. 243: 81-85 (1990).

S23 Savage, J.R.K. and D.G. Papworth. The effect of variable G_2 duration upon the interpretation of yield-time curves of radiation-induced chromatid aberrations. J. Theor. Biol. 38: 17-38 (1973).

S24 Stewart, A., J. Webb and D. Hewitt. A survey of childhood malignancies. Br. Med. J. 1: 1495-1508 (1958).

S25 Suzuki, M., K. Nakano, K. Suzuki et al. Effect of chronic low dose irradiation of ^{60}Co gamma rays in human embryo cells. p. 307-310 in: Low Dose Irradiation and Biological Defense Mechanisms. (T. Sugahara, L.A. Sagan and T. Aoyama, eds.) Elsevier Science Publishers, The Netherlands, 1992.

S26 Smith, B.P., K.L. Gale, M. Einspenner et al. Stimulated human fibroblast cell survival/ clonogenicity in response to low doses of ionizing radiation. p. 299-302 in: Low Dose Irradiation and Biological Defense Mechanisms. (T. Sugahara, L.A. Sagan and T. Aoyama, eds.) Elsevier Science Publishers, The Netherlands, 1992.

S27 Shapiro, R. Damage to DNA caused by hydrolysis. p. 3-18 in: Chromosome Damage and Repair. (E. Seeberg and K. Kleepe, eds.) Plenum Press, New York, 1981.

S28 Studzinski, G.P. Invited review - oncogenes, growth, and the cell cycle: an overview. Cell Tissue Kinet. 22: 405-424 (1989).

S29 Sariban, E., T. Mitchell and D. Kufe. Expression of the c-fms proto-oncogene during human monocytic differentiation. Nature 316: 64-66 (1985).

S30 Shimizu, Y., H. Kato, W.J. Schull et al. Dose-response analysis among atomic-bomb survivors exposed to low-level radiation. p. 71-74 in: Low Dose Irradiation and Biological Defense Mechanisms. (T. Sugahara, L.A. Sagan and T. Aoyama, eds.) Elsevier Science Publishers, The Netherlands, 1992.

S31 Sciandra, J.J., J.R. Subjeck and C.S. Hughes. Induction of glucose-regulated proteins during anaerobic exposure and of heat-shock proteins after re-oxygenation. Proc. Natl. Acad. Sci. (USA) 81: 4843-4847 (1984).

S32 Semenetc, T.N., O.V. Semina, A.M. Poverenny et al. Adaptive response in vivo: the reasons of CFU-S radioresistance increase. Radiatsionnaya Biologia Radioekologia 33: 812-815 (1993).

S33 Singh, S.P. and M.F. Lavin. DNA-binding protein activated by gamma radiation in human cells. Mol. Cell. Biol. 10: 5279-5285 (1990).

S34 Sies, H. (ed.). Oxidative Stress. Academic Press, London, 1985.

S35 Slater, T.F. and K.H. Cheesemann. Free radical mechanism and tissue injury and mechanism of protection. p. 1-14 in: Reactive Oxygen Species in Chemistry, Biology and Medicine. (A. Quintaniha, ed.) Plenum Press, New York, 1988.

S36 Shortman K., M. Egerton, G.J. Spangrude et al. The generation and fate of thymocytes. Semin. Immunol. 2: 3-13 (1990).

S37 Schwartz, R.H. A cell culture model for T lymphocyte clonal energy. Science 248: 1342-1347 (1990).

S38 Samelson, L.E., A.F. Philips, E.T. Loung et al. Association of the fyn protein tyrosine kinase with the T cell antigen receptor. Proc. Natl. Acad. Sci. (USA) 87: 4358-4362 (1990).

S39 Sado, T., H. Kamisaku, Y. Ikarashi et al. Immediate and long-term effects of radiation on the immune system of specific-pathogen-free mice. Int. J. Radiat. Biol. 53: 177-187 (1988).

S40 Sato, K., J.F. Flood and T. Makinodan. Influence of conditioned psychological stress on immunological recovery in mice exposed to low-dose X irradiation. Radiat. Res. 98: 381-388 (1984).

S41 Suzuki, K., M. Watanabe and J. Miyoshi. Differences in effects of oncogenes on resistance to gamma rays, ultraviolet light, and heat shock. Radiat. Res. 129: 157-162 (1992).

S42 Sato, F., N. Kawashima, K. Fukutsu et al. Influence of physical and biological factors on radiation carcinogenesis: some delayed effects of continuous irradiations in mice. p. 29-35 in: Annual Report April 1982 - March 1983. National Institute of Radiological Sciences, Japan (1983).

S43 Stepanyan, N.S., G.F. Krupnova and V. Zhestyanikov. Reparation of cytogenetical damage in plant cells upon an unequal fractionation of injurious action. I. Treatment by nitrogen mustard. Tsitologiya 25: 958-964 (1983). (in (Russian)

S44 Shadley, J.D. and G. Dai. Evidence that the adaptive response of human lymphocytes to ionizing radiation sets on lethal damage in nonaberrant cells. Mutat. Res. 301: 171-176 (1993).

S45 Sklar, M.D. The ras oncogenes increase the intrinsic resistance of NIH 3T3 cells to ionizing radiation. Science 239: 645-647 (1988).

S46 Sierra-Rivera, E., G.J. Vourhees and M.L. Freeman. Gamma irradiation increases hsp-70 in CHO cells. Radiat. Res. 135: 40-45 (1993).

S47 Schreck, R., P. Rieber and P.A. Baeuerle. Reactive oxygen intermediates as apparently widely used messengers in the activation of the NF-kappa B transcription factor and HIV-1. EMBO J. 10: 2247-2258 (1991).

S48 Sherman, M.L., R. Datta, D.E. Hallahan et al. Ionizing radiation regulates expression of the c-jun protooncogene. Proc. Natl. Acad. Sci. (USA) 87: 5663-5666 (1990).

S49 Sherman, M.L., R. Datta, D.E. Hallahan et al. Regulation of tumor necrosis factor gene expression by ionizing radiation in human myeloid leukemia cells and peripheral blood monocytes. J. Clin. Invest. 87: 1794-1797 (1991).

S50 Smith, K.A. Interleukin 2. Sci. Am. 262 (3): 26-33 (1990).

S51 Schmidtke, J.R. and F.J. Dixon. Effects of sublethal irradiation on the plaque-forming cell response in mice. J. Immunol. 111: 691-697 (1973).

S52 Sacher, G.A. and G. Grahn. Survival of mice under duration-of-life exposure to gamma rays. I. The dosage-survival relation and lethality function. J. Natl. Cancer Inst. 32: 277 (1964).

S53 Sato, F., S. Sasaki and N. Kawashima. Late effects of whole- or partial-body x-irradiation on mice. Causes of death. p. 59-70 in: Late Biological Effects of Ionizing Radiation. Volume II. IAEA-SM-224/210. IAEA, Vienna, 1978.

S54 Sasaki, S. and T. Kasuga. Life shortening and carcinogenesis in mice irradiated at the perinatal period with gamma rays. 357-367 in: Life-Span Radiation Effects Studies in Animals: What Can They Tell Us? (R.C. Thompson and J.A. Mahaffey, eds.) CONF-830951 (1986).

S55 Storer, J.B. Radiation resistance with age in normal and irradiated populations of mice. Radiat. Res. 25: 435-459 (1965).

T1 Tabocchini, M.A., J.B. Little and H.L. Liber. Mutation induction in human lymphoblasts after protracted exposure to low doses of tritiated water. p. 439-445 in: Low Dose Radiation: Biological Bases of Risk Assessment. (K.F. Baverstock and J.W. Stather, eds.) Taylor and Frances, London, 1989.

T2 Tuschl, H., R. Kovac and H. Altmann. UDS and SCE in lymphocytes of persons occupationally exposed to low levels of ionizing radiation. Health Phys. 45: 1-7 (1983).

T3 Tuschl, H., H. Altmann, R. Kovac et al. Effects of low-dose radiation on repair processes in human lymphocytes. Radiat. Res. 81: 1-9 (1980).

T4 Thompson Roy, C. Life-span effects of ionizing radiation in the beagle dog. A summary account of four decades of research funded by the US Department of Energy and its predecessor agencies. p. 1-323 in: PNL-6822 UC-408 (1989).

T5 Treisman, R. Identification of a protein-binding site that mediates transcriptional response of the c-fos gene to serum factors. Cell 46: 567-574 (1986).

T6 Taliaferro, W.H., L.G. Taliaferro and B.N. Jaroslow. Radiation and Immune Mechanisms. Academic Press, New York, 1964.

T7 Taliaferro, W.H. and L.G. Taliaferro. Effects of radiation on the initial and anamnestic IgM hemolysin responses in rabbits: antigen injection after x-rays. J. Immunol. 103: 559-569 (1969).

T8 Taliaferro, W.H. and L.G. Taliaferro. Effects of irradiation on initial and anamnestic hemolysin responses in rabbits: antigen injection before x-rays. J. Immunol. 104: 1364-1376 (1970).

T9 Troup, G.M. and R.E. Anderson. Radiation-induced augmentation of mitogenic responsiveness in murine spleen cells. Fed. Proc., Fed. Am. Soc. Exp. Biol. 4: 335 (1982).

T10 Tuschl, H., R. Kovac and A. Wottawa. T-lymphocyte subsets in occupationally exposed persons. Int. J. Radiat. Biol. 58: 651-659 (1990).

T11 Takizawa, Y., J. Yamashita, A. Wakizaka et al. Background radiation can stimulate the proliferation of mouse-L-5178Y cells. p. 234-236 in: International Conference on Radiation Effects and Protection, Mito, 1992. Japan Atomic Energy Research Institute, Tokyo, 1992.

T12 Tamura, K., Y. Kanaoka, S. Jinno et al. Cyclin G: a new mammalian cyclin with homology to fission yeast Cig 1. Oncogene 8: 2113-2118 (1993).

T13 Thompson, L.H., K.W. Brookman, N.J. Jones et al. Molecular cloning of the human XRCCI gene, which corrects defective DNA strand break repair and sister chromatid exchange. Mol. Cell. Biol. 12: 6160-6171 (1990).

T14 Tomita, Y., S. Hashimoto, S. Sawadi et al. Regulatory role for T cell activation. Nihon Univ. J. Med. 32: 207-221 (1990).

T15 Thomson, J.F. and D. Grahn. Life shortening in mice exposed to fission neutrons and γ radiation. Radiat. Res. 118: 151-160 (1989).

T16 Takai, Y., S. Yamada, K. Nemoto et al. Anti-tumour effect of low dose total (or half) body irradiation and changes of the functional subset of peripheral lymphocytes in non-Hodgkin's lymphoma patients after TBI (HBI). p. 113-116 in: International Conference on Low Dose Irradiation and Biological Defense Mechanisms, Abstracts, Kyoto, Japan, 1992.

U1 United Nations. Sources and Effects of Ionizing Radiation. United Nations Scientific Committee on the Effects of Atomic Radiation, 1993 Report to the General Assembly, with scientific annexes. United Nations sales publication E.94.IX.2. United Nations, New York, 1993.

U12 Upton, A.C., M.L. Randolph, J.W. Conklin et al. Late effects of fast neutrons and gamma rays in mice as influenced by the dose rate of irradiation: life shortening. Radiat. Res. 32: 493-509 (1967).

U13 Upton, A.C., M.L. Randolph, J.W. Conklin et al. Late effects of fast neutrons and gamma-rays in mice as influenced by the dose rate of irradiation: induction of neoplasia. Radiat. Res. 41: 467-491 (1970).

U14 Ullrich, R.L. and J.B. Storer. Influence of dose, dose rate and radiation quality on radiation carcinogenesis and life shortening in RFM and BALB/c mice. p. 95-113 in: Late Biological Effects of Ionizing Radiation. Volume II. IAEA-SM-224/210. IAEA, Vienna, 1978.

U15 Ullrich, R.L. Tumor induction in BALB/c mice after fractionated or protracted exposures to fission-spectrum neutrons. Radiat. Res. 97: 587-597 (1984).

V1 Vichi, P. and T.R. Tritton. Stimulation of growth in human and murine cells by adriamycin. Cancer Res. 49: 2679-2682 (1989).

V2 Vijayalaxmi and W. Burkart. Resistance and cross-resistance to chromosome damage in human blood lymphocytes adapted to bleomycin. Mutat. Res. 211: 1-5 (1989).

V3 Vilenchik, M.J. Spontaneous DNA damage and repair. p. 266-267 in: Mechanisms of DNA Damage and Repair. Biological Centre of the Academy of Sciences, Puschcino, 1971.

V4 Vilenchik, M.M. DNA Instability and Late Radiation Effect. Energoatomizdat, Moscow, 1987.

V5 von Boehmer, H. The developmental biology of T lymphocytes. Annu. Rev. Immunol. 6: 309-326 (1988).

V6 Veillette, A., M.A. Bookman, E.M. Horak et al. The CD4 and CD8 T cell surface antigens are associated with the internal membrane tyrosine-protein kinase $p56^{lck}$. Cell 55: 301-308 (1988).

W1 Wilson, V.L. and P.A. Jones. Inhibition of DNA methylation by chemical carcinogens in vitro. Cell 32: 239-246 (1983).

W2 Walker, G.C. Inducible DNA repair systems. Annu. Rev. Biochem. 54: 425-457 (1985).

W3 Wiencke, J.K., V. Afzal, G. Olivieri et al. Evidence that the [^{3}H] thymidine-induced adaptive response of human lymphocytes to subsequent doses of X-rays involves the induction of a chromosomal repair mechanism. Mutagenesis 1: 375-380 (1986).

W4 Wolff, S., V. Afzal, J.K. Wiencke et al. Human lymphocytes exposed to low doses of ionizing radiation become refractory to high doses of radiation as well as chemical mutagens that induce double-strand breaks in DNA. Int. J. Radiat. Biol. 53: 39-48 (1988).

W5 Wojcik, A., K. Bonk, W.U. Müller et al. Absence of adaptive response to low doses of X-rays in pre-implantation embryos and spleen lymphocytes of an inbred mouse strain as compared to human peripheral lymphocytes: a cytogenetic study. Int. J. Radiat. Biol. 62 (2): 177-186 (1992).

W6 Wolff, S. Low-dose exposures and the induction of adaptation. p. 21-28 in: Low Dose Irradiation and Biological Defense Mechanisms. (T. Sugahara, L.A. Sagan and T. Aoyama, eds.) Elsevier Science Publishers, The Netherlands, 1992.

W7 Wang, Z.Q., S. Saigusa and M.S. Sasaki. Adaptive response to chromosome damage in cultured human lymphocytes primed with low doses of X-rays. Mutat. Res. 246: 179-186 (1991).

W8 Wolff, S., J.K. Wiencke, V. Afzal et al. The adaptive response of human lymphocytes to very low doses of ionizing radiation: a case of induced chromosomal repair with the induction of specific proteins. p. 446-454 in: Low Dose Radiation: Biological Bases of Risk Assessment. (K.F. Baverstock and J.W. Stather, eds.) Taylor and Frances, London, 1989.

W9 Wojcik, A. and H. Tuschl. Indications of an adaptive response in C57BL mice pre-exposed in vivo to low doses of ionizing radiation. Mutat. Res. 243: 67-73 (1990).

W10 Watanabe, M., M. Suzuki, K. Suzuki et al. Effect of multiple radiation with low doses of gamma rays on growth ability of human embryo cells in vitro. p. 303-306 in: Low Dose Irradiation and Biological Defense Mechanisms. (T. Sugahara, L.A. Sagan and T. Aoyama, eds.) Elsevier Science Publishers, The Netherlands, 1992.

W11 Watanabe, M., M. Suzuki, K. Suzuki et al. Effect of multiple irradiation with low doses of gamma-rays on morphological transformation and growth ability of human embryo cells in vitro. Int. J. Radiat. Biol. 62 (6): 711-718 (1992).

W12 Ward, J.F. Some biological consequences of the spatial distribution of ionizing radiation-produced free radicals. Radiat. Res. 86: 185-195 (1981).

W13 Ward, J.F. Biochemistry of DNA lesions. Radiat. Res. 104: S103-S111 (1985).

W14 Ward, J.F. DNA damage produced by ionising radiation in mammalian cells: identies, mechanisms of formation and repairability. Prog. Nucleic Acid Res. Mol. Biol. 35: 95-125 (1988).

W15 Ward, J.F. Response to commentary by D. Billen. Radiat. Res. 126: 385-387 (1991).

W16 Woloschak, G.E. and C.M. Chang-Liu. Differential modulation of specific gene expression following high- and low-LET radiations. Radiat. Res. 124: 183-187 (1990).

W17 Woloschak, G.E., C.M. Chang-Liu and P. Shearin-Jones. Regulation of protein kinase C by ionizing radiation. Cancer Res. 50: 3963-3967 (1990).

W18 Woloschak, G.E. and C.M. Chang-Liu. Effects of low-dose radiation on gene expression in Syrian hamster embryo cells: comparison of JANUS neutrons and gamma rays. p. 239-242 in: Low Dose Irradiation and Biological Defense Mechanisms. (T. Sugahara, L.A. Sagan and T. Aoyama, eds.) Elsevier Science Publishers, The Netherlands, 1992.

W19 Wilson, R.E. and R.M. Sutherland. Enhanced synthesis of specific proteins, RNA, and DNA caused by hypoxia and reoxygenation. Int. J. Radiat. Oncol. Biol. Phys. 16: 957-961 (1989).

W20 Wyllie, A.H., J.F.R. Kerr and A.R. Currie. Cell death: the significance of apoptosis. Int. Rev. Cytol. 68: 251-306 (1980).

W21 Wyllie, A.H., R.G. Morris, A.L. Smith et al. Chromatin cleavage in apoptosis: association with condensed chromatin morphology and dependence on macromolecular synthesis. J. Pathol. 142: 67-77 (1984).

W22 Weiss, A., J. Imboden, K. Hardy et al. The role of T3/antigen receptor complex in T cell activation. Annu. Rev. Immunol. 4: 593-619 (1986).

W23 Wojcik, A., K. Bonk and C. Streffer. Indications of strain specificity for the induction of adaptive response to ionizing radiation in mice. p. 311-314 in: Low Dose Irradiation and Biological Defense Mechanisms. (T. Sugahara, L.A. Sagan and T. Aoyama, eds.) Elsevier Science Publishers, The Netherlands, 1992.

W24 Walker, G.C. Mutagenesis and inducible responses to deoxyribonucleic acid damage in Escherichia coli. Microbiol. Rev. 48: 60-93 (1984).

W25 Weichselbaum, R.R., D.E. Hallahan, V. Sukhatme et al. Biological consequences of gene regulation after ionizing radiation exposure. J. Natl. Cancer Inst. 83: 480-484 (1991).

W26 Woloschak, G.E., C.M.L. Liu, S.P. Jones et al. Modulation of gene expression in Syrian hamster embryo cells following ionizing radiation. Cancer Res. 50: 339-344 (1990).

W27 Wojcik, A., K. Bonk and C. Streffer. Is there an adaptive response in spleen lymphocytes of C57Bl/6 mice as assessed by chromosomal aberrations. Radiat. Res. 135 (2): 249-256 (1993).

W28 Wang, Z., J.D. Boice Jr., L. Wei et al. Thyroid nodularity and chromosome aberrations among women in areas of high background radiation in China. J. Natl. Cancer Inst. 82: 478-485 (1990).

W29 Wei, L. Re-analysis of some data related to the relationship between dose and cancer mortality rate in the high background radiation areas of Yangjiang, China. p. 1-11 in: Proceedings of Workshop on Radiological Biology, Medicine and Protection. (Zhao Ru and Tao Zufan, eds.) China-Japan Medical Conference, Beijing, 1992.

W30 Wei, L., Y. Zha, Z. Tao et al. Epidemiological investigation of radiological effects in high background radiation areas of Yangjiang, China. J. Radiat. Res. 31: 119-136 (1990).

X1 Xiong, Y., H. Zhang and D. Beach. D type cyclins associate with multiple protein kinases and the DNA replication and repair factor PCNA. Cell 71: 505-514 (1992).

Y1 Youngblom, J.H., J.K. Wiencke and S. Wolff. Inhibition of the adaptive response of human lymphocytes to very low doses of ionizing radiation by the protein synthesis inhibitor cycloheximide. Mutat. Res. 227: 257-261 (1989).

Y2 Yu, W.R., M.D. Wang, L. Cai et al. Pre-exposure of mice to low-dose or low-dose-rate ionizing radiations reduces chromosome aberrations by subsequent high dose of radiation or mitomycin C. p. C2-10 in: International Conference on Low Dose Irradiation and Biological Defense Mechanisms, Abstracts, Kyoto, Japan, 1992.

Y3 Yezzi, M.J., E.A. Blakely and C.A. Tobias. Split-dose recovery and protein synthesis in X-irradiated CHO cells. Radiat. Res. 114: 281-296 (1988).

Y4 Yamaoka, K., R. Edamatsu and M. Akitane. Effects of low dose x-ray irradiation on old rats - SOD activity, liquid peroxide level and membrane fluidity. p. 419-422 in: Low Dose Irradiation and Biological Defense Mechanisms. (T. Sugahara, L.A. Sagan and T. Aoyama, eds.) Elsevier Science Publishers, The Netherlands, 1992.

Y5 Yi, P.N. and C. Rha. Kinetics of repair in mouse lymphoma cells in the low dose rate range. p. 431-434 in: Low Dose Irradiation and Biological Defense Mechanisms. (T. Sugahara, L.A. Sagan and T. Aoyama, eds.) Elsevier Science Publishers, The Netherlands, 1992.

Y6 Yi, P.N., W.S. Stanley and W. Lee. Relationship between mitotic delay and the minimum dose rate of x irradiation required to stop cell proliferation. Radiat. Res. 133: 163-169 (1993).

Y7 Yonezowa, M., J. Misonoh and Y. Hosokawa. Radioresistance acquired after low doses of x-rays in mice. p. 48 in: International Symposium on Biological Effects of Low Level Exposures to Radiation and Related Agents (Isbelles '93), Abstracts, Changchun, China, 1993. [To be published in the Chinese Medical Journal (1994)].

Y8 Yao, J., Y.R. Cha and Z.X. Lin. Effects of long term low-level radiation on the immune system in human beings. p. 64 in: International Symposium on Biological Effects of Low Level Exposures to Radiation and Related Agents (Isbelles '93), Abstracts, Changchun, China, 1993. [To be published in the Chinese Medical Journal (1994)].

Y9 Yonezawa, M., A. Takeda and J. Misonoh. Acquired radioresistance after low dose x-irradiation in mice. J. Radiat. Res. 31: 256-262 (1990).

Z1 Zamboglou, N., W. Porschen, H. Mühlensiepen et al. Low dose effect of ionizing radiation on incorporation of iododeoxyuridine into bone marrow cells. Int. J. Radiat. Biol. 39: 83-93 (1981).

Z2 Zhivotovsky, B.D., K. Hanson, P. Nicotera et al. Involvement of protein synthesis and intracellular Ca^{2+} changes in the early stages of radiation-induced interphase death of rat thymocytes. p. 251-254 in: Low Dose Irradiation and Biological Defense Mechanisms. (T. Sugahara, L.A. Sagan and T. Aoyama, eds.) Elsevier Science Publishers, The Netherlands, 1992.

Z3 Zhivotovsky, B.D., L. Perlaky, A. Fonagy et al. Nuclear protein synthesis in thymocytes of X-irradiated rats. Int. J. Radiat. Biol. 54: 999-1006 (1988).

Z4 Zhang, Q.M., S. Yonel and M. Kato. Multiple pathways for repair of oxidative DNA damages caused by x rays and hydrogen perioxide in escherichia coli. Radiat. Res. 132: 334-338 (1992).

Z5 Zhou, P.K., X.Y. Liu, W.Z. Sun et al. Cultured mouse SR-1 cells exposed to low dose of gamma-rays become less susceptible to the induction of mutagenesis by radiation as well as bleomycin. Mutagenesis 8 (2): 109-111 (1993).

Z6 Zhou, P.K., W.Z. Sun, X.Y. Liu et al. Adaptive response to mutagenesis induced by low dose γ-rays in human HL-60 cells and its molecular basis. p. 67 in: International Symposium on Biological Effects of Low Level Exposures to Radiation and Related Agents (Isbelles '93), Abstracts, Changchun, China, 1993. [To be published in the Chinese Medical Journal (1994)].

Z7 Zhang, L.S., N.H. Wang and J.C. Miao. Effect of low dose radiation on IL-2 production and DNA synthesis by human lymphocytes. p. 51 in: International Symposium on Biological Effects of Low Level Exposures to Radiation and Related Agents (Isbelles '93), Abstracts, Changchun, China, 1993. [To be published in the Chinese Medical Journal (1994)].